PERGAMON INTERNATIONAL LIBRARY
of Science, Technology, Engineering and Social Studies
The 1000-volume original paperback library in aid of education, industrial training and the enjoyment of leisure
Publisher: Robert Maxwell, M.C.

REFRIGERATION PROCESSES

A practical handbook on the physical properties of refrigerants and their applications

INTERNATIONAL SERIES IN

HEATING, VENTILATION AND REFRIGERATION

VOLUME 12

GENERAL EDITORS: N. S. BILLINGTON AND E. OWER

COMPLETE LIST OF TITLES IN THE SERIES

Vol 1 OSBORNE
Fans, 2nd Edition

Vol 2 EDE
An Introduction to Heat Transfer Principles and Calculations

Vol 3 KUT
Heating and Hot Water Services in Buildings

Vol 4 ANGUS
The Control of Indoor Climate

Vol 5 DOWN
Heating and Cooling Load Calculations

Vol 6 DIAMANT
Total Energy

Vol 7 KUT
Warm Air Heating

Vol 8 BATURIN
Fundamentals of Industrial Ventilation

Vol 9 DORMAN
Dust Control and Air Cleaning

Vol 10 CROOME & ROBERTS
Airconditioning and Ventilation of Buildings

Vol 11 CROOME
Noise, Buildings and People

Vol 12 MEACOCK
Refrigeration Processes.
A practical handbook on the physical properties of refrigerants and their applications

Vol 13 MacKENZIE-KENNEDY
District Heating - Thermal Generation and Distribution.
A practical guide to centralised generation and distribution of heat services.

REFRIGERATION PROCESSES

A practical handbook on the physical properties of refrigerants and their applications

by

H. M. MEACOCK

Lewes, Sussex

PERGAMON PRESS

OXFORD · NEW YORK · TORONTO · SYDNEY · PARIS · FRANKFURT

U.K.	Pergamon Press Ltd., Headington Hill Hall, Oxford OX3 0BW, England
U.S.A.	Pergamon Press Inc., Maxwell House, Fairview Park, Elmsford, New York 10523, U.S.A.
CANADA	Pergamon of Canada, Suite 104, 150 Consumers Road, Willowdale, Ontario M2J 1P9, Canada
AUSTRALIA	Pergamon Press (Aust.) Pty. Ltd., P.O. Box 544, Potts Point, N.S.W. 2011, Australia
FRANCE	Pergamon Press SARL, 24 rue des Ecoles, 75240 Paris, Cedex 05, France
FEDERAL REPUBLIC OF GERMANY	Pergamon Press GmbH, 6242 Kronberg-Taunus, Pferdstrasse 1, Federal Republic of Germany

First edition 1979

British Library Cataloguing in Publication Data

Meacock, H. M.
Refrigeration processes. - (International series
in heating, ventilation and refrigeration; vol. 12).
1. Refrigeration and refrigerating machinery
I. Title II. Series
621.5'6 TP492 78-41287

ISBN 0 08 024211 1 hard cover

ISBN 0 08 024234 0 flexi cover

In order to make this volume available as economically and as rapidly as possible the author's typescript has been reproduced in its original form. This method unfortunately has its typographical limitations but it is hoped that they in no way distract the reader.

Printed and bound at William Clowes & Sons Limited
Beccles and London

Contents

For detailed list of tables and charts see page 69.

List of Figures

Acknowledgements

The approach to the treatment of the properties of refrigerants which I have used was much influenced by the work of Dr B J Eiseman Jnr who, as far back as 1952, published the results of his work on the correlation of the pressure–temperature–volume relationship of a wide variety of refrigerants. He was kind enough to give me the originals of the figures in his article which are now reproduced as figures 4, 8, 10 and 11.

Again, the importance of the concept of flow-work is put so well by Professor F W Hutchinson of Berkeley University , that with his permission I have drawn heavily on his writing.

E I DuPont de Nemours, as befits the largest manufacturers of fluorinated hydrocarbons, were exceptionally helpful with material for the compilation of the Data Section.

The Danish company Danfoss and the Gulf Publishing Company of Houston, Texas have helped by giving permission to reproduce Mollier charts and tables, as have the Allied Chemical Corporation.

I must record my thanks to Mr E Ower for his advice and interest and my most grateful thanks to Michael Meacock and Lawrence Perry without whose help this book would never have emerged from the indecipherable note stage.

Walton on Thames
May, 1978

Introduction

Refrigeration is one of the newer branches of engineering and, because of this, still retains some of the barriers to full knowledge which arise in any new form of art or science. These barriers have been erected from time immemorial by those possessing or claiming to possess some specialised knowledge in order to ensure the continuity of recourse to the fountainhead by seekers. It is the basic principle of priesthood that knowledge should be disseminated only to the chosen few and then only in small doses. In social life it ensures the continuity of caste and the creation of an elite group whether they be economists, men of religion, weather forecasters, doctors, lawyers or one of many other groups. The pattern is followed in Industry where specialised knowledge is a strong commercial asset and has even given birth to a new industry of Industrial Espionage.

The Refrigeration Industry in recent years has been most forthcoming with its information and in particular with facts concerning refrigerant gases. To some extent this is due to the fact that it is the Chemical Industry which produces refrigerants and which has no interest in preserving ignorance concerning its products. Nonetheless even here the tendency exists in another form and whilst manufacturers of refrigerants are open handed with accurate and profuse information concerning their products there seems to have been little attempt to offer instruction on basic principles.

Moreover the figures supplied by refrigerant manufacturers are in a number of cases simply computer printouts of input formulae which when given to six, seven and sometimes eight figures tend to be confusing. It is for this reason that all the tables have been edited and are given to the same four figure pattern that was prevalent in the pre-computer era.

The refrigeration cycle is a simple one. A refrigerant is a gaseous/liquid substance which will evaporate from the liquid state and condense from the gaseous state at combinations of temperature and pressure suitable to the designer's requirements. The temperatures at which evaporation and condensation take place vary as the pressure, an increase in one corresponding to an increase in the other.

If a liquid refrigerant with a boiling point of say −28° F is enclosed in a vessel as in Figure 1 it will absorb heat from its surroundings and evaporate, the gas will escape to atmosphere and the surroundings will be cooled. However if the vessel is closed the pressure will increase as more liquid evaporates. As the pressure rises so will the temperature until a state of equilibrium is reached and no further cooling takes place. If the rate of escape of the gas is controlled as in Figure 2 the temperature and concomitantly the rate of cooling of the surroundings may be controlled.

Fig. a

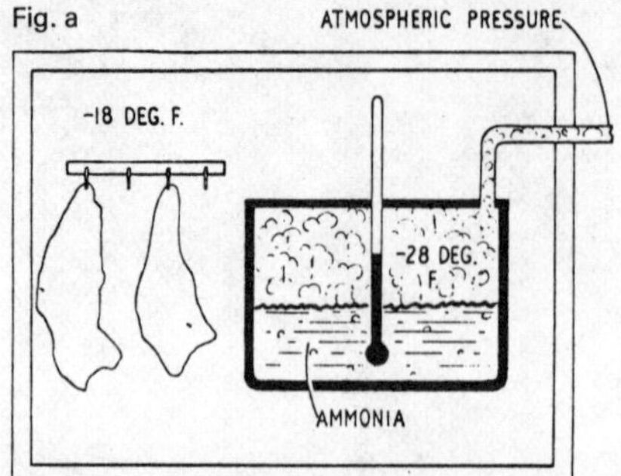

Fig. b

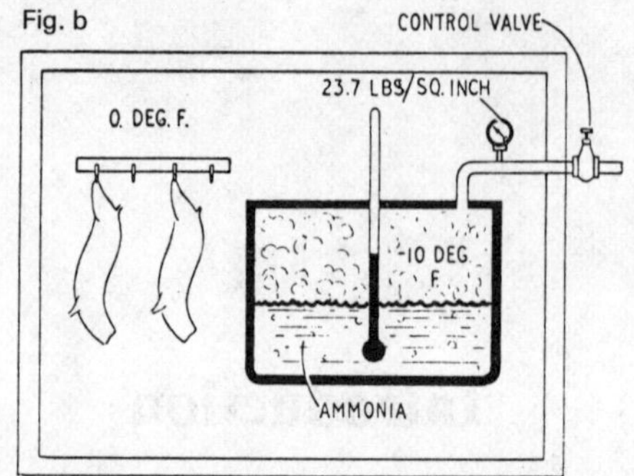

If we now arrange for a compressor to carry off the gas and compress it to a pressure which will result in a temperature higher than the surrounding atmosphere, it may be allowed to cool and liquify thus providing a source of liquid refrigerant still at a higher pressure than in the evaporator for re-evaporation. The addition of a throttling device to reduce the pressure of the hot liquid to that of the evaporating vessel provides the basis of a closed cycle continuous refrigeration plant.

The withdrawal and compression of the gas from the evaporating vessel can be accomplished by several means; The most common is the use of a reciprocating compressor although for large loads, particularly in the air conditioning field, centrifugal axial flow and screw type compressors are used. The absorption of the gas by a liquid or adsorption by a solid to be subsequently driven off by heat is another common method. The entrainment of the low pressure gas by a high velocity jet of the same 'substance' is a further variant.

It will be clear from the foregoing brief description of the refrigeration cycle that a vitally important factor in any cycle is the pressure–temperature relationship of the refrigerant. Normally the chapter devoted to 'Refrigerants' in most textbooks on refrigeration is prefaced by a note on the diversity of physical characteristics and illustrated by a diagram of pressure/temperature relationships something like Figure c, and the general indication that the whole matter is best derived from tables and Mollier charts many of which are readily available.

When a chart such as Figure d is presented covering the pressure/temperature relationship of virtually all the common refrigerants the study becomes much clearer; add to this the knowledge that the whole pressure–temperature–volume–enthalpy tables can be calculated with relative ease from the slenderest information and the study becomes infinitely more interesting.

Fig. c

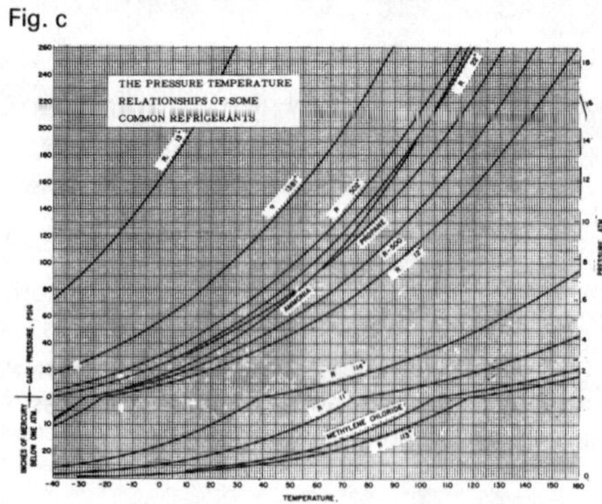

Fig. d

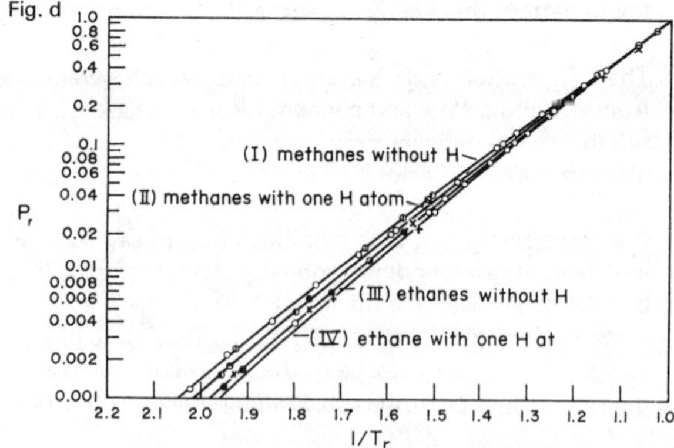

The calculation of liquid heat is taken fom a base temperature, this temperature when tables are in Imperial units is almost universally −40° in all tables. The greater part of the development of refrigerants in recent years has been done by countries using Imperial units and the majority of the metric or S.I. tables are merely straight conversions from the original works. Different countries have used different base points for the Enthalpy calculations when converting, NH_3 tables for instance sometimes use the base of 100 kcal/kg as the base for liquid heat in metric units whilst 200 kJ/kg seems popular with the fluorinated hydrocarbons in S.I. units.

It is largely because of the difficulties of obtaining original calculations of the properties of refrigerants in metric or S.I. units that the greater part of the data present in this book is in Imperial units.

Further confusion has been caused by these differences of bases since it has destroyed one of the few simple relationships in metrication i.e. that entropy expressed in units of kcal/kg ° K is the same as if expressed in units of Btu/lb ° R. The introduction of S.I. units as opposed to true metric units has also helped to destroy this relationship.

This book is in two parts, the first part deals with the definitions of the physical properties and outlines the methods of their calculation. The second part is devoted to calculated data on a range of refrigerants. The treatment takes the form of a data sheet, one for each of about thirty refrigerants; this data sheet gives the essential information from which can be calculated close approximations of the main main characteristics of the fluid. Following this is a set of tables of the saturation properties in Imperial units and Metric or S.I. units where they are available. Pressure Enthalpy charts follow the tables.

The refrigerants are arranged in the order of the now almost universally accepted numerical classification introduced by the American Standards Association and adopted by the British Standards Institution.

CHAPTER 1

Temperature

Definition of Temperature

The idea of temperature comes to us at an early age: fire is hot, ice is cold, and our senses recognise a wide variety of perceptions between. Terms which are used to describe temperature however are subjective terms and may well mean all things to all men. A hot day to an Eskimo is quite different from a hot day to a desert Arab. In order to express temperature in such a way that identical conditions may be described by a number of people it is necessary to quantify the property in some way and to provide some sort of numerical scale.

Some states of certain systems can be specified by a single physical quantity known as a 'state co-ordinate' – length, pressure, and electrical resistance are some – temperature is another. Temperature is a state co-ordinate which determines whether or not systems are in thermal equilibrium. If a number of systems are in thermal equilibrium they are said to have the same temperature, or to put it another way, they would have the same numerical value when measured on the same temperature scale.

Two words which occur frequently in thermo-dynamics are 'adiabatic' and 'isothermal'. The adiabatic process is one in which no heat is added to or taken away from a substance undergoing a process. An isothermal process indicates a change taking place at constant temperature. When two systems are separated by a thermal partition which allows their thermal state co-ordinates to vary over a large range of temperature independently it is said to be an adiabatic wall. When two systems are separated by a wall which allows one system to influence the other then it is called a diathermic wall. Neither of these two walls is obtainable perfectly in practice but a thick wall of cork or asbestos insulation approximates to an adiabatic wall and thin sheet of platinum approximates to a diathermic wall. When two systems are in contact through a diathermic wall, if their related thermal state co-ordinates differ they will change, coming closer together until after a time no further change takes place. When a state exists when all changes in the co-ordinates have ceased they are said to be in thermal equilibrium. Two systems in thermal equilibrium with a third are also in thermal equilibrium with one another and they are all said to have the same temperature.

One characteristic of a change of temperature in a substance is that it is normally accompanied by a change of length or volume and in most substances an increase in temperature results in an increase in length or volume. This characteristic has been known for many hundreds of years, the contraction and expansion of a liquid in a tube to indicate a change in temperature was applied in the middle of the seventeenth century. The earliest known instrument was owned by Ferdinand II of Tuscany in 1654 but there was no scale to the instrument and a movable cursor was used which indicated changes in temperature without assigning a numerical value to fixed points on the scale.

Temperature Scales

The earliest scale was devised by Gabriel Daniel Fahrenheit who produced a mercury-in-glass thermometer of very great accuracy. Fahrenheit was born in Danzig, or Gdansk as it then was, in l686. His business was that of a meteorological instrument maker although his study of natural science and particularly his contributions to the scientific journals of the time gained for him a great reputation as a physicist. Gdansk had passed through many political vicissitudes and in Fahrenheit's time was part of Poland. Fahrenheit left with his parents as a boy to live with them in Holland and later in England where he spent most of his life with frequent visits to Holland. He introduced his temperature scale in 1714 when he was twenty eight, and ten years later was elected a fellow of the Royal Society. He died in England in 1736.

The obvious requirement of a thermometric scale is some fixed or set point at which a commonly occuring substance exhibits some phenomenon which is always at the same temperature. Water being one of the commonest substances and which exhibits two thermometric phenomena frequently in everyday life was a natural choice by Fahrenheit for the earliest set point on a temperature scale. Water boils and water freezes. In Fahrenheit's time the belief was commonly held that when water froze, this was the lowest temperature obtainable. Fahrenheit however observed that lower temperatures were obtainable by using a mixture of ice and common salt and he took this as his 'absolute zero' and gave it as the base of his scale assigning it the value 0°. Seeking an upper fixed point for the scale he took the body temperature of a healthy man and then divided the distance between into twelve parts. Finding these divisions too big he divided each into quarters giving a degree approximately equal to a degree Centigrade. These were obviously still too big for practical use in whole numbers so Fahrenheit then divided each of his twelve degrees into eight parts in stead of four. Thus, the upper set point became 96° and from his scale he derived the normal melting point of ice as 32° and the normal boiling point of water as 212°.

The accuracy of Fahrenheit's instruments must have been exceptionally high since these two values have remained almost completely unchallenged up to the present day. His assessment of 96° as the normal human body temperature is also very close in view of its subjective nature and the diurnal cyclic variations.

A Swedish astronomer, Anders Celsius (1701 – 1744), divided the distance between the two set points of water into 100 divisions and the scale, known by his name and alternatively as Centigrade, was in general use in Europe.

Set Points

A system composed of a solid and a liquid of the same material and maintained at constant pressure will remain in phase equilibrium only at one constant temperature. Likewise, a liquid will remain in phase equilibrium with its vapour only at one temperature under constant pressure conditions.

When a liquid and its vapour are in phase equilibrium, this is said to be the boiling point; the boiling point however was found to vary with the atmospheric pressure and so a set of conditions had to be established to ensure uniformity. This is known as Standard Atmospheric Pressure (S.A.P.). S.A.P. is defined internationally and is represented by conditions which will support a column of mercury having a mass of 13.59 grams per cubic centimetre 760 millimetres high when subject to a gravitational acceleration of 980.7 centimetres per second per second. This is equivalent to 1013 milibars or 14.696 lbs per square inch or 29.921 inches of mercury.

A liquid and its vapour in contact at this standard pressure is said to be at its normal boiling point. Again, a liquid and its solid form in similar conditions is said to be at its normal melting point. At one definite pressure and temperature a substance can exist in all three states and this is known as the triple point. Water vapour can exist in equilibrium with water or ice, and ice and water can exist at 32° Fahrenheit at S.A.P. and since this is the only point at which it can exist in all three states – the triple point, it was taken for many years as the standard.

Refinements in thermometry in recent years resulted in the finding that the triple point of water differs from the equilibrium temperature of ice, water and air-saturated water vapour by 0.01° C.

At the Seventh General Conference of Weights and Measures held in Paris in September 1927, it was agreed to preserve the accepted temperature scale on which the temperature of equilibrium of ice and air-saturated water and the temperature of condensing water vapour both under the pressure of one Standard Atmosphere were numbered 0° and 100° respectively.

As a corollory to this, the triple point of water was fixed at 0.01° Centigrade. At the General Conference this was agreed as the only Standard Fixed Point in thermometry and it was given the arbitrary value of 273.16° Kelvin (° K). The Celsius scale had its zero point shifted so as to make the Celsius temperature of the triple point of water 0.01° Centigrade. Understandably some confusion remains in both the lay and the engineering mind.

By the time that the General Conference of 1927 was held it had become clear that the increasingly low temperatures being used in science and industry required a scale with its zero point much lower than that of melting ice, and the value of 273.16° Kelvin which the Conference assigned to the Fixed Point is based on the hypothesis that 0° on the Kelvin scale is the lowest temperature possible and is known as the Absolute Zero.

At the same time, the temperature of 273.16 was accepted as the sole fixed set point. Other basic set points were agreed and are shewn in Figure 1 in relation to the Kelvin and Centigrade scales.

A further set of calculations were agreed and ratified at a later conference when the Kelvin and Centigrade scales were related to Fahrenheit's scale and a similarly based scale known as the Rankine (R) scale had the value of 459.6° alloted to the fixed set point. Figure 1 also shews the relationship of the basic fixed points to the Fahrenheit and Rankine scales.

The Absolute Zero

As advances in low temperature physics are made, theories which seemed tenable and provided a working hypothesis for the phenomena of the period are outdated by further discoveries leading to a better insight into the laws which govern them.

Early experiments on the rate of expansion and contraction of gases indicated that all gases altered in volume by $^1/_{273}$ part of their volume at 0° Centigrade for each degree Centigrade that they were heated or cooled. There is some doubt as to the origin of the discovery and it is referred to as Gay Lussac's Law by some and as a modification of Charles' Law by others. Both however drew the same conclusion, that if a gas were cooled to –273° Centigrade, it would have no volume and that this point therefore must be the absolute zero of temperature.

Some philosophers believed that if they could attain the absolute zero they would have discovered the secret of creation, the underlying idea was that if a substance could be cooled until it contracted to a

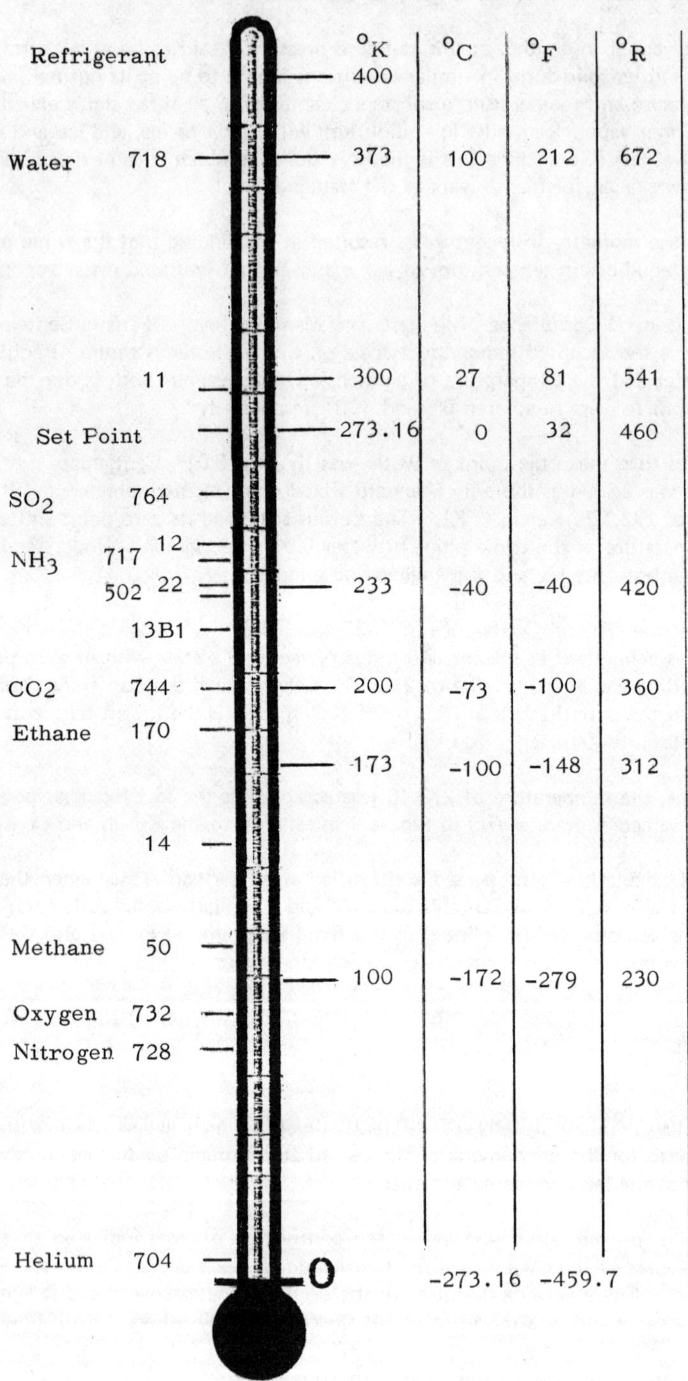

Fig. 1 Set Points on Thermoscales

point of non—existance there would still remain the Phlogistan, that mysterious all—purpose substance which provided so many useful explanations for otherwise inexplicable phenomena.

The next step after having isolated the phlogistan would be to warm it up again and the substance which had been made to disappear by cooling could conveniently be re-created into some substance — preferably gold — of the philosopher's choice. The phlogistan theory was widely accepted nonetheless, and many distinguished scientists such as Henry Cavendish (1731—1814) and Joseph Priestly (1733—1804) subscribed to it.

Again, much more recently, it having been observed that molecular activity decreased with a drop in temperature, the hypothesis that the absolute zero was the point at which all molecular activity ceased was and still is widely accepted. This theory too, has been discredited since it has been shewn that the atoms of a solid have a store of kinetic energy known as 'zero point energy'. The zero point energy of liquid helium for instance is almost three times as large as the heat of vaporisation.

Looking at Boltzmann's equation in the field of stastical thermo-dynamics we find that, so long as T_{abs} is positive, as by definition it must be, then all calculations are orderly and the results are what are to be expected.

By making use of more recent knowledge of energy states, it is possible to evaluate the total energy of a system and to subtract various terms that refer to potential energy. The results are indicated roughly in Figure 2.

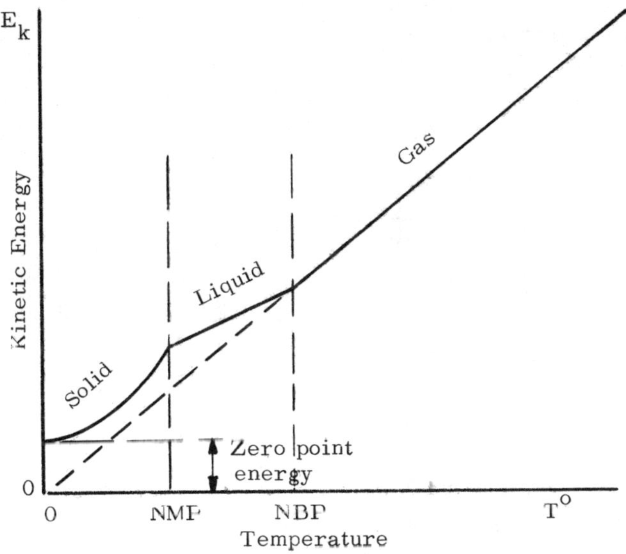

Fig. 2 Kinetic energy as a function of temperature and zero point energy

It is seen that the kinetic energy is proportional to T_{abs} only when the substance is a gas. It is not true of the liquid and solid areas, and at the absolute zero the kinetic energy is not zero but a finite quantity. Bearing in mind that zero point energy was finite, E.M. Purcell and R.V. Pound conducted a series of experiments at Harvard in 1951. They found as might be expected, that in a system with an infinite number of energy levels, if T_{abs} were to be negative then the system would require a quantity of energy greater than infinity and thus negative temperatures on the absolute scale were an absurdity.

However, with a system in which there were a finite number of levels, the picture changes and it was this new view that led them to carry out their experiments. Using the spin of Lithium ions in a Lithium fluoride solution they had a system in which there were four finite energy levels when immersed in a magnetic field. From the results of these experiments they postulated the possibility of the existence of negative temperatures on the absolute scale but the suggestion now is that the approach to the lowest possible temperature is asymptotic and that the numeral value of the Absolute Zero would therefore be $1/\infty$ ° K.

A new quantity, called a 'negcitemp' equal to the negative reciprocal of the Kelvin temperature has been postulated as a means of simplifying the complex issues involved in the mathematical solution of multiple finite energy levels. It may be that the ridiculed phlogistan theory was in fact an early attempt to explain zero point energy in terms appropriate to the times and the extent of then existing knowledge.

The Kelvin Scale

If the values of pressure P and temperature T of a gas are plotted when it undergoes a change of condition, a pressure–temperature graph may be drawn as in Figure 3 with the ideal gas scale shewn on the Y axis. A constant quantity of gas allowed to expand adiabatically from A to B would provide a smooth set of readings. By setting the temperature of the gas before cooling at a higher point, a second curve C–D would be obtained.

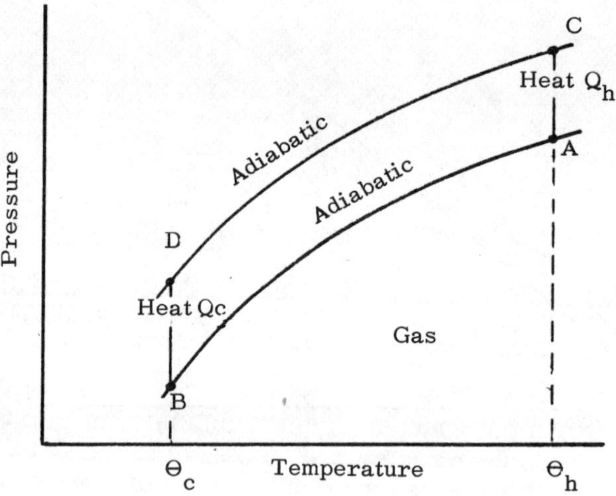

Fig. 3 The ratio of the Kelvin temperatures T_h / T_c is defined as being equal to the ratio of the heats Q_h / Q_c.

If the quantity of heat required to enter or leave the gas to enable the change from A to C and D to B in either direction is measured and these quantities are designated Q_h and Q_c then the ratio Q_h/Q_c equals the ratio of two identical functions of the temperatures h and c. Thus, $Q_h/Q_c = T_h/T_c$ and this latter was chosen to be the ratio of the Kelvin scale. This can be re-stated as:

$$\frac{T_h}{T_c} = \frac{\text{Isothermal process between two adiabatics}}{\text{Isothermal process between the same two adiabatics}}$$

If one of the isothermal processes is performed at the triple point of water and the value arbitrarily assigned to it is 273.16° K, then the Kelvin temperature for any other process is given by: $T = 273.16\ Q/Q_t$. Kelvin demonstrated that, within the limits of use, the ideal gas scale is identical with the scale which he postulated and which was ultimately given his name.

Since the smaller the value of Q, the lower the value of T, then clearly when Q = 0 the corresponding value of T = 0 or the absolute zero.

The Absolute Zero may be defined as 'the temperature at which a reversible isothermal process takes place in a system without the transfer of heat'.

CHAPTER 2

Pressure and Flow Work

The second most important state coordinate in refrigeration is pressure. This can be taken to mean, in the context of this book, the force exerted upon a body expressed by the equivalent of the weight upon a unit area.

Like temperature, some scale or scales must be evolved to allow measurement and comparison. But unlike temperature, pressure is of a much more complex nature. Pressure always exists as atmospheric pressure against which variable condition comparisons must be made and which provides for positive and negative values in many everyday readings.

Pressure moreover has significance in three quite separate and distinct ways: the static pressure as of the atmosphere, the velocity pressure as of the wind, and an energy significance derived from the static pressure of a fluid in motion but independent of its velocity.

Static Pressure

One of the simplest measurements of pressure is by measurement of the height of a column of liquid in a 'U' tube. A vertical U tube containing a liquid subject to pressure at both ends (one of which may be atmospheric) , will indicate a difference in pressure, if it exists, by a difference in height of the two legs of the tube. If the weight of the liquid in the column is known, the pressure is given by the weight of the difference. For the purposes of convenience this weight is usually expressed either in terms of unit area eg grams per square centimetre or pounds per square inch or in terms of the height of the column of liquid eg 760mm of mercury.

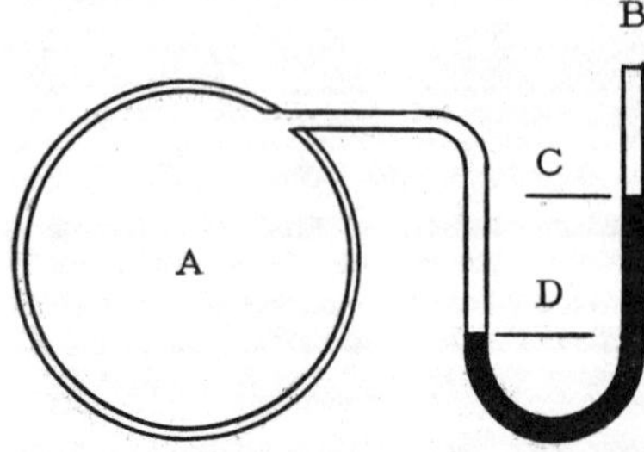

If one end of the tube containing a liquid is connected to a closed vessel A having a pressure within it greater than that of the atmosphere, and the other end B is left open, then the pressure above atmosphere will be shewn by the difference in height C–D of the two legs of the liquid column. This will be the static pressure in the vessel.

Velocity Pressure

Supposing that the vessel A (diagram p8) which is shewn as being cylindrical were to be taken as part of a conduit through which a fluid was flowing; providing that the pressure condition within did not change, the pressure indication would remain constant. If however, instead of a connection at E which simply butts on the perimeter, a connection is made which extends into the conduit, turns at 90° and has an open end facing the direction of flow, then there will be an increase in pressure related to the velocity of the fluid flow.

The pressure shewn by the U tube will then indicate the sum of the static pressure and the velocity pressure, and is referred to as the total pressure. If we then connected the atmospheric end of the U tube to the conduit in the original way this will carry the static pressure to the open end and the height of the column will indicate the velocity pressure only.

The velocity pressure will depend, apart from the velocity, on the molecular weight of the fluid and its significance will depend on its relationship to the total pressure. Thus in the case of air ducts in an air conditioning plant, the velocity pressure is high in relation to the total pressure in the duct so that the pressure is commonly used as a measure of velocity. Contrariwise the velocity pressure of a refrigerant in the cooling circuit of a compression plant is generally so low in relation to the total pressure that it is usually neglected in normal operational work. However in the centrifugal or axial flow compressor the velocities sometimes in excess of Mach 1 are sufficiently high to produce high ratios of compression when converted to static pressure.

The Energy Significance of Pressure

Pressure is of significance as an energy term indicative of the rate of transmission of work, per unit volume, through a fluid which is in steady and continuous motion.

This role of pressure requires special care in interpretation. Pressure as a property always exists and can always be expressed in units of say pounds per square foot absolute or Pascals. Pressure as an energy term however is of significance only when the fluid is in steady and continuous motion. Its comparable units would then be foot-pounds per cubic foot or Pascals per cubic metre; but though the absolute value of the pressure may not change when motion ceases, the meaning of the energy units will at that time entirely disappear. The energy significance of pressure depends on motion, but is independent of the magnitude of the motion; whether the flow velocity is very large or infinitesimally small, the amount of flow work associated with unit volume remains the same, but when motion ceases, this energy quantity instantly becomes zero, even though the pressure, as a property, remains unchanged.

In some fields of engineering, as in the design of piping systems, convention dictates the use of pressure as a property even though the conditions of use indicate that it is the energy significance that is actually of interest. Thus in piping problems, pressure drop is used as indicative of the loss of energy due to turbulence or to skin friction in a pipeline. Actually the decrease of energy results not from loss of pressure as a property but from loss of flow work, hence from decrease in pressure when expressed as an energy term with units of Pascals per cubic metre or ft-lbs per ft^3. Since the decrease in pressure is proportional to the decrease in energy, it is obviously of no practical concern whether the pressure be expressed as a force per unit area or as work per unit volume, but if the engineer is to develop a correct visualization of the relationships involved in any problem, he should at all times be aware as to which of these two concepts is the one significant to the context.

Flow Work

The subject of flow work has been brought in at this point because of its relevance in the appreciation of the energy significance of pressure and it is convenient also to insert a note on circumferential force as a manifestation of flow work.

Work is the most desirable of all energy forms, since a conversion from work to any other energy form proceeds more readily than conversions from other forms to work and for this reason a good deal of space is devoted to its analysis.

In the majority of processes and cycles the end product is shaft work. In one respect work and heat are similar: neither the presence, nor absence, influences the state at any point in the working substance. If one were to examine the state of the metal at any point in a copper rod, one could not determine whether or not a transfer of heat was occurring through the rod. Neither work nor heat influence the state of the material through which they are flowing, although the addition or removal of either work or heat from any substance (in contrast to flow through the substance) does affect the state of the material. The flow of water under steady state conditions through a pipe does not lead to any alteration of conditions within the pipe, whereas the flow of water to or from a storage tank has a definite effect on conditions within the tank.

Shaft Work

A shaft is essentially a conduit through which energy flows as work from the point of origin to the point of dissipation. Consider an object which will slide across a rough surface only when a force is applied to overcome friction. If the force is applied directly to the object, work is done in causing a movement. However if the force is applied at the end of a rigid rod the work required to cause motion is as before, and the fact that the force acts through the rod in no way influences the situation or leads to any change in the rod. The transit of work through the rod occurs in an infinitesimal time; that is to say, at the instant a force is applied at the end, it is immediately transmitted through the rod and becomes active against the object. Work which is transmitted through the rod is energy in transition. It does not come from storage in the rod, but is supplied from some exterior energy source and passes through the rod without changing conditions within the rod.

Irrespective of the cross-sectional area of the rod, the force acting at every cross-section is necessarily the same, but the stress, S (force per unit area) is less for a larger rod area. If we choose to consider a length of rod, L (of uniform cross-section), such that its weight is 1Kg, then the work transmitted as this length L moves past the original marked position is: $W=FL=SAL=S\bar{V}$ kGm/kg where $\bar{V}$ is the reciprocal of the density or specific volume, m^3/Kg., and A is the cross sectional area.

One could therefore say the $S\bar{V}$ units of work flow through the rod for each unit weight of rod that is displaced over the mark on the ground, but the total weight of the rod would have work significance only if the total length of rod moved past the mark. The $S\bar{V}$ term does not represent energy stored in unit weight of rod, but merely indicates the energy that would pass through this piece of the rod if displacement past the ground mark were to occur.

Let the rod through which the force is delivered to the object be replaced by a metal tube, filled with liquid and with a frictionless piston, or long plug, partially inserted in each end. If now a force is applied to the plug at one end of the tube, this force will cause the plug to move in, the fluid to move, and the plug at the other end to move out and the object to slide. In every respect, the situation is identical with that previously considered, except that the column of fluid now serves in place of the rod as a device for transmitting energy as work. The fluid will be no more influenced by the passage of energy through it than was the rod.

Returning to the solid rod, if the force is 10kg and the distance the object moves is 5m the total work delivered is 10 x 5 = 50kGm and this quantity of energy must pass every cross-section of the rod. If now, one chooses to take a position beside the rod and make a mark on the ground and another on the rod to identify its original position, one can say that the work required to move the rod one metre past the mark on the ground is 10 x 1 = 10kGm. As visualized from the fixed reference position beside the rod, the work associated with each metre of rod length that passes the section in question is 10kGm. Such a statement has meaning however, only when applied to conditions as they exist at the fixed reference point; the work associated with the entire rod is not 10 times its length, but if an infinite number of reference points were set up an infinitesimal distance apart, the work associated with passage of one foot of rod past each reference point would still be 10kGm. As the rod moved, the same 10kGm would pass each cross-section in turn; one could not add the energy passing the various cross-sections because it would in every case be exactly the same energy.

When dealing with the transmission of work through a fluid, the force applied per unit area of cross-section is the pressure, P, rather than the stress, and the work which flows through the fluid as unit weight of fluid passes a given cross-section of the tube is then $P\bar{V}$kGm/kg. Here again the term $P\bar{V}$ represents energy associated with, rather than stored in, the fluid, and this term has energy significance only when motion is occurring and work is being done. Thus, although the product of the two properties P and $\bar{V}$ is itself a property, it is a term which signifies energy only when the fluid is in steady and continuous flow; for this reason $P\bar{V}$ is frequently referred to as the flow work associated with any fluid which is in a state of steady and continuous motion. When motion does not exist, the numerical value of the $P\bar{V}$ product is of course unchanged and its dimensions remain kGm/kg, but this value is meaningless except insofar as it shows how much energy would flow through the fluid if it were in motion. In every real sense, flow work is identical with shaft work and represents energy in transition.

Since the work delivered through the fluid depends only on the force applied and the distance through which that force acts, it is evident that the result would be the same whether the connecting section of fluid were very thick or very thin, and whether it were long or short. Irrespective of length or thickness of the fluid section, the same amount of energy will be passing every cross-section and, in literal fact, exactly the same energy will be passing each cross-section since the transfer of work can be visualized as occurring instantly. Hence, the work entering one end is simultaneously leaving the other end.

Irrespective of the form in which it may manifest itself, work is always equivalent to, and expressed as, the action of force through distance. Thus all expressions applicable to special manifestations of work must come from the one fundamental equation: W = Fd = FdS, where F is force, d is distance, and dS is the infinitesimal distance over which the force acts. The integration of this equation obviously requires knowledge of the way in which the force varies as a function of distance. Since the relationship between force and distance is sometimes rather complex, it is evident that integrable forms of the equation may be diverse, but it is important to remember that all expressions for energy in transition as work must be some form of this one basic equation.

Circumferential Force

Another method of transmitting energy through a solid rod is by applying a circumferential force. The distance through which that force moves is equal to the distance travelled by a point on the circumference. Since π d is equal to the length of travel per revolution, the distance per unit time is π dn where n is the number of revolutions in unit time. In this case, the work done at the other end of the shaft will likewise be πdn and this will be entirely independent of the shaft length. Once again, the energy in transit as shaft work flows through the shaft but does not in any way affect the state of the shaft itself; the shaft acts merely as a defining path along which energy flows.

CHAPTER 3

Other Thermodynamic Properties

Specific Heat

The specific heat of a substance is the ratio of the thermal capacity of the substance to the thermal capacity of an equal weight of water. Being a ratio it is strictly a dimensionless quantity but because most units of heat are referred to a water basis, the specific heat is often given in terms of heat/mass/degree temperature.

Although there are slight variations in the specific heat of solids, they are very small and the specific heat is accepted as being substantially constant.

The specific heat of a liquid will vary with the temperature and usually does so considerably. It tends to rise slowly and uniformly from very low temperatures and to rise rapidly as it approaches the critical temperature. The value of the reduced temperature at which the rapid increase commences is approximately 0.8.

In the case of gases the value of the specific heat is again substantially constant providing that either the pressure or the volume remains constant. Thus a gas has two specific heats, one at constant volume, the other at constant pressure. For all working substances which expand as they undergo a temperature increase, the numerical value of the specific heat at constant pressure (Cp) exceeds that of the specific heat at constant volume (Cv).

Under constant volume conditions a change in temperature and pressure takes place when heat is added or taken away. When heat is added to a gas at constant volume the temperature and the pressure increase, and since no work is done the gas retains all its energy. When heat is added at constant pressure an increase occurs in temperature and volume and some energy must be expended in increasing the volume against the pressure surrounding the container. Thus more heat is necessary to raise the temperature of a volume of gas at constant pressure than to raise the temperature of the same weight of gas at constant volume.

The difference between the two specific heats — at Cp and at Cv — represents either energy going into internal potential storage or work done on the fluid during compression. In rare cases where material decreases in volume while its temperature is being raised, eg water between the temperatures of 0° and 4° Centigrade, work is done on the fluid; hence the work term is negative, and for such cases the specific heat at constant pressure is less than the specific heat at constant volume. When heat addition takes place without change in volume and no external work is done, all energy added as heat remains stored in the fluid as internal kinetic energy. Thus it is evident that a valued desirable characteristic of a working refrigerant should be a considerable difference between Cp and Cv since this difference largely represents the conversion of work into thermal energy as the result of the process of compression.

ENTHALPY

Neglecting mechanical energy and assuming flow through an insulated pipe, the two forms of energy stored in or associated with a fluid will be: internal energy in storage, and flow work in transition. Since the expression for flow work, $P\overline{V}$, is a product of properties, it must itself be a property and it will remain a property irrespective of whether it is possessed of energy significance. The expression for internal energy, u, is likewise a property and since any combination of properties must itself be a property, it follows that we can define a new property, enthalpy, by the equation $H = u + AP\overline{V}$, where H = the enthalpy, and A is the coefficient changing units of the $P\overline{V}$ product from work units to heat units.

Since enthalpy is defined by the above equation it follows that it is an arbitrarily selected term having units usually of Btu per lb or Calories per gram, but only having valid energy significance if and when the $AP\overline{V}$ product has such significance. Thus enthalpy, like flow work, is sometimes an energy term and sometimes not, depending on whether the fluid is in steady and continuous motion. By far the greater number of processes considered in refrigeration or air conditioning involve the consideration of steady motion, hence permit use of enthalpy as representing the sum of internal energy and associated flow work, but in those cases where motion ceases the energy significance of this term must also cease even though, as is usually the case, its numerical value remains unchanged.

Two Energy Quantities considered.

Although dimensionally consistent, enthalpy is somewhat of an anomaly since it represents, for a fluid in motion, the sum of two energy quantities, one of which is in storage and the other in transition. This distinction must always be remembered since the treatment of stored and transferred energy must often be different and in such cases some confusion might exist as to the method of handling enthalpy. Thus, if w units of fluid are flowing in a pipe, the product wu is equal to the total internal energy stored in the fluid, and the term wH is, by definition, the total enthalpy. But whereas wu corresponds to an actual energy quantity, wH does not. The total enthalpy, wH, is no more than a number, without reference to any existing energy quantity, which is indicative of the energy that would be delivered, either in stored form or flow work, if w units of fluid were forced across a given cross–section of the pipe.

There is a wide variety of ways in which the property enthalpy finds use in engineering, but the most important and most common use is in cases where steady flow exists and where H is, therefore, equal to the sum of internal energy and associated flow work. Since internal energy and flow work are the only two practical means of transferring large quantities of energy from the source to the place of utilization, it follows that the property enthalpy finds widespread use in all varieties of engineering applications.

If a liquid refrigerant is placed in a vertical vessel of extended length with a frictionless piston and supposing the temperature to be –40° C, then, if this temperature is raised to say 0° C the refrigerant would expand and the piston would be forced out against the pressure of the atmosphere and energy would be expended. The heat equivalent of the work done by the expansion of the liquid is so small as to be negligible. This is not the case when a change of state from liquid to gas takes place when the increase in volume is very much greater and the heat equivalent of the work done is correspondingly greater. These two quantities represent the energy required to perform External Work or Flow Energy.

The energy of the liquid is the sum of: the internal work which is calculated from an arbitrary base (usually −40° C), and the external work of expansion against the atmosphere. Because of the negligible values of the latter there is no point in separating them and the liquid heat quoted in the tables is given as a single quantity.

The latent heat of a vapour is the sum of the energy required to overcome the pressure exerted during the expansion in addition to the energy required to cause the change of state. All the energy cannot have originated as thermal energy even though the external work content is expressed in heat units. For this reason the term Enthalpy is used although in some older tables the term 'heat content' appears.

Enthalpy is the internal energy (u) plus the value representing flow energy of the fluid. The symbol generally in use is H, which equals $u + {}^{pv}/_J$ where u = Internal Energy, p = pressure, v = specific volume, and J = Joules equivalent in appropriate units.

The three commonly used phases of Enthalpy are (1) the liquid heat, (2) the latent heat, and (3) the vapour heat, which is the sum of (1) and (2).

Latent Heat of Vaporization.

When a liquid is used as a working substance, work of expansion is done during the process of evaporation. For a constant pressure evaporative process, the temperature remains constant, but a large amount of energy goes into internal potential storage and, at the same time, the volume increase associated with the change of phase from liquid to vapour causes delivery of work of expansion to whatever the fluid expands against.

In any expansion process work must be done in pushing back the atmosphere. It is evident therefore, that the effectiveness of the process will increase as the pressure increases, since the ratio of absolute pressure to atmospheric pressure will be equal to the ratio of total work to work done on the atmosphere. On the other hand, the amount of expansion during evaporation decreases with increasing pressure, hence investigation is necessary to determine where greatest advantage is to be found; but it will be seen that the effectiveness of phase change as a means of transforming thermal energy to work increases with pressure.

The thermal energy input to a liquid to achieve evaporation is defined as the latent heat of vaporization. As already noted, the latent heat goes partially into potential storage and is partially dissipated as work done during expansion; but since the process occurs at constant temperature, it is not possible to express the latent heat of vaporization in terms of specific heat.

During the reverse process of condensation of a vapour at atmospheric pressure, an equal amount of thermal energy is liberated, but again one must note that this energy comes only in part from storage in the vapour; the remaining fraction is a transformation of external potential energy stored in the atmosphere to work done in reducing the volume of the vapour as it undergoes phase change to a liquid. This work is then transformed to thermal energy and leaves the fluid as heat.

Latent Heat of Fusion.

The primary energy difference between a phase change from liquid to vapour and from solid to liquid is that in the latter case the change in volume is very much less than in the former, hence by far the greater part of the energy which must be added to achieve the phase change goes into internal potential storage. For most solids there is a volume increase as liquefaction occurs. In some cases however, of which the phase change from ice to water is an example, there is a volume decrease during the phase change, hence work is done on the material and the change of internal potential energy then exceeds the latent heat of fusion. Such special cases can be readily treated, provided the ΔV term in the usual equation is considered to represent an increase rather than merely a change in volume; on this basis the $AP\Delta V$ term of the equation would be negative for materials which shrink during melting.

ENTROPY

Inconceivably large quantities of heat energy are present on the Earth but since, as is stated in the second law of thermodynamics, "heat must always flow from a hot body to a colder one", much of it is not available for use by humanity.

In any study of the relationship between heat and mechanical energy, some parameter is needed to define the availability or the unavailability of this energy. Entropy is a factor devised by engineers to provide such an indicator.

Entropy according to the Oxford Dictionary, is the "measure of the unavailability of a system's thermal energy for conversion into mechanical work" which is not very helpful. Sir James Jeans uses entropy to define the second law of thermodynamics thus: "the entropy of a natural system always increases until the final stage is reached in which entropy can increase no farther" which again takes us very little nearer to understanding.

In some ways entropy and temperature are much alike; as we saw earlier in discussing temperature, the first measuring instrument only indicated changes in temperature and it is this property which is commonly of most interest. If the night is bitterly cold, it is of little interest to know the temperature whilst it may be vital to know whether it is rising or falling and by how much. In the same way with entropy, the numerical value is of little use but the change in entropy gives a great deal of useful information.

Entropy cannot be measured but must be calculated, and is a useful composite property the physical significance of which is not important in refrigeration problems. The numerical value as we said has little significance; it is the change in entropy which is of concern to the engineer.

Entropy change is the heat energy transfer to or from a substance per degree of average absolute temperature. The fundamental property of the entropy of a system (S) is that, in a reversible process the product of the temperature and the entropy change is equal to the heat transfer.

The base used for the calculation of entropy varies as between authorities but in most refrigerant tables the base used is $-40°$ and is usually given as a footnote if otherwise. This base has been generally adopted for tables in Imperial units but where conversion to S.I. units has been made, the base used has been at the discretion of the convertor and the results are somewhat confusing.

The characteristics of the entropies of practically all solids, liquids and gases are:

(1) If only the temperature is varied – the higher the temperature the greater the entropy.

(2) If only the volume is varied – the larger the volume the greater the entropy.

If you consider the isentropic expansion of a gas, an increase in volume provides a contribution to an entropy change. To keep the entropy constant there must be a compensating negative contribution to the entropy change provided by a decrease in temperature. The temperature drop which takes place when a gas expands adiabatically may therefore be regarded as an effect that is needed to offset the effect of the rise in volume in order to keep the entropy constant. Hence when an isolated system undergoes an irreversible process on its way toward equilibrium its entropy must always increase and the equilibrium state is the state of maximum entropy.

Despite the complex nature of the term however, the actual calculation of entropy values is quite simple, as will be seen in the section relating to calculations.

The change in entropy in passing from one state to another is the same, irrespective of the reversible path which the substance follows. It is fixed when the state of the substance is fixed and is a property of the fluid.

Entropy and Random Distribution

Atoms comprise the basis of all substances, a molecule is a combination of atoms, and a compound is a mass of material made up of molecules comprising two or more different kinds of atom. The atomic structure comprises a positive nucleus with one or more negative electrons travelling at high speed in orbit round the nucleus. The number of electrons and their balancing positive charges constitutes the difference between the various elements. Atoms and molecules are in constant agitation, collision and random movement; the electron orbits are irregular. An increase in internal kinetic energy usually results in an increase in the random motion of the molecules and vice versa, in some cases, electrical charges attract and act as cohesive forces holding the molecules together.

Entropy may be considered as a measure of the degree of molecular disorder. If a gas occupies a very small volume it is spatially set apart, all molecules are in a restricted region and no molecules are anywhere else which is a state of low disorder. If the gas expands so that the molecules occupy a greater space they reach a state of greater disorder and as the system proceeds towards equilibrium the array of disorder becomes greater.

Again, the similarity between entropy and temperature appears, for temperature may be considered as a measure of molecular agitation of a random kind.

The word 'random' is something of a paradox since it is normally taken to mean haphazard or disorderly; and yet the greater the state of disorder, the more closely will any parameter conform to the laws of probability thus making it easier to forecast its behaviour accurately. The more closely the distribution approaches true randomness, the more closely it conforms with the laws of probability.

A further factor contributing towards conformity with the laws of probability is the size of the sample; the larger the sample, the greater the conformity. When dealing with molecules, the sample size is enormous so that great accuracy in behaviour forecasting is possible. This whole aspect of Physics has developed into a sub-branch known as Statistical Mechanics.

CHAPTER 4

Energy Transformation

The Laws of Energy Transformation

All refrigeration processes have as their purpose the removal, supply, transformation, transfer or storage of energy. Energy is the ability of a substance to move or to change other things. Heat is the energy transfer that is connected exclusively with temperature differences between two systems. The refrigeration cycle is concerned primarily with heat and:

(a) its storage and concomitantly the evaluation of the storage quantity,
(b) its transfer and concomitantly the measure of the rate of transfer, and
(c) the transformation and concomitantly the conditions and extent of its transformation.

Storage

A refrigerant has stored within it a certain amount of internal molecular energy. This is of two kinds – kinetic molecular, and potential molecular. Changes in the kinetic molecular energy result in changes of temperature. Changes in potential molecular energy produce changes in its state. Internal potential molecular energy is the energy required to offset the cohesion required between the molecules rather than to change the temperature. Thus, molecules of water have a certain amount of cohesion which retains them in a liquid form. The addition of energy to water 100° C at atmospheric pressure will break down this cohesion and change the liquid to a gas. This gas will have acquired an additional amount of potential energy equal to the amount required to produce vaporization, that is to say the latent heat.

Transfer of Energy

The transfer of heat energy depends on two basic laws of thermodynamics. These were promulgated by scientists in early thermodynamics, and expressed relationship between the properties of thermal systems. The first law of thermodynamics is a statement of the law of conservation of energy, and it denies the possibility of a device from which power work can be drawn continuously without replenishment. It requires at least as much energy to be supplied to the device as is drawn from it. Conversely, in respect of a refrigeration system, whatever energy is put in must either stay in or come out. It has been stated in a possibly more precise but less direct form, eg: "Energy can neither be created from nothing nor annihilated, and the different forms of energy are equivalent; consequently the sum of heat and other forms of energy received from the surroundings in different parts of a closed cycle without natural generation of energy, without accumulation of energy and without loss of previously conserved energy is the sum of the heat and other energy given off to other parts of the cycle". However the law is so patently simple as to require no further explanation. Thus, in a refrigeration plant, heat absorbed in the evaporator plus equivalent of work done in compression equals the heat rejected to the condenser.

The second law of thermodynamics states that in sensible heat flow, heat only flows from a hot body to cooler one. This has also been stated in other ways, eg: "The entropy of a natural system always increases until the final stage is reached in which entropy can increase no further" (Sir James Jeans). "No cyclic process in possible whose sole result is the flow of heat from a single heat reservoir and the performance of an equivalent amount of work on a work reservoir" (Kelvin Plank statement).

The two laws may be co-related in the statement that whilst the first law says the work produced can never be greater than the energy supplied, the second law goes farther and says that it must always be less.

Transformation of Energy

It is a relatively simple matter to transform shaft work into either of the forms of energy in storage which have already been considered. Thus if the shaft drove a propellor (as in a blender) submerged in a fluid, the result would be rotational motion (external kinetic energy) of the fluid and eventually under the influence of fluid friction, a rise in fluid temperature in an amount proportional to the increase in kinetic energy of the fluid. This was one of the methods used by James Prescott Joule in his classic experiment to determine the mechanical equivalent of heat which he carried out in Manchester in 1850.

Reversible Processes

A reversible process is one in which the system and its surroundings are without complication such as acceleration, eddies, turbulence, friction. Every property like temperature and pressure is uniform throughout the system so that one value of the property holds for the entire system. The system is very close to equilibrium at all times.

When a change of state occurs without fluid friction and with uniformity of all properties throughout the mass, it becomes possible to express the state of the fluid at any point along the path. Such a process is said to be reversible. In practice, friction, turbulence and heat transfer within the body of the fluid cause variation throughout the mass, and it is sometimes impossible to define the state of the fluid at successive intervals during the process; such a process is called irreversible.

All real processes shew some degree of lack of homogeneity caused by mechanical or fluid frictions, heat gains or losses, but fortunately for most heat engine processes the degree of irreversibility is very small and allows valid analysis to be carries out as for a reversible process.

GAS LAWS

Robert Boyle

Robert Boyle was born into a wealthy and influential Irish family on the 25th January 1627 at Lismore in County Waterford. Although he was the seventh child there was no lack of money and he went to Eton in 1635 at the age of eight which is very young by present day standards. After four years there he was sent to the Continent with a tutor and lived mainly in Switzerland, but also in France and Italy, until 1644. He returned to England during the Civil War and was reunited with his sister Katharine Lady Ranelagh, to whom he had always been devoted; he lived with her when in London and spent a proportion of his time in Stalbridge, Dorset, the manor which he had inherited from his father and where he wrote many of his moral essays. His life was a strange mixture of science and religion; he wrote widely on religion and was the author of 'The Christian Virtuoso' seeking to prove that the study of nature was the religious duty of everyone. He revived, revised, and financially assisted through printing, the Irish Old Testament. He wrote a number of sermons on the theme of God and Nature. On the scientific side he developed the theory of primary particles in which all natural phenomena were explained by the motion and organization of primary particles, something very advanced for its time.

At the house of his sister Katharine in London he met many interesting people, among them Samuel Hartlib with whom he constructed the first air pump. It was directly as a result of the work with this pump that he made the experiments which eventually led to his postulating the law which carries his name. He was interested in trade and manufacture, and with one of his assistants owned an interest in a chemical manufacturing company. Robert Boyle had little time for social life other than that which brought him into contact with other scientists and theologians. He never married, for his life was too full. He was a fellow of the Royal Society and a director of the New England Company. In his will he endowed a trust to provide for the Boyle Lectures 'to prove Christian Religion against Notable Infidels' which continue even now. He died on the last day of 1691 and is buried in the Church of St Martins-in-the-Fields.

Boyle's Law may be stated as follows:

"with constant temperature, the volume of a perfect gas varies inversely as the pressure applied" — or: PV = constant.

Joseph Louis Gay Lussac

Gay Lussac was born in the early morning of the 6th December 1778 in the small town of St Leonard in the Department of Haut Vienne. He was the son of an important citizen in the town; his father was a civil servant and a judge. But life for young Joseph was not easy, for his father was imprisoned for his sympathy with the aristocrats during the rebellion and times were difficult for the family. However, he received a sound education at the town school supplemented by his mother's teaching, and at the age of nineteen he entered the Ecole Polytechnique and graduated in 1800. Though he was offered a post in Paris he refused it in favour of the opportunity to become assistant to the then famous Berthollet at Arceuil. His natural curiosity had taught him a great deal, from his youth he was always seeking to know 'why', and Arceuil had become the centre of a group of young scientists who spent hours in discourse after their days' work was over.

He was adventurous and undertook several hazardous tasks in the pursuit of knowledge; among other things he made three balloon ascents to study the air and the magnetic fields above the earth, and whilst he undertook these ascents to further his knowledge, his third ascent was to a height of 23,018 feet which was nearly 3,000 feet higher than the record at that time.

His work was not unrecognised and he became a professor in Paris and was elected to the Institut de France.

The great year in Gay Lussac's life was 1808; in the spring he was married and was settling down very happily and finalizing his most important work. On the last day of the year he announced the publication of the results and postulated the law which bears his name.

Gay Lussac was a highly respected member of the university community in Paris; he had accepted the Chair of Chemistry at the Muséum National d'Histoire Naturelle in Paris in 1832 and was elected to the Chamber of Deputies in 1831, 1834 and 1837, and in 1839 was given a peerage by Louis Phillipe. In the early spring of 1850 he fell ill and died on the 9th May 1850 at the age of 71.

Gay Lussac's Law states that:

''the volume of a specific weight of gas varies in direct proportion to its absolute temperature at constant pressure''.

Jaques-Alexander-César Charles

Beaugency today is probably much the same as it was in 1746 when Charles was born on the 12th November. It is a quiet provincial town in the Orleannais where fishermen sit all day along the embankment of the Loire and life moves gently.

His education was that of the other middle class children of small provincial French towns – somewhat sketchy and without possibility to develop any particular aptitude. He was good at arithmetic however and without difficulty obtained a position in the local office of the Ministry of Finance where he quickly rose in esteem. His work gave him time to satisfy his enquiring mind and turning to science he experimented with electricity. His first practical works of note were not so much original as improving the work of others. He produced a reflecting goniometer and improved on Fahrenheit's aerometer. Branching out on his own he invented a hydrometer and in the course of his experiments came into contact with the brothers Nickolas and Anne-Jean Robert, and together with them built the first hydrogen balloon and with Robert made the first recorded ascent to more than 5,000 feet.

In 1785 he was elected to the Academie des Sciences in Paris and later became Professor of Physics at the Conservatoire Arts et Métiers.

In 1787 he anticipated Gay Lussac's work on the expansion of gases which resulted in that part of the gas laws now known by his name.

Although he was known for his contributions to physics, not a great many of his published works are in that field; his main written contributions were purely mathematical in content. He lived something of a spartan life despite his comfortable financial situation, and this may well have had something to do with his living to a ripe old age. He was 77 when he died in Paris on the 7th April 1823.

Charles' Law states that:

"the pressure of a specific volume of gas is directly proportional to its absolute temperature".

James Prescott Joule

James Prescott Joule was one of a middle class Salford family and received the usual middle class education of the period – he was born in 1818. From the age of thirteen he attended Manchester New College and came under the influence of John Dalton who was then at the height of his career. It is generally said that Joule was largely self taught and this is basically true, but it was mainly Dalton's influence that led him to think for himself and to know how to "teach" himself. At the age of nineteen he published in "The Annals of Electricity" a description of an electro magnetic engine. He devoted his time to chemical and physical research and obtained the post of Library Secretary of the Manchester Society which greatly facilitated his research work. He became President of the Manchester Society and was elected FRS at the age of 32 in 1850. In contrast to Dalton who was a quiet retiring man and a devout Quaker, Joule enjoyed the honours he received with obvious pleasure but without ostentation. He received the Copley Medal from the Royal Society in 1860 and the Albert Medal of the Society of Arts in 1880. He was given honorary doctorates from Oxford, Dublin and Edinburgh Universities. He died in 1889 at the age of 69.

Joule's first discovery was connected with the production of heat by voltaic electricity, but the most important one was the equivalence of heat and energy. By the use of weights, cords and pulleys he rotated paddles within a calorimeter. Heat was generated which raised the temperature of the water in the calorimeter and by this means was able to establish the relationship between heat and mechanical energy. He published the results of several years' work in 1850 under the title "A New Theory of Heat".

He expressed the mechanical equivalent of one British Thermal Unit as 778 foot pounds which is known as Joule's Equivalent. The unit of energy which has been given his name reflects his electrical achievements rather than the thermodynamic.

CHAPTER 5

Refrigerants

"What are the most commonly occurring substances used as refrigerants?" It is very rare that the answer to this question, even from informed audiences is "air and water". These two are not only more commonly occurring but, in the case of air, probably the most commonly used refrigerant.

At this point it is necessary to differentiate between primary refrigerants and secondary refrigerants and these conform, for the purposes of this book, to the following definitions.

A primary refrigerant is one which is used in a recirculating cycle and is accompanied by changes in state. A secondary refrigerant is one which is used as a heat transfer medium without a change of state but with a change in temperature.

Examples of primary refrigerants are:

1. The ammonia in a conventional cooling plant which is compressed, liquified, evaporated and recompressed.

2. The water in a lithium bromide absorption plant which is evaporated, absorbed in the lithium bromide solution, driven off in the generator and recondensed.

Examples of a secondary refrigerant are:

1. Chilled water used in a conventional air conditioning plant and circulated through the air cooling coils.

2. Air in a cold room, having a direct expansion cooler, by which the incoming heat from the cold room and its contents are transferred to the primary refrigerant in the cooler. In this context it should be noted that in the case of the air conditioning plant employing a secondary refrigerant such as chilled water, the recirculating air is a tertiary refrigerant transferring the fabric and other heat gains from the building to the chilled water.

Air and water are not so commonly found as other refrigerants in use as primary refrigerants and attention will therefore first be given to the more common primary refrigerants e.g., ammonia, halocarbons and similar substances used in conventional refrigerating plants.

CONVENTIONAL REFRIGERANTS

These fall naturally into four groups:

1. The Classic Refrigerants

Ether, ammonia, sulphur dioxide and carbon dioxide. Methyl chloride may be considered as a bridge between the classic refrigerants and the halocarbons since it was the earliest refrigerant to be formed by displacing the hydrogen atoms in the hydrocarbon group in the search for new and better refrigerants.

2. Halocarbons

Were formed after considerable research in extending the process of displacement of hydrogen atoms referred to in the last paragraph.

3. Hydrocarbons

Are all interesting to the refrigeration engineer primarily for the problems which they present in the compression and condensation for storage and transport and revaporisation for external consumption rather than for their use as refrigerants.

4. Azeotropes

Among the commonest examples of an azeotropic mixture is air but the term is generally used by refrigeration engineers to cover the conventional refrigerant mixtures.

DEVELOPMENT & HISTORY

In the early days when mechanical refrigeration was limited to a few applications of an industrial nature, there were very few refrigerants known or used.

Ether of course deserves mention for its place in Jacob Perkins historic plants but it was soon displaced, as were some other similar substances.

Ammonia and Carbon Dioxide were the only two of the early refrigerants which retained their popularity over a long period. Carbon Dioxide because of its safe properties was used in almost all marine installations and Ammonia because of its excellent thermal properties was to become widely used for many years and still retains its popularity.

The greater use of refrigeration in retail business such as butcher's shops and the realisation that there was big money in domestic refrigeration called for a low pressure refrigerant to allow for lighter, smaller, cheaper plants and in the 1920's Sulphur Dioxide was introduced as a refrigerant and for a time was used almost universally in small commercial refrigeration. It formed Sulphurous acid with very small quantities of moisture – attacking all copper in pipelines, etc., and it was also very toxic indeed.

Its main reason for lasting so long (10 years or so) was that it had one great advantage – it was cheap – about 2½p per pound. Even at today's values it is still among the cheapest gases to be used as a refrigerant.

The search for a completely safe refrigerant with good thermal properties led chemists in the 1920's to examine the gas Methane (CH_4) and they found that one of the hydrogen atoms could be synthetically replaced with one of Chlorine. This produced a compound known as Methyl Chloride.

Thus

```
      H
      |
  H — C — H = CH4 = Methane
      |
      H
```

Replace one of the Hydrogen atoms with Chlorine.

```
      H
      |
 Cl. — C — H  CH3 Cl = Methyl Chloride
      |
      H
```

This gas was used extensively before and during the last war mainly in the commercial and domestic field. Although Methyl Chloride only came into general use as a refrigerant in the 1930's it is worth mentioning that the French firm of Douane was awarded a gold medal at the Paris Exhibition of 1879 for a refrigerating machine using Methyl Chloride. Now, whilst Methyl Chloride was suitable as a refrigerant in many ways, it had some disadvantages. It is very similar in chemical structure to Chloroform ($CH\ Cl_3$), which itself can be used as a refrigerant, and in large concentrations it has anaesthetic properties. It can also be explosive or inflamable if mixed with air in certain concentr-ations and it can be dangerous to servicemen when using leak detectors in confined spaces.

In contact with moisture it will displace copper from the interior of say a copper pipe and deposit it on surfaces of iron. Thus some moving parts of a compressor may become "copper plated" to the extent that they will seize completely.

So, the American Chemical firm E.I. du Pont de Nemours continued the search by initiating a research programme under Dr. I. Midgeley and the first commercial results of his team work was a fluorine derivation of the parafin family in 1930. The compound was marketed commercially as Freon 12 and it had many of the properties ascribed to the hypothetical ideal refrigerant; apart from its thermal properties it was safe and it was stable.

The development of Freon 12 was a continuation of the process which had produced Methyl Chloride ($CH_3\ Cl$). As was shown previously, the hydrogen atoms of Methane (CH_4) can be synthetically replaced by other substances. By substituting Chlorine for all four hydrogen atoms carbon tetra chloride is obtained thus:

```
      H                        Cl
      |                        |
  H — C — H Methane      Cl — C — Cl — Carbon tetrachloride
      |                        |
      H                        Cl
```

Starting with carbon tetrachloride and using a catalyst and hydrofluoric acid, one of the chlorine atoms can be replaced by one fluorine atom and subsequently a second change can be made to take place. This is more conveniently shown by the following equations:

<u>Phase 1</u>

CCl_4	+	HF	=	CCl_3F	+	HCl
Carbon tetra Chloride	+	Hydrofluoric acid	=		+	Hydrochloric acid

Phase 2

CCl_3F	+	HF	=	CCl_2F_2 Freon 12	+	HCL

The resultant compound, dichlorodifluoromethane was the substance given the name of Freon 12.

Fluorine is highly reactive and its strong affinity for some other elements results in combinations which are stable and non reactive.

Many other related compounds followed as a result of this work, probably the next best known being Freon 22. Freon 22 was prepared exactly in the same way as Freon 12 but starting with Chloroform ($CH\ Cl_3$) the two phase reaction was obtained thus:

Phase 1

$CH\ Cl_3$ Chloroform	+	HF Hydrofluoric Acid	=	$CH\ Cl_2F$	+	HCl Hydrochloric Acid

Phase 2

$CH\ Cl_2F$	+	HF	=	$CH\ ClF_2$ Freon 22	+	HCl

Freon 22 has a higher pressure than Freon 12 and thus requires a smaller swept volume for a given refrigerating effect.

Manufacture of the halocarbon refrigerants was taken up by other companies and in other countries and a wide range of names and numbers for the same substance became available on the market; Genetron, Isceon, Arcton and a number of other names were applied to these substances. It was largely as a result of the conflict of particulars attatched to the trade names that the American Standard 879 – 1 was introduced. From the list of halocarbon compounds given in the extracts from the above document it will be seen that a very wide variety of substances now available, are suitable as refrigerants. This has now been adopted by the British Standards Institution and published as BS 4580.

Azeotropes

The continued search for the ideal refrigerant was rapidly exhausting the halocarbon series and little else appeared to be useful as a line of investigation in the field of single compounds. The next logical step was towards mixtures

It is comparatively easy to exhaust the possibilities where single compounds are concerned but quite a different matter when it comes to mixtures and much work lies ahead in the study of the properties of azeotropes.

The Hydrocarbons

These substances hold an important place in the list refrigerants for two reasons. First, that as has just been described, Methane is the main basis from which the halocarbons are derived – though not now directly. Secondly, the wide use of "natural" gas has created a whole new field of problems closely related to refrigeration in the liquifaction and subsequent vaporisation for industrial and domestic consumption.

CLASSIFICATION

It was said earlier that it was largely as a result of the heterogeneous collection of trade names for the same halocarbon substances that the American Standards Association issued their classification which covered all refrigerants or substances which were closely related. A similar list has been issued by the British Standards Institution (No. 4850).

A numerical system of classification was devised and internationally accepted. The work was initiated by the American Society of Heating, Refrigerating & Air Conditioning Engineers, and was approved by the American Standards Association and issued as B79.1 – 1960 from which the following is summarised.

An identifying number is assigned to each refrigerant.

The identifying numbers assigned to azeotropes (500 series), miscellaneous organic compounds (600 series), and inorganic compounds (700 series) are purely arbitrary.

Within the inorganic 700 series, the molecular weights of the compounds have been added to 700 to arrive at the identifying refrigerant number.

The identifying numbers assigned to the hydrocarbons and halocarbons of the methane, ethane, propane and cyclobutane series are such that the refrigerant compound may be deduced from the refrigerant number, and vice versa, without ambiguity.

Another group of unsaturated compounds has been located in the 1000 series.

The greatest attention of the classifiers was devoted to the halocarbon compounds, very logically, since this was the field most interesting to refrigerating engineers.

The system of numbering these compounds is that, the first digit from the right indicates the number of fluorine atoms, the second digit from the right is one less than the number of hydrogen atoms in the compound and the third digit from the right is one less than the number of carbon atoms in the compound (this is omitted when zero).

The additional suffix B followed by a number indicates the number of bromine atoms present. The number of chlorine (Cl) atoms in the compound is found by subtracting the sum of the fluorine (F) and hydrogen (H) atoms from the total number of atoms which can be connected to the carbon (C) atoms.

When only 1 carbon atom is involved, the total number of attached atoms is 4. When 2 carbon atoms are present, the total number of attached atoms is 6, unless the compound is unsaturated; in this case, the total number of attached atoms is 4.

For saturated refrigerants, the total number of attached atoms is:

4 when $C = 1$
6 when $C = 2$
8 when $C = 3$
10 when $C = 4$
$2n + 2$ when $C = n$

For unsaturated and cyclic refrigerants, the total number of attached atoms is:

4 when C = 2
6 when C = 3
8 when C = 4
10 when C = 5
2n when C = n

For cyclic derivatives the letter C is used before the identifying refrigerant number.

In those instances where bromine is present in place of part or all of the chlorine, the same rules apply except that the letter B after the designation for the parent chloro-fluoro compound shows the presence of bromine (Br). The number following the letter B shows the number of bromine atoms present.

In the case of isomers, each has the same number, and the most symmetrical one is indicated by the number without any letter following it. As the isomers become more and more unsymmetrical, the letters a, b, c, etc are appended. Symmetry is determined by adding the atomic weights of the groups attached to each carbon atom and subtracting one sum from the other. The smaller the difference, the more symmetrical the product.

If the compound is unsaturated, the above rules apply, except that the number 1 is used as the fourth digit from the right.

The following Table is reproduced from the American Standard B79.1 – 1960 and lists the substances covered by the original specification.

REFRIGERANT NUMBERING SYSTEM

Refrigerant No. Designation	Chemical Name	Chemical Formula	Molecular Weight	Boiling Point, F.
Halocarbon Compounds				
10	Carbontetrachloride	CCl_4	153.8	170.2
11	Trichloromonofluoromethane	CCl_3F	137.4	74.8
12	Dichlorodifluoromethane	CCl_2F_2	120.9	– 21.6
13	Monochlorotrifluoromethane	$CClF_3$	104.5	–114.6
13B1	Monobromotrifluoromethane	$CBrF_3$	148.9	– 72.0
14	Carbontetrafluoride	CF_4	88.0	–198.4
20	Chloroform	$CHCl_3$	119.4	142
21	Dichloromonofluoromethane	$CHCl_2F$	102.9	48.1
22	Monochlorodifluoromethane	$CHClF_2$	86.5	– 41.4
23	Trifluoromethane	CHF_3	70.0	–119.9
30	Methylene chloride	CH_2Cl_2	84.9	105.2
31	Monochloromonofluoromethane	CH_2ClF	68.5	48.0
32	Methylene fluoride	CH_2F_2	52.0	– 61.4
40	Methyl chloride	CH_3Cl	50.5	– 10.8
41	Methyl fluoride	CH_3F	34.0	–109
50	Methane †	CH_4	16.0	–259
110	Hexachloroethane	CCl_3CCl_3	236.8	365
111	Pentachloromonofluoroethane	CCl_3CCl_2F	220.3	279

Refrigerant No. Designation	Chemical Name	Chemical Formula	Molecular Weight	Boiling Point, F.
Halocarbon Compounds (continued)				
112	Tetrachlorodifluoroethane	CCl_2FCCl_2F	203.8	199.0
112a	Tetrachlorodifluoroethane	CCl_3CClF_2	203.8	195.8
113	Trichlorotrifluoroethane	CCl_2FCClF_2	187.4	117.6
113a	Trichlorotrifluoroethane	CCl_3CF_3	187.4	114.2
114	Dichlorotetrafluoroethane	$CClF_2CClF_2$	170.9	38.4
114a	Dichlorotetrafluoroethane	CCl_2FCF_3	170.9	38.5
114B2	Dibromotetrafluoroethane	$CBrF_2CBrF_2$	259.9	117.5
115	Monochloropentafluoroethane	$CClF_2CF_3$	154.5	– 37.7
116	Hexafluoroethane	CF_3CF_3	138	–108.8
120	Pentachloroethane	$CHCl_2CCl_3$	202.3	324
123	Dichlorotrifluoroethane	$CHCl_2CF_3$	153	83.7
124	Monochlorotetrafluoroethane	$CHClFCF_3$	136.5	10.4
124a	Monochlorotetrafluoroethane	CHF_2CClF_2	136.5	14
125	Pentafluoroethane	CHF_2CF_3	120	– 55
133a	Monochlorotrifluoroethane	CH_2ClCF_3	118.5	43
140a	Trichloroethane	CH_3CCl_3	133.4	165
142b	Monochlorodifluoroethane	CH_3CClF_2	100.5	12.2
143a	Trifluoroethane	CH_3CF_3	84	– 53.5
150a	Dichloroethane	CH_3CHCl_2	98.9	140
152a	Difluoroethane	CH_3CHF_2	66	– 12.4
160	Ethyl chloride	CH_3CH_2Cl	64.5	54
170	Ethane†	CH_3CH_3	30	–127.5
218	Octafluoropropane	$CF_3CF_2CF_3$	188	– 36.4
290	Propane†	$CH_3CH_2CH_3$	44	– 44.2
Cyclic Organic Compounds				
C316	Dichlorohexafluorocyclobutane	$C_4Cl_2F_6$	233	140
C317	Monochloroheptafluorocyclobutane	C_4ClF_7	216.5	77
C318	Octafluorocyclobutane	C_4F_8	200	21.1
Azeotropes				
500	Refrig. 12/152a 73.8/26.2 wt % *	CCl_2F_2/CH_3CHF_2	99.29	– 28.0
501	Refrig. 22/12 75/25 wt %	$CHClF_2/CCl_2F_2$	93.1	– 42
502	Refrig. 22/115 48.8/51.2 wt %	$CHClF_2/CClF_2CF_3$	112	– 50.1
503	Refrig. 23/13 40.1/59.9 %	$CHF_3/CClF_3$	87.5	–126.1
504	Refrig. 32/115 48.2/57.8 %	CH_2F_2/CCl_2CF_3	79.9	– 71

† The compounds methane, ethane, and propane appear in the Halocarbon section in their proper numerical positions, but these products are not halocarbons.

* Carrier Corp. Document 2-D-127, p 1.

Refrigerant No. Designation	Chemical Name	Chemical Formula	Molecular Weight	Boiling Point, F.
Miscellaneous Organic Compounds:				
Hydrocarbons				
50	Methane	CH_4	16	−259
170	Ethane	CH_3CH_3	30	−127.5
290	Propane	$CH_3CH_2CH_3$	44	− 44.2
600	Butane	$CH_3CH_2CH_2CH_3$	58.1	31.3
601	Isobutane	$CH(CH_3)_3$	58.1	14
1150	Ethylene ††	$CH_2=CH_2$	28	−155
1270	Propylene ††	$CH_3CH=CH_2$	42.1	− 53.7
Oxygen Compounds				
610	Ethyl ether	$C_2H_5OC_2H_5$	74.1	94.3
611	Methyl formate	$HCOOCH_3$	60	89.2
Sulphur Compounds				
620				
Nitrogen Compounds				
630	Methyl amine	CH_3NH_2	31.1	20.3
631	Ethyl amine	$C_2H_5NH_2$	45.1	61.8
Inorganic Compounds				
717	Ammonia	NH_3	17	− 28
718	Water	H_2O	18	212
729	Air		29	−318
744	Carbon dioxide	CO_2	44	−109 (sublime)
744A	Nitrous oxide	N_2O	44	−127
764	Sulphur dioxide	SO_2	64	14
Unsaturated Organic Compounds				
1112a	Dichlorodifluoroethylene	$CCl=CF_2$	133	67
1113	Monochlorotrifluoroethylene	$CClF=CF_2$	116.5	− 18.2
1114	Tetrafluoroethylene	$CF_2=CF_2$	100	−105
1120	Trichloroethylene	$CHCl=CCl_2$	131.4	187
1130	Dichloroethylene	$CHCl=CHCl$	96.9	118
1132a	Vinylidene fluoride	$CH_2=CF_2$	64	−119
1140	Vinyl chloride	$CH_2=CHCl$	62.5	7
1141	Vinyl fluoride	$CH_2=CHF$	46	− 98
1150	Ethylene	$CH_2=CH_2$	28	−155
1270	Propylene	$CH_3CC=CH_2$	42.1	− 53.7

†† The compounds ethylene and propylene appear in the Hydrocarbon section in order to indicate that these compounds are hydro-carbons. Ethylene and propylene are properly identified under Unsaturated Organic Compounds.

GENERAL CHARACTERISTICS OF CONVENTIONAL REFRIGERANTS

Some notes follow on the general properties of the substances mentioned under the foregoing categories. First, the classics.

Classic Refrigerants

Carbon Dioxide. Carbon dioxide has many properties similar to sulphur dioxide. It can be prepared by roasting carbonates just as sulphur dioxide is obtained by roasting sulphides.

It dissolves in water giving an acid solution which combines with bases to give carbonates.

Carbon dioxide is a colourless gas at ordinary temperatures with no definite smell, but if inhaled a sharp acid taste will be noticed. It requires considerable pressure to liquify it at ordinary temperatures (1066psi at 87.8° F – the critical point) and is then obtained as a colourless, mobile liquid.

Carbon dioxide is a thermally stable gas and does not decompose until temperatures of well over I,000° C are obtained. Its reactivity is of a much lower order than sulphur dioxide; for example, its water solution is only a weak acid. For this reason it does not attack the normal materials of construction as vigorously as in the case of sulphur dioxide. Nevertheless, moisture must be excluded as far as possible in a refrigeration system.

The harmful effects of carbon dioxide are mainly due to suffocation since the compound is not particularly poisonous. The gas is heavier than air, and in enclosed spaces, especially those with only a top entrance, it is possible for dangerous concentrations to develop.

There is no fire or explosive hazard with carbon dioxide; it is, in fact, a fire extinguisher.

Ammonia. Ammonia is still without doubt the most important industrial refrigerant of the present day because its thermodynamic properties are good and it is cheap.

Ammonia is a pungent smelling, colourless gas readily liquifying at atmospheric temperatures when compressed to give a clear, colourless liquid. It is only feebly combustible in air, the heat of combustion being insufficient to maintain a flame. However, air and ammonia in certain proportions may form an explosive mixture which can be ignited by a naked flame. The reported explosive range is between 16 and 27% by volume with air, and the explosive mixture may be ignited with difficulty by a flame. Since this condition exists, suitable precautions must be taken. Rooms where ammonia plant operates should be well ventilated and no naked lights should be permitted.

Recent "safety" measures introduced by the electrical industry may have widespread and expensive repercussions among users of industrial ammonia refrigeration plant.

Ammonia is a reactive chemical and attacks many metals and alloys. Fortunately, mild steel and cast iron are not attacked by it, and they should be used wherever possible. Truly anhydrous ammonia does not attack copper, but traces of moisture are sufficient to give rise to corrosion and it is therefore always essential to avoid copper and copper alloys. Zinc also is attacked by ammonia in the presence of moisture and for this reason galvanized iron and zinc should also be avoided.

Nickel bearing alloys and light metals are attacked and cannot be used.

Sulphur Dioxide. Sulphur dioxide is at ordinary temperatures a colourless gas having a pungent, suffocating odour. It may readily be liquified to give a clear, colourless, mobile liquid. It is stable and is not dissociated until relatively high temperatures (1,200° C) are attained. In contact with air in the presence of a catalyst the compound unites with oxygen at much lower temperatures to give sulphur trioxide (SO_3) which, when dissolved in water, gives sulphuric acid.

Sulphur dioxide readily dissolves in water giving sulphurous acid solutions which are corrosive and which on exposure to air will oxidize slowly to give traces of sulphuric acid.

Sulphur dioxide is a reactive chemical and in the presence of moisture its reactivity is enhanced, the moist sulphur dioxide readily attacking many metals. It is because of the affinity of sulphur dioxide for moisture and its corrosive action when moist that sulphur dioxide is not a really satisfactory refrigerant.

When inhaled, the gas attacks the mucous membrane and is a general irritant. It has, however, an intolerable odour when present in relatively low concentrations which serves to act as a warning agent and serious injury is therefore rare. Sulphur dioxide does not burn, neither does it support combustion.

Methyl Chloride. The gas is heavier than air and will flow along floors collecting in any pits or depressions. Since it is flammable and can give explosive mixtures with air within the concentration range of 8.1 to 17.2 parts by volume, it is important to avoid using naked flames. In this connection it is worth noting that the halide lamp used for detecting leaks of methyl chloride is hazardous.

Dry methyl chloride is stable at normal temperatures and at those encountered in compression machines, but it does decompose when heated to fairly high temperatures. In contact with metals such as copper, iron, galvanized iron, brass, bronze, and solder, decomposition starts at about 795° F.

In the presence of moisture slow hydrolysis takes place, forming hydrochloric acid. For this reason methyl chloride is supplied to the refrigeration industry with a limiting moisture content of 75 p.p.m.

When absolutely dry, methyl chloride is non-corrosive to most of the common engineering metals. In the presence of traces of moisture however, some metals are attacked, notably zinc, magnesium, alloys of aluminium, magnesium alloys and die castings. It is of particular importance never to use aluminium, as with this material methyl chloride reacts to give a spontaneously explosive/ flammable compound.

Copper plating is sometimes encountered in refrigeration units in which methyl chloride is used as the refrigerant. The phenomenon is an indication of decomposition of the refrigerant and is associated with corrosion of the unit. It only occurs when moisture gives an electrolyte. Brass or bronze bearings form a couple with the cast-iron shell of the compressor. A little copper dissolves in the acid and is deposited electrolytically on the bright parts replacing some of the iron in doing so

If leaks of methyl chloride develop there is the risk of fire or explosion, or danger of the personnel being overcome, particularly in enclosed spaces , and the practice of adding warning agents was therefore introduced in an effort to reduce the possibility of mishap. The warning agent used was a lachrymatory, the most common one being acrolein. The substance is not very stable and is rather a reactive chemical. It was found, after using it for some time, that some compressors in which it was employed seized up. Investigation of the cause of the seizure showed that methyl chloride containing acrolein in the presence of lubricating oil, traces of moisture, and a catalyst (which could be finely divided metal), reacted together to give a greenich, tacky substance. When conditions were favourable this could be formed within the unit, and was deposited in the cylinder where it set when the machine stopped and effectively prevented restarting.

Hydrocarbons

This family of compounds, which includes the paraffins, is of interest to the refrigeration engineer in a rather different way. Most of the substances are highly flammable and not suitable for general use in refrigeration plants, although propane has been used for refrigeration purposes in the oil industry where the precautions against fire are in any case of an extremely high calibre.

With the introduction of methane and other natural gas products as fuel into industry in larger quantities the problems of transport and storage have arisen. The liquefaction and transport in liquid form together with the revaporization for distribution to industry are essentially problems for the refrigeration industry, although not directly with the production of cold.

Propane and butane have already been used extensively, distributed in high pressure cylinders, as domestic fuel. The liquefaction of these substances is at relatively low pressures and presents no problems. Methane, requiring a pressure of around 1000 lbs to liquefy at atmospheric temperature presents on the other hand problems to which the industry is unaccustomed, and as a result a large portion of this work has gone outside the conventional refrigeration industry. The properties of the main compounds in this family are given below. Those in which the refrigerating engineer is primarily interested are: methane, ethane, propane, and butane. These have been allocated numbers in the classification system already mentioned, and are 50, 170, 290, and 600 respectively.

It is convenient to set out the structure of this family where the logical pattern may be clearly seen.

Name	Chemical Formula	Structure	BP °F
Methane	CH_4	H–C(H)(H)–H	–259
Ethane	C_2H_6	H–C(H)(H)–C(H)(H)–H	–127
Propane	C_3H_8	H–C(H)(H)–C(H)(H)–C(H)(H)–H	– 43.7

Name	Chemical Formula	Structure	BP °F
Butane	C_4H_{10}	H H H H H – C – C – C – C – H H H H H	– 31.1
Iso Butane	C_4H_{10}	H H C C C H H – C – H H	10.4
Pentane	C_5H_{12}	H H H H H H – C – C – C – C – C – H H H H H H	96.9

Air and Water

Air. The use of air as a primary refrigerant under the conditions laid down in the beginning of this note is almost non-existent since there is no system in general commercial use where air is liquefied and used as a refrigerant. However, many years ago air was used as a primary refrigerant in which compression and expansion of the air without a change of state took place. This was commercially accepted in the Bell-Coleman cold air machine which was later manufactured by Haslam and Lightfoot. This cold air system was in use on many marine installations. Owing to the weight and volume of the machinery required, it became displaced by CO_2 refrigerating plants of a conventional nature.

The system fell into disuse for many years until it was reintroduced for use in aircraft cooling. The old piston-type compressor was replaced by high-speed turbo compressors followed by the turbo expander, with the result that using aluminium alloy the weight and volume of the plant could be reduced to a size suitable for use in aircraft.

The use of air as a secondary refrigerant is universal, but since this employs air/water vapour mixtures it is not appropriate to this book.

Water. The use of water as a refrigerant is growing with the popularity of lithium bromide absorption plant. This type of plant is particularly useful where quantities of waste steam are available.

It was used quite extensively in steam jet refrigeration where high pressure steam serves as an impellent compressing the low pressure water vapour by entrainment.

The Halocarbons

The substitution of chlorine or fluorine for hydrogen in the hydrocarbons has been described earlier but the effects of the substitutions are now detailed. This group of halogen compounds is divided into two, the first based on methane, the second on ethane. The properties of the compounds are controlled by the replacement of the hydrogen atoms with chlorine, fluorine, bromine, or a combination of all three.

The Methane Series. In the first group, methane, the basic fluid has a N.B.P. of −162° C (−260° F), and as each atom of chlorine is added the N.B.P. rises, thus:

	Compound	NBP °C	NBP °F	Ref'g No.
Methane	CH_4	−162	−259	50
Methyl chloride	CH_3Cl	− 24	− 11	40
Methylene chloride	CH_2Cl_2	40	104	30
Chloroform	$CHCl_3$	61	142	20
Carbon tetrachloride	CCl_4	77	170	10

Looking now at the effect of substituting fluorine in place of chlorine, it is seen that the NBP falls with the increase, thus:

CCl_4	77	170	10
CCl_3F	24	75	11
CCl_2F_2	−30	−22	12
$CClF_3$	−81	− 114	13
CF_4	−128	− 198	14

In the same way, the compound with one hydrogen atom is affected by the substitution of fluorine for chlorine thus:

$CHCl_3$	61	142	20
$CHCl_2F$	9	48	21
$CHClF_2$	−41	−42	22
CHF_3	−84	−120	23

The pattern is followed in the other compounds, and the methane series may be summarized thus:

Fluorine atoms	Series Number NBP: °C (°F)					
0	**10** 77 (170)	**20** 61 (142)	**30** 40 (104)	**40** −24 (−11)	**50** −162 (−259)	
1	**11** 24 (75)	**21** 9 (48)	**31** −9 (16)	**41** −78 (−109)		
2	**12** −30 (−22)	**22** −41 (−42)	**32** −53 (−61)			
3	**13** −81 (−114)	**23** −84 (−120)				
4	**14** −128 (−198)					

The Ethane Series. This series follows very much the same general pattern, the NBP rising with the number of chlorine atoms, thus:

	Compound	°C NBP	°F	Ref'g No.
Ethane	C_2H_6	−89	−128	170
	C_2H_5Cl	13	56	160
	$C_2H_4Cl_2$	60	140	150
	$C_2H_3Cl_3$	113	236	140a
	$C_2H_2Cl_4$	146	295	130c
	C_2HCl_5	162	324	120
	C_2Cl_6	sublime		110

Taking the 110 group to see the effect of increasing the number of fluorine atoms by substitution for chlorine, we get:

Compound	°C	°F	Ref'g No.
C_2Cl_6	sublime		110
C_2Cl_5F	138	279	111
$C_2Cl_4F_2$	93	199	112
$C_2Cl_3F_3$	48	118	113
$C_2Cl_2F_4$	4	38	114a
C_2ClF_5	−3	−36	115
C_2F_6	−78	−109	116

These variations of properties take place in a similar pattern throughout each group — summed up thus:

Fluorine atoms	Series Number NBP °C (°F)						
0	110 Sublimes	120 162 (324)	130 146 (295)	140 113 (236)	150a 60 (140)	160 13 (56)	170 −89 (−128)
1	111 137 (279)	121 117 (242)	131 103 (217)	141 74 (165)	151 35 (95)	161 −37 (−35)	
2	112 93 (199)	122 72 (162)	132 60 (140)	142 35 (95)	152a −25 (−13)		
3	113 48 (118)	123 29 (84)	133a 6 (43)	143 5 (41)			
4	114 3 (38)	124 −12 (10)	134 −23 (−9)				
5	115 −39 (−38)	125 −48 (−55)					
6	116 −78 (−109)						

General. It will be seen from the foregoing that an enormous variety of compounds is available, but the number in general use as refrigerants is quite small. Some notes follow on the most commonly used. There are in addition to those mentioned, one or two brominated compounds of which 13B1 Bromotrifluoromethane (CF_3Br) is the only one in common use.

R11 – CCl_3F MW 137.4* This is a low pressure type refrigerant in general use in centrifugal compressors. Due to its molecular weight it is frequently used for solvent purposes in flushing out plants using a different refrigerant. Its low viscosity and freezing point have also led to its use as a brine.

R12 – CCl_2F_2 MW 120.9. This was the first of the halo-carbons to go into general use, and its development has been described in the introductory section. It is miscible with oil in all proportions although an excess will tend to reduce the heat transfer rate. Care must be taken to avoid the use of magnesium zinc and natural rubber as well as some of the synthetic types. It has a low discharge temperature even at low suction temperatures.

R13 – $CClF_3$ MW 104.5. Is sometimes used in cascade plants with R12 or R22 on high side.

R13B1 – CF_3Br MW 148.9. Useful and gaining popularity in low temperature plants. It offers higher capacity at low temperatures than most of the other refrigerants.

R22 – $CHClF_2$ MW 86.5. Has become very widely used and in many cases superseded R12; this is mainly due to the increase in capacity for the same swept volume which approximates to that of ammonia. It is advantageous in plants with low evaporating temperatures although the discharge temperature may be high in such cases and liquid injection may be necessary. R22 is miscible with oil at high temperatures but only partly miscible on the low temperature side and some difficulties have been encountered.

R113 – CCl_2FCClF_2 MW 187.4. This is an ethane and not a methane derivative; it is a low pressure refrigerant suitable for use in centrifugal compressors due to its high molecular weight.

R114 – $CClF_2CClF_2$ MW 70.9. A low pressure refrigerant again, sometimes used in centrifugal compressors operating on low temperature applications.

Azeotopes

It has already been pointed out that much work is required in the field of azeotropy. As the first step, it is necessary to study azeotropy to try to classify the knowledge. The early work leading to a more scientific classification covers a very broad scope. Many of the sideline binary systems investigated to establish principles of azeotropy are reported elsewhere.

On considering single compounds, little thought need be given to the constancy of conditions which should exist in different parts of the refrigeration machine. In the evaporator for example, the temperature will be practically constant throughout the entire length of the coil if a single compound is used. On the other hand, most mixtures will not have a constant boiling point and, therefore, will not give a constant temperature in the evaporator unless it is the flooded type. Fortunately there are some mixtures which are characterized by a constant boiling composition. Such a constant boiling mixture, where the constituents are completely miscible, is called a homoazeotrope, or by the more general classification, azeotrope. An advantage of an azeotrope over an ordinary mixture is that there will not be trouble with fractionation if there is a leak.

---

* MW = Molecular Weight in all cases.

Ready references are found in several books on azeotropy, and a fairly good list of azeotropes is in the International Critical Tables.

There are two types of homoazeotropes that must be considered. One has a pressure lower and a boiling point higher than either pure constituent; the other has a pressure higher and a boiling point lower than either constituent. One is called a minimum pressure or maximum temperature azeotrope; the other is the maximum pressure or minimum temperature type.

If the attraction between the unlike molecules is greater than that between the like molecules, it will be more difficult for the molecules to escape from the solution, and therefore such an azeotrope will be of the minimum pressure type. Much more often however, the unlike molecules have a lesser attraction for each other and consequently most homoazeotropes are of the maximum pressure type. As a matter of fact, unless the solution is ionic there will probably need to be some real chemical affinity between the two compounds to get the minimum pressure type.

Seeking an azeotrope therefore, which will be satisfactory as a refrigerant means that a great deal of work has to be done on measuring the properties of various binary systems before it is possible to classify them satisfactorily.

Owing to the large number of combinations, one cannot hope to discover a desirable azeotropic refrigerant just by pouring together haphazardly all pairs of compounds; there must be some limitations in the search or the problem could be almost insurmountable.

The search is made very much less arduous in that the objectives for properties are usually fairly limited and thus the limitations of performance can be laid down from the start. One result of this is to eliminate all but a few of the possible substances.

The first step would be to set an upper limit not too far above the lower limit of the boiling point; such an azeotrope is not likely to be more than a few degrees lower than that of the lower boiling constituent. The difference is very likely to be less than 10 degrees and just an outside chance that it would be as much as 20 degrees.

If the boiling point is very much higher the azeotrope will probably not contain a high percentage of the second substance. An azeotrope containing only a small amount of the second substance cannot boil at a much lower temperature.

The series number 500 has been allocated to azeotropes, and at present there are four compounds in this series – 500, 502, 503 and 504, 502 being the most commonly used.

R500 is composed of 73.8% R12 and 26.2% R152a. Its NBP is –33° C (–28° F).

R502 is composed of 48.8% R22 and 51.2% R115 with a resultant NBP of –46° C (–50° F).

R503 is composed of 40.1% R23 and 59.9% R13 with a NBP of –89° C (–127° F).

R504 is composed of 48.2% R32 and 51.8% R115 with a NBP of –57.6° C (–71° F).

Mixtures

The properties of azeotropes have been dealt with fairly extensively in current literature but the subject of non-azeotropic mixtures has been neglected to a large extent. It was examined tentatively by

V.F. Chaikovsky and A.P. Kutznetsov of the Odessa Technological Institute who feel that non-azeotropic mixtures may possess the following advantages in some conditions.

1. The power characteristics of the machine improve when effecting non isothermal processes of heat transfer.

2. Low temperature cycles are effected with single stage compression.

3. Control of the composition provides for optimum working conditions in a wide range of compositions and temperatures in the evaporator.

CHAPTER 6

Calculations

PROPERTIES OF REFRIGERANTS

The ideal refrigerant has been specified many times: it should be suitably free from corrosive action, non flammable, non toxic and non irritant; but of course it is probable that no known refrigerant fills these requirements in all circumstances. Stability is closely associated with corrosive properties or the reaction between the refrigerant and the materials of construction of the plant. The presence of impurities which may act as catalysts or can produce a reaction which in turn will produce the corrosive materials must be considered and electrolytic action must also not be overlooked.

It is however the thermodynamic properties of refrigerants which concern the refrigeration engineer most, and in two different ways: first, the comparative properties of a variety of different refrigerants, and second, the properties of a single refrigerant as they vary from one phase and condition to another.

The information which is of interest to the refrigeration engineer concerning refrigerants is the interrelationship between pressure, temperature, volume, entropy, and enthalpy.

A set of tables of the properties of refrigerants is usually set out like this:

Temp. °F.	PRESSURE lb./sq. in.		VOLUME cu. ft./lb.		DENSITY lb./cu. ft.		ENTHALPY B.t.u /lb.			ENTROPY B.t.u./lb. °R.	
	Absolute	Gauge	Liquid v_f	Vapour v_g	Liquid $1/v_f$	Vapour $1/v_g$	Liquid h_f	Latent h_{fg}	Vapour h_g	Liquid s_f	Vapour s_g
25	39·310	24·614	0·011366	1·0039	87·981	0·99613	13·958	65·946	79·904	0·030772	0·16683
26	40·056	25·360	·011380	0·98612	87·870	1·0141	14·178	65·829	80·007	·031221	·16676
27	40·813	26·117	·011395	·96874	87·760	1·0323	14·398	65·713	80·111	·031670	·16669
28	41·580	26·884	·011409	·95173	87·649	1·0507	14·618	65·596	80·214	·032118	·16662
29	42·359	27·663	·011424	·93509	87·537	1·0694	14·838	65·478	80·316	·032566	·16655
30	43·148	28·452	0·011438	0·91880	87·426	1·0884	15·058	65·361	80·419	0·033013	0·16648
31	43·948	29·252	·011453	·90286	87·314	1·1076	15·279	65·243	80·522	·033460	·16642
32	44·760	30·064	·011468	·88725	87·202	1·1271	15·500	65·124	80·624	·033905	·16635
33	45·583	30·887	·011482	·87197	87·090	1·1468	15·720	65·006	80·726	·034351	·16629
34	46·417	31·721	·011497	·85702	86·977	1·1668	15·942	64·886	80·828	·034796	·16622

or else like this:

TEMP °C	PRESSURE bar	VOLUME m³/kg · 10³		DENSITY kg/m³		ENTHALPY kcal kg			ENTROPY kcal (kg) (°K)		TEMP °C
		LIQUID v_f	VAPOR v_g	LIQUID $1/v_f$	VAPOR $1/v_g$	LIQUID h_f	LATENT h_{fg}	VAPOR h_g	LIQUID s_f	VAPOR s_g	
-60	0.2262	0.63689	637.91	1.57014	0.0015	146.463	177.77	324.236	0.7797	1.6137	-60
-59	0.2396	0.63795	604.68	1.56753	0.0016	147.331	177.37	324.702	0.7838	1.6120	-59
-58	0.2537	0.63902	573.51	1.56491	0.0017	148.199	176.96	325.168	0.7878	1.6103	-58
-57	0.2683	0.64009	544.27	1.56228	0.0018	149.068	176.56	325.635	0.7918	1.6087	-57
-56	0.2837	0.64117	516.80	1.55964	0.0019	149.938	176.16	326.101	0.7959	1.6070	-56
-55	0.2998	0.64226	491.00	1.55699	0.0020	150.808	175.75	326.567	0.7999	1.6055	-55
-54	0.3166	0.64336	466.73	1.55434	0.0021	151.679	175.35	327.033	0.8038	1.6039	-54
-53	0.3341	0.64446	443.91	1.55168	0.0022	152.550	174.94	327.500	0.8078	1.6024	-53
-52	0.3524	0.64557	422.42	1.54901	0.0023	153.422	174.54	327.966	0.8117	1.6009	-52
-51	0.3715	0.64669	402.18	1.54634	0.0024	154.295	174.13	328.431	0.8157	1.5995	-51

In both cases the format and the information given are the same. Sometimes the tables are condensed, the columns most often omitted being: the volume of the liquid, the density of the vapour, and the latent heat.

The usual accompaniment to sets of tables is a set of Mollier charts, of which this is a typical example:

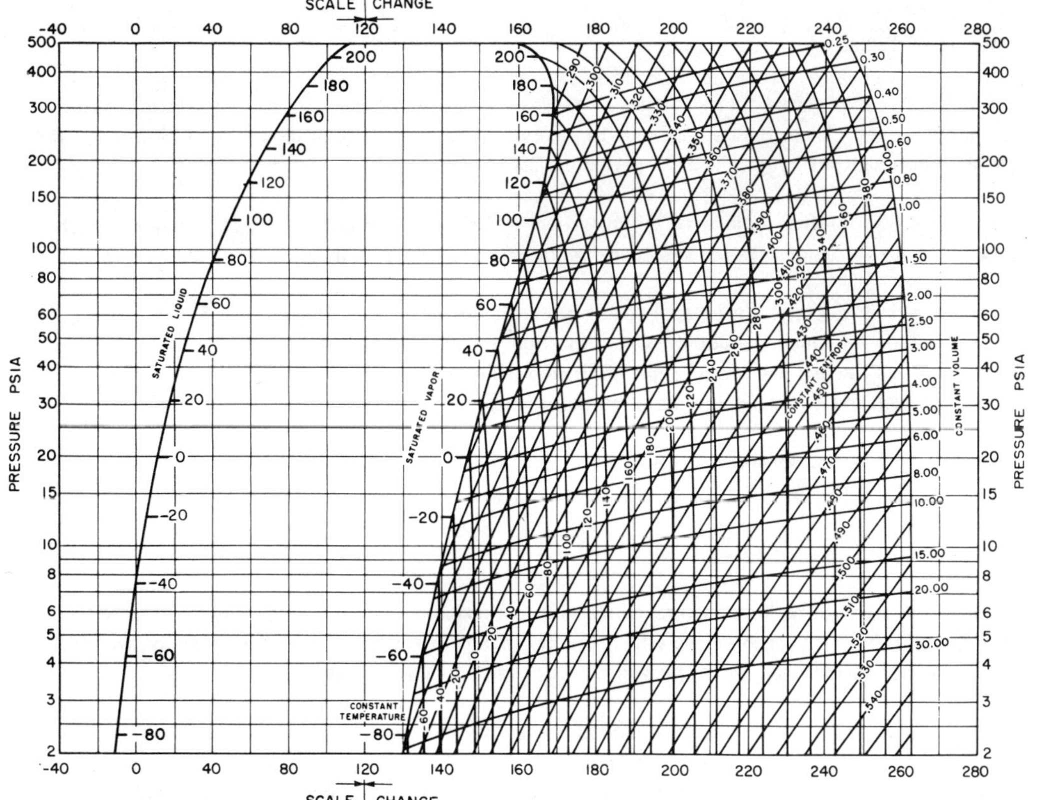

These tables and charts are generally felt to be indispensable to the refrigeration engineer, and so they are. What is not so often recognized however, is that given only a minimum of information about a refrigerant, the construction of a set of tables from which the Mollier chart can be drawn, is a relatively simple if tedious task. It is strongly recommended that anyone studying the subject of refrigeration should work out the framework of at least one set of thermodynamic properties of a refrigerant, preferably in the conventional form as depicted at the beginning of this chapter. This will create a familiarity with them far better than any amount of ordinary use.

It will have been noticed that the first example of the tables is in Imperial units followed by the example in S.I. units. This is because the greatest part of the development of the fluorinated hydrocarbons has been in the U.S.A. and much of the refrigeration equipment also has its origin there. The result has been that, whilst there is a wealth of data available, all the original work was produced in Imperial units and in general terms the few available sets of data in S.I. units are merely conversions of the original work.

With the early enthusiasm for metrication some tables were produced in metric units and indeed had been in use by some of the European countries for many years. The change to metric units would have been fairly general, since much of the data was already available in that form. The change to S.I. units however meets some resistance from both Imperial and Metric users and little movement has so far been made towards the general use of these units by the greater part of the Refrigeration Industry internationally.

Evolution of Tables

It must first be emphasized that all tables of refrigerant properties are built up empirically from experimental data. These data are subject to errors in reading and conditions so that almost invariably they are smoothed out by graphical or mathematical means.

Comparisons of published data by different writers will show inconsistencies in early results which gradually tend to disappear as the substance becomes more common in use.

This point is well illustrated in a paper read before the Institute of Refrigeration on The Properties of Methyl Chloride, in which the author, G.C. Hodsdon, quotes no less than ten differing sets of data for Methyl Chloride ranging over twenty-five years and giving a table showing the differences even then existing between the two generally accepted authorities.

Part of the table is reproduced opposite.

Perspective

An important point to be kept in view concerning the accuracy of the data on thermodynamic properties of refrigerants, is the place which the data takes in the context of the other information used in the calculations leading to the selection or design of equipment.

The discharge pressure, for instance, depends in nearly all cases on the air temperature which, the weather data shows, is hardly if ever constant to 1° F (½° C) for any period of 20 minutes in any period of 24 hours. Still less is it likely that such a steady period would coincide precisely with the predicted design conditions. The possiblity of the design conditions being attained are even more

° F.	Authority	Absolute pressure in lbs. persq.in.	Liquid weight in lbs. per c. ft.	Vapour volume in c. ft. per lb.	Heat in B.T.U. per lb.			Entropy	
					of Liquid	of Evaporation	of Vapour	of Liquid	of Vapour
−22	Shorthose ...	11.20	63.33	8.458	6.44	186.83	193.27	0.0150	0.4321
	T. B. and M.	11.13	63.29	8.136	6.427	186.78	193.21	0.0150	0.4418
+ 5	Shorthose ...	20.82	61.61	4.680	16.25	180.89	197.14	0.0367	0.4261
	T. B. and M.	21.15	61.65	4.471	16.21	180.70	196.92	0.0367	0.4257
32	Shorthose ...	36.56	59.91	2.691	26.24	174.41	200.65	0.0576	0.4125
	T. B. and M.	37.11	59.92	2.640	26.18	174.08	200.26	0.0576	0.4117
59	Shorthose ...	60.70	58.14	1.634	36.42	167.57	203.99	0.0777	0.4010
	T. B. and M.	60.94	58.07	1.651	36.33	166.90	203.23	0.0776	0.3995
86	Shorthose ...	95.51	56.30	1.043	46.76	160.18	206.94	0.0972	0.3909
	T. B. and M.	94.70	56.24	1.081	46.67	159.13	205.80	0.0970	0.3887
104	Shorthose ...	126.2	55.07	0.8186	53.76	155.15	208.91	0.1098	0.3852
	T. B. and M.	123.6	55.01	0.8331	53.65	153.60	207.25	0.1097	0.3822

remote if a cooling tower or an evaporative condenser is involved, since wind force and direction and relative humidity are further variants.

Again, the suction pressure is dependent on load conditions and only on very rare occasions is it possible to design a plant for a set of constant and continuous load conditions, although this does happen sometimes in chemical process applications.

Consider then that a variation of 5° C in the suction temperature can result in a variation of 10° to 25° in the plant capacity and that a variation in the saturated discharge temperature of 5° C may give a capacity variation of 6° . Add to this the fact that the majority of plants are not designed to operate on a 24 hours a day basis but on a purely arbitrary x hours a day and it will be apparent that the use of six and eight figure property tables for compression calculations is rather like giving the distance between two villages as say 8km 352 metres and 16 mm. Some of the methods of calculation quoted hereafter are approximations and apart from study should not be used when published tabulated data are available; they will nonetheless produce results well in line with the many approximations made in refrigerating plant and equipment design.

The outline of the treatment of the subject of refrigerants followed in this book takes into account, first, the neccessity of a clear understanding of the meaning and measurement of temperature, pressure and the definitions of entropy, enthalpy, and other thermo dynamic parameters. Secondly, the background of refrigerants and their development is given, followed by the methods of calculation of the relevant physical properties. The latter part of the book contains data sheets for more than 30 refrigerants in Metric and Imperial units.

TEMPERATURE/PRESSURE RELATIONSHIP

The most commonly used physical relationship in a refrigerant is undoubtedly the saturated pressure/temperature relation.

Most service men carry a conversion rule on which these figures are given for a variety of the most commonly used refrigerants, and pressure guages for refrigeration work are frequently calibrated in temperature as well as pressure for a given refrigerant.

Let us then, first look at the way in which this relationship can be calculated with only a bare amount of information available.

Any attempt to co-relate pressure-temperature relationship of a wide variety of substances in general use as refrigerants would produce no discernable pattern; this relationship for some common refrigerants which has been plotted on Fig. c, illustrates this point.

Seeking a means of clarifying these relationships, Dr. B.J. Eiseman, in early 1952, gave consideration to the use of the "reduced form" with some remarkable results which helped to simplify much of the calculation of saturated pressure/temperature relations.

Reduced Form

The "reduced form" of a substance is the relationship between some particular value of its temperature, pressure or volume to the critical temperature, pressure or volume of that substance.

Thus (1) The reduced pressure is the actual absolute pressure divided by the critical pressure: $P_r = P/P_c$.

(2) The reduced temperature is the ratio of the actual absolute temperature to the critical temperature: $T_r = T/T_c$.

In a number of calculations the reciprocal of the reduced temperature is more conveniently used than the reduced temperature itself. The reduced volume applies equally with the pressure temperature relationship, thus the three properties may be calculated by the same means.

(3) The reduced volume is the actual value per unit mass V divided by the critical volume V_c per unit mass: $V_r = V/V_c$. Two values, however, apply for the reduced volume, one for the *vapour* volume, the other for the *liquid* volume.

In dealing with reduced pressures we must remember that we are dealing with pressure as a state point in a static system — it has no energy or velocity significance in this case.

From the principle of corresponding states it is shown that a third quantity may be derived when any two of the reduced quantities are given and this third quantity has approximately the same value for any substance.

In the broadest form there may well be some variation, but in any groups of closely related compounds the agreement is very close, and a very good approximation may be made.

The reduced form thus produces an entirely different picture from that shown on Fig. c, as will be seen from Fig. 4.

One group of related substances covering a wide range of properties are the fluorinated hydrocarbons, where good pressure-temperature-volume data are available for a number of compounds.

Table: NUMERICAL VALUES FOR THE REDUCED VAPOUR PRESSURE CURVES

Reduced pressure, Pr	*Reciprocal of Reduced Temperature 1/T* **Curve 1** Methanes without H atom	**Curve 2** Methanes having one H atom	**Curve 3** Ethanes without H atom	**Curve 4** Ethanes having one H atom
1.0	1.000	1.000	1.000	1.000
.9	1.017	1.016	1.014	1.014
.8	1.037	1.034	1.032	1.032
.7	1.058	1.055	1.053	1.051
.6	1.083	1.079	1.075	1.074
.5	1.112	1.107	1.103	1.101
.4	1.148	1.141	1.136	1.134
.35	1.169	1.162	1.156	1.154
.3	1.194	1.186	1.179	1.176
.25	1.222	1.215	1.206	1.202
.2	1.258	1.250	1.240	1.235
.15	1.303	1.294	1.283	1.277
.1	1.366	1.355	1.343	1.336
.09	1.383	1.371	1.368	1.351
.08	1.401	1.389	1.375	1.368
.07	1.422	1.409	1.395	1.387
.06	1.446	1.433	1.417	1.409
.05	1.475	1.460	1.443	1.434
.04	1.509	1.493	1.475	1.465
.035	1.530	1.513	1.495	1.483
.03	1.553	1.536	1.516	1.505
.025	1.581	1.562	1.542	1.531
.02	1.615	1.597	1.573	1.561
.015	1.658	1.637	1.614	1.601
.01	1.718	1.695	1.670	1.657
.009	1.734	1.710	1.685	1.671
.008	1.752	1.727	1.702	1.687
.007	1.771	1.746	1.720	1.706
.006	1.794	1.769	1.742	1.727
.005	1.820	1.794	1.766	1.751
.004	1.852	1.825	1.795	1.781
.0035	1.872	1.844	1.814	1.799
.003	1.894	1.866	1.835	1.820
.0025	1.919	1.892	1.860	1.845
.002	1.951	1.923	1.890	1.874
.0015	1.992	1.963	1.929	1.912
.001	2.049	2.017	1.984	1.965

The above table gives the numerical values for reduced vapour pressures and the reciprocal of the reduced temperature and the following Figs. 4 to 7 give the curves derived from them. By reference to one or the other, unknown values of saturated pressures or temperatures may be found.

There is a very slight but uniform divergence between the values for reduced pressure above and below .01 and for accuracy these values have been plotted on the separate charts following.

REDUCED VAPOUR PRESSURE
OF SOME REFRIGERANTS
REDUCED PRESSURE
1/REDUCED TEMPERATURE
1·0
1·1
1·2
1·3
1·4
1·5
1·6
1·7
1·8
1·9
2·0
1·0 ·8 ·6 ·5 ·4 ·3 ·2 ·1 ·08 ·06 ·05 ·04 ·03 ·02 ·01 ·008 ·006 ·005 ·004 ·003 ·002 ·001

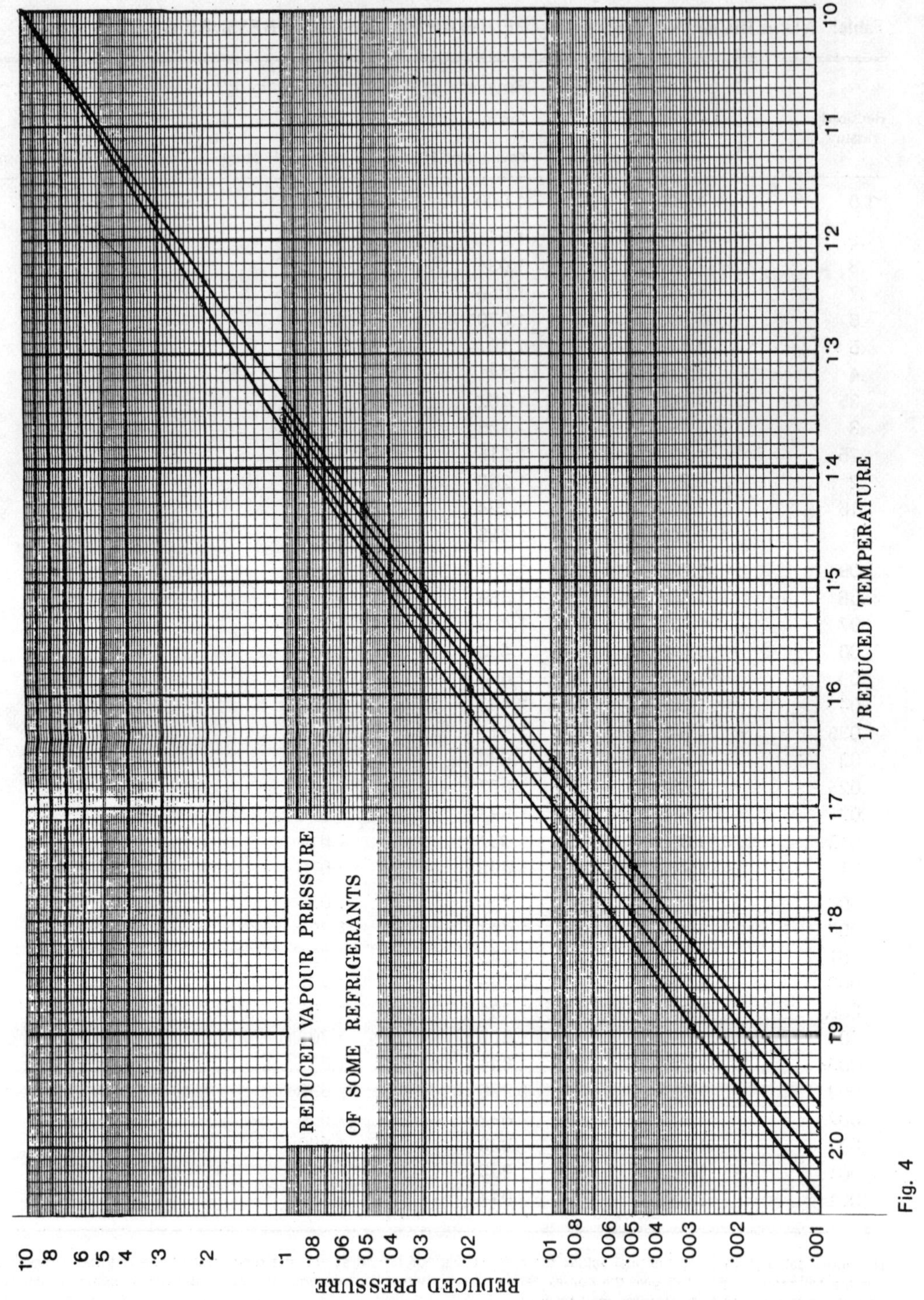

Fig. 4

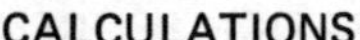

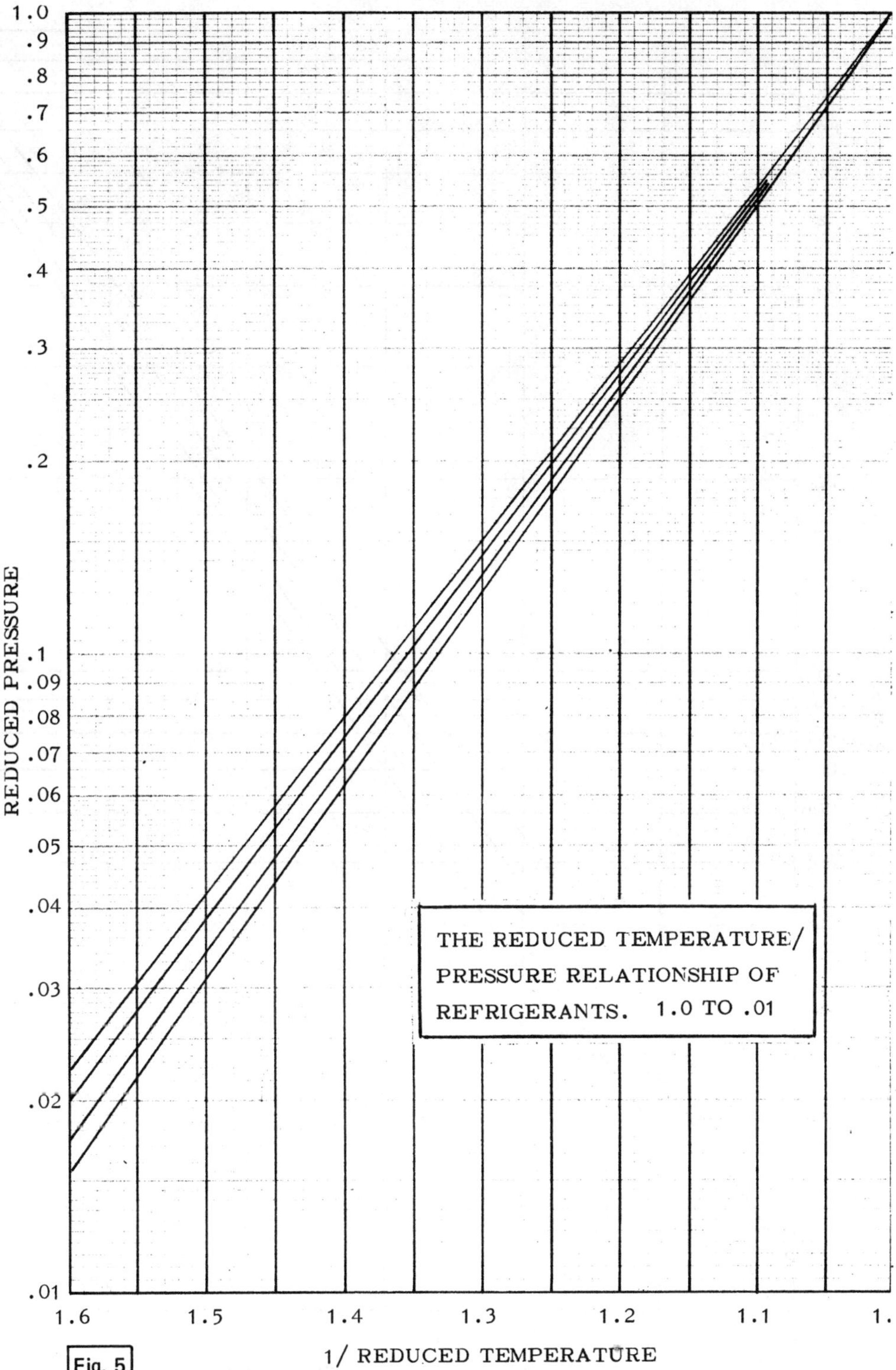

REDUCED PRESSURE

1.0
.9
.8
.7
.6
.5
.4
.3
.2
.1
.09
.08
.07
.06
.05
.04
.03
.02
.01

THE REDUCED TEMPERATURE/
PRESSURE RELATIONSHIP OF
REFRIGERANTS. 1.0 TO .01

1.6 1.5 1.4 1.3 1.2 1.1 1.

1/ REDUCED TEMPERATURE

Fig. 5

.1
.09
.08
.07
.06
.05
.04
.03
.02
.01
.009
.008
.007
.006
.005
.004
.003
.002
.001
REDUCED PRESSURE
2.0
1.9
1.8
1.7
1.6
1.5
1.4
1/REDUCED TEMPERATURE
THE REDUCED TEMPERATURE/
PRESSURE RELATIONSHIP OF
REFRIGERANTS. 0.1 TO .001

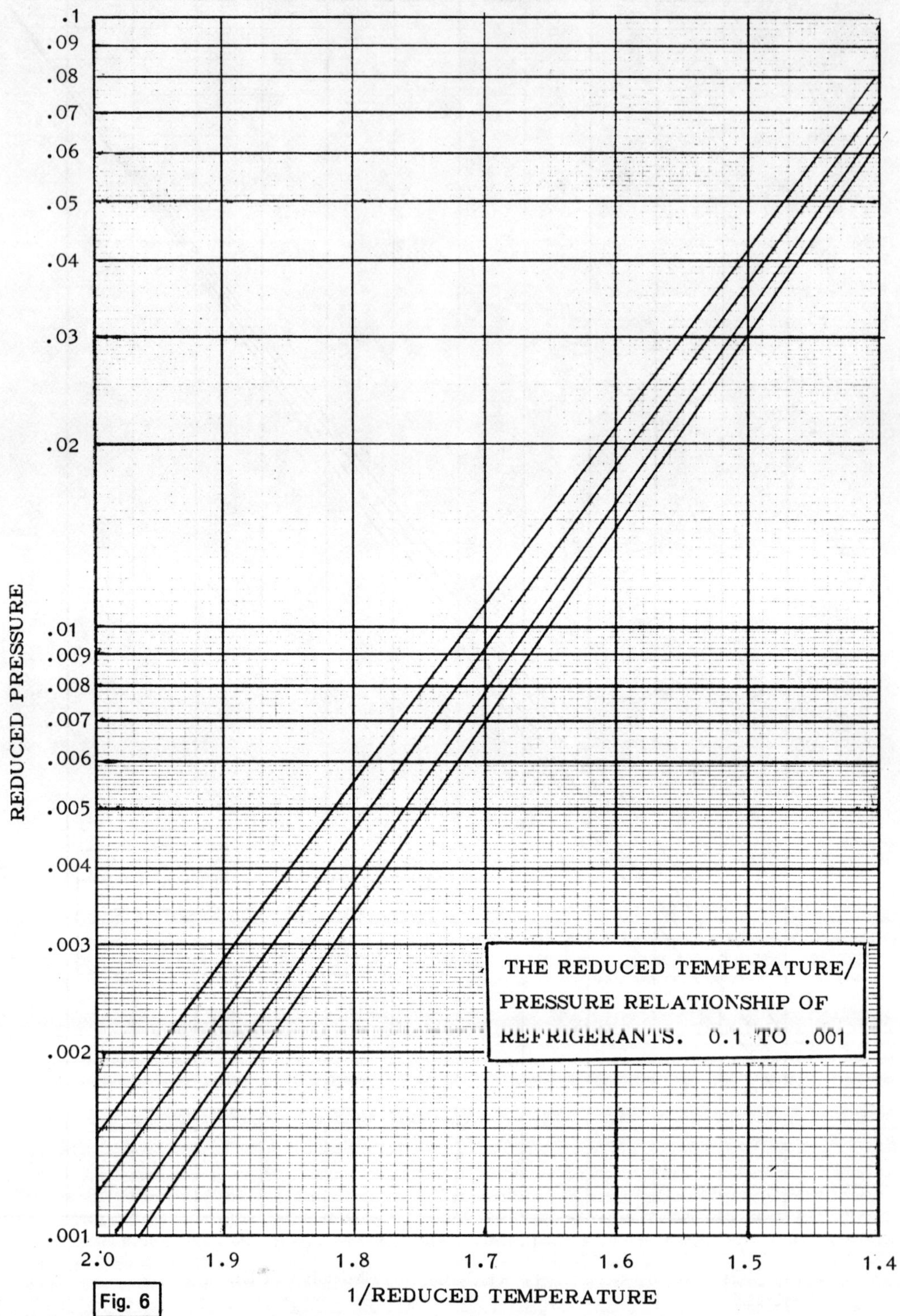

Fig. 6

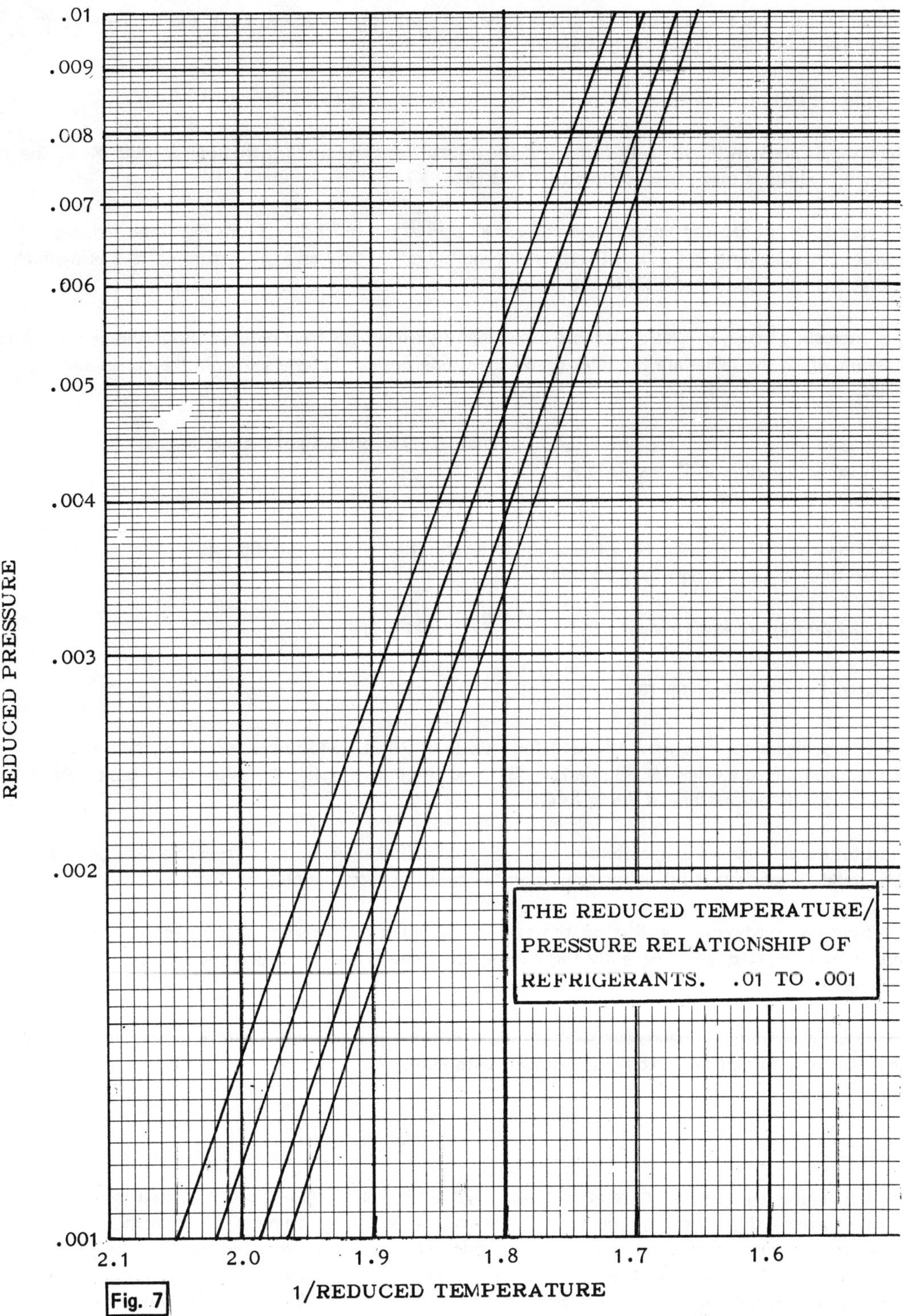

Fig. 7

Take as an example the case of R22; this is a methane with one hydrogen atom and so the second curve from the left is applicable (Group II).

Suppose we require the saturated vapour pressure corresponding to 40° C.

The critical pressure (Pcrit) is 48.7 atm and the critical temperature (Tcrit) is 96° C (369° K). The pressure required is that corresponding to 40° C (313° K).

The reciprocal of reduced temperature at 313° K is $369/313 = 1.179$. From Fig. 5 the reduced pressure corresponding to 1.179 is 0.315 which multiplied by 48.7 gives a pressure of 15.3 atm which accords with the reading of 15.3 in the tables.

It is interesting however that substances other than hydrocarbons have reduced properties very close to those on which the tables are formulated and the properties can be assessed from the corresponding values.

As an example, let us take the properties of ammonia. The critical point conditions are:
– critical temperature 271° F = 731° R,
– critical pressure 1657 psia.

It is known that 85° F (545° R) corresponds to a saturation pressure of 166 psia. What is the temperature corresponding to say 83 psia?

Reduced pressure of known value $\frac{166}{1657} = 0.1$

Reciprocal of reduced Temperature $\frac{731}{545} = 1.341$

The closest figure corresponding to 1.341 for a value of 0.1 is 1.343 in column III of the Table. The reduced pressure value for 83 psia is 0.05; using column III again, the value of the temperature ratio for a reduced pressure of .05 is 1.443.

Substituting, this gives a temperature of $\frac{731}{1,443} = 506.6°$ R or 46.6° F It will be seen on checking with standard tables that this is the saturation temperature corresponding to 83 psia.

The foregoing examples show that the pressure–temperature relationships of most refrigerants can be very easily calculated, and frequently this can be done when only the critical pressure and pemperature are known.

The calculation of the pressure/temperature values in the superheat area is rather more complex, but since it is only infrequently required it is dealt with later.

Volume.

The temperature–volume relationship is an equally simple one. The reduced temperature can be applied against reduced density, and the result shown in Fig. 8 has been derived from the numerical values of reduced density curves shown in the Table. It will be seen that for reduced temperatures between .6 and .8 or even .9, there is very slight variation in the reduced density on the standard vapour side, and because of this the results obtainable from calculations using this method will not necessarily be as close as desirable.

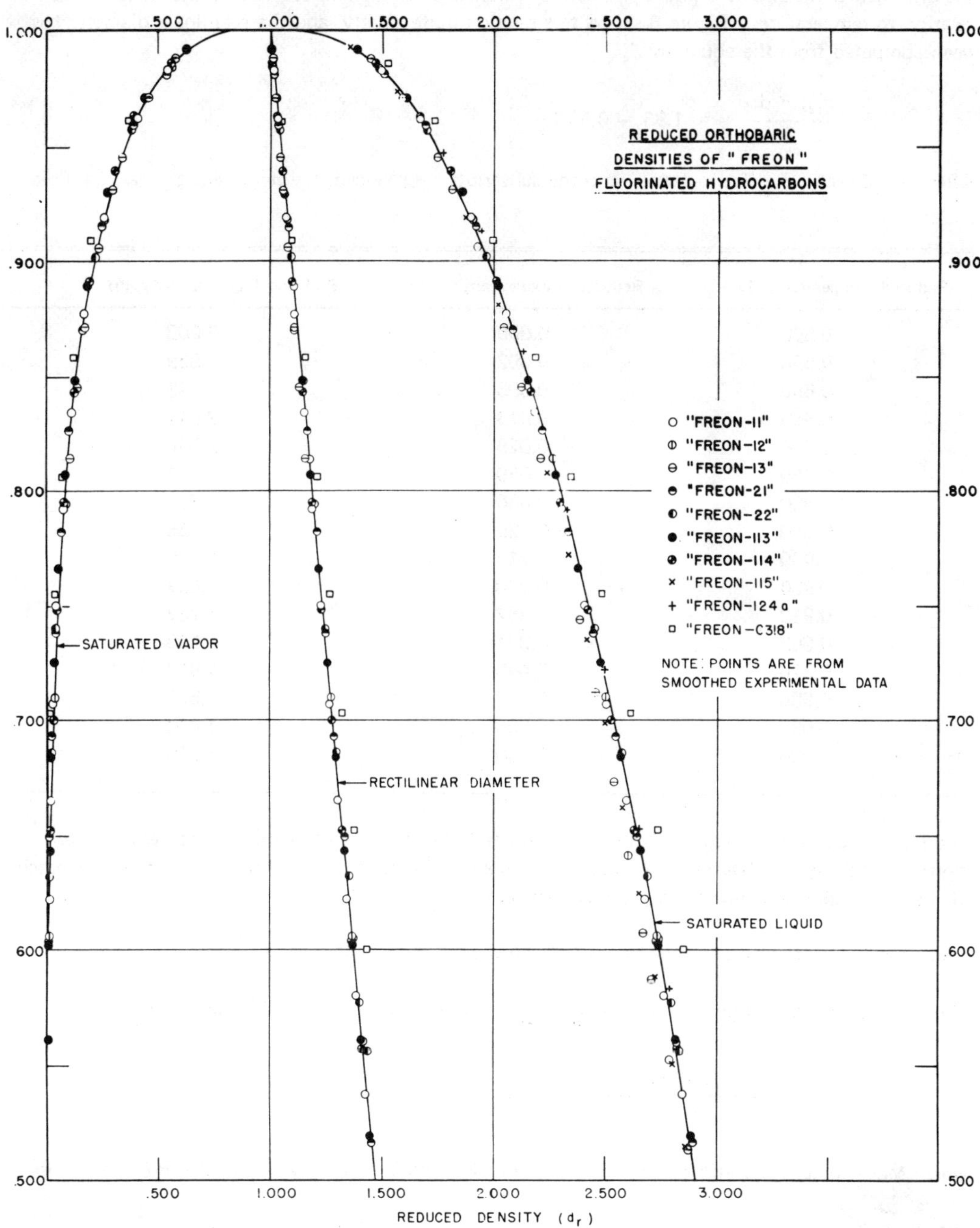

Fig.8: Rectilinear Diameter

The hypothesis of Rectilinear Diameter was put forward by Cailletet and Mathias, and postulates that the arithmetical mean of the liquid and vapour volumes of a free unassociated substance is a linear relation to temperature. Figure 8 shows this relation quite clearly, and the rectilinear diameter has been computed from the equation:

$$\frac{d_{rsg} + d_{rf}}{2} = 1.93 - 0.93T$$

where d = density, T = Temperature, and the subscripts r = reduced, s = saturated, g = gas, f = fluid.

Reduced Temperature, T_r	Reduced Vapour Density, d_{rsg}	Reduced Liquid Density, d_{rf}
0.500	0.000	2.903
0.550	0.002	2.829
0.600	0.005	2.742
0.650	0.013	2.644
0.700	0.028	2.538
0.750	0.050	2.421
0.800	0.080	2.300
0.850	0.129	2.155
0.900	0.210	1.970
0.930	0.290	1.837
0.950	0.352	1.737
0.960	0.391	1.683
0.970	0.445	1.617
0.980	0.514	1.526
0.990	0.605	1.400
1.000	1.000	1.000

The figures obtained for liquid densities are good and easy to compute with fair accuracy, but as already mentioned the figures for vapour densities are difficult to obtain owing to the close approach of the saturated vapour line to the vertical ordinate.

Two other ways of calculating the density of the vapour are available to us.

The first method gives a good approximation of vapour density at the NBP by using the formula:

$$\overline{V} = \frac{M_w \times 12.66 \times P}{B_k}$$

where M_w = molecular weight; B_k = the NBP in ° K; and $\overline{V}$ = density in g/l. P = Pressure in atm.

Alternatively:

$$\overline{V}_1 = \frac{M_w \times 1.42 \times P}{B_R}$$

where B_R = NBP in ° R, and $\overline{V}_1$ = density in lbs/ft^3. P = Pressure in atm.

These formulae give the density at reduced pressures up to .5 if one value higher than NBP is known. From this known value the correction value at the known reduced pressure can be found and its position noted on Figure 9 below. This point then sets the location of the line from which interpolation can be made.

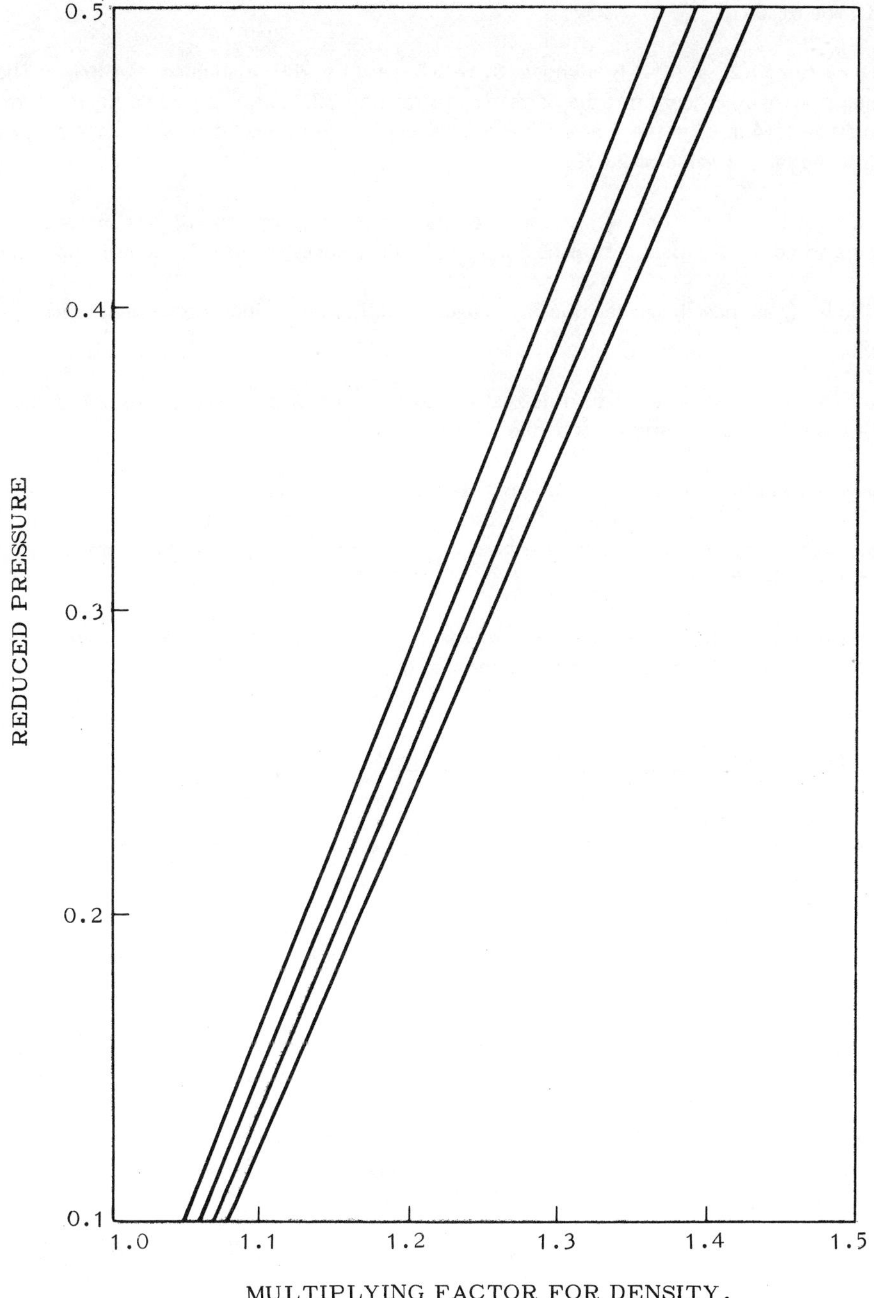

Fig. 9: Correction Factors for Volume.

Thus the density of R22 at 52° C is known to be .0902Kg x 10^3 / m^3 at a pressure of 20.3 bar = .4 P_{red}. The calculated density is .0684 so the correction factor is 1.32 which locates it between lines II and III. The corresponding correction factor for a reduced pressure of .2 = 9.95 bar = say 23° C is 1.14. The calculated density from the formula is .0368 which, times 1.14 = .0419 against the tabulated value of .0418.

Again, if the density of R22 at 124° F is known to be 5.5 lbs/ft^3 = 289 psia which is .4 P_{red}. The calculated value is 4.14 which confirms the correction factor of 1.32. Using the same figure as for the previous example ie 1.14 at .2 = 144.7 psia = 74° F, this, applied to a calculated figure of 2.29 gives 2.611 as against the tabular value of 2.635.

The correction factors are based on the fact that the values between .1 and about .5 or .6 for the reduced pressure so nearly approach a straight line that the differences are of no practical importance.

As with specific heat, the near linear relationship changes rapidly as P_{red} increases commencing at about .7 or .8.

It is of interest (though of little value) to note that the correction factor is almost constant at about 3.6 for all the commonly used refrigerants at the critical point.

The other method available is an extension of the first method in combination with the graphs of Figure 10. It is complex, and the figures reproduced are mainly for interest since this book is essentially for practical application and the methods of producing the graphs involve corrections using the compressibility factor.

It will be seen that Figure 10 is a general picture of the P—V—T relationship over a wide range, and that Figure 11 is an enlarged part of the low density area (boxed).

ENTHALPY

Calculation of Enthalpy

The columns in Property Tables under the heading "Enthalpy" are usually three in number headed respectively "Liquid", "Latent", and "Vapour". In the older tables enthalpy is referred to as Heat Content, and vapour heat sometimes as total heat.

The tables could logically be divided into five columns for enthalpy and some of the early tables do this. The additional columns would separate the internal energy from the external. The latent heat could be divided between the heat required to change the liquid to vapour, and the energy required to perform the work of expansion. Similarly the energy of the liquid could be divided, but the external work performed by the expanding liquid is extremely small and separating it would serve no useful purpose.

The calculation of Latent Heat is dealt with first, followed by that of Liquid Heat. The Vapour Heat is the sum of the two.

REDUCED PROPERTIES OF GASEOUS
REFRIGERANTS
REDUCED PRESSURE (P_r)
REDUCED DENSITY (d_r)
T_r = 1.75
T_r = 1.65
T_r = 1.55
T_r = 1.50
T_r = 1.45
T_r = 1.40
T_r = 1.35
T_r = 1.30
T_r = 1.25
T_r = 1.20
T_r = 1.15
T_r = 1.10
T_r = 1.05
T_r = 1.00
T_r = .95
T_r = .90
0
0.4
0.8
1.2
1.6
2.0
2.4
2.8
3.2
3.6

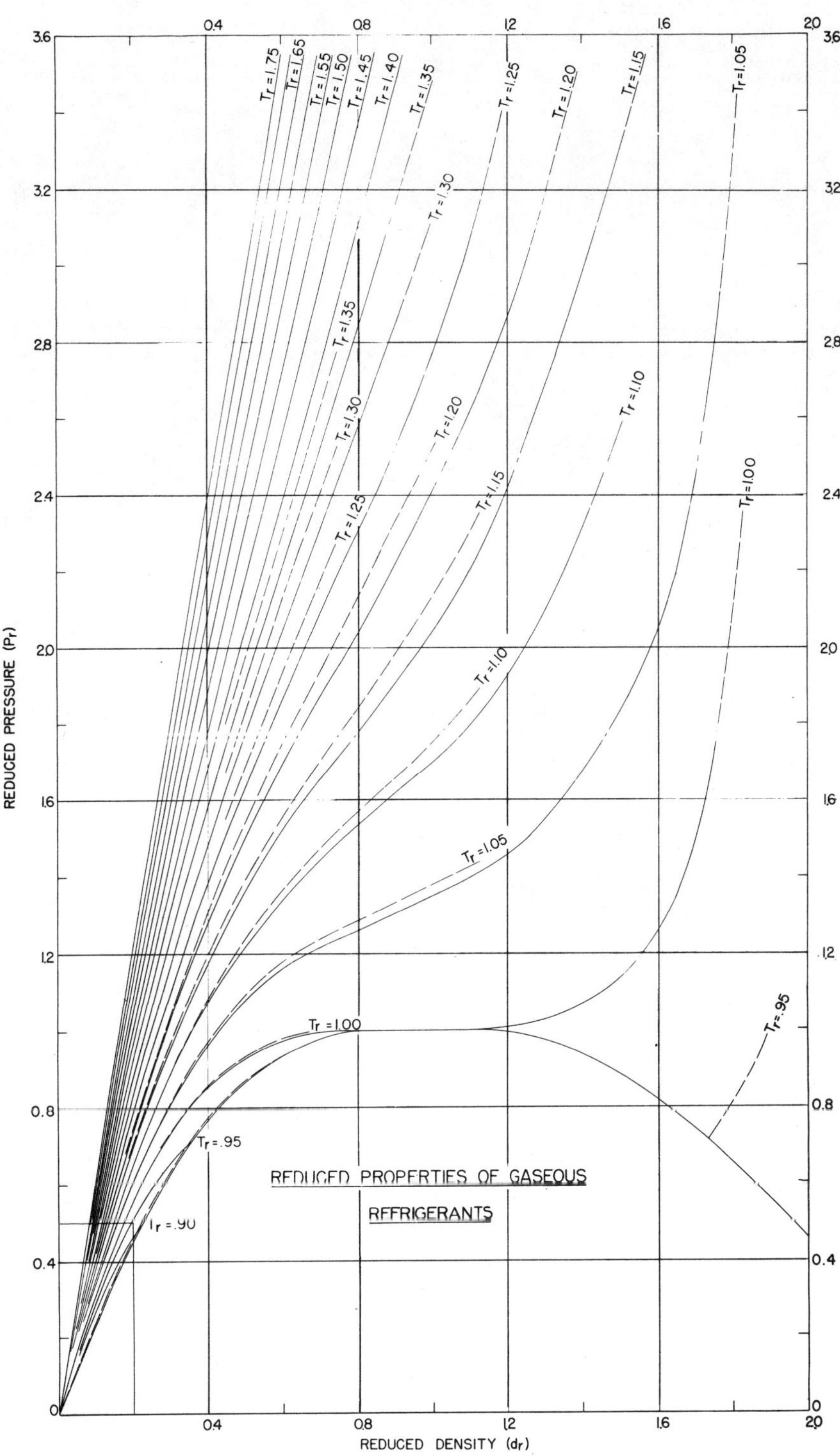

Fig. 10

REDUCED PROPERTIES OF GASEOUS
"FREON" COMPOUNDS AT LOW DENSITIES
REDUCED PRESSURE (P_r)
REDUCED DENSITY (d_r)
0
0.04
0.08
0.12
0.16
0.20
0.24
0.28
0.32
0.36
0.40
0.44
0.48
0.52
T_r=1.70
T_r=1.60
T_r=1.50
T_r=1.40
T_r=1.30
T_r=1.25
T_r=1.20
T_r=1.15
T_r=1.10
T_r=1.05
T_r=1.00
T_r=0.95
T_r=0.90
T_r=.85
T_r=.80
T_r=.75
T_r=.70
T_r=.65

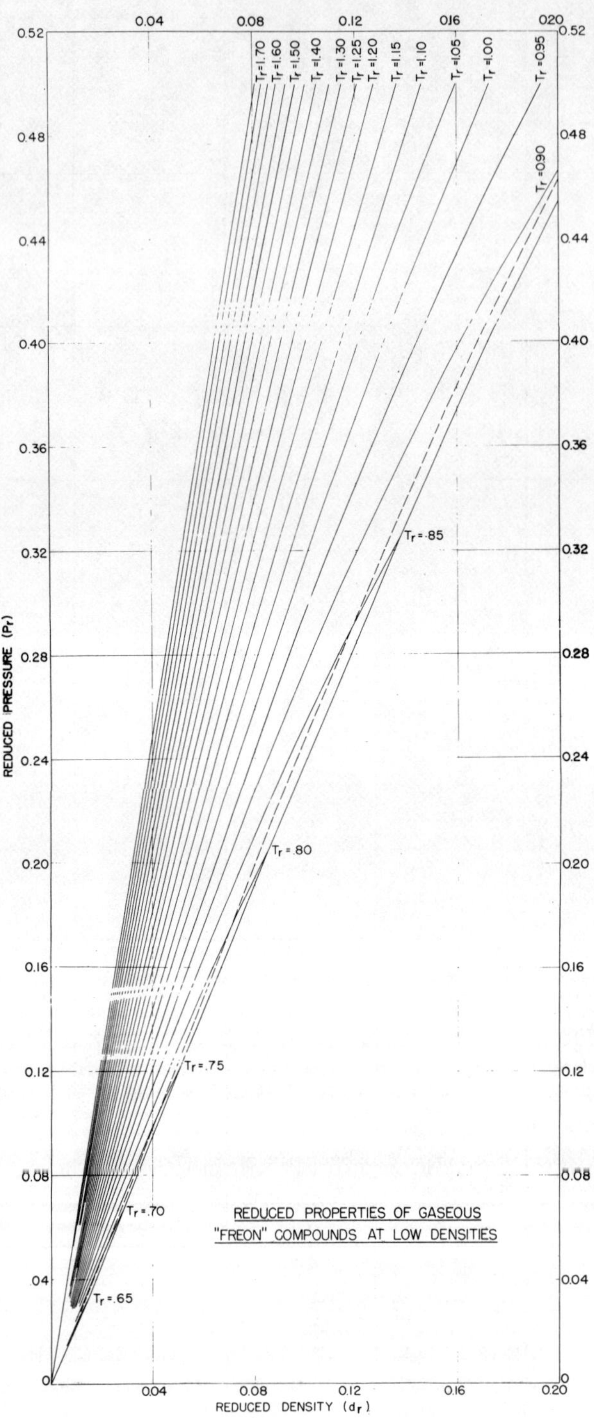

Fig. 11

Calculation of Latent Heat. The calculation of latent heat by means of the Clausius–Clapyron equation is somewhat cumbersome and time consuming, and does not lend itself to the unification of a method applicable to all refrigerants.

Some attempts have been made to correlate differing substances by seeking a common factor. One such attempt was made by Trouton who investigated latent heat in depth and postulated Trouton's Number. Trouton's Number is obtained by dividing the molar latent heat at NBP by the absolute temperature at NBP. The resulting number usually falls between the limits of 18 and 23 with most substances close to 20. It will be seen from Figure 12 that in general the molar latent heat increases with the NBP of the substance. Trouton's Number however only applies to the NBP.

Since the latent heat is affected by pressure rather more than any other factor, it seemed logical to seek some connection between latent heat and pressure which could be used for calculation, or at least close approximation of latent heat.

The first step was to extend Trouton's Number as a series dividing the molar latent heat at pressures other than at NBP; but this step offered little encouragement.

The application of the reduced form to the extended Trouton series however, produces a clear relationship. If the extended Trouton series is plotted against the reduced pressure corresponding to the temperature, the results – shown in Figure 13 – are a clear family of curves covering a large number of liquefiable gases in much the same way as the reduced temperature/pressure curves of Figure 4, p46.

The latent heat at the critical point must obviously be 0, and from that origin a group of curves is obtained from which a very close approximation of the latent heat may be obtained.

It is necessary to know one value of latent heat, and since this is usually available at the NBP, that suffices to place the point among the family of curves in Figure 13.

For example, the latent heat of R22 at +10° C is required. The following is known:

P_{crit} = 50.3 Kg/cm²; T_{crit} = 96° C; Mol Wt = 86.5;

Latent Heat at NBP = 56 cal/g; NBP = –41° C (232° K).

Reduced pressure at NBP = $^1/_{50.3}$ = .0199.

Trouton's Number = $\frac{56 \times 86.5}{232}$ = 20.88 which puts a point slightly to the right of the centre curve.

Reduced pressure at 10° C = $^7/_{50.3}$ = 0.139.

Extended Trouton Number corresponding to 0.139 from Figure 13 = 14.2 at 283° K.

Therefore: latent heat = $\frac{14.2 \times 283}{86.5}$ = 46.7 cal/g.

This result compares with the value from Tables of 47.34 cal/g.

A great advantage of using this method is that the system of units chosen is irrelevant, providing that it is constant.

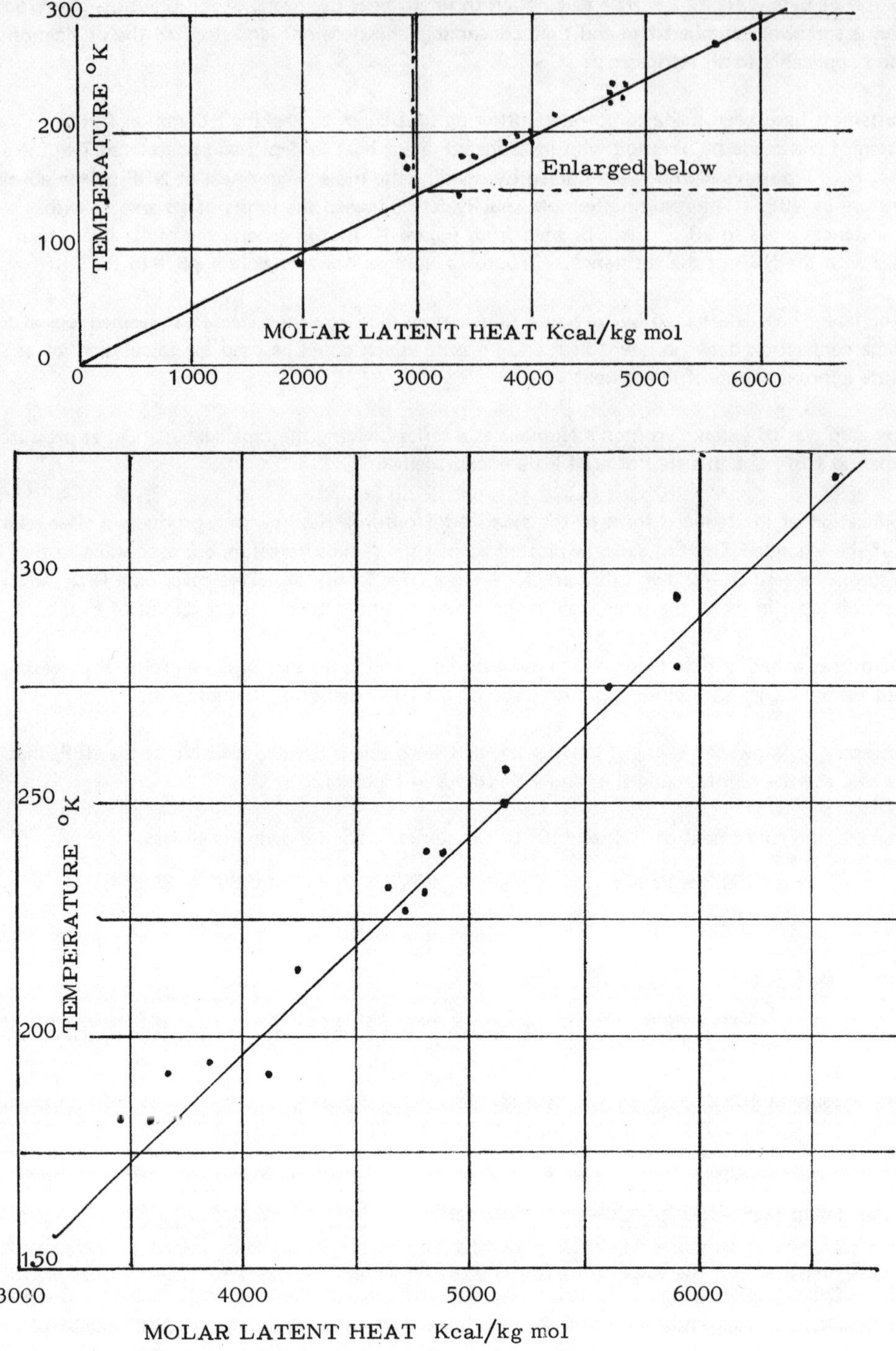

Fig. 12 THE MOLAR LATENT HEAT OF SOME REFRIGERANTS

REDUCED PRESSURE
1·0
·8
·6
·4
·2
·1
·08
·06
·04
·02
·01
·008
·006
·004
·002
TROUTON'S NUMBER
0
2
4
6
8
10
12
14
16
18
20
22
24
26
28
30

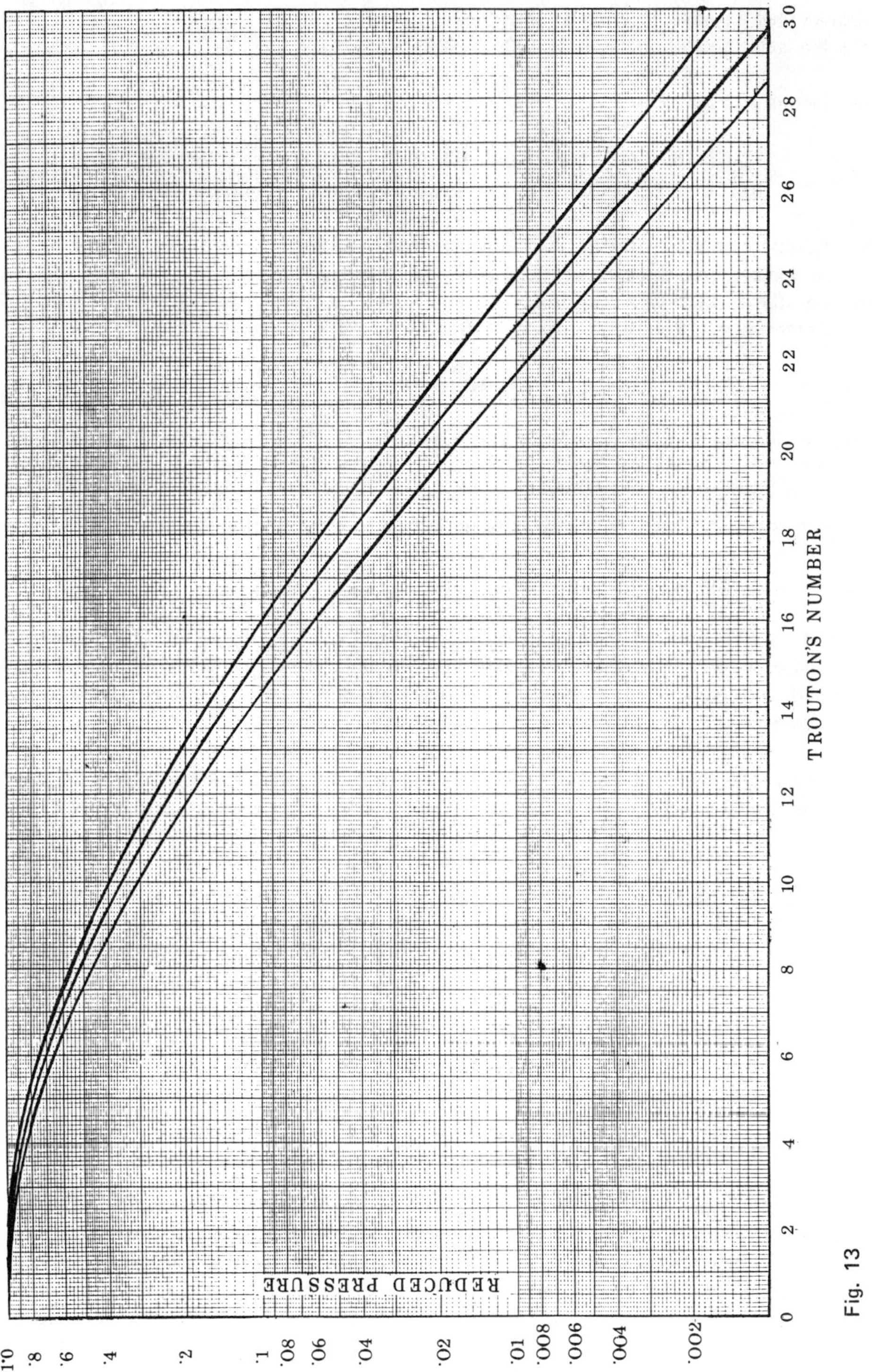

Fig. 13

To take another example, in Imperial units:
what is the latent heat of ammonia at 100° F (560° R) ?

Known factors are:

P_{crit} = 1657; T_{crit} = 271° F = 731° R; Mol Wt = 86.5;
NBP = −28° F (432° R); Latent heat at NBP = 589 BTU/lb.

Reduced pressure at NBP = .0088. Trouton's Number = 23.2.
Trouton's Number from the chart at .0088 is on the centre line.
Pressure at 100° F (560° R) = 212 psia.
Reduced pressure = .128 x Trouton No at .128 = 14.3.

$$\text{Latent heat at } 100^\circ \text{ F} = \frac{14.3 \times 560}{17 \text{ (Mol Wt)}} = 471 \text{ BTU/lb.}$$

This result compares with the value from Tables of 477.

The Liquid Heat. The Liquid Heat is taken from a base temperature, usually −40°, and is the product of the mean specific heat of the liquid between the temperature under consideration and the base temperature, and the temperature range.

No way of correlating specific heats of a range of refrigerants to a common base has presented itself and the values are derived basically from experimental data. This wide variation in value and form shows clearly in Figures 14 to 19; however the charts are sufficiently accurate to use for normal practice purposes.

Take the case of ammonia at say −33° C. The mean specific heat is 1.064 so that the liquid heat is 7.45 to a base of 0 at −40°, or, if a base of 100 cal/g at 0° is used the liquid heat would be 64.21.

The Enthalpy of the Vapour. The latent heat and the liquid heat having already been found, the vapour heat is simply the sum of these two. For this reason, when condensing tables of properties, the latent heat is frequently omitted since it can be found by taking the difference between the vapour heat and the liquid heat.

ENTROPY

Calculation of Entropy

The entropy of a liquid is the change in the heat of the liquid from the base point, divided by the mean temperature between the temperature of consideration and the base temperature which is purely arbitrary. This base temperature is almost always −40° for the Imperial units tables. S.I. tables however, although primarily conversions from the original Imperial tables, have used a number of different bases for enthalpy from which entropy must be calculated. Some tables use a figure of 200 kJ/Kg for the liquid at 0° C as a basis for the tables, whilst using 100 Kcal/Kg as a basis for the Mollier Chart; yet other tables re-set the entropy value to 1 at 0° C.

SPECIFIC HEAT OF LIQUID REFRIGERANTS
R 11, R 21, R 22, R 40 AND R 113
Degrees Centigrade
-80
-60
-40
-20
0
20
40
60
80
100
120
140
Specific Heat.
0.45
0.4
0.35
0.3
0.25
0.2
R 22
R 40
R 21
R 113
R 11
-100
-50
0
50
100
150
200
250
300
Degrees Fahrenheit

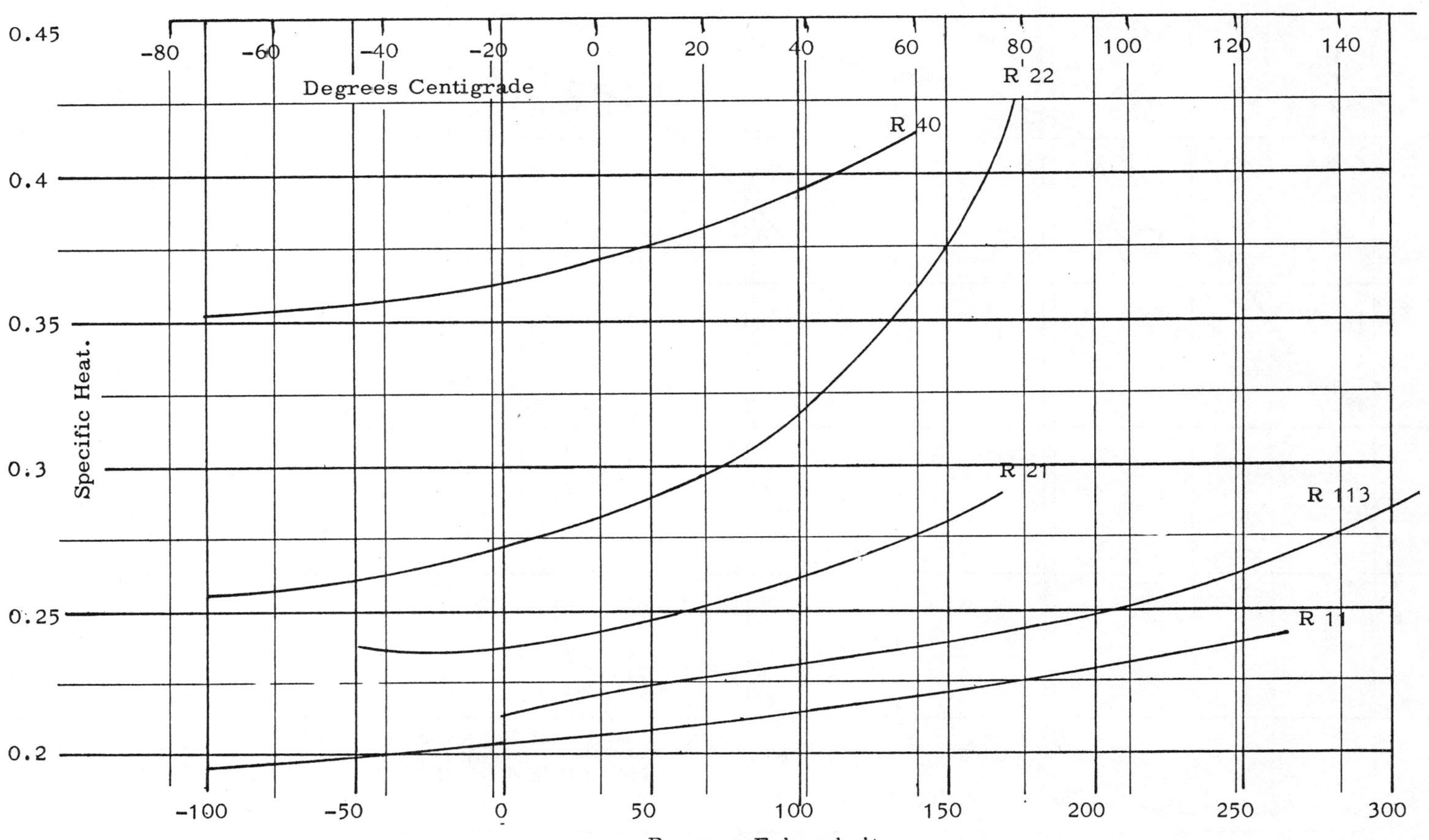

Fig. 14

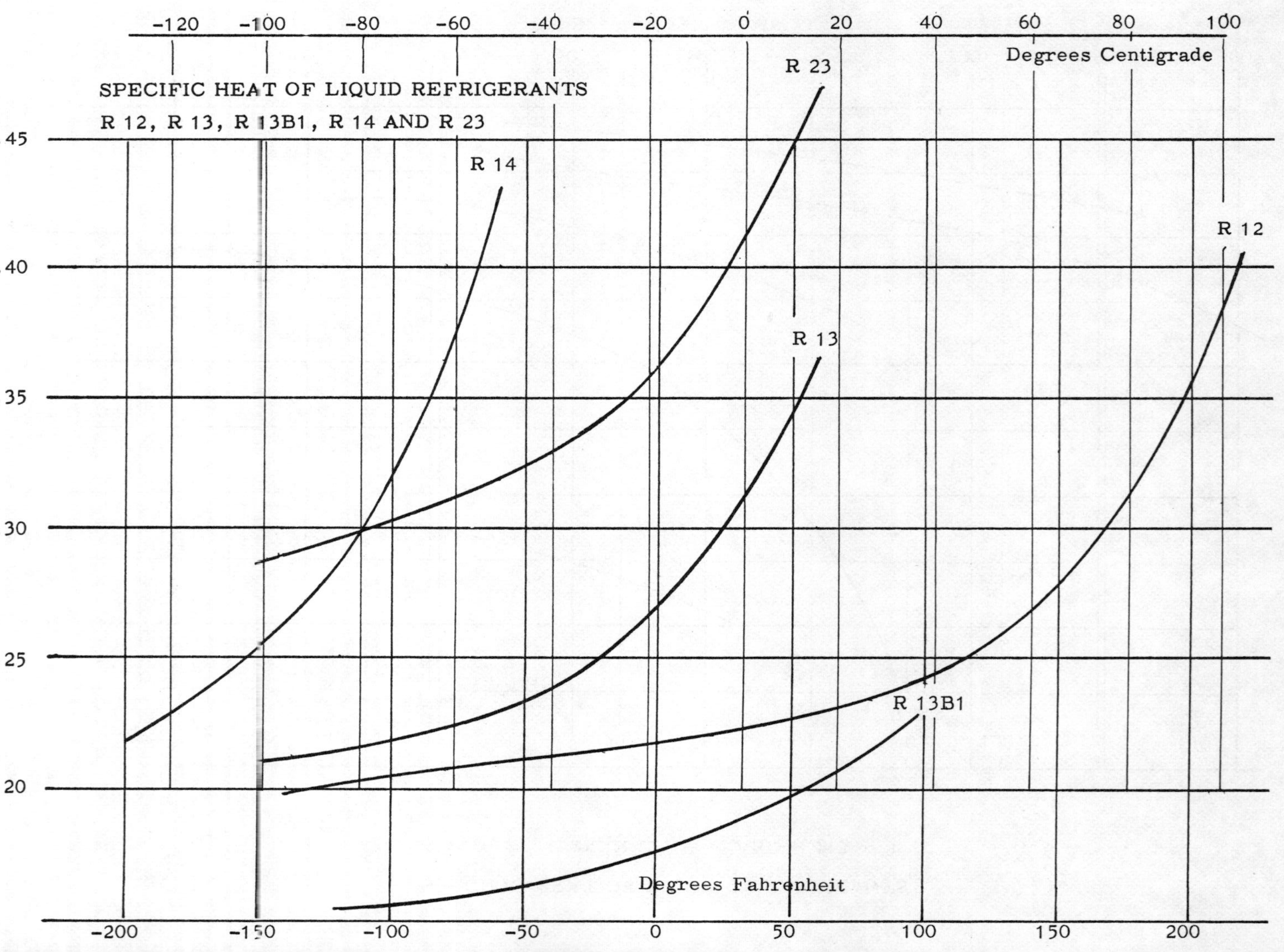

SPECIFIC HEAT OF LIQUID REFRIGERANTS
R 12, R 13, R 13B1, R 14 AND R 23
Degrees Centigrade
-120
-100
-80
-60
-40
-20
0
20
40
60
80
100
Degrees Fahrenheit
-200
-150
-100
-50
0
50
100
150
200
.45
.40
.35
.30
.25
.20
R 23
R 14
R 12
R 13
R 13B1

SPECIFIC HEAT OF LIQUID REFRIGERANTS
R 114, R 115, R 142b, R 318(C), AND R 744.
Degrees Centigrade
-80
-60
-40
-20
0
20
40
60
80
100
120
140
Specific Heat
.50
.45
.40
.35
.30
.25
.20
R 744
R 115
R 142b
R C318
R114
Degrees Fahrenheit
-100
-50
0
50
100
150
200
250

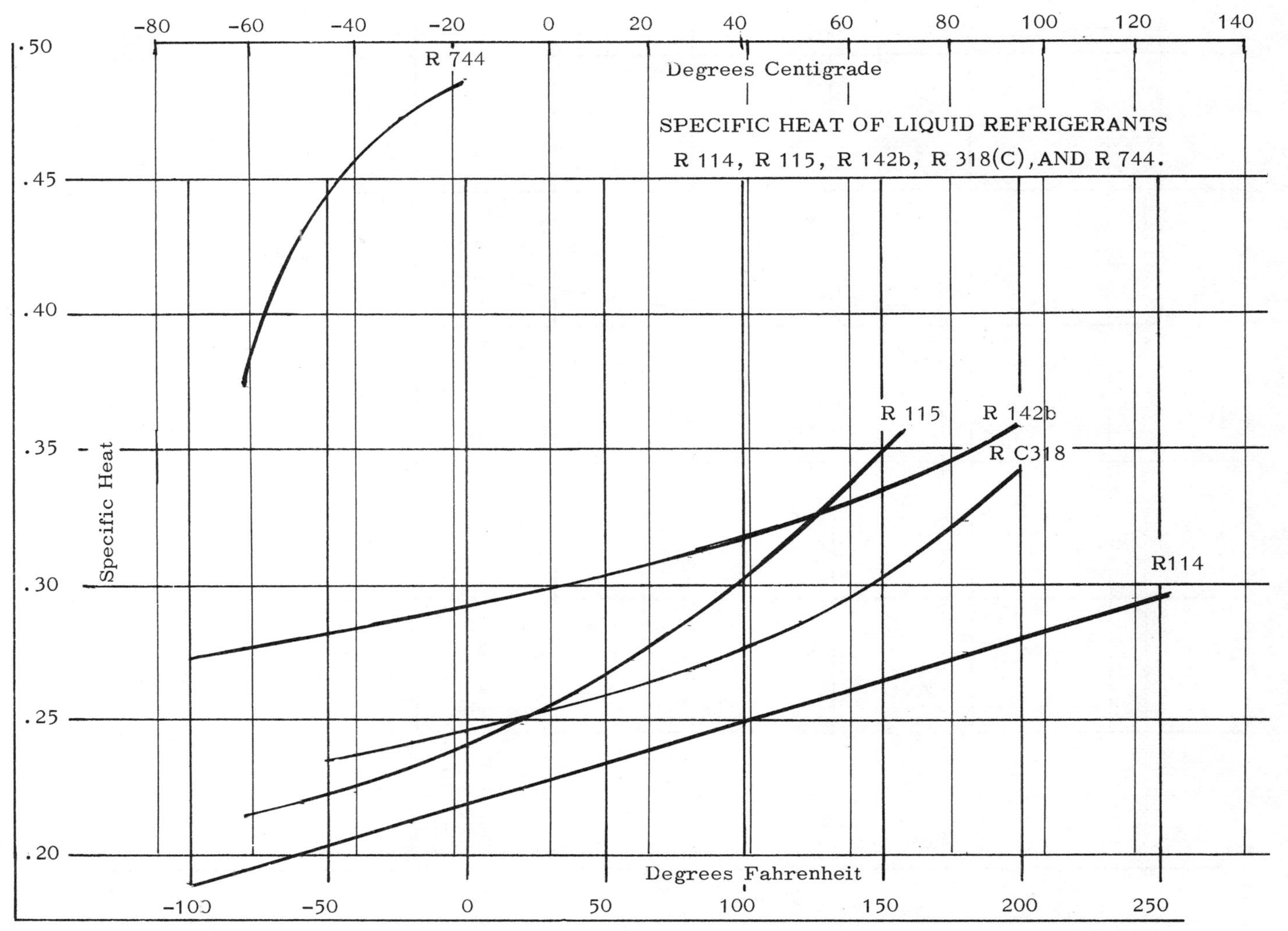

Fig. 16

SPECIFIC HEAT OF LIQUID AZEOTROPES
R 500, R 502, R 503 AND R 504
Degrees Centigrade
-120
-100
-80
-60
-40
-20
0
20
40
60
80
100
120
Degrees Fahrenheit
-200
-150
-100
-50
0
50
100
150
200
250
Specific Heat
.60
.50
.40
.30
.20
.10
R 500
R.502
R 503
R 504

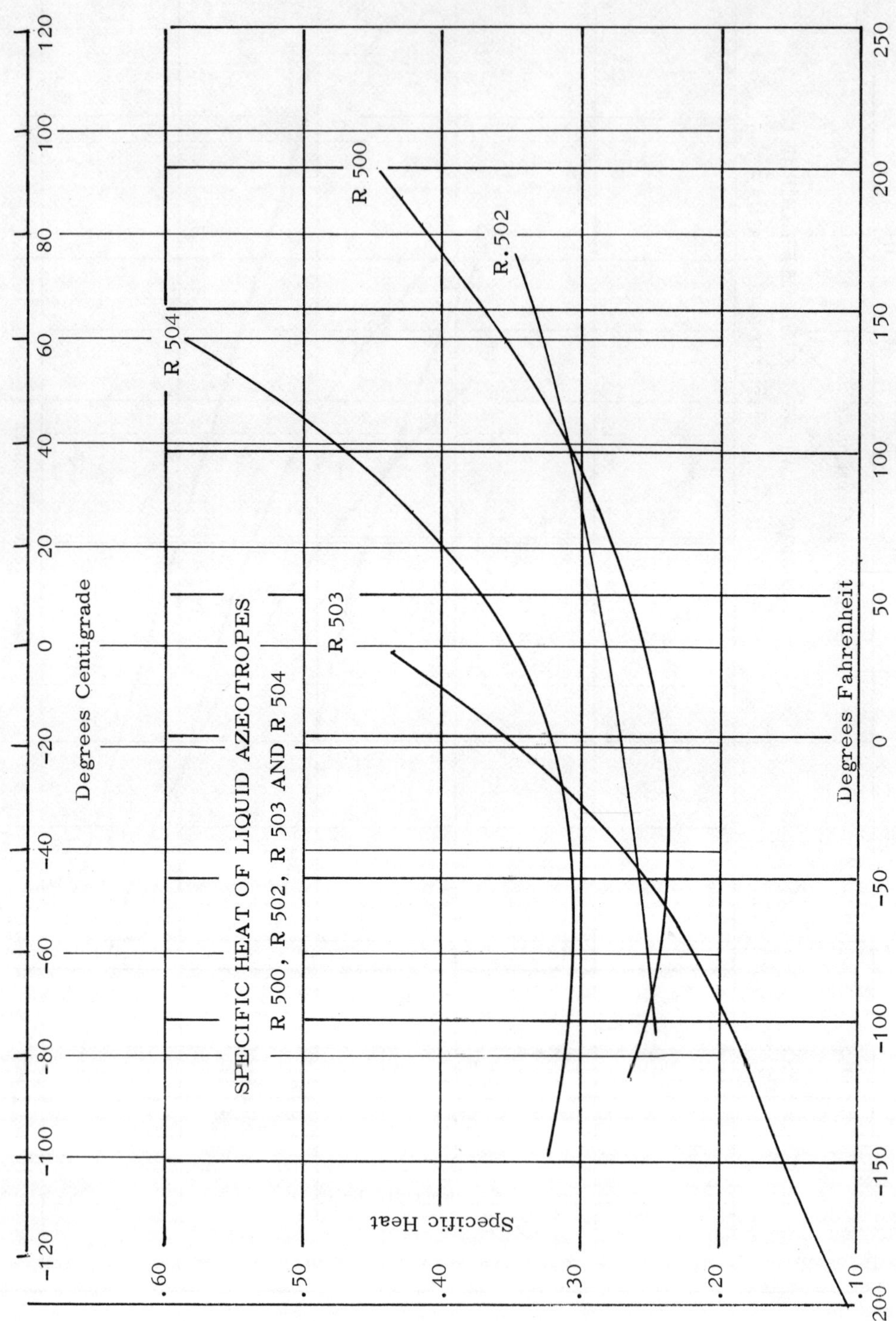

Fig. 17

Degrees Centigrade
-80 -60 -40 -20 0 20 40 60 80 100 120
SPECIFIC HEAT OF LIQUID AMMONIA
Refrigerant 717
Specific Heat
1.5 1.4 1.3 1.2 1.1 1.0
-100 -50 0 50 100 150 200 250 300
Degrees Fahrenheit

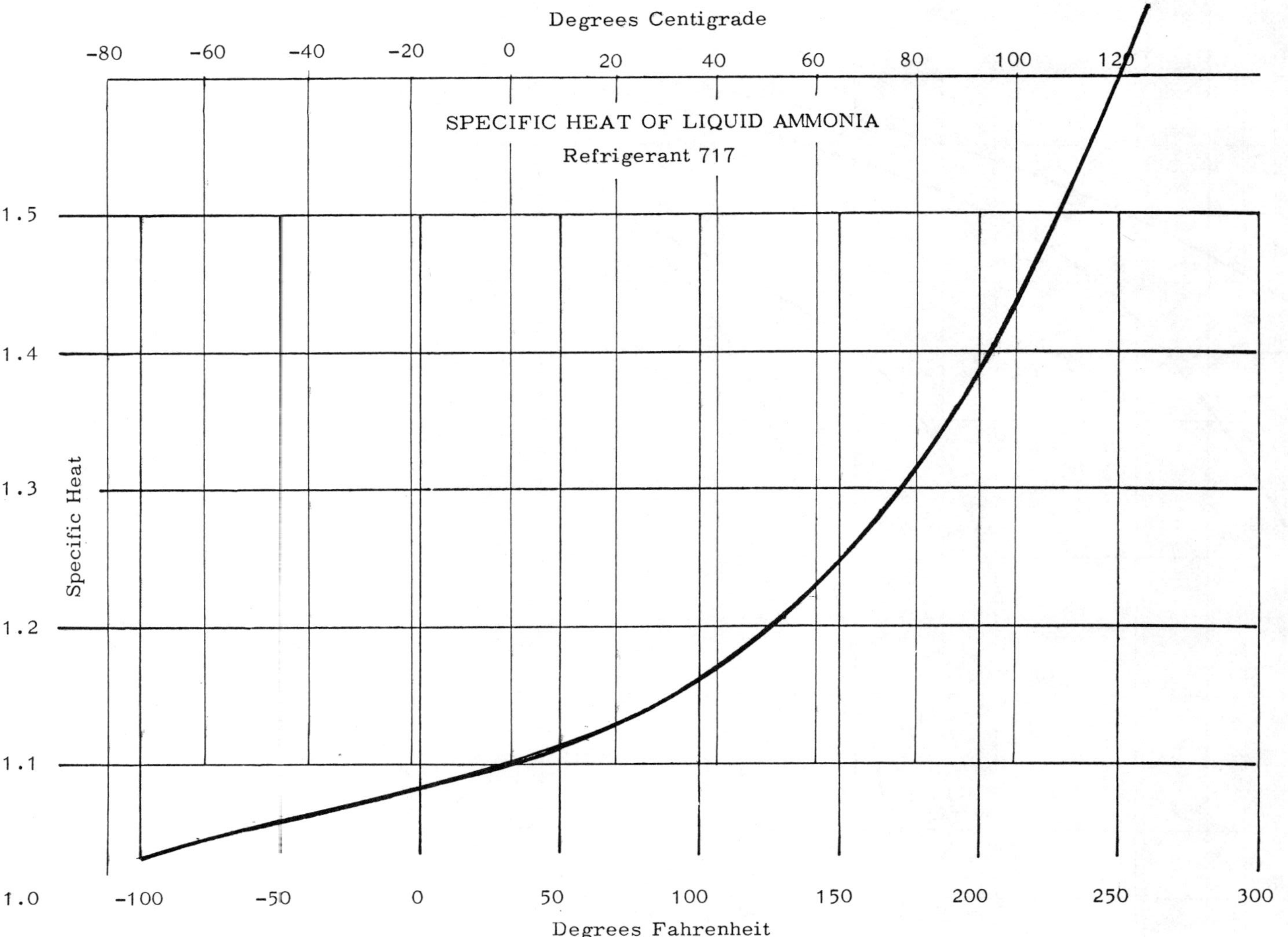

Fig. 18

Degrees Centigrade
-160
-140
-120
-100
-80
-60
-40
-20
0
20
40
60
80
Methane
R 50
Ethylene
R 1150
Ethane
R 170
Propane
R290
Butane
R 600
SPECIFIC HEAT OF
LIQUID HYDROCARBONS
1.10
1.00
0.90
0.80
0.70
0.60
0.50
Degrees Fahrenheit
-250
-200
-150
-100
-50
0
50
100
150

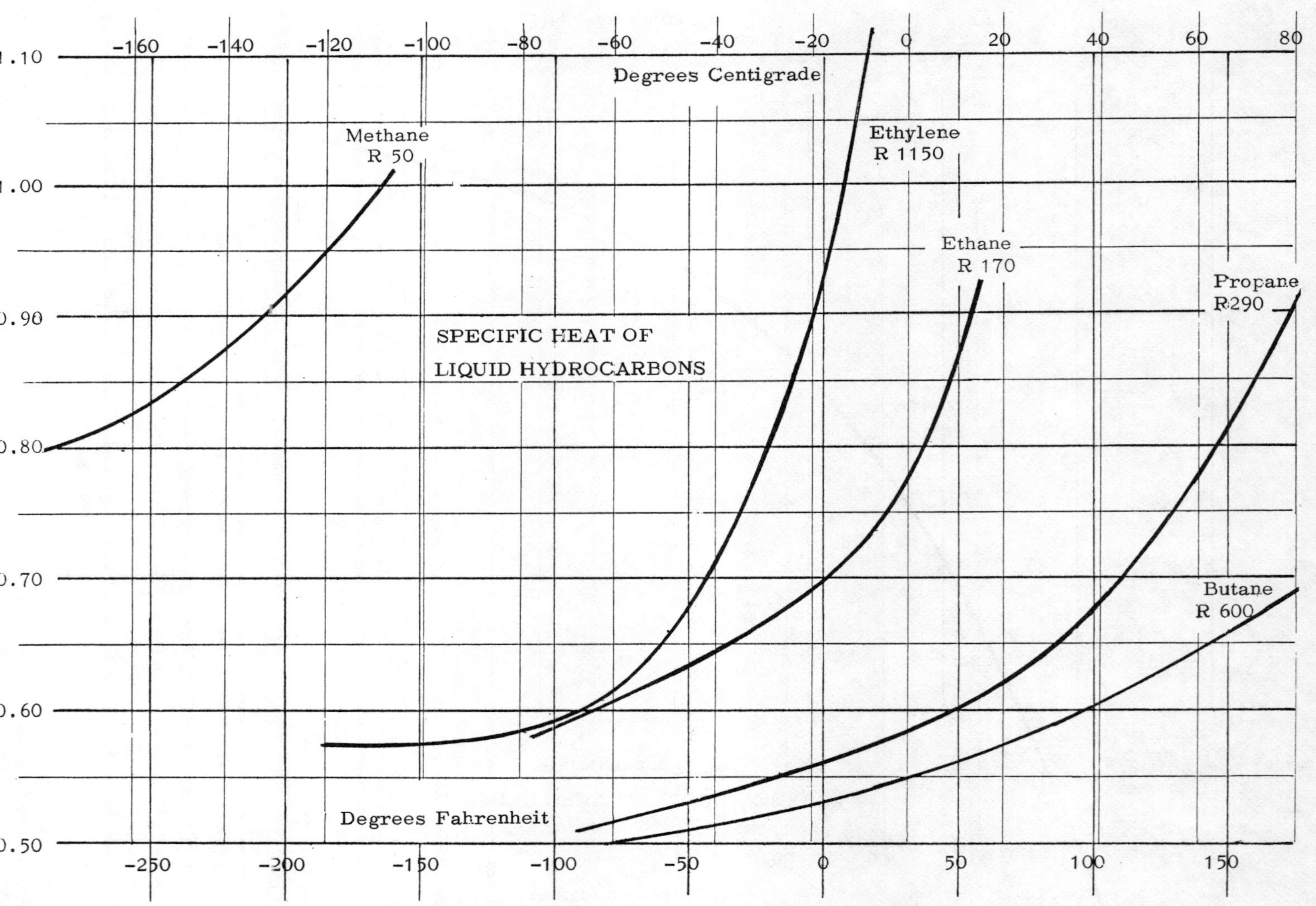

Fig. 19

Entropy of Liquid. This is calculated by taking the change in the heat of the liquid, and dividing it by the average absolute temperature at which the change occurs. Thus from the Table of Properties: the change in the liquid heat between say 30° C and the base temperature of 0° C is 228.5 – 200 = 28.5. The average absolute temperature is 273 + $^{30}/_{2}$ = 288°.K = .097 which when added to the base value of 1.0 gives the 1.097 in the Table.

Taking an example from the Imperial tables for R22 (Table of Properties):
the entroly of the liquid at say –80° F is found by taking the liquid heat –9.84 and dividing it by the average temperature between –80° F and the base temperature which is –40° F.

Thus, the average absolute temperature of the change is 460 – 60 = 400.
–9.84 divided by 400 = –.0246.

Entropy of the Vapour. This is the sum of the entropy of the liquid and the entropy of evaporation. This latter is simply the latent heat divided by the absolute temperature; so, for the first example above, having already found the entropy of the liquid at 30° tp be 1.097, we take the latent heat at 30° = $^{135}/_{303}$ = .446, and add 1.097 to obtain 1.543.

Similarly in the second example; the latent heat is $^{105.5}/_{380}$ = .278 + the liquid entropy –.0246 = .2534.

CONCLUSION

From the foregoing it can be seen that the compilation of tables of the saturation properties of refrigerants within workable limits of accuracy is not a difficult operation. It is not the intention to advocate that readers should do this for themselves since the following chapter gives a wide selection of tables of saturated properties of refrigerants which have already been compiled. Nonetheless, it is a highly profitable exercise to compile a skeleton set of one particular substance and, only on completion, to check the results by comparison with standard tables.

CHAPTER 7

Data on Refrigerants

The second part of this book is devoted to information on the properties of some refrigerants. They are placed in the order of the now almost universally accepted standard numbering system described on pages 28 et seq.

There is a data sheet for each refrigerant which gives the properties from which close approximations of pressure, temperature, volume and enthalpy can be made for any predicated conditions. The critical temperature and pressure, and the normal boiling point constitute four known points enabling the appropriate curve on figures 4 to 7 to be located so that other conditions of pressure/temperature relationship may be determined.

Given the pressure and the molecular weight, the volume can be obtained by the procedure outlined on page 52. The molecular latent heat for predicated conditions can be found by the use of the extended Trouton Number using the chart, Figure 13. Again, given the latent heat and the specific heat from one of the figures 14 to 19, the enthalpy may be found, and from this the entropy.

However, values have been tabulated for the properties of many refrigerants, and each data sheet is followed by tables of the saturation properties in Imperial units and where available in metric or SI units.

The enthalpy and the entropy may be shown as the abscissae of a chart with the temperature and pressure as families of curves. This form of chart was devised by Prof. Mollier who worked with steam tables and put forward many ways of showing properties in chart form. The original Mollier form is sometimes useful in refrigeration but one of Mollier's later forms in which pressure and enthalpy form the coordinates with volume, temperature and entropy shown as families of curves is the form generally accepted in the industry. This pressure—enthalpy chart is in fact a true Mollier chart and is sometimes referred to as such in the text; however, it is now more generally known as a Pressure Enthalpy chart and this term is used throughout the Data Section. Pressure Enthalpy charts follow the tables of properties in most cases.

Refrigerant No.	Data Sheet	Saturation Properties		Pressure Enthalpy Diagram	
		Imperial	*Metric*	*Imperial*	*Metric*
11	70	71	–	79	80
12	81	82	87	91	92
13	93	94	–	100	101
13 BI	102	103	–	109	–
14	110	111	–	114	–
21	115	116	–	118	119
22	120	121	128	132	133
23	134	135	–	–	–
30	137	138	138	–	–
40	139	140	–	142	143
50	144	145	–	146	–
113	147	148	–	150	151
114	152	153	–	154	155
115	156	157	–	159	–
142b	160	161	–	162	–
152a	163	164	–	165	–
170	166	167	–	168	–
216	169	170	–	171	–
290	172	173	–	174	–
C 318	175	176	–	–	–
500	180	181	–	183	–
502	184	185	191	194	195
503	196	197	–	199	–
504	200	201	–	203	–
600	204	205	–	206	–
601	207	208	–	209	–
717	210	211	–	215	217
744	219	220	–	223	224
764	225	226	–	–	–

DATA SHEET FOR R 11

$CFCl_3$	MONOFLUOROTRICHLOROMETHANE	Molecular Wt. 137.4

Pressure		Temperature				Volume		Density	
43.2	kg/cm²	198	°C	388	°F	1.80	l/kg	.554	kg/l.
635	psia	471	°K	848	°R	.0289	ft³/lb	34.6	lbs/ft³
								4.74	Air=1

AT CRITICAL POINT

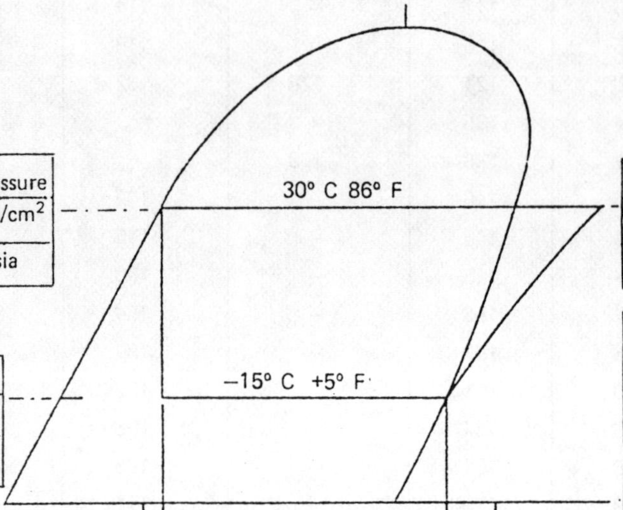

Discharge Pressure	
1.28	kg/cm²
18.3	psia

Inlet Pressure	
.205	kg/cm²
2.92	psia

Discharge Temperature			
44	°C	111	°F
317	°K	571	°R

Normal Boiling Point			
23.8	°C	74.8	°F
297	°K	534	°R

Triple Point			
-111	°C	-168	°F
162	°K	292	°R

Refrigerating Effect	
48.6 kcal/m³	5.47 Btu/ft³

Latent Heat at NBP	43.5 kcal/kg	5977 /kg mol	78.3 Btu/lb	10758 /lb mol

Trouton's No. 20.13	Gas Constant 6.17 kg.m/kg/° K	11.25 ft.lbs/lb/° R

Specific Heat Liquid at 30° C 86° F .209 Fig 13	Gas C_p .135	C_p/C_v 1.136 at 30°C

Liquid at 30° C 86° F Density 1.46 kg/l.	91.4 lbs/ft³ Viscosity .405 cp

Reduced Form at NBP Pressure .023	1 / Temperature 1.61

Reduced value of 1/T	1.92	1.71	1.359
when Reduced Pressure is	a) .001	b) .01	c) .1

TEMP. °F	PRESSURE PSIA	PRESSURE PSIG	VOLUME cu ft/lb LIQUID v_f	VOLUME cu ft/lb VAPOR v_g	DENSITY lb/cu ft LIQUID $1/v_f$	DENSITY lb/cu ft VAPOR $1/v_g$	ENTHALPY Btu/lb LIQUID h_f	ENTHALPY Btu/lb LATENT h_{fg}	ENTHALPY Btu/lb VAPOR h_g	ENTROPY Btu/(lb)(°R) LIQUID s_f	ENTROPY Btu/(lb)(°R) VAPOR s_g	TEMP °F
−168	0.000844	29.91 *	0.009055	26985	110.4	0.0000371	−26.06	99.17	73.11	−0.07414	0.2658	−168
−167	0.000915	29.91 *	0.009061	24985	110.3	0.0000400	−25.85	99.07	73.21	−0.07343	0.2650	−167
−166	0.000991	29.91 *	0.009067	23150	110.2	0.0000432	−25.64	98.96	73.31	−0.07272	0.2642	−166
−165	0.001072	29.91 *	0.009072	21463	110.2	0.0000466	−25.44	98.85	73.41	−0.07202	0.2634	−165
−164	0.001160	29.91 *	0.009078	19911	110.1	0.0000502	−25.23	98.74	73.51	−0.07132	0.2626	−164
−163	0.001254	29.91 *	0.009084	18482	110.0	0.0000541	−25.02	98.64	73.61	−0.07062	0.2618	−163
−162	0.001355	29.91 *	0.009089	17166	110.0	0.0000583	−24.82	98.53	73.71	−0.06992	0.2611	−162
−161	0.001462	29.91 *	0.009095	15953	109.9	0.0000627	−24.61	98.43	73.81	−0.06923	0.2603	−161
−160	0.001578	29.91 *	0.009101	14835	109.8	0.0000674	−24.40	98.32	73.91	−0.06854	0.2595	−160
−159	0.001702	29.91 *	0.009107	13802	109.8	0.0000725	−24.20	98.21	74.01	−0.06785	0.2588	−159
−158	0.001834	29.91 *	0.009112	12849	109.7	0.0000778	−23.99	98.11	74.11	−0.06717	0.2580	−158
−157	0.001975	29.91 *	0.009118	11969	109.6	0.0000836	−23.78	98.00	74.22	−0.06648	0.2573	−157
−156	0.002127	29.91 *	0.009124	11155	109.6	0.0000896	−23.58	97.90	74.32	−0.06580	0.2565	−156
−155	0.002288	29.91 *	0.009130	10402	109.5	0.0000961	−23.37	97.79	74.42	−0.06512	0.2558	−155
−154	0.002460	29.91 *	0.009136	9705	109.4	0.0001030	−23.16	97.69	74.52	−0.06445	0.2551	−154
−153	0.002644	29.91 *	0.009141	9059	109.3	0.0001104	−22.96	97.59	74.62	−0.06378	0.2544	−153
−152	0.002840	29.91 *	0.009147	8462	109.3	0.0001182	−22.75	97.48	74.73	−0.06310	0.2537	−152
−151	0.003049	29.91 *	0.009153	7907	109.2	0.0001265	−22.55	97.38	74.83	−0.06244	0.2530	−151
−150	0.003272	29.91 *	0.009159	7393	109.1	0.0001352	−22.34	97.28	74.93	−0.06177	0.2523	−150
−149	0.003508	29.91 *	0.009165	6916	109.1	0.0001446	−22.13	97.17	75.03	−0.06111	0.2516	−149
−148	0.003761	29.91 *	0.009171	6473	109.0	0.0001545	−21.93	97.07	75.14	−0.06045	0.2510	−148
−147	0.004029	29.91 *	0.009176	6062	108.9	0.0001650	−21.72	96.97	75.24	−0.05979	0.2503	−147
−146	0.004314	29.91 *	0.009182	5679	108.9	0.0001761	−21.52	96.87	75.34	−0.05913	0.2496	−146
−145	0.004617	29.91 *	0.009188	5324	108.8	0.0001878	−21.31	96.76	75.45	−0.05847	0.2490	−145
−144	0.004938	29.91 *	0.009194	4993	108.7	0.0002003	−21.11	96.66	75.55	−0.05782	0.2484	−144
−143	0.005280	29.91 *	0.009200	4684	108.6	0.0002135	−20.90	96.56	75.66	−0.05717	0.2477	−143
−142	0.005642	29.90 *	0.009206	4397	108.6	0.0002274	−20.70	96.46	75.76	−0.05652	0.2471	−142
−141	0.006027	29.90 *	0.009212	4130	108.5	0.0002421	−20.49	96.36	75.86	−0.05588	0.2465	−141
−140	0.006434	29.90 *	0.009218	3880	108.4	0.0002577	−20.28	96.26	75.97	−0.05524	0.2458	−140
−139	0.006866	29.90	0.009224	3648	108.4	0.0002741	−20.08	96.16	76.07	−0.05460	0.2452	−139
−138	0.007323	29.90 *	0.009230	3431	108.3	0.0002914	−19.87	96.06	76.18	−0.05396	0.2446	−138
−137	0.007807	29.90 *	0.009236	3228	108.2	0.0003097	−19.67	95.96	76.28	−0.05332	0.2440	−137
−136	0.008319	29.90 *	0.009242	3039	108.2	0.0003290	−19.46	95.86	76.39	−0.05269	0.2434	−136
−135	0.008860	29.90 *	0.009248	2862	108.1	0.0003494	−19.26	95.76	76.49	−0.05205	0.2428	−135
−134	0.009433	29.90 *	0.009254	2696	108.0	0.0003708	−19.05	95.66	76.60	−0.05142	0.2423	−134
−133	0.01004	29.90	0.009260	2541	107.9	0.0003934	−18.85	95.56	76.70	−0.05080	0.2417	−133
−132	0.01068	29.89	0.009266	2396	107.9	0.0004172	−18.64	95.46	76.81	−0.05017	0.2411	−132
−131	0.01135	29.89	0.009272	2261	107.8	0.0004422	−18.44	95.36	76.92	−0.04955	0.2406	−131
−130	0.01206	29.89 *	0.009278	2134	107.7	0.0004686	−18.24	95.26	77.02	−0.04892	0.2400	−130
−129	0.01282	29.89 *	0.009284	2015	107.7	0.0004962	−18.03	95.16	77.13	−0.04830	0.2395	−129
−128	0.01361	29.89 *	0.009290	1903	107.6	0.0005254	−17.83	95.07	77.24	−0.04769	0.2389	128
−127	0.01444	29.89 *	0.009296	1798	107.5	0.0005559	−17.62	94.97	77.34	−0.04707	0.2384	−127
−126	0.01532	29.89 *	0.009302	1700	107.5	0.0005881	−17.42	94.87	77.45	−0.04646	0.2378	−126
−125	0.01625	29.88 *	0.009308	1608	107.4	0.0006218	−17.21	94.77	77.56	−0.04585	0.2373	−125
−124	0.01723	29.88 *	0.009314	1521	107.3	0.0006572	−17.01	94.68	77.66	−0.04524	0.2368	−124
−123	0.01826	29.88 *	0.009321	1440	107.2	0.0006943	−16.80	94.58	77.77	−0.04463	0.2363	−123
−122	0.01934	29.88 *	0.009327	1363	107.2	0.0007333	−16.60	94.48	77.88	−0.04402	0.2358	−122
−121	0.02048	29.87 *	0.009333	1291	107.1	0.0007741	−16.40	94.39	77.99	−0.04342	0.2352	−121
−120	0.02167	29.87 *	0.009339	1224	107.0	0.0008169	−16.19	94.29	78.10	−0.04282	0.2347	−120
−119	0.02293	29.87 *	0.009345	1160	107.0	0.0008618	−15.99	94.20	78.20	−0.04222	0.2342	−119
−118	0.02425	29.87 *	0.009351	1100	106.9	0.0009087	−15.78	94.10	78.31	−0.04162	0.2338	−118
−117	0.02563	29.86 *	0.009358	1044	106.8	0.0009579	−15.58	94.01	78.42	−0.04103	0.2333	−117
−116	0.02709	29.86 *	0.009364	990.7	106.7	0.001009	−15.38	93.91	78.53	−0.04043	0.2328	−116
−115	0.02861	29.86 *	0.009370	940.6	106.7	0.001063	−15.17	93.81	78.64	−0.03984	0.2323	−115
−114	0.03022	29.85 *	0.009376	893.3	106.6	0.001119	−14.97	93.72	78.75	−0.03925	0.2318	−114

*Inches of mercury below one atmosphere

TEMP.	PRESSURE		VOLUME cu ft/lb		DENSITY lb/cu ft		ENTHALPY Btu/lb			ENTROPY Btu/(lb)(°R)		TEMP.
°F	PSIA	PSIG	LIQUID v_f	VAPOR v_g	LIQUID $1/v_f$	VAPOR $1/v_g$	LIQUID h_f	LATENT h_{fg}	VAPOR h_g	LIQUID s_f	VAPOR s_g	°F
−113	0.03190	29.85*	0.009382	848.7	106.5	0.001178	−14.77	93.63	78.86	−0.03866	0.2314	−113
−112	0.03366	29.85*	0.009389	806.6	106.5	0.001239	−14.56	93.53	78.96	−0.03808	0.2309	−112
−111	0.03550	29.84*	0.009395	766.8	106.4	0.001304	−14.36	93.44	79.07	−0.03749	0.2305	−111
−110	0.03744	29.84*	0.009401	729.3	106.3	0.001371	−14.16	93.34	79.18	−0.03691	0.2300	−110
−109	0.03946	29.84*	0.009408	693.9	106.2	0.001441	−13.95	93.25	79.29	−0.03633	0.2296	−109
−108	0.04158	29.83*	0.009414	660.4	106.2	0.001514	−13.75	93.16	79.40	−0.03575	0.2291	−108
−107	0.04380	29.83*	0.009420	628.7	106.1	0.001590	−13.55	93.06	79.51	−0.03517	0.2287	−107
−106	0.04612	29.82*	0.009426	598.8	106.0	0.001670	−13.34	92.97	79.62	−0.03460	0.2282	−106
−105	0.04854	29.82*	0.009433	570.4	106.0	0.001752	−13.14	92.88	79.73	−0.03402	0.2278	−105
−104	0.05108	29.81*	0.009439	543.6	105.9	0.001839	−12.94	92.78	79.84	−0.03345	0.2274	−104
−103	0.05373	29.81*	0.009445	518.2	105.8	0.001929	−12.73	92.69	79.96	−0.03288	0.2270	−103
−102	0.05650	29.80*	0.009452	494.2	105.7	0.002023	−12.53	92.60	80.07	−0.03231	0.2265	−102
−101	0.05939	29.80*	0.009458	471.4	105.7	0.002121	−12.33	92.51	80.18	−0.03174	0.2261	−101
−100	0.06241	29.79*	0.009465	449.8	105.6	0.002222	−12.12	92.42	80.29	−0.03118	0.2257	−100
− 99	0.06557	29.78*	0.009471	429.4	105.5	0.002328	−11.92	92.32	80.40	−0.03062	0.2253	− 99
− 98	0.06886	29.78*	0.009477	410.0	105.5	0.002438	−11.72	92.23	80.51	−0.03005	0.2249	− 98
− 97	0.07229	29.77*	0.009484	391.6	105.4	0.002553	−11.51	92.14	80.62	−0.02949	0.2245	− 97
− 96	0.07587	29.76*	0.009490	374.1	105.3	0.002672	−11.31	92.05	80.73	−0.02894	0.2241	− 96
− 95	0.07961	29.75*	0.009497	357.5	105.2	0.002796	−11.11	91.96	80.85	−0.02838	0.2238	− 95
− 94	0.08350	29.75*	0.009503	341.8	105.2	0.002925	−10.91	91.87	80.96	−0.02782	0.2234	− 94
− 93	0.08755	29.74*	0.009510	326.8	105.1	0.003059	−10.70	91.78	81.07	−0.02727	0.2230	− 93
− 92	0.09178	29.73*	0.009516	312.6	105.0	0.003198	−10.50	91.69	81.18	−0.02672	0.2226	− 92
− 91	0.09617	29.72*	0.009523	299.1	105.0	0.003342	−10.30	91.60	81.29	−0.02617	0.2222	− 91
− 90	0.1007	29.71*	0.009529	286.3	104.9	0.003492	−10.10	91.51	81.41	−0.02562	0.2219	− 90
− 89	0.1055	29.70*	0.009536	274.1	104.8	0.003647	− 9.898	91.42	81.52	−0.02507	0.2215	− 89
− 88	0.1104	29.69*	0.009542	262.5	104.7	0.003809	− 9.696	91.33	81.63	−0.02453	0.2212	− 88
− 87	0.1156	29.68*	0.009549	251.4	104.7	0.003976	− 9.493	91.24	81.75	−0.02398	0.2208	− 87
− 86	0.1210	29.67*	0.009556	240.9	104.6	0.004149	− 9.291	91.15	81.86	−0.02344	0.2205	− 86
− 85	0.1265	29.66*	0.009562	230.9	104.5	0.004329	− 9.089	91.06	81.97	−0.02290	0.2201	− 85
− 84	0.1323	29.65*	0.009569	221.4	104.5	0.004516	− 8.886	90.97	82.09	−0.02236	0.2198	− 84
− 83	0.1384	29.63*	0.009575	212.3	104.4	0.004709	− 8.684	90.88	82.20	−0.02182	0.2194	− 83
− 82	0.1446	29.62*	0.009582	203.6	104.3	0.004909	− 8.482	90.79	82.31	−0.02129	0.2191	− 82
− 81	0.1511	29.61*	0.009589	195.4	104.2	0.005116	− 8.279	90.71	82.43	−0.02075	0.2187	− 81
− 80	0.1579	29.59*	0.009595	187.5	104.2	0.005331	− 8.077	90.62	82.54	−0.02022	0.2184	− 80
− 79	0.1649	29.58*	0.009602	180.0	104.1	0.005553	− 7.875	90.53	82.65	−0.01969	0.2181	− 79
− 78	0.1721	29.57*	0.009609	172.9	104.0	0.005783	− 7.673	90.44	82.77	−0.01916	0.2178	− 78
− 77	0.1797	29.55*	0.009615	166.0	104.0	0.006021	− 7.471	90.35	82.88	−0.01863	0.2174	− 77
− 76	0.1875	29.53*	0.009622	159.5	103.9	0.006267	− 7.269	90.27	83.00	−0.01810	0.2171	− 76
− 75	0.1956	29.52*	0.009629	153.3	103.8	0.006522	− 7.067	90.18	83.11	−0.01758	0.2168	− 75
− 74	0.2040	29.50*	0.009635	147.3	103.7	0.006785	− 6.864	90.09	83.23	−0.01705	0.2165	− 74
− 73	0.2128	29.48*	0.009642	141.7	103.7	0.007057	− 6.662	90.00	83.34	−0.01653	0.2162	− 73
− 72	0.2218	29.46*	0.009649	136.2	103.6	0.007338	− 6.460	89.92	83.46	−0.01601	0.2159	− 72
− 71	0.2312	29.45*	0.009656	131.0	103.5	0.007628	− 6.258	89.83	83.57	−0.01549	0.2156	− 71
− 70	0.2409	29.43*	0.009662	126.1	103.4	0.007929	− 6.056	89.74	83.69	−0.01497	0.2153	− 70
− 69	0.2509	29.41*	0.009669	121.3	103.4	0.008238	− 5.854	89.66	83.80	−0.01445	0.2150	− 69
− 68	0.2613	29.38*	0.009676	116.8	103.3	0.008558	− 5.652	89.57	83.92	−0.01393	0.2147	− 68
− 67	0.2721	29.36*	0.009683	112.5	103.2	0.008889	− 5.451	89.48	84.03	−0.01342	0.2144	− 67
− 66	0.2832	29.34*	0.009690	108.3	103.2	0.009230	− 5.249	89.40	84.15	−0.01290	0.2141	− 66
− 65	0.2947	29.32*	0.009697	104.3	103.1	0.009581	− 5.047	89.31	84.26	−0.01239	0.2139	− 65
− 64	0.3066	29.29*	0.009703	100.5	103.0	0.009944	− 4.845	89.22	84.38	−0.01188	0.2136	− 64
− 63	0.3190	29.27*	0.009710	96.9	102.9	0.01031	− 4.643	89.14	84.49	−0.01137	0.2133	− 63
− 62	0.3317	29.24*	0.009717	93.4	102.9	0.01070	− 4.441	89.05	84.61	−0.01086	0.2130	− 62
− 61	0.3449	29.21*	0.009724	90.0	102.8	0.01110	− 4.239	88.97	84.73	−0.01036	0.2128	− 61
− 60	0.3585	29.19*	0.009731	86.8	102.7	0.01151	− 4.037	88.88	84.84	−0.00985	0.2125	− 60
− 59	0.3726	29.16*	0.009738	83.7	102.6	0.01193	− 3.835	88.79	84.96	−0.00935	0.2122	− 59

*Inches of mercury below one atmosphere

TEMP. °F	PRESSURE PSIA	PRESSURE PSIG	VOLUME cu ft/lb LIQUID v_f	VOLUME cu ft/lb VAPOR v_g	DENSITY lb/cu ft LIQUID $1/v_f$	DENSITY lb/cu ft VAPOR $1/v_g$	ENTHALPY Btu/lb LIQUID h_f	ENTHALPY Btu/lb LATENT h_{fg}	ENTHALPY Btu/lb VAPOR h_g	ENTROPY Btu/(lb)(°R) LIQUID s_f	ENTROPY Btu/(lb)(°R) VAPOR s_g	TEMP. °F
– 58	0.38715	29.13297*	0.0097455	80.831	102.61	0.012371	– 3.634	88.714	85.080	–0.008848	0.21201	– 58
– 57	0.40215	29.10241*	0.0097524	78.002	102.54	0.012820	– 3.432	88.628	85.197	–0.008346	0.21176	– 57
– 56	0.41765	29.07086*	0.0097594	75.288	102.46	0.013282	– 3.230	88.543	85.313	–0.007845	0.21150	– 56
– 55	0.43365	29.03829*	0.0097664	72.684	102.39	0.013758	– 3.028	88.458	85.430	–0.007346	0.21125	– 55
– 54	0.45016	29.00466*	0.0097735	70.185	102.32	0.014248	– 2.826	88.373	85.547	–0.006848	0.21100	– 54
– 53	0.46720	28.96997*	0.0097805	67.785	102.24	0.014752	– 2.624	88.288	85.664	–0.006351	0.21075	– 53
– 52	0.48479	28.93417*	0.0097875	65.482	102.17	0.015271	– 2.423	88.203	85.781	–0.005855	0.21050	– 52
– 51	0.50292	28.89724*	0.0097946	63.269	102.10	0.015806	– 2.221	88.119	85.898	–0.005361	0.21026	– 51
– 50	0.52163	28.85916*	0.0098017	61.144	102.02	0.016355	– 2.019	88.034	86.015	–0.004868	0.21002	– 50
– 49	0.54092	28.81988*	0.0098088	59.102	101.95	0.016920	– 1.817	87.950	86.133	–0.004376	0.20979	– 49
– 48	0.56080	28.77940*	0.0098159	57.139	101.88	0.017501	– 1.615	87.865	86.250	–0.003885	0.20955	– 48
– 47	0.58130	28.73766*	0.0098230	55.253	101.80	0.018099	– 1.413	87.781	86.368	–0.003395	0.20932	– 47
– 46	0.60243	28.69466*	0.0098301	53.439	101.73	0.018713	– 1.211	87.697	86.485	–0.002907	0.20909	– 46
– 45	0.62419	28.65034*	0.0098373	51.695	101.65	0.019344	– 1.010	87.612	86.603	–0.002419	0.20886	– 45
– 44	0.64661	28.60469*	0.0098444	50.017	101.58	0.019993	– 0.808	87.528	86.721	–0.001933	0.20864	– 44
– 43	0.66970	28.55768*	0.0098516	48.404	101.51	0.020660	– 0.606	87.444	86.839	–0.001448	0.20842	– 43
– 42	0.69348	28.50926*	0.0098588	46.851	101.43	0.021344	– 0.404	87.360	86.957	–0.000964	0.20820	– 42
– 41	0.71797	28.45941*	0.0098660	45.356	101.36	0.022048	– 0.202	87.276	87.075	–0.000482	0.20798	– 41
– 40	0.74317	28.40809*	0.0098733	43.917	101.28	0.022770	0.000	87.193	87.193	0.000000	0.20776	– 40
– 39	0.76911	28.35527*	0.0098805	42.532	101.21	0.023511	0.202	87.109	87.311	0.000481	0.20755	– 39
– 38	0.79581	28.30092*	0.0098878	41.198	101.14	0.024273	0.404	87.025	87.429	0.000960	0.20734	– 38
– 37	0.82327	28.24500*	0.0098951	39.914	101.06	0.025054	0.606	86.942	87.548	0.001438	0.20713	– 37
– 36	0.85153	28.18748*	0.0099023	38.676	100.99	0.025856	0.808	86.858	87.666	0.001915	0.20693	– 36
– 35	0.88059	28.12832*	0.0099097	37.483	100.91	0.026679	1.010	86.775	87.784	0.002392	0.20673	– 35
– 34	0.91047	28.06748*	0.0099170	36.333	100.84	0.027523	1.212	86.691	87.903	0.002867	0.20652	– 34
– 33	0.94119	28.00493*	0.0099243	35.225	100.76	0.028389	1.414	86.608	88.022	0.003341	0.20633	– 33
– 32	0.97277	27.94063*	0.0099317	34.157	100.69	0.029277	1.616	86.524	88.140	0.003814	0.20613	– 32
– 31	1.0052	27.8745*	0.0099391	33.127	100.61	0.030187	1.818	86.441	88.259	0.004285	0.20593	– 31
– 30	1.0386	27.8066*	0.0099464	32.133	100.54	0.031120	2.020	86.358	88.378	0.004756	0.20574	– 30
– 29	1.0729	27.7369*	0.0099539	31.175	100.46	0.032077	2.222	86.275	88.497	0.005226	0.20555	– 29
– 28	1.1081	27.6652*	0.0099613	30.250	100.39	0.033058	2.425	86.192	88.616	0.005695	0.20536	– 28
– 27	1.1442	27.5915*	0.0099687	29.357	100.31	0.034063	2.627	86.108	88.735	0.006162	0.20518	– 27
– 26	1.1814	27.5159*	0.0099762	28.496	100.24	0.035093	2.829	86.025	88.854	0.006629	0.20499	– 26
– 25	1.2195	27.4383*	0.0099837	27.664	100.16	0.036148	3.031	85.942	88.974	0.007095	0.20481	– 25
– 24	1.2586	27.3586*	0.0099911	26.861	100.09	0.037228	3.233	85.859	89.093	0.007559	0.20463	– 24
– 23	1.2988	27.2768*	0.0099987	26.086	100.01	0.038335	3.436	85.776	89.212	0.008023	0.20446	– 23
– 22	1.3401	27.1928*	0.010006	25.337	99.938	0.039468	3.638	85.694	89.331	0.008485	0.20428	– 22
– 21	1.3824	27.1067*	0.010014	24.613	99.863	0.040628	3.840	85.611	89.451	0.008947	0.20411	– 21
– 20	1.4258	27.0183*	0.010021	23.914	99.788	0.041816	4.043	85.528	89.570	0.009408	0.20393	– 20
– 19	1.4703	26.9276*	0.010029	23.239	99.712	0.043032	4.245	85.445	89.690	0.009867	0.20376	– 19
– 18	1.5100	26.8346*	0.010036	22.586	99.637	0.044276	4.448	85.362	89.810	0.010326	0.20360	18
– 17	1.5629	26.7392*	0.010044	21.954	99.561	0.045549	4.650	85.279	89.929	0.010783	0.20343	– 17
– 16	1.6109	26.6414*	0.010052	21.344	99.486	0.046852	4.852	85.197	90.049	0.011240	0.20327	– 16
– 15	1.6602	26.5411*	0.010059	20.754	99.410	0.048184	5.055	85.114	90.169	0.011696	0.20310	– 15
– 14	1.7106	26.4383*	0.010067	20.183	99.334	0.049547	5.258	85.031	90.289	0.012151	0.20294	– 14
– 13	1.7624	26.3330*	0.010075	19.630	99.259	0.050941	5.460	84.948	90.408	0.012605	0.20279	– 13
– 12	1.8154	26.2250*	0.010082	19.096	99.183	0.052367	5.663	84.866	90.528	0.013058	0.20263	– 12
– 11	1.8697	26.1144*	0.010090	18.579	99.107	0.053824	5.865	84.783	90.648	0.013510	0.20247	– 11
– 10	1.9254	26.0011*	0.010098	18.079	99.031	0.055314	6.068	84.700	90.768	0.013961	0.20232	– 10
– 9	1.9824	25.8850*	0.010106	17.594	98.955	0.056837	6.271	84.617	90.888	0.014411	0.20217	– 9
– 8	2.0408	25.7661*	0.010113	17.125	98.879	0.058393	6.474	84.535	91.008	0.014860	0.20202	– 8
– 7	2.1006	25.6443*	0.010121	16.671	98.803	0.059984	6.677	84.452	91.129	0.015308	0.20187	– 7
– 6	2.1618	25.5197*	0.010129	16.231	98.726	0.061609	6.879	84.369	91.249	0.015756	0.20173	– 6
– 5	2.2245	25.3920*	0.010137	15.805	98.650	0.063270	7.082	84.287	91.369	0.016202	0.20158	– 5
– 4	2.2887	25.2614*	0.010145	15.393	98.574	0.064966	7.285	84.204	91.489	0.016648	0.20144	– 4

*Inches of mercury below one atmosphere

TEMP. °F	PRESSURE		VOLUME cu ft/lb		DENSITY lb/cu ft		ENTHALPY Btu/lb			ENTROPY Btu/(lb)(°R)		TEMP. °F
	PSIA	PSIG	LIQUID v_f	VAPOR v_g	LIQUID $1/v_f$	VAPOR $1/v_g$	LIQUID h_f	LATENT h_{fg}	VAPOR h_g	LIQUID s_f	VAPOR s_g	
− 3	2.35	25.12 *	0.01015	14.99	98.49	0.06669	7.48	84.12	91.60	0.01709	0.2013	− 3
− 2	2.42	24.99 *	0.01016	14.60	98.42	0.06846	7.69	84.03	91.73	0.01753	0.2011	− 2
− 1	2.49	24.85 *	0.01016	14.23	98.34	0.07027	7.89	83.95	91.85	0.01797	0.2010	− 1
0	2.56	24.70 *	0.01017	13.86	98.26	0.07211	8.09	83.87	91.97	0.01842	0.2008	0
1	2.63	24.56 *	0.01018	13.51	98.19	0.07400	8.30	83.79	92.09	0.01886	0.2007	1
2	2.70	24.41 *	0.01019	13.17	98.11	0.07592	8.50	83.70	92.21	0.01930	0.2006	2
3	2.78	24.25 *	0.01020	12.83	98.03	0.07788	8.70	83.62	92.33	0.01974	0.2004	3
4	2.85	24.10 *	0.01020	12.51	97.96	0.07989	8.91	83.54	92.45	0.02018	0.2003	4
5	2.93	23.94 *	0.01021	12.20	97.88	0.08193	9.11	83.45	92.57	0.02061	0.2002	5
6	3.01	23.77 *	0.01022	11.90	97.80	0.08401	9.31	83.37	92.69	0.02105	0.2001	6
7	3.10	23.60 *	0.01023	11.60	97.73	0.08614	9.52	83.29	92.81	0.02149	0.1999	7
8	3.18	23.43 *	0.01024	11.32	97.65	0.08831	9.72	83.21	92.93	0.02192	0.1998	8
9	3.27	23.26 *	0.01024	11.04	97.57	0.09052	9.92	83.12	93.05	0.02236	0.1997	9
10	3.35	23.08 *	0.01025	10.77	97.49	0.09278	10.13	83.04	93.17	0.02279	0.1996	10
11	3.44	22.90 *	0.01026	10.51	97.42	0.09508	10.33	82.96	93.29	0.02322	0.1994	11
12	3.53	22.71 *	0.01027	10.26	97.34	0.09743	10.53	82.87	93.41	0.02366	0.1993	12
13	3.63	22.52 *	0.01028	10.01	97.26	0.09982	10.74	82.79	93.53	0.02409	0.1992	13
14	3.72	22.33 *	0.01028	9.778	97.18	0.1022	10.94	82.71	93.65	0.02452	0.1991	14
15	3.82	22.13 *	0.01029	9.546	97.11	0.1047	11.15	82.62	93.77	0.02495	0.1990	15
16	3.92	21.92 *	0.01030	9.321	97.03	0.1072	11.35	82.54	93.89	0.02537	0.1989	16
17	4.02	21.71 *	0.01031	9.101	96.95	0.1098	11.55	82.46	94.02	0.02580	0.1988	17
18	4.13	21.50 *	0.01032	8.889	96.87	0.1125	11.76	82.37	94.14	0.02623	0.1986	18
19	4.23	21.29 *	0.01033	8.682	96.80	0.1151	11.96	82.29	94.26	0.02666	0.1985	19
20	4.34	21.07	0.01033	8.481	96.72	0.1179	12.17	82.21	94.38	0.02708	0.1984	20
21	4.45	20.84 *	0.01034	8.285	96.64	0.1206	12.37	82.12	94.50	0.02751	0.1983	21
22	4.56	20.61 *	0.01035	8.095	96.56	0.1235	12.57	82.04	94.62	0.02793	0.1982	22
23	4.68	20.3 *	0.01036	7.910	96.48	0.1264	12.78	81.96	94.74	0.02835	0.1981	23
24	4.80	20.1 *	0.01037	7.730	96.40	0.1293	12.98	81.87	94.86	0.02878	0.1980	24
25	4.92	19.90 *	0.01038	7.556	96.33	0.1323	13.19	81.79	94.98	0.02920	0.1979	25
26	5.04	19.65 *	0.01038	7.385	96.25	0.1353	13.39	81.71	95.10	0.02962	0.1978	26
27	5.16	19.40 *	0.01039	7.220	96.17	0.1385	13.60	81.62	95.22	0.03004	0.1977	27
28	5.29	19.14 *	0.01040	7.059	96.09	0.1416	13.80	81.54	95.34	0.03046	0.1976	28
29	5.42	18.87 *	0.01041	6.902	96.01	0.1448	14.00	81.45	95.46	0.03088	0.1975	29
30	5.55	18.61 *	0.01042	6.750	95.93	0.1481	14.21	81.37	95.58	0.03129	0.1974	30
31	5.69	18.33 *	0.01043	6.601	95.86	0.1514	14.41	81.29	95.70	0.03171	0.1973	31
32	5.82	18.05 *	0.01044	6.457	95.78	0.1548	14.62	81.20	95.83	0.03213	0.1973	32
33	5.96	17.77 *	0.01044	6.316	95.70	0.1583	14.82	81.12	95.95	0.03254	0.1972	33
34	6.10	17.48 *	0.01045	6.179	95.62	0.1618	15.03	81.03	96.07	0.03296	0.1971	34
35	6.25	17.18 *	0.01046	6.046	95.54	0.1653	15.23	80.95	96.19	0.03337	0.1970	35
36	6.40	16.88 *	0.01047	5.916	95.46	0.1690	15.44	80.86	96.31	0.03379	0.1969	36
37	6.55	16.57 *	0.01048	5.790	95.38	0.1727	15.64	80.78	96.43	0.03420	0.1968	37
38	6.70	16.26 *	0.01049	5.666	95.30	0.1764	15.85	80.70	96.55	0.03461	0.1967	38
39	6.86	15.94 *	0.01050	5.546	95.22	0.1802	16.05	80.61	96.67	0.03502	0.1966	39
40	7.02	15.62 *	0.01051	5.429	95.14	0.1841	16.26	80.53	96.79	0.03543	0.1966	40
41	7.18	15.29 *	0.01051	5.315	95.06	0.1881	16.46	80.44	96.91	0.03584	0.1965	41
42	7.34	14.95 *	0.01052	5.204	94.98	0.1921	16.67	80.36	97.03	0.03625	0.1964	42
43	7.51	14.61 *	0.01053	5.096	94.90	0.1962	16.88	80.27	97.15	0.03666	0.1963	43
44	7.68	14.26 *	0.01054	4.990	94.82	0.2003	17.08	80.19	97.27	0.03707	0.1962	44
45	7.86	13.91 *	0.01055	4.887	94.74	0.2045	17.29	80.10	97.39	0.03748	0.1962	45
46	8.04	13.55 *	0.01056	4.787	94.66	0.2088	17.49	80.02	97.51	0.03788	0.1961	46
47	8.22	13.18 *	0.01057	4.689	94.58	0.2132	17.70	79.93	97.63	0.03829	0.1960	47
48	8.40	12.80 *	0.01058	4.594	94.50	0.2176	17.90	79.84	97.75	0.03869	0.1959	48
49	8.59	12.42 *	0.01059	4.501	94.42	0.2221	18.11	79.76	97.87	0.03910	0.1959	49
50	8.78	12.03 *	0.01059	4.410	94.34	0.2267	18.32	79.67	97.99	0.03950	0.1958	50
51	8.97	11.64 *	0.01060	4.321	94.26	0.2313	18.52	79.59	98.11	0.03990	0.1957	51

*Inches of mercury below one atmosphere

TEMP.	PRESSURE		VOLUME cu ft/lb		DENSITY lb/cu ft		ENTHALPY Btu/lb			ENTROPY Btu/(lb)(°R)		TEMP.
°F	PSIA	PSIG	LIQUID v_f	VAPOR v_g	LIQUID $1/v_f$	VAPOR $1/v_g$	LIQUID h_f	LATENT h_{fg}	VAPOR h_g	LIQUID s_f	VAPOR s_g	°F
52	9.1733	11.2444*	0.010617	4.2355	94.186	0.23610	18.732	79.506	98.238	0.040310	0.19570	52
53	9.3737	10.8362*	0.010626	4.1512	94.106	0.24089	18.938	79.420	98.358	0.040711	0.19563	53
54	9.5776	10.4211*	0.010635	4.0690	94.025	0.24576	19.144	79.334	98.478	0.041112	0.19556	54
55	9.7850	9.9988*	0.010645	3.9887	93.944	0.25071	19.350	79.248	98.598	0.041512	0.19549	55
56	9.9960	9.5693*	0.010654	3.9103	93.864	0.25573	19.556	79.162	98.718	0.041912	0.19542	56
57	10.210	9.133*	0.010663	3.8338	93.783	0.26083	19.763	79.075	98.838	0.042311	0.19536	57
58	10.429	8.688*	0.010672	3.7592	93.702	0.26602	19.969	78.989	98.958	0.042709	0.19529	58
59	10.650	8.237*	0.010681	3.6863	93.620	0.27128	20.176	78.902	99.078	0.043107	0.19523	59
60	10.876	7.778*	0.010691	3.6151	93.539	0.27662	20.382	78.816	99.198	0.043504	0.19517	60
61	11.105	7.311*	0.010700	3.5456	93.458	0.28204	20.589	78.729	99.317	0.043900	0.19511	61
62	11.338	6.836*	0.010709	3.4777	93.377	0.28754	20.795	78.642	99.437	0.044295	0.19505	62
63	11.575	6.354*	0.010719	3.4114	93.295	0.29313	21.002	78.555	99.557	0.044690	0.19499	63
64	11.816	5.864*	0.010728	3.3467	93.213	0.29880	21.209	78.468	99.676	0.045085	0.19493	64
65	12.061	5.366*	0.010737	3.2834	93.132	0.30456	21.416	78.380	99.796	0.045478	0.19487	65
66	12.309	4.859*	0.010747	3.2216	93.050	0.31040	21.623	78.293	99.916	0.045872	0.19481	66
67	12.562	4.345*	0.010756	3.1612	92.968	0.31633	21.830	78.205	100.035	0.046264	0.19475	67
68	12.819	3.822*	0.010766	3.1022	92.886	0.32235	22.037	78.118	100.154	0.046656	0.19470	68
69	13.080	3.291*	0.010775	3.0445	92.804	0.32846	22.244	78.030	100.274	0.047047	0.19464	69
70	13.345	2.752*	0.010785	2.9882	92.722	0.33465	22.451	77.942	100.393	0.047437	0.19459	70
71	13.614	2.204*	0.010795	2.9331	92.640	0.34094	22.658	77.854	100.512	0.047827	0.19454	71
72	13.887	1.647*	0.010804	2.8792	92.557	0.34731	22.866	77.766	100.631	0.048217	0.19448	72
73	14.165	1.081*	0.010814	2.8266	92.475	0.35378	23.073	77.678	100.751	0.048605	0.19443	73
74	14.447	0.507*	0.010823	2.7751	92.392	0.36034	23.280	77.589	100.870	0.048993	0.19438	74
75	14.733	0.037	0.010833	2.7248	92.310	0.36700	23.488	77.501	100.989	0.049381	0.19433	75
76	15.024	0.328	0.010843	2.6756	92.227	0.37375	23.696	77.412	101.108	0.049768	0.19428	76
77	15.319	0.623	0.010853	2.6275	92.144	0.38059	23.903	77.323	101.226	0.050154	0.19423	77
78	15.619	0.923	0.010862	2.5804	92.061	0.38754	24.111	77.234	101.345	0.050540	0.19419	78
79	15.924	1.228	0.010872	2.5344	91.978	0.39458	24.319	77.145	101.464	0.050925	0.19414	79
80	16.233	1.537	0.010882	2.4893	91.895	0.40172	24.527	77.056	101.583	0.051309	0.19409	80
81	16.546	1.850	0.010892	2.4453	91.812	0.40895	24.735	76.967	101.701	0.051693	0.19405	81
82	16.865	2.169	0.010902	2.4021	91.728	0.41629	24.943	76.877	101.820	0.052076	0.19400	82
83	17.188	2.492	0.010912	2.3600	91.645	0.42374	25.151	76.787	101.938	0.052459	0.19396	83
84	17.516	2.820	0.010922	2.3187	91.561	0.43128	25.359	76.698	102.057	0.052841	0.19391	84
85	17.848	3.152	0.010932	2.2783	91.477	0.43893	25.567	76.608	102.175	0.053222	0.19387	85
86	18.186	3.490	0.010942	2.2388	91.394	0.44668	25.776	76.518	102.293	0.053603	0.19383	86
87	18.529	3.833	0.010952	2.2001	91.310	0.45453	25.984	76.427	102.411	0.053983	0.19379	87
88	18.876	4.180	0.010962	2.1622	91.226	0.46250	26.192	76.337	102.529	0.054363	0.19375	88
89	19.229	4.533	0.010972	2.1251	91.142	0.47057	26.401	76.246	102.647	0.054742	0.19371	89
90	19.587	4.891	0.010982	2.0888	91.057	0.47874	26.610	76.156	102.765	0.055121	0.19367	90
91	19.950	5.254	0.010992	2.0532	90.973	0.48703	26.818	76.065	102.883	0.055499	0.19363	91
92	20.318	5.622	0.011002	2.0184	90.889	0.49543	27.027	75.974	103.001	0.055876	0.19359	92
93	20.691	5.995	0.011013	1.9844	90.804	0.50394	27.236	75.883	103.119	0.056253	0.19355	93
94	21.070	6.374	0.011023	1.9510	90.719	0.51256	27.445	75.791	103.236	0.056629	0.19352	94
95	21.454	6.758	0.011033	1.9183	90.635	0.52130	27.654	75.700	103.354	0.057005	0.19348	95
96	21.843	7.147	0.011044	1.8863	90.550	0.53015	27.863	75.608	103.471	0.057380	0.19345	96
97	22.238	7.542	0.011054	1.8549	90.465	0.53912	28.072	75.516	103.588	0.057755	0.19341	97
98	22.638	7.942	0.011064	1.8241	90.379	0.54820	28.281	75.424	103.706	0.058129	0.19338	98
99	23.044	8.348	0.011075	1.7940	90.294	0.55740	28.491	75.332	103.823	0.058502	0.19334	99
100	23.456	8.760	0.011085	1.7645	90.209	0.56672	28.700	75.240	103.940	0.058875	0.19331	100
101	23.873	9.177	0.011096	1.7356	90.123	0.57616	28.909	75.147	104.057	0.059248	0.19328	101
102	24.296	9.600	0.011106	1.7073	90.038	0.58572	29.119	75.055	104.174	0.059620	0.19325	102
103	24.724	10.028	0.011117	1.6795	89.952	0.59540	29.329	74.962	104.290	0.059991	0.19322	103
104	25.159	10.463	0.011128	1.6523	89.866	0.60521	29.538	74.869	104.407	0.060362	0.19319	104
105	25.599	10.903	0.011138	1.6256	89.780	0.61514	29.748	74.776	104.524	0.060732	0.19316	105
106	26.045	11.349	0.011149	1.5995	89.694	0.62519	29.958	74.682	104.640	0.061102	0.19313	106

*Inches of mercury below one atmosphere

TEMP.	PRESSURE		VOLUME cu ft/lb		DENSITY lb/cu ft		ENTHALPY Btu/lb			ENTROPY Btu/(lb)(°R)		TEMP.
°F	PSIA	PSIG	LIQUID v_f	VAPOR v_g	LIQUID $1/v_f$	VAPOR $1/v_g$	LIQUID h_f	LATENT h_{fg}	VAPOR h_g	LIQUID s_f	VAPOR s_g	°F
107	26.497	11.801	0.011160	1.5739	89.608	0.63538	30.168	74.589	104.757	0.061471	0.19310	107
108	26.956	12.260	0.011170	1.5487	89.522	0.64569	30.378	74.495	104.873	0.061840	0.19307	108
109	27.420	12.724	0.011181	1.5241	89.435	0.65612	30.588	74.401	104.989	0.062208	0.19304	109
110	27.890	13.194	0.011192	1.4999	89.349	0.66669	30.798	74.307	105.105	0.062575	0.19301	110
111	28.367	13.671	0.011203	1.4763	89.262	0.67739	31.009	74.213	105.221	0.062942	0.19299	111
112	28.850	14.154	0.011214	1.4530	89.175	0.68822	31.219	74.118	105.337	0.063309	0.19296	112
113	29.339	14.643	0.011225	1.4302	89.088	0.69919	31.429	74.023	105.453	0.063675	0.19294	113
114	29.834	15.138	0.011236	1.4079	89.001	0.71029	31.640	73.929	105.569	0.064041	0.19291	114
115	30.336	15.640	0.011247	1.3860	88.914	0.72152	31.851	73.833	105.684	0.064406	0.19289	115
116	30.844	16.148	0.011258	1.3645	88.826	0.73289	32.061	73.738	105.800	0.064770	0.19286	116
117	31.359	16.663	0.011269	1.3434	88.739	0.74440	32.272	73.643	105.915	0.065134	0.19284	117
118	31.880	17.184	0.011280	1.3227	88.651	0.75605	32.483	73.547	106.030	0.065498	0.19281	118
119	32.408	17.712	0.011291	1.3024	88.564	0.76784	32.694	73.451	106.145	0.065861	0.19279	119
120	32.943	18.247	0.011303	1.2824	88.476	0.77977	32.905	73.355	106.260	0.066223	0.19277	120
121	33.484	18.788	0.011314	1.2629	88.388	0.79184	33.116	73.259	106.375	0.066585	0.19275	121
122	34.032	19.336	0.011325	1.2437	88.300	0.80406	33.328	73.162	106.490	0.066947	0.19273	122
123	34.587	19.891	0.011336	1.2249	88.211	0.81642	33.539	73.066	106.605	0.067308	0.19271	123
124	35.149	20.453	0.011348	1.2064	88.123	0.82893	33.751	72.969	106.719	0.067669	0.19269	124
125	35.718	21.022	0.011359	1.1882	88.034	0.84159	33.962	72.872	106.834	0.068029	0.19267	125
126	36.294	21.598	0.011371	1.1704	87.946	0.85440	34.174	72.774	106.948	0.068388	0.19265	126
127	36.877	22.181	0.011382	1.1529	87.857	0.86735	34.385	72.677	107.062	0.068747	0.19263	127
128	37.467	22.771	0.011394	1.1358	87.768	0.88046	34.597	72.579	107.176	0.069106	0.19261	128
129	38.064	23.368	0.011405	1.1189	87.679	0.89372	34.809	72.481	107.290	0.069464	0.19259	129
130	38.668	23.972	0.011417	1.1024	87.589	0.90714	35.021	72.383	107.404	0.069822	0.19257	130
131	39.280	24.584	0.011429	1.0861	87.500	0.92071	35.233	72.284	107.518	0.070179	0.19256	131
132	39.899	25.203	0.011440	1.0702	87.410	0.93443	35.446	72.186	107.631	0.070536	0.19254	132
133	40.525	25.829	0.011452	1.0545	87.321	0.94832	35.658	72.087	107.745	0.070892	0.19252	133
134	41.159	26.463	0.011464	1.0391	87.231	0.96237	35.870	71.988	107.858	0.071248	0.19251	134
135	41.801	27.105	0.011476	1.0240	87.141	0.97657	36.083	71.888	107.971	0.071604	0.19249	135
136	42.450	27.754	0.011488	1.0091	87.051	0.99094	36.296	71.789	108.084	0.071959	0.19248	136
137	43.106	28.410	0.011499	0.99456	86.960	1.0055	36.508	71.689	108.197	0.072313	0.19246	137
138	43.771	29.075	0.011511	0.98023	86.870	1.0202	36.721	71.589	108.310	0.072667	0.19245	138
139	44.443	29.747	0.011523	0.96615	86.779	1.0350	36.934	71.489	108.423	0.073021	0.19243	139
140	45.123	30.427	0.011536	0.95232	86.689	1.0501	37.147	71.388	108.535	0.073374	0.19242	140
141	45.810	31.114	0.011548	0.93873	86.598	1.0653	37.360	71.287	108.648	0.073727	0.19241	141
142	46.506	31.810	0.011560	0.92538	86.507	1.0806	37.574	71.186	108.760	0.074079	0.19239	142
143	47.210	32.514	0.011572	0.91226	86.415	1.0962	37.787	71.085	108.872	0.074431	0.19238	143
144	47.921	33.225	0.011584	0.89937	86.324	1.1119	38.001	70.983	108.984	0.074783	0.19237	144
145	48.641	33.945	0.011597	0.88670	86.232	1.1278	38.214	70.881	109.096	0.075134	0.19236	145
146	49.369	34.673	0.011609	0.87424	86.141	1.1439	38.428	70.779	109.207	0.075484	0.19235	146
147	50.105	35.409	0.011621	0.86200	86.049	1.1601	38.642	70.677	109.319	0.075835	0.19233	147
148	50.850	36.154	0.011634	0.84996	85.957	1.1765	38.856	70.574	109.430	0.076184	0.19232	148
149	51.603	36.907	0.011646	0.83813	85.864	1.1931	39.070	70.472	109.542	0.076534	0.19231	149
150	52.364	37.668	0.011659	0.82650	85.772	1.2099	39.284	70.368	109.653	0.076883	0.19230	150
151	53.133	38.437	0.011671	0.81507	85.680	1.2269	39.499	70.265	109.764	0.077231	0.19229	151
152	53.912	39.216	0.011684	0.80383	85.587	1.2440	39.713	70.161	109.874	0.077579	0.19228	152
153	54.698	40.002	0.011697	0.79278	85.494	1.2614	39.928	70.057	109.985	0.077927	0.19227	153
154	55.494	40.798	0.011709	0.78191	85.401	1.2789	40.142	69.953	110.096	0.078275	0.19227	154
155	56.298	41.602	0.011722	0.77122	85.308	1.2966	40.357	69.849	110.206	0.078622	0.19226	155
156	57.111	42.415	0.011735	0.76071	85.214	1.3146	40.572	69.744	110.316	0.078968	0.19225	156
157	57.933	43.237	0.011748	0.75037	85.121	1.3327	40.787	69.639	110.426	0.079314	0.19224	157
158	58.763	44.067	0.011761	0.74021	85.027	1.3510	41.003	69.533	110.536	0.079660	0.19223	158
159	59.603	44.907	0.011774	0.73021	84.933	1.3695	41.218	69.428	110.646	0.080006	0.19223	159
160	60.451	45.755	0.011787	0.72037	84.839	1.3882	41.433	69.322	110.755	0.080351	0.19222	160
161	61.309	46.613	0.011800	0.71070	84.745	1.4071	41.649	69.215	110.865	0.080695	0.19221	161

TEMP. °F	PRESSURE PSIA	PRESSURE PSIG	VOLUME cu ft/lb LIQUID v_f	VOLUME cu ft/lb VAPOR v_g	DENSITY lb/cu ft LIQUID $1/v_f$	DENSITY lb/cu ft VAPOR $1/v_g$	ENTHALPY Btu/lb LIQUID h_f	ENTHALPY Btu/lb LATENT h_{fg}	ENTHALPY Btu/lb VAPOR h_g	ENTROPY Btu/(lb)(°R) LIQUID s_f	ENTROPY Btu/(lb)(°R) VAPOR s_g	TEMP. °F
162	62.17	47.48	0.01181	0.7011	84.65	1.426	41.86	69.10	110.9	0.08104	0.1922	162
163	63.05	48.35	0.01182	0.6918	84.55	1.445	42.08	69.00	111.0	0.08138	0.1922	163
164	63.93	49.24	0.01184	0.6826	84.46	1.465	42.29	68.89	111.1	0.08172	0.1921	164
165	64.83	50.13	0.01185	0.6735	84.36	1.484	42.51	68.78	111.3	0.08207	0.1921	165
166	65.73	51.04	0.01186	0.6646	84.27	1.504	42.72	68.68	111.4	0.08241	0.1921	166
167	66.64	51.95	0.01188	0.6558	84.17	1.524	42.94	68.57	111.5	0.08275	0.1921	167
168	67.57	52.87	0.01189	0.6472	84.07	1.545	43.16	68.46	111.6	0.08309	0.1921	168
169	68.50	53.80	0.01190	0.6387	83.98	1.565	43.37	68.35	111.7	0.08344	0.1921	169
170	69.44	54.75	0.01192	0.6303	83.88	1.586	43.59	68.24	111.8	0.08378	0.1921	170
171	70.39	55.70	0.01193	0.6221	83.79	1.607	43.81	68.13	111.9	0.08412	0.1921	171
172	71.36	56.66	0.01194	0.6140	83.69	1.628	44.03	68.02	112.0	0.08446	0.1921	172
173	72.33	57.63	0.01196	0.6060	83.59	1.650	44.24	67.91	112.1	0.08480	0.1921	173
174	73.31	58.61	0.01197	0.5981	83.50	1.671	44.46	67.80	112.2	0.08514	0.1921	174
175	74.30	59.60	0.01199	0.5904	83.40	1.693	44.68	67.69	112.3	0.08548	0.1921	175
176	75.30	60.61	0.01200	0.5828	83.30	1.715	44.90	67.58	112.4	0.08582	0.1921	176
177	76.31	61.62	0.01201	0.5753	83.21	1.738	45.11	67.47	112.5	0.08616	0.1921	177
178	77.34	62.64	0.01203	0.5679	83.11	1.760	45.33	67.36	112.6	0.08650	0.1921	178
179	78.37	63.67	0.01204	0.5607	83.01	1.783	45.55	67.24	112.8	0.08684	0.1921	179
180	79.41	64.71	0.01206	0.5535	82.91	1.806	45.77	67.13	112.9	0.08717	0.1921	180
181	80.46	65.77	0.01207	0.5465	82.81	1.829	45.99	67.02	113.0	0.08751	0.1921	181
182	81.52	66.83	0.01208	0.5396	82.72	1.853	46.21	66.90	113.1	0.08785	0.1921	182
183	82.60	67.90	0.01210	0.5327	82.62	1.877	46.43	66.79	113.2	0.08819	0.1921	183
184	83.68	68.99	0.01211	0.5260	82.52	1.901	46.65	66.68	113.3	0.08852	0.1921	184
185	84.78	70.08	0.01213	0.5194	82.42	1.925	46.86	66.56	113.4	0.08886	0.1921	185
186	85.88	71.19	0.01214	0.5129	82.32	1.949	47.08	66.45	113.5	0.08920	0.1921	186
187	87.00	72.30	0.01216	0.5064	82.22	1.974	47.30	66.33	113.6	0.08953	0.1921	187
188	88.13	73.43	0.01217	0.5001	82.12	1.999	47.52	66.22	113.7	0.08987	0.1921	188
189	89.26	74.57	0.01219	0.4939	82.02	2.024	47.74	66.10	113.8	0.09020	0.1921	189
190	90.41	75.72	0.01220	0.4877	81.92	2.050	47.96	65.98	113.9	0.09054	0.1921	190
191	91.57	76.88	0.01222	0.4817	81.82	2.075	48.18	65.87	114.0	0.09087	0.1921	191
192	92.74	78.05	0.01223	0.4757	81.72	2.101	48.41	65.75	114.1	0.09121	0.1921	192
193	93.92	79.23	0.01225	0.4698	81.62	2.128	48.63	65.63	114.2	0.09154	0.1921	193
194	95.12	80.42	0.01226	0.4640	81.52	2.154	48.85	65.51	114.3	0.09188	0.1921	194
195	96.32	81.63	0.01228	0.4583	81.41	2.181	49.07	65.40	114.4	0.09221	0.1921	195
196	97.54	82.84	0.01229	0.4527	81.31	2.208	49.29	65.28	114.5	0.09255	0.1921	196
197	98.76	84.07	0.01231	0.4472	81.21	2.236	49.51	65.16	114.6	0.09288	0.1921	197
198	100.0	85.31	0.01232	0.4417	81.11	2.263	49.73	65.04	114.7	0.09321	0.1921	198
199	101.2	86.56	0.01234	0.4363	81.01	2.291	49.96	64.92	114.8	0.09355	0.1921	199
200	102.5	87.82	0.01236	0.4310	80.90	2.319	50.18	64.80	114.9	0.09388	0.1921	200
201	103.7	89.10	0.01237	0.4258	80.80	2.348	50.40	64.68	115.0	0.09421	0.1921	201
202	105.0	90.38	0.01239	0.4207	80.70	2.377	50.62	64.55	115.1	0.09455	0.1921	202
203	106.3	91.68	0.01240	0.4156	80.59	2.406	50.85	64.43	115.2	0.09488	0.1921	203
204	107.6	92.99	0.01242	0.4106	80.49	2.435	51.07	64.31	115.3	0.09521	0.1921	204
205	109.0	94.31	0.01243	0.4056	80.39	2.465	51.29	64.19	115.4	0.09554	0.1921	205
206	110.3	95.64	0.01245	0.4008	80.28	2.494	51.52	64.06	115.5	0.09587	0.1921	206
207	111.6	96.99	0.01247	0.3960	80.18	2.525	51.74	63.94	115.6	0.09620	0.1921	207
208	113.0	98.35	0.01248	0.3912	80.07	2.555	51.97	63.82	115.7	0.09654	0.1921	208
209	114.4	99.72	0.01250	0.3866	79.97	2.586	52.19	63.69	115.8	0.09687	0.1921	209
210	115.7	101.10	0.01252	0.3820	79.86	2.617	52.41	63.56	115.9	0.09720	0.1921	210
220	130	115.6	0.01269	0.3393	78.79	2.946	54.68	62.28	116.9	0.1005	0.1921	220
230	146.1	131.4	0.01287	0.3021	77.69	3.309	56.96	60.94	117.9	0.1037	0.1921	230
240	163	148.6	0.01306	0.2695	76.55	3.710	59.29	59.54	118.8	0.1070	0.1921	240
250	182	167.3	0.01326	0.2407	75.38	4.153	61.64	58.06	119.7	0.1103	0.1921	250

TEMP. °F	PRESSURE		VOLUME cu ft/lb		DENSITY lb/cu ft		ENTHALPY Btu/lb			ENTROPY Btu/(lb)(°R)		TEMP. °F
	PSIA	PSIG	LIQUID v_f	VAPOR v_g	LIQUID $1/v_f$	VAPOR $1/v_g$	LIQUID h_f	LATENT h_{fg}	VAPOR h_g	LIQUID s_f	VAPOR s_g	
260	202.24	187.55	0.013484	0.21538	74.163	4.6430	64.047	56.508	120.555	0.11364	0.19216	260
270	224.09	209.39	0.013719	0.19284	72.893	5.1858	66.494	54.856	121.350	0.11694	0.19212	270
280	247.63	232.94	0.013973	0.17274	71.565	5.7889	68.997	53.096	122.093	0.12027	0.19205	280
290	272.98	258.28	0.014251	0.15476	70.170	6.4615	71.563	51.212	122.776	0.12363	0.19194	290
300	300.21	285.51	0.014557	0.13860	68.696	7.2152	74.205	49.183	123.387	0.12703	0.19177	300
310	329.42	314.73	0.014897	0.12399	67.128	8.0651	76.933	46.980	123.914	0.13049	0.19153	310
320	360.71	346.02	0.015280	0.11072	65.443	9.0315	79.767	44.570	124.337	0.13404	0.19120	320
330	394.1	379.4	0.01572	0.09858	63.61	10.14	82.72	41.90	124.6	0.1376	0.1907	330
340	429.96	415.26	0.016236	0.087394	61.591	11.442	85.851	38.911	124.763	0.14149	0.19015	340
350	468.14	453.44	0.016861	0.076940	59.310	12.997	89.183	35.486	124.669	0.14548	0.18931	350
360	508.84	494.15	0.017654	0.066989	56.643	14.928	92.807	31.442	124.249	0.14977	0.18813	360
370	552.21	537.52	0.018750	0.057172	53.333	17.491	96.892	26.399	123.291	0.15455	0.18636	370
380	598.40	583.70	0.020570	0.046570	48.614	21.473	101.912	19.249	121.160	0.16036	0.18328	380
381	603.18	588.48	0.020843	0.045365	47.978	22.044	102.518	18.290	120.809	0.16106	0.18282	381
382	607.99	593.29	0.021148	0.044104	47.285	22.674	103.160	17.250	120.410	0.16180	0.18229	382
383	612.83	598.13	0.021495	0.042769	46.522	23.381	103.844	16.106	119.950	0.16259	0.18170	383
384	617.70	603.00	0.021898	0.041334	45.666	24.193	104.585	14.825	119.410	0.16344	0.18102	384
385	622.60	607.90	0.022382	0.039753	44.679	25.156	105.405	13.349	118.754	0.16439	0.18020	385
386	627.53	612.84	0.022994	0.037943	43.489	26.356	106.344	11.574	117.918	0.16548	0.17916	386
387	632.50	617.80	0.023851	0.035696	41.928	28.014	107.500	9.240	116.740	0.16682	0.17773	387
388	637.49	622.79	0.025444	0.032168	39.302	31.087	109.253	5.285	114.538	0.16886	0.17510	388
388.40	639.50	624.80	0.028927	0.028927	34.570	34.570	112.080	0.000	112.080	0.17219	0.17219	388.40

From Published data of E.I. du Pont de Nemours & Co. Inc. Used by permission.

ENTHALPY (Btu/lb above saturated liquid at −40° F)

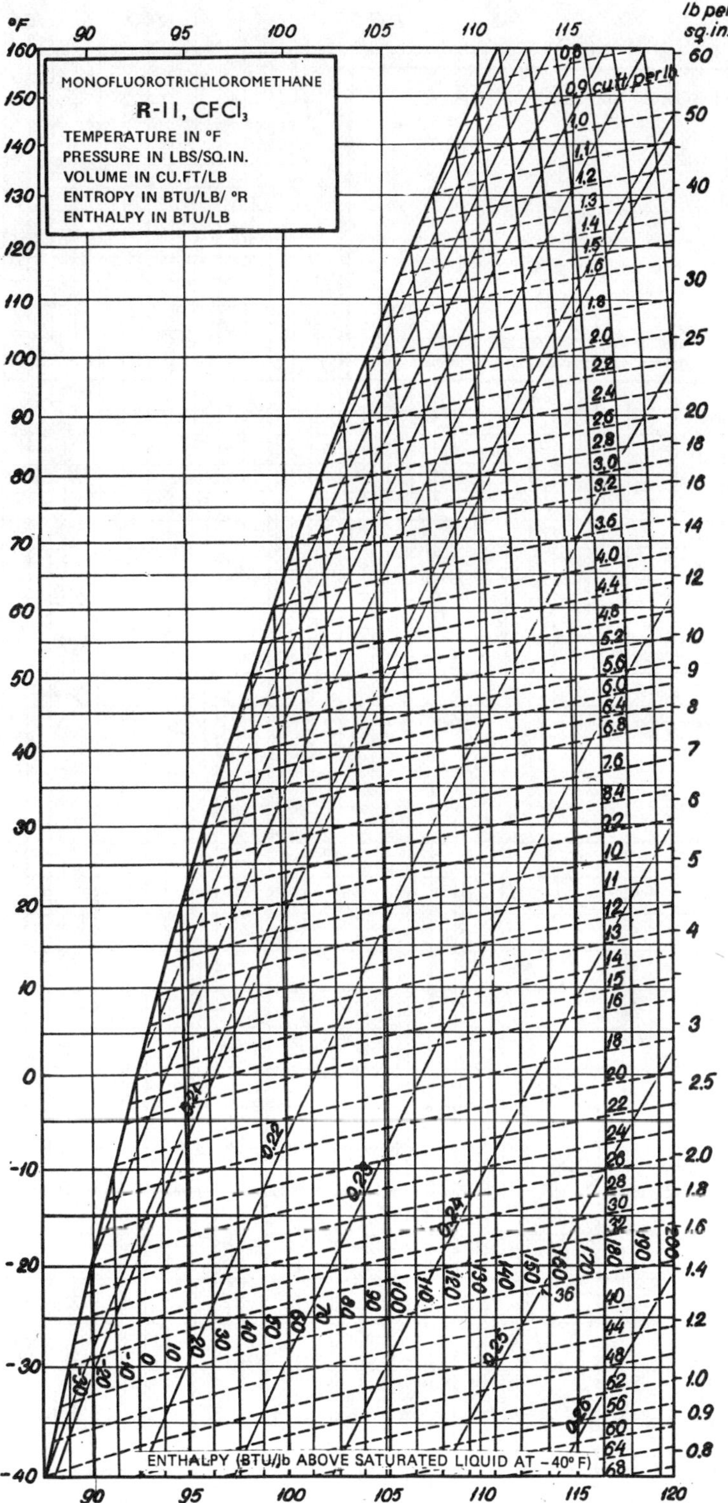
MONOFLUOROTRICHLOROMETHANE
R-11, $CFCl_3$
TEMPERATURE IN °F
PRESSURE IN LBS/SQ.IN.
VOLUME IN CU.FT/LB
ENTROPY IN BTU/LB/ °R
ENTHALPY IN BTU/LB
°F
lb per sq.in.
cu.ft/per lb
ENTHALPY (BTU/lb ABOVE SATURATED LIQUID AT −40° F)

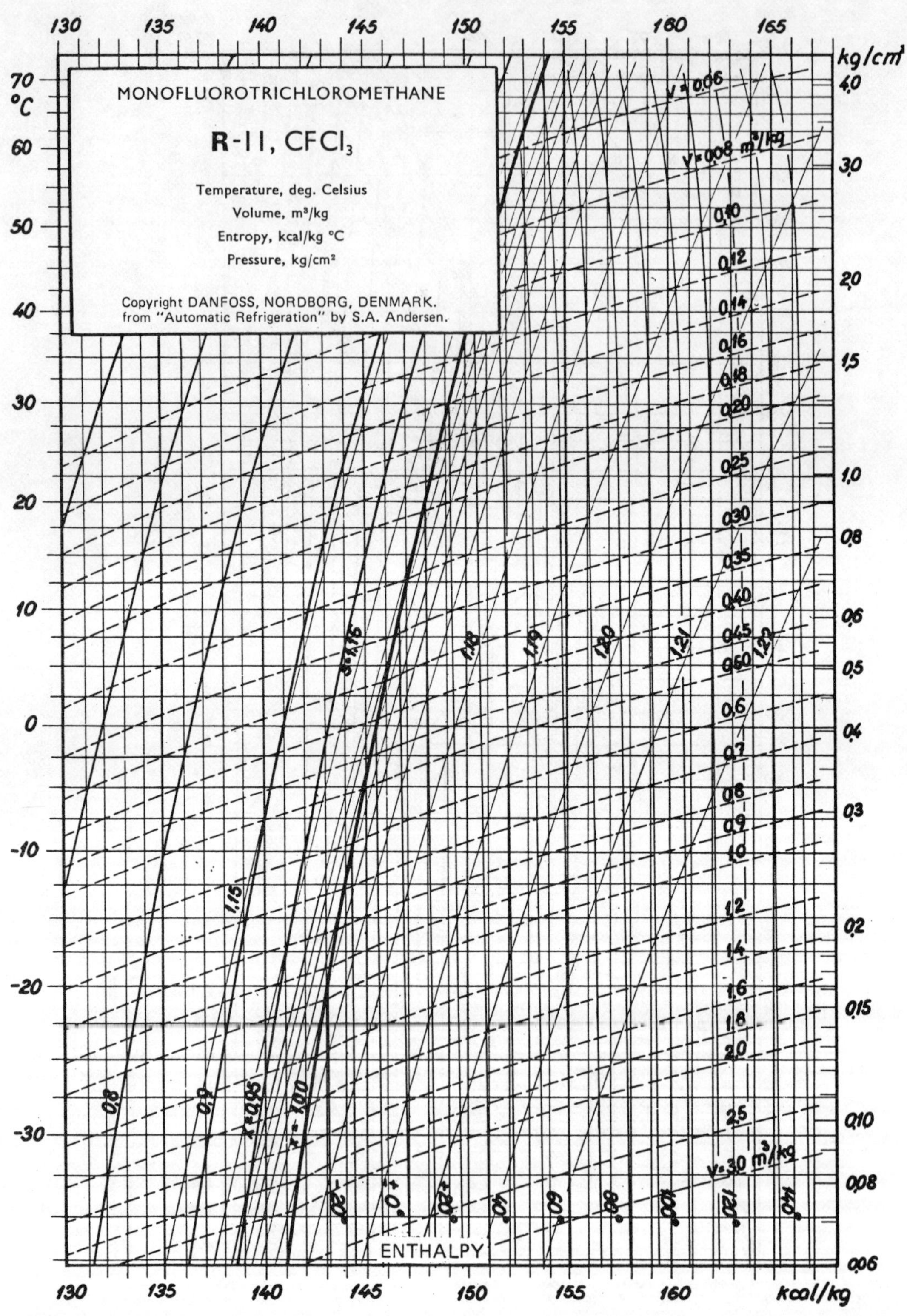

from "Automatic Refrigeration" by S.A. Andersen.

DATA SHEET FOR R 12

CCl_2F_2	DICHLORODIFLUOROMETHANE	Molecular Wt.	120.9

Pressure		Temperature				Volume		Density	
40.9	kg/cm²	112	°C	233.6	°F	1.792	l/kg	.558	kg/l.
582	psia	385	°K	644	°R	.0287	ft³/lb	34.8	lbs/ft³
								4.17	Air=1

AT CRITICAL POINT

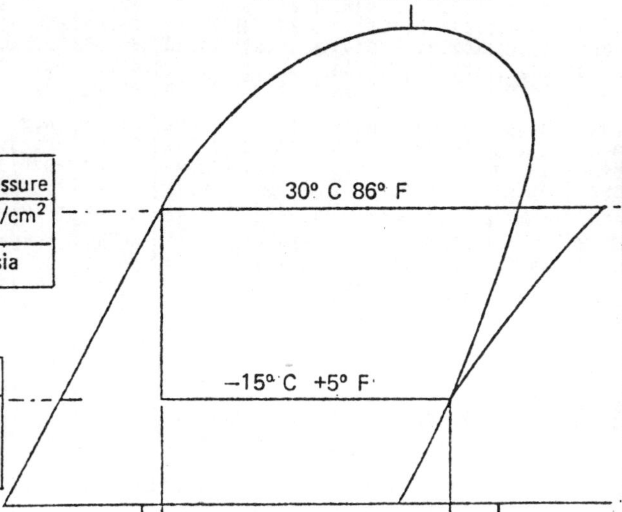

Discharge Pressure	
7.6	kg/cm²
108	psia

Inlet Pressure	
1.86	kg/cm²
26.5	psia

Discharge Temperature			
38	°C	100	°F
311	°K	560	°R

Normal Boiling Point			
-29.8	°C	-216	°F
243	°K	438	°R

Triple Point			
-158	°C	-252	°F
115	°K	208	°R

Refrigerating Effect			
304	kcal/m³	34.2	Btu/ft³

Latent Heat at NBP	39.4 kcal/kg	4775 /kg mol	71 Btu/lb	8584 /lb mol

Trouton's No. 19.65	Gas Constant 7.014 kg.m/kg/° K	12.8 ft.lbs/lb/° R

Specific Heat Liquid at 30° C 86° F Fig 14	Gas C_p .148	C_p/C_v 1.137 at 38°C

Liquid at 30° C 86° F Density 1.29 kg/l.	80.7 lbs/ft³ Viscosity .251 cp

Reduced Form at NBP Pressure .0246	1 / Temperature 1.584

Reduced value of 1/T	2.05	1.47	1.17
when Reduced Pressure is	a) .001	b) .01	c) .1

TEMP. °F	PRESSURE PSIA	PRESSURE PSIG	VOLUME cu ft/lb LIQUID v_f	VOLUME cu ft/lb VAPOR v_g	DENSITY lb/cu ft LIQUID $1/v_f$	DENSITY lb/cu ft VAPOR $1/v_g$	ENTHALPY Btu/lb LIQUID h_f	ENTHALPY Btu/lb LATENT h_{fg}	ENTHALPY Btu/lb VAPOR h_g	ENTROPY Btu/(lb)(°R) LIQUID s_f	ENTROPY Btu/(lb)(°R) VAPOR s_g	TEMP. °F
-150	0.153	29.60	.00958	178.64	104.36	.0055	-22.69	83.53	60.83	-0.062	.2071	-150
-146	0.189	29.53	.00961	146.73	104.03	.0068	-21.87	83.13	61.25	-0.059	.2050	-146
-142	0.232	29.44	.00964	121.25	103.71	.0082	-21.06	82.74	61.68	-0.057	.2030	-142
-138	0.282	29.34	.00967	100.76	103.38	.0099	-20.24	82.35	62.10	-0.054	.2011	-138
-134	0.342	29.22	.00970	84.20	103.05	.0118	-19.42	81.96	62.53	-0.052	.1993	-134
-130	0.412	29.08	.00973	70.73	102.71	.0141	-18.60	81.57	62.96	-0.049	.1976	-130
-126	0.493	28.91	.00976	59.71	102.38	.0167	-17.79	81.19	63.40	-0.047	.1959	-126
-122	0.588	28.72	.00980	50.65	102.04	.0197	-16.97	80.80	63.83	-0.044	.1943	-122
-118	0.698	28.49	.00983	43.17	101.70	.0231	-16.15	80.42	64.27	-0.042	.1928	-118
-114	0.825	28.24	.00986	36.96	101.36	.0270	-15.33	80.04	64.70	-0.040	.1914	-114
-110	0.970	27.94	.00989	31.77	101.02	.0314	-14.51	79.66	65.14	-0.037	.1900	-110
-106	1.136	27.60	.00993	27.43	100.67	.0364	-13.69	79.28	65.58	-0.035	.1887	-106
-102	1.324	27.22	.00996	23.77	100.33	.0420	-12.87	78.90	66.02	-0.033	.1874	-102
-100	1.428	27.01	.00998	22.16	100.15	.0451	-12.46	78.71	66.24	-0.032	.1868	-100
-98	1.538	26.79	.01000	20.68	99.97	.0483	-12.05	78.52	66.46	-0.030	.1862	-98
-96	1.655	26.55	.01002	19.31	99.80	.0517	-11.64	78.33	66.69	-0.029	.1856	-96
-94	1.779	26.29	.01003	18.05	99.62	.0553	-11.23	78.14	66.91	-0.028	.1850	-94
-92	1.911	26.03	.01005	16.89	99.45	.0591	-10.82	77.95	67.13	-0.027	.1845	-92
-90	2.050	25.74	.01007	15.82	99.27	.0632	-10.40	77.76	67.35	-0.026	.1839	-90
-88	2.198	25.44	.01009	14.82	99.09	.0674	-9.99	77.57	67.57	-0.025	.1834	-88
-86	2.355	25.12	.01010	13.90	98.91	.0719	-9.58	77.38	67.79	-0.024	.1829	-86
-84	2.521	24.78	.01012	13.05	98.74	.0765	-9.17	77.19	68.02	-0.023	.1824	-84
-82	2.696	24.43	.01014	12.26	98.56	.0815	-8.75	77.00	68.24	-0.021	.1819	-82
-80	2.880	24.05	.01016	11.53	98.38	.0867	-8.34	76.81	68.46	-0.020	.1814	-80
-78	3.075	23.65	.01018	10.85	98.20	.0921	-7.93	76.62	68.68	-0.019	.1809	-78
-76	3.281	23.24	.01020	10.21	98.02	.0978	-7.51	76.42	68.91	-0.018	.1804	-76
-74	3.497	22.80	.01022	9.62	97.83	.1038	-7.10	76.23	69.13	-0.017	.1800	-74
-72	3.725	22.33	.01024	9.08	97.65	.1101	-6.68	76.04	69.35	-0.016	.1796	-72
-70	3.965	21.84	.01025	8.56	97.47	.1167	-6.27	75.85	69.58	-0.015	.1791	-70
-68	4.217	21.33	.01027	8.09	97.29	.1235	-5.85	75.66	69.80	-0.014	.1787	-68
-66	4.481	20.79	.01029	7.64	97.10	.1307	-5.44	75.46	70.02	-0.013	.1783	-66
-64	4.759	20.23	.01031	7.23	96.92	.1383	-5.02	75.27	70.24	-0.012	.1779	-64
-62	5.051	19.63	.01033	6.84	96.73	.1461	-4.60	75.08	70.47	-0.011	.1775	-62
-60	5.357	19.01	.01035	6.47	96.55	.1543	-4.19	74.88	70.69	-0.010	.1771	-60
-59	5.515	18.69	.01036	6.30	96.46	.1586	-3.98	74.78	70.80	-0.009	.1769	-59
-58	5.678	18.36	.01037	6.13	96.36	.1629	-3.77	74.69	70.91	-0.009	.1767	-58
-57	5.843	18.02	.01038	5.97	96.27	.1673	-3.56	74.59	71.02	-0.008	.1765	-57
-56	6.013	17.67	.01039	5.81	96.18	.1718	-3.35	74.49	71.13	-0.008	.1763	-56
-55	6.187	17.32	.01040	5.66	96.08	.1765	-3.14	74.39	71.25	-0.007	.1762	-55
-54	6.365	16.96	.01041	5.51	95.99	.1812	-2.93	74.29	71.36	-0.007	.1760	-54
-53	6.546	16.59	.01042	5.37	95.89	.1860	-2.72	74.20	71.47	-0.006	.1758	-53
-52	6.732	16.21	.01043	5.23	95.80	.1909	-2.52	74.10	71.58	-0.006	.1756	-52
-51	6.922	15.82	.01044	5.10	95.71	.1959	-2.31	74.00	71.69	-0.005	.1755	-51
-50	7.116	15.43	.01045	4.97	95.61	.2010	-2.10	73.90	71.80	-0.005	.1753	-50
-49	7.315	15.02	.01046	4.84	95.52	.2062	-1.89	73.80	71.91	-0.004	.1751	-49
-48	7.518	14.61	.01047	4.72	95.42	.2115	-1.68	73.70	72.02	-0.004	.1750	-48
-47	7.725	14.19	.01049	4.60	95.33	.2169	-1.47	73.60	72.13	-0.003	.1748	-47
-46	7.937	13.76	.01050	4.49	95.23	.2225	-1.26	73.51	72.24	-0.003	.1746	-46
-45	8.154	13.31	.01051	4.38	95.14	.2281	-1.05	73.41	72.35	-0.002	.1745	-45
-44	8.375	12.86	.01052	4.27	95.04	.2339	-0.84	73.31	72.47	-0.002	.1743	-44
-43	8.601	12.40	.01053	4.17	94.94	.2397	-0.63	73.21	72.58	-0.001	.1741	-43
-42	8.831	11.94	.01054	4.06	94.85	.2457	-0.42	73.11	72.69	-0.001	.1740	-42
-41	9.067	11.46	.01055	3.97	94.75	.2518	-0.21	73.01	72.80	-0.000	.1738	-41

TEMP.	PRESSURE		VOLUME cu ft/lb		DENSITY lb/cu ft		ENTHALPY Btu/lb			ENTROPY Btu/(lb)(° R)		TEMP.
°F	PSIA	PSIG	LIQUID v_f	VAPOR v_g	LIQUID $1/v_f$	VAPOR $1/v_g$	LIQUID h_f	LATENT h_{fg}	VAPOR h_g	LIQUID s_f	VAPOR s_g	°F
−40	9.307	10.9	0.01056	3.875	94.66	0.2580	0	72.91	72.91	0	0.1737	−40
−39	9.553	10.4	0.01057	3.782	94.56	0.2643	0.2107	72.81	73.02	0.000500	0.1735	−39
−38	9.803	9.96	0.01058	3.692	94.46	0.2708	0.4215	72.71	73.13	0.001000	0.1734	−38
−37	10.05	9.44	0.01059	3.604	94.37	0.2774	0.6324	72.61	73.24	0.001498	0.1732	−37
−36	10.32	8.90	0.01060	3.519	94.27	0.2841	0.8434	72.51	73.35	0.001995	0.1731	−36
−35	10.58	8.36	0.01061	3.437	94.17	0.2909	1.054	72.40	73.46	0.002492	0.1729	−35
−34	10.85	7.81	0.01062	3.357	94.08	0.2978	1.265	72.30	73.57	0.002988	0.1728	−34
−33	11.13	7.25	0.01064	3.279	93.98	0.3049	1.477	72.20	73.68	0.003482	0.1727	−33
−32	11.41	6.67	0.01065	3.203	93.88	0.3121	1.688	72.10	73.79	0.003976	0.1725	−32
−31	11.70	6.08	0.01066	3.130	93.78	0.3194	1.900	72.00	73.90	0.004469	0.1724	−31
−30	11.99	5.49	0.01067	3.058	93.69	0.3269	2.112	71.90	74.01	0.004961	0.1722	−30
−29	12.29	4.88	0.01068	2.989	93.59	0.3345	2.323	71.80	74.12	0.005452	0.1721	−29
−28	12.60	4.25	0.01069	2.921	93.49	0.3423	2.535	71.69	74.23	0.005942	0.1720	−28
−27	12.91	3.62	0.01070	2.855	93.39	0.3501	2.747	71.59	74.34	0.006431	0.1718	−27
−26	13.23	2.97	0.01071	2.791	93.29	0.3582	2.960	71.49	74.45	0.006919	0.1717	−26
−25	13.55	2.32	0.01073	2.729	93.19	0.3663	3.172	71.39	74.56	0.007407	0.1716	−25
−24	13.88	1.64	0.01074	2.669	93.09	0.3746	3.384	71.28	74.67	0.007894	0.1715	−24
−23	14.22	0.96	0.01075	2.610	92.99	0.3831	3.597	71.18	74.78	0.008379	0.1713	−23
−22	14.56	0.27	0.01076	2.552	92.89	0.3917	3.810	71.08	74.89	0.008864	0.1712	−22
−21	14.91	0.21	0.01077	2.497	92.79	0.4004	4.022	70.97	75.00	0.009348	0.1711	−21
−20	15.26	0.57	0.01078	2.442	92.69	0.4093	4.235	70.87	75.11	0.009831	0.1710	−20
−19	15.62	0.93	0.01079	2.390	92.59	0.4183	4.448	70.77	75.21	0.01031	0.1709	−19
−18	15.99	1.30	0.01081	2.338	92.49	0.4275	4.661	70.66	75.32	0.01079	0.1707	−18
−17	16.37	1.67	0.01082	2.288	92.39	0.4369	4.875	70.56	75.43	0.01127	0.1706	−17
−16	16.75	2.05	0.01083	2.239	92.29	0.4464	5.088	70.45	75.54	0.01175	0.1705	−16
−15	17.14	2.44	0.01084	2.192	92.19	0.4561	5.302	70.35	75.65	0.01223	0.1704	−15
−14	17.53	2.84	0.01085	2.146	92.09	0.4659	5.515	70.24	75.76	0.01271	0.1703	−14
−13	17.93	3.24	0.01087	2.101	91.99	0.4759	5.729	70.14	75.87	0.01319	0.1702	−13
−12	18.34	3.65	0.01088	2.057	91.89	0.4861	5.943	70.03	75.97	0.01366	0.1701	−12
−11	18.76	4.06	0.01089	2.014	91.79	0.4964	6.157	69.93	76.08	0.01414	0.1699	−11
−10	19.18	4.49	0.01090	1.972	91.68	0.5069	6.371	69.82	76.19	0.01461	0.1698	−10
− 9	19.62	4.92	0.01091	1.932	91.58	0.5175	6.585	69.71	76.30	0.01509	0.1697	− 9
− 8	20.05	5.36	0.01093	1.892	91.48	0.5284	6.800	69.61	76.41	0.01556	0.1696	− 8
− 7	20.50	5.81	0.01094	1.853	91.38	0.5394	7.014	69.50	76.52	0.01603	0.1695	− 7
− 6	20.96	6.26	0.01095	1.816	91.28	0.5506	7.229	69.39	76.62	0.01650	0.1694	− 6
− 5	21.42	6.72	0.01096	1.779	91.17	0.5619	7.444	69.29	76.73	0.01697	0.1693	− 5
− 4	21.89	7.19	0.01098	1.743	91.07	0.5735	7.659	69.18	76.84	0.01744	0.1692	− 4
− 3	22.36	7.67	0.01099	1.708	90.97	0.5852	7.874	69.07	76.95	0.01791	0.1691	− 3
− 2	22.85	8.15	0.01100	1.674	90.86	0.5971	8.089	68.96	77.05	0.01838	0.1690	− 2
− 1	23.34	8.65	0.01101	1.641	90.76	0.6092	8.305	68.85	77.16	0.01885	0.1689	− 1
0	23.84	9.15	0.01103	1.608	90.65	0.6215	8.520	68.75	77.27	0.01932	0.1688	0
1	24.35	9.66	0.01104	1.577	90.55	0.6340	8.736	68.64	77.37	0.01978	0.1687	1
2	24.87	10.1	0.01105	1.546	90.45	0.6467	8.952	68.53	77.48	0.02025	0.1686	2
3	25.40	10.7	0.01106	1.516	90.34	0.6595	9.168	68.42	77.59	0.02071	0.1686	3
4	25.93	11.2	0.01108	1.486	90.24	0.6726	9.384	68.31	77.69	0.02118	0.1685	4
5	26.48	11.7	0.01109	1.458	90.13	0.6858	9.600	68.20	77.80	0.02164	0.1684	5
6	27.03	12.3	0.01110	1.429	90.03	0.6993	9.816	68.09	77.91	0.02211	0.1683	6
7	27.59	12.9	0.01112	1.402	89.92	0.7130	10.03	67.98	78.01	0.02257	0.1682	7
8	28.16	13.4	0.01113	1.375	89.81	0.7268	10.25	67.87	78.12	0.02303	0.1681	8
9	28.74	14.0	0.01114	1.349	89.71	0.7409	10.46	67.76	78.22	0.02349	0.1680	9
10	29.33	14.6	0.01116	1.324	89.60	0.7552	10.68	67.65	78.33	0.02395	0.1679	10
11	29.93	15.2	0.01117	1.299	89.49	0.7697	10.90	67.53	78.44	0.02441	0.1679	11
12	30.53	15.8	0.01118	1.274	89.39	0.7844	11.11	67.42	78.54	0.02487	0.1678	12
13	31.15	16.4	0.01120	1.251	89.28	0.7993	11.33	67.31	78.65	0.02532	0.1677	13
14	31.78	17.0	0.01121	1.227	89.17	0.8144	11.55	67.20	78.75	0.02578	0.1676	14

TEMP.	PRESSURE		VOLUME cu ft/lb		DENSITY lb/cu ft		ENTHALPY Btu/lb			ENTROPY Btu/(lb)(° R)		TEMP.
°F	PSIA	PSIG	LIQUID v_f	VAPOR v_g	LIQUID $1/v_f$	VAPOR $1/v_g$	LIQUID h_f	LATENT h_{fg}	VAPOR h_g	LIQUID s_f	VAPOR s_g	°F
15	32.415	17.719	0.011227	1.2050	89.070	0.82986	11.771	67.090	78.861	0.026243	0.16758	15
16	33.060	18.364	0.011241	1.1828	88.962	0.84544	11.989	66.977	78.966	0.026699	0.16750	16
17	33.714	19.018	0.011254	1.1611	88.854	0.86125	12.207	66.864	79.071	0.027154	0.16742	17
18	34.378	19.682	0.011268	1.1399	88.746	0.87729	12.426	66.750	79.176	0.027608	0.16734	18
19	35.052	20.356	0.011282	1.1191	88.637	0.89356	12.644	66.636	79.280	0.028062	0.16727	19
20	35.736	21.040	0.011296	1.0988	88.529	0.91006	12.863	66.522	79.385	0.028515	0.16719	20
21	36.430	21.734	0.011310	1.0790	88.419	0.92679	13.081	66.407	79.488	0.028968	0.16712	21
22	37.135	22.439	0.011324	1.0596	88.310	0.94377	13.300	66.293	79.593	0.029420	0.16704	22
23	37.849	23.153	0.011338	1.0406	88.201	0.96098	13.520	66.177	79.697	0.029871	0.16697	23
24	38.574	23.878	0.011352	1.0220	88.091	0.97843	13.739	66.061	79.800	0.030322	0.16690	24
25	39.310	24.614	0.011366	1.0039	87.981	0.99613	13.958	65.946	79.904	0.030772	0.16683	25
26	40.056	25.360	0.011380	0.98612	87.870	1.0141	14.178	65.829	80.007	0.031221	0.16676	26
27	40.813	26.117	0.011395	0.96874	87.760	1.0323	14.398	65.713	80.111	0.031670	0.16669	27
28	41.580	26.884	0.011409	0.95173	87.649	1.0507	14.618	65.596	80.214	0.032118	0.16662	28
29	42.359	27.663	0.011424	0.93509	87.537	1.0694	14.838	65.478	80.316	0.032566	0.16655	29
30	43.148	28.452	0.011438	0.91880	87.426	1.0884	15.058	65.361	80.419	0.033013	0.16648	30
31	43.948	29.252	0.011453	0.90286	87.314	1.1076	15.279	65.243	80.522	0.033460	0.16642	31
32	44.760	30.064	0.011468	0.88725	87.202	1.1271	15.500	65.124	80.624	0.033905	0.16635	32
33	45.583	30.887	0.011482	0.87197	87.090	1.1468	15.720	65.006	80.726	0.034351	0.16629	33
34	46.417	31.721	0.011497	0.85702	86.977	1.1668	15.942	64.886	80.828	0.034796	0.16622	34
35	47.263	32.567	0.011512	0.84237	86.865	1.1871	16.163	64.767	80.930	0.035240	0.16616	35
36	48.120	33.424	0.011527	0.82803	86.751	1.2077	16.384	64.647	81.031	0.035683	0.16610	36
37	48.989	34.293	0.011542	0.81399	86.638	1.2285	16.606	64.527	81.133	0.036126	0.16604	37
38	49.870	35.174	0.011557	0.80023	86.524	1.2496	16.828	64.406	81.234	0.036569	0.16598	38
39	50.763	36.067	0.011573	0.78676	86.410	1.2710	17.050	64.285	81.335	0.037011	0.16592	39
40	51.667	36.971	0.011588	0.77357	86.296	1.2927	17.273	64.163	81.436	0.037453	0.16586	40
41	52.584	37.888	0.011603	0.76064	86.181	1.3147	17.495	64.042	81.537	0.037893	0.16580	41
42	53.513	38.817	0.011619	0.74798	86.066	1.3369	17.718	63.919	81.637	0.038334	0.16574	42
43	54.454	39.758	0.011635	0.73557	85.951	1.3595	17.941	63.796	81.737	0.038774	0.16568	43
44	55.407	40.711	0.011650	0.72341	85.836	1.3823	18.164	63.673	81.837	0.039213	0.16562	44
45	56.373	41.677	0.011666	0.71149	85.720	1.4055	18.387	63.550	81.937	0.039652	0.16557	45
46	57.352	42.656	0.011682	0.69982	85.604	1.4289	18.611	63.426	82.037	0.040091	0.16551	46
47	58.343	43.647	0.011698	0.68837	85.487	1.4527	18.835	63.301	82.136	0.040529	0.16546	47
48	59.347	44.651	0.011714	0.67715	85.371	1.4768	19.059	63.177	82.236	0.040966	0.16540	48
49	60.364	45.668	0.011730	0.66616	85.254	1.5012	19.283	63.051	82.334	0.041403	0.16535	49
50	61.394	46.698	0.011746	0.65537	85.136	1.5258	19.507	62.926	82.433	0.041839	0.16530	50
51	62.437	47.741	0.011762	0.64480	85.018	1.5509	19.732	62.800	82.532	0.042276	0.16524	51
52	63.494	48.798	0.011779	0.63444	84.900	1.5762	19.957	62.673	82.630	0.042711	0.16519	52
53	64.563	49.867	0.011795	0.62428	84.782	1.6019	20.182	62.546	82.728	0.043146	0.16514	53
54	65.646	50.950	0.011811	0.61431	84.663	1.6278	20.408	62.418	82.826	0.043581	0.16509	54
55	66.743	52.047	0.011828	0.60453	84.544	1.6542	20.634	62.290	82.924	0.044015	0.16504	55
56	67.853	53.157	0.011845	0.59495	84.425	1.6808	20.859	62.162	83.021	0.044449	0.16499	56
57	68.977	54.281	0.011862	0.58554	84.305	1.7078	21.086	62.033	83.119	0.044883	0.16494	57
58	70.115	55.419	0.011879	0.57632	84.185	1.7352	21.312	61.903	83.215	0.045316	0.16489	58
59	71.267	56.571	0.011896	0.56727	84.065	1.7628	21.539	61.773	83.312	0.045748	0.16484	59
60	72.433	57.737	0.011913	0.55839	83.944	1.7909	21.766	61.643	83.409	0.046180	0.16479	60
61	73.613	58.917	0.011930	0.54967	83.823	1.8193	21.993	61.512	83.505	0.046612	0.16474	61
62	74.807	60.111	0.011947	0.54112	83.701	1.8480	22.221	61.380	83.601	0.047044	0.16470	62
63	76.016	61.320	0.011965	0.53273	83.580	1.8771	22.448	61.248	83.696	0.047475	0.16465	63
64	77.239	62.543	0.011982	0.52450	83.457	1.9066	22.676	61.116	83.792	0.047905	0.16460	64
65	78.477	63.781	0.012000	0.51642	83.335	1.9364	22.905	60.982	83.887	0.048336	0.16456	65
66	79.729	65.033	0.012017	0.50848	83.212	1.9666	23.133	60.849	83.982	0.048765	0.16451	66
67	80.996	66.300	0.012035	0.50070	83.089	1.9972	23.362	60.715	84.077	0.049195	0.16447	67
68	82.279	67.583	0.012053	0.49305	82.965	2.0282	23.591	60.580	84.171	0.049624	0.16442	68
69	83.576	68.880	0.012071	0.48555	82.841	2.0595	23.821	60.445	84.266	0.050053	0.16438	69

TEMP. °F	PRESSURE		VOLUME cu ft/lb		DENSITY lb/cu ft		ENTHALPY Btu/lb			ENTROPY Btu/(lb)(° R)		TEMP. °F
	PSIA	PSIG	LIQUID v_f	VAPOR v_g	LIQUID $1/v_f$	VAPOR $1/v_g$	LIQUID h_f	LATENT h_{fg}	VAPOR h_g	LIQUID s_f	VAPOR s_g	
70	84.88	70.1	0.01208	0.4781	82.71	2.091	24.05	60.30	84.35	0.05048	0.1643	70
71	86.21	71.5	0.01210	0.4709	82.59	2.123	24.28	60.17	84.45	0.05091	0.1642	71
72	87.55	72.8	0.01212	0.4638	82.46	2.155	24.51	60.03	84.54	0.05133	0.1642	72
73	88.91	74.2	0.01214	0.4568	82.34	2.188	24.74	59.89	84.63	0.05176	0.1642	73
74	90.29	75.5	0.01216	0.4500	82.21	2.222	24.97	59.75	84.73	0.05219	0.1641	74
75	91.68	76.9	0.01218	0.4432	82.08	2.256	25.20	59.62	84.82	0.05262	0.1641	75
76	93.08	78.3	0.01220	0.4366	81.96	2.290	25.43	59.48	84.91	0.05304	0.1640	76
77	94.50	79.8	0.01222	0.4301	81.83	2.324	25.66	59.34	85.00	0.05347	0.1640	77
78	95.94	81.2	0.01223	0.4237	81.70	2.359	25.89	59.20	85.10	0.05390	0.1640	78
79	97.40	82.7	0.01225	0.4175	81.57	2.395	26.13	59.05	85.19	0.05432	0.1639	79
80	98.87	84.1	0.01227	0.4113	81.45	2.431	26.36	58.91	85.28	0.05475	0.1639	80
81	100.3	85.6	0.01229	0.4053	81.32	2.467	26.59	58.77	85.37	0.05517	0.1638	81
82	101.8	87.1	0.01231	0.3993	81.19	2.504	26.83	58.63	85.46	0.05560	0.1638	82
83	103.3	88.6	0.01233	0.3935	81.06	2.541	27.06	58.48	85.55	0.05602	0.1638	83
84	104.9	90.2	0.01235	0.3877	80.93	2.578	27.30	58.34	85.64	0.05645	0.1637	84
85	106.4	91.7	0.01237	0.3821	80.80	2.617	27.53	58.19	85.73	0.05687	0.1637	85
86	108.0	93.3	0.01239	0.3765	80.67	2.655	27.76	58.05	85.82	0.05730	0.1636	**86**
87	109.6	94.9	0.01241	0.3711	80.53	2.694	28.00	57.90	85.91	0.05772	0.1636	87
88	111.2	96.5	0.01243	0.3657	80.40	2.734	28.24	57.75	85.99	0.05814	0.1636	88
89	112.8	98.1	0.01245	0.3604	80.27	2.774	28.47	57.60	86.08	0.05857	0.1635	89
90	114.4	99.7	0.01247	0.3552	80.14	2.814	28.71	57.46	86.17	0.05899	0.1635	90
91	116.1	101.4	0.01249	0.3501	80.00	2.855	28.95	57.31	86.26	0.05942	0.1634	91
92	117.8	103.1	0.01252	0.3451	79.87	2.897	29.18	57.16	86.34	0.05984	0.1634	92
93	119.5	104.8	0.01254	0.3402	79.74	2.939	29.42	57.00	86.43	0.06026	0.1634	93
94	121.2	106.5	0.01256	0.3354	79.60	2.981	29.66	56.85	86.52	0.06069	0.1633	94
95	122.9	108.2	0.01258	0.3306	79.47	3.024	29.90	56.70	86.60	0.06111	0.1633	95
96	124.7	110.0	0.01260	0.3259	79.33	3.068	30.14	56.55	86.69	0.06153	0.1633	96
97	126.4	111.7	0.01262	0.3213	79.19	3.112	30.38	56.39	86.77	0.06195	0.1632	97
98	128.2	113.5	0.01264	0.3167	79.06	3.156	30.61	56.24	86.86	0.06238	0.1632	98
99	130.0	115.3	0.01267	0.3123	78.92	3.201	30.85	56.08	86.94	0.06280	0.1631	99
100	131.8	117.1	0.01269	0.3079	78.78	3.247	31.10	55.92	87.02	0.06322	0.1631	100
101	133.7	119.0	0.01271	0.3036	78.64	3.293	31.34	55.77	87.11	0.06364	0.1631	101
102	135.5	120.8	0.01273	0.2993	78.50	3.340	31.58	55.61	87.19	0.06407	0.1630	102
103	137.4	122.7	0.01276	0.2951	78.36	3.387	31.82	55.45	87.27	0.06449	0.1630	103
104	139.3	124.6	0.01278	0.2910	78.22	3.435	32.06	55.29	87.36	0.06491	0.1630	104
105	141.2	126.5	0.01280	0.2870	78.08	3.484	32.31	55.13	87.44	0.06533	0.1629	105
106	143.1	128.4	0.01282	0.2830	77.94	3.533	32.55	54.97	87.52	0.06576	0.1629	106
107	145.1	130.4	0.01285	0.2791	77.80	3.582	32.79	54.80	87.60	0.06618	0.1629	107
108	147.1	132.4	0.01287	0.2752	77.66	3.633	33.04	54.64	87.68	0.06660	0.1628	108
109	149.1	134.4	0.01290	0.2714	77.51	3.684	33.28	54.47	87.76	0.06702	0.1628	109
110	151.1	136.4	0.01292	0.2676	77.37	3.735	33.53	54.31	87.84	0.06745	0.1627	110
111	153.1	138.4	0.01294	0.2640	77.23	3.787	33.77	54.14	87.92	0.06787	0.1627	111
112	155.1	140.4	0.01297	0.2603	77.08	3.840	34.02	53.97	88.00	0.06829	0.1627	112
113	157.2	142.5	0.01299	0.2568	76.94	3.894	34.27	53.80	88.07	0.06871	0.1626	113
114	159.3	144.6	0.01302	0.2532	76.79	3.948	34.51	53.63	88.15	0.06914	0.1626	114
115	161.4	146.7	0.01304	0.2498	76.64	4.002	34.76	53.46	88.23	0.06956	0.1626	115
116	163.6	148.9	0.01307	0.2464	76.50	4.058	35.01	53.29	88.31	0.06998	0.1625	116
117	165.7	151.0	0.01309	0.2430	76.35	4.114	35.26	53.12	88.38	0.07041	0.1625	117
118	167.9	153.2	0.01312	0.2397	76.20	4.171	35.51	52.94	88.46	0.07083	0.1624	118
119	170.1	155.4	0.01314	0.2364	76.05	4.228	35.76	52.77	88.53	0.07125	0.1624	119
120	172.3	157.6	0.01317	0.2332	75.90	4.287	36.01	52.59	88.61	0.07168	0.1624	120
121	174.5	159.8	0.01320	0.2301	75.75	4.345	36.26	52.42	88.68	0.07210	0.1623	121
122	176.8	162.1	0.01322	0.2269	75.60	4.405	36.51	52.24	88.75	0.07252	0.1623	122
123	179.1	164.4	0.01325	0.2239	75.45	4.466	36.76	52.06	88.83	0.07295	0.1623	123
124	181.4	166.7	0.01328	0.2208	75.29	4.527	37.02	51.88	88.90	0.07337	0.1622	124

TEMP. °F	PRESSURE PSIA	PRESSURE PSIG	VOLUME cu ft/lb LIQUID v_f	VOLUME cu ft/lb VAPOR v_g	DENSITY lb/cu ft LIQUID $1/v_f$	DENSITY lb/cu ft VAPOR $1/v_g$	ENTHALPY Btu/lb LIQUID h_f	ENTHALPY Btu/lb LATENT h_{fg}	ENTHALPY Btu/lb VAPOR h_g	ENTROPY Btu/(lb)(°R) LIQUID s_f	ENTROPY Btu/(lb)(°R) VAPOR s_g	TEMP. °F
125	183.7	169.0	0.01330	0.2179	75.14	4.589	37.27	51.69	88.97	0.07380	0.1622	125
126	186.1	171.4	0.01333	0.2149	74.99	4.651	37.52	51.51	89.04	0.07422	0.1621	126
127	188.4	173.7	0.01336	0.2120	74.83	4.715	37.78	51.33	89.11	0.07465	0.1621	127
128	190.8	176.1	0.01339	0.2092	74.68	4.779	38.04	51.14	89.18	0.07507	0.1621	128
129	193.2	178.5	0.01341	0.2064	74.52	4.844	38.29	50.95	89.25	0.07550	0.1620	129
130	195.7	181.0	0.01344	0.2036	74.36	4.910	38.55	50.76	89.32	0.07592	0.1620	130
131	198.1	183.4	0.01347	0.2009	74.20	4.977	38.81	50.57	89.38	0.07635	0.1619	131
132	200.6	185.9	0.01350	0.1982	74.05	5.045	39.06	50.38	89.45	0.07677	0.1619	132
133	203.1	188.4	0.01353	0.1955	73.89	5.113	39.32	50.19	89.52	0.07720	0.1618	133
134	205.6	190.9	0.01356	0.1929	73.72	5.182	39.58	50.00	89.58	0.07763	0.1618	134
135	208.2	193.5	0.01359	0.1903	73.56	5.253	39.84	49.80	89.65	0.07806	0.1618	135
136	210.7	196.0	0.01362	0.1878	73.40	5.324	40.11	49.60	89.71	0.07848	0.1617	136
137	213.3	198.6	0.01365	0.1853	73.24	5.396	40.37	49.40	89.78	0.07891	0.1617	137
138	216.0	201.3	0.01368	0.1828	73.07	5.469	40.63	49.21	89.84	0.07934	0.1616	138
139	218.6	203.9	0.01371	0.1803	72.91	5.543	40.89	49.00	89.90	0.07977	0.1616	139
140	221.3	206.6	0.01374	0.1779	72.74	5.618	41.16	48.80	89.96	0.08020	0.1615	140
141	224.0	209.3	0.01377	0.1756	72.58	5.694	41.42	48.60	90.02	0.08063	0.1615	141
142	226.7	212.0	0.01381	0.1732	72.41	5.771	41.69	48.39	90.08	0.08106	0.1615	142
143	229.4	214.7	0.01384	0.1709	72.24	5.849	41.95	48.18	90.14	0.08149	0.1614	143
144	232.2	217.5	0.01387	0.1686	72.07	5.928	42.22	47.97	90.20	0.08192	0.1614	144
145	235.0	220.3	0.01390	0.1664	71.90	6.008	42.49	47.76	90.26	0.08236	0.1613	145
146	237.8	223.1	0.01394	0.1642	71.73	6.089	42.76	47.55	90.31	0.08279	0.1613	146
147	240.6	225.9	0.01397	0.1620	71.55	6.171	43.03	47.33	90.37	0.08322	0.1612	147
148	243.5	228.8	0.01400	0.1598	71.38	6.255	43.30	47.12	90.42	0.08366	0.1612	148
149	246.4	231.7	0.01404	0.1577	71.21	6.339	43.57	46.90	90.48	0.08409	0.1611	149
150	249.3	234.6	0.01407	0.1556	71.03	6.425	43.85	46.68	90.53	0.08453	0.1611	150
151	252.2	237.5	0.01411	0.1535	70.85	6.512	44.12	46.46	90.58	0.08496	0.1610	151
152	255.2	240.5	0.01414	0.1515	70.67	6.600	44.39	46.23	90.63	0.08540	0.1609	152
153	258.1	243.4	0.01418	0.1494	70.50	6.689	44.67	46.01	90.68	0.08584	0.1609	153
154	261.2	246.5	0.01422	0.1475	70.31	6.779	44.95	45.78	90.73	0.08628	0.1608	154
155	264.2	249.5	0.01425	0.1455	70.13	6.871	45.22	45.55	90.78	0.08671	0.1608	155
156	267.3	252.6	0.01429	0.1435	69.95	6.964	45.50	45.32	90.83	0.08715	0.1607	156
157	270.3	255.6	0.01433	0.1416	69.77	7.059	45.78	45.08	90.87	0.08760	0.1607	157
158	273.5	258.8	0.01437	0.1397	69.58	7.155	46.06	44.85	90.92	0.08804	0.1606	158
159	276.6	261.9	0.01441	0.1378	69.39	7.252	46.35	44.61	90.96	0.08848	0.1605	159
160	279.8	265.1	0.01444	0.1360	69.20	7.350	46.63	44.37	91.00	0.08892	0.1605	160
161	283.0	268.3	0.01448	0.1342	69.01	7.451	46.91	44.13	91.04	0.08937	0.1604	161
162	286.2	271.5	0.01452	0.1324	68.82	7.552	47.20	43.88	91.08	0.08981	0.1604	162
163	289.4	274.7	0.01457	0.1306	68.63	7.655	47.48	43.63	91.12	0.09026	0.1603	163
164	292.7	278.0	0.01461	0.1288	68.44	7.760	47.77	43.38	91.16	0.09071	0.1602	164
165	296.0	281.3	0.01465	0.1271	68.24	7.866	48.06	43.13	91.19	0.09115	0.1602	165
166	299.4	284.7	0.01469	0.1254	68.04	7.974	48.35	42.87	91.23	0.09160	0.1601	166
167	302.7	288.0	0.01473	0.1237	67.85	8.083	48.64	42.62	91.26	0.09205	0.1600	167
168	306.1	291.4	0.01478	0.1220	67.64	8.195	48.93	42.36	91.29	0.09251	0.1600	168
169	309.5	294.8	0.01482	0.1203	67.44	8.308	49.23	42.09	91.33	0.09296	0.1599	169
170	313.0	298.3	0.01487	0.1187	67.24	8.422	49.52	41.83	91.35	0.09341	0.1598	170
171	316.4	301.7	0.01491	0.1171	67.03	8.539	49.82	41.56	91.38	0.09387	0.1597	171
172	319.9	305.2	0.01496	0.1155	66.83	8.657	50.12	41.29	91.41	0.09433	0.1596	172
173	323.5	308.8	0.01501	0.1139	66.62	8.778	50.42	41.01	91.43	0.09478	0.1596	173
174	327.0	312.3	0.01505	0.1123	66.41	8.900	50.72	40.73	91.46	0.09524	0.1595	174
175	330.6	315.9	0.01510	0.1108	66.19	9.025	51.02	40.45	91.48	0.09570	0.1594	175
176	334.2	319.5	0.01515	0.1092	65.98	9.151	51.33	40.17	91.50	0.09617	0.1593	176
177	337.9	323.2	0.01520	0.1077	65.76	9.280	51.63	39.88	91.51	0.09663	0.1592	177
178	341.5	326.8	0.01525	0.1062	65.54	9.411	51.94	39.59	91.53	0.09710	0.1591	178
179	345.2	330.5	0.01530	0.1047	65.32	9.544	52.25	39.29	91.54	0.09756	0.1591	179
180	349.0	334.3	0.01536	0.1033	65.10	9.680	52.56	38.99	91.56	0.09803	0.1590	180

From Published data of E. I. du Pont de Nemours & Co. Inc. Used by permission.

ENTHALPY (Btu/lb above saturated liquid at −40° F)

TEMP. °C	PRESSURE kg/cm²	PRESSURE ATM	VOLUME LIQUID l/kg v_f	VOLUME VAPOR m³/kg v_g	DENSITY LIQUID kg/l $1/v_f$	DENSITY VAPOR kg/m³ $1/v_g$	ENTHALPY kcal/kg LIQUID h_f	ENTHALPY kcal/kg LATENT h_{fg}	ENTHALPY kcal/kg VAPOR h_g	ENTROPY kcal/(kg) (°K) LIQUID s_f	ENTROPY kcal/(kg) (°K) VAPOR s_g	TEMP. °C
−100	0.0120	0.0116	0.599114	10.09988	1.6691	0.09901	79.007	46.298	125.305	0.90479	1.17215	−100
−99	0.0132	0.0128	0.599959	9.24977	1.6668	0.10811	79.211	46.199	125.410	0.90597	1.17123	−99
−98	0.0145	0.0140	0.600810	8.48149	1.6644	0.11790	79.416	46.100	125.516	0.90714	1.17032	−98
−97	0.0158	0.0153	0.601664	7.78628	1.6621	0.12843	79.620	46.002	125.622	0.90830	1.16943	−97
−96	0.0173	0.0168	0.602524	7.15639	1.6597	0.13974	79.825	45.904	125.728	0.90946	1.16856	−96
−95	0.0189	0.0183	0.603387	6.58499	1.6573	0.15186	80.029	45.806	125.835	0.91061	1.16771	−95
−94	0.0207	0.0200	0.604256	6.06601	1.6549	0.16485	80.233	45.708	125.942	0.91175	1.16687	−94
−93	0.0225	0.0218	0.605129	5.59408	1.6525	0.17876	80.438	45.611	126.049	0.91289	1.16605	−93
−92	0.0245	0.0237	0.606006	5.16445	1.6501	0.19363	80.642	45.514	126.156	0.91402	1.16525	−92
−91	0.0267	0.0258	0.606889	4.77285	1.6477	0.20952	80.846	45.417	126.263	0.91515	1.16446	−91
−90	0.0290	0.0281	0.607776	4.41554	1.6453	0.22647	81.051	45.321	126.371	0.91627	1.16369	−90
−89	0.0315	0.0304	0.608668	4.08914	1.6429	0.24455	81.255	45.224	126.479	0.91738	1.16294	−89
−88	0.0341	0.0330	0.609565	3.79066	1.6405	0.26381	81.459	45.128	126.587	0.91848	1.16220	−88
−87	0.0370	0.0358	0.610467	3.51742	1.6381	0.28430	81.664	45.032	126.696	0.91958	1.16147	−87
−86	0.0400	0.0387	0.611373	3.26702	1.6357	0.30609	81.868	44.936	126.804	0.92068	1.16076	−86
−85	0.0432	0.0418	0.612285	3.03731	1.6332	0.32924	82.073	44.840	126.913	0.92177	1.16007	−85
−84	0.0467	0.0452	0.613202	2.82638	1.6308	0.35381	82.277	44.744	127.022	0.92285	1.15939	−84
−83	0.0504	0.0488	0.614124	2.63249	1.6283	0.37987	82.482	44.649	127.131	0.92393	1.15872	−83
−82	0.0543	0.0526	0.615051	2.45409	1.6259	0.40748	82.686	44.554	127.240	0.92500	1.15807	−82
−81	0.0585	0.0566	0.615983	2.28980	1.6234	0.43672	82.891	44.458	127.349	0.92607	1.15743	−81
−80	0.0630	0.0609	0.616921	2.13834	1.6210	0.46765	83.096	44.363	127.459	0.92713	1.15680	−80
−79	0.0677	0.0655	0.617864	1.99859	1.6185	0.50035	83.301	44.268	127.569	0.92819	1.15618	−79
−78	0.0727	0.0704	0.618812	1.86953	1.6160	0.53489	83.506	44.173	127.679	0.92924	1.15558	−78
−77	0.0780	0.0755	0.619766	1.75023	1.6135	0.57135	83.711	44.078	127.789	0.93029	1.15499	−77
−76	0.0837	0.0810	0.620725	1.63985	1.6110	0.60981	83.916	43.983	127.899	0.93133	1.15441	−76
−75	0.0896	0.0868	0.621690	1.53765	1.6085	0.65034	84.121	43.888	128.009	0.93237	1.15384	−75
−74	0.0960	0.0929	0.622660	1.44293	1.6060	0.69303	84.326	43.793	128.119	0.93340	1.15329	−74
−73	0.1027	0.0994	0.623636	1.35507	1.6035	0.73797	84.532	43.698	128.230	0.93443	1.15274	−73
−72	0.1097	0.1062	0.624618	1.27351	1.6010	0.78523	84.737	43.603	128.341	0.93546	1.15221	−72
−71	0.1172	0.1134	0.625605	1.19773	1.5985	0.83491	84.943	43.509	128.451	0.93647	1.15169	−71
−70	0.1251	0.1211	0.626599	1.12728	1.5959	0.88709	85.148	43.414	128.562	0.93749	1.15117	−70
−69	0.1334	0.1291	0.627598	1.06171	1.5934	0.94187	85.354	43.319	128.673	0.93850	1.15067	−69
−68	0.1422	0.1376	0.628603	1.00066	1.5908	0.99934	85.560	43.224	128.784	0.93951	1.15018	−68
−67	0.1514	0.1465	0.629614	0.94376	1.5883	1.05959	85.766	43.129	128.895	0.94051	1.14970	−67
−66	0.1611	0.1559	0.630632	0.89069	1.5857	1.12272	85.972	43.033	129.006	0.94150	1.14923	−66
−65	0.1713	0.1658	0.631655	0.84116	1.5831	1.18883	86.179	42.938	129.117	0.94250	1.14876	−65
−64	0.1821	0.1762	0.632685	0.79490	1.5806	1.25801	86.385	42.843	129.228	0.94349	1.14831	−64
−63	0.1934	0.1872	0.633721	0.75166	1.5780	1.33038	86.592	42.748	129.339	0.94447	1.14787	−63
−62	0.2052	0.1986	0.634763	0.71122	1.5754	1.40604	86.799	42.652	129.451	0.94545	1.14743	−62
−61	0.2170	0.2106	0.635812	0.67336	1.5728	1.48508	87.006	42.556	129.562	0.94643	1.14701	−61
−60	0.2307	0.2233	0.636868	0.63791	1.5702	1.56762	87.213	42.461	129.673	0.94740	1.14659	−60
−59	0.2443	0.2365	0.637930	0.60468	1.5676	1.65376	87.420	42.365	129.785	0.94837	1.14618	−59
−58	0.2587	0.2503	0.638998	0.57351	1.5649	1.74363	87.627	42.269	129.896	0.94933	1.14578	−58
−57	0.2736	0.2648	0.640074	0.54427	1.5623	1.83732	87.835	42.173	130.007	0.95029	1.14539	−57
−56	0.2893	0.2800	0.641156	0.51680	1.5597	1.93496	88.042	42.076	130.119	0.95125	1.14500	−56
−55	0.3057	0.2959	0.642245	0.49100	1.5570	2.03666	88.250	41.980	130.230	0.95221	1.14463	−55
−54	0.3228	0.3124	0.643341	0.46673	1.5544	2.14254	88.458	41.883	130.342	0.95316	1.14426	−54
−53	0.3407	0.3297	0.644444	0.44390	1.5517	2.25271	88.666	41.786	130.453	0.95410	1.14390	−53
−52	0.3594	0.3478	0.645555	0.42242	1.5491	2.36731	88.875	41.689	130.564	0.95505	1.14354	−52
−51	0.3789	0.3667	0.646672	0.40218	1.5464	2.48645	89.083	41.592	130.675	0.95598	1.14320	−51
−50	0.3992	0.3864	0.647797	0.38310	1.5437	2.61025	89.292	41.495	130.787	0.95692	1.14286	−50
−49	0.4204	0.4069	0.648929	0.36511	1.5410	2.73885	89.501	41.397	130.898	0.95785	1.14252	−49
−48	0.4425	0.4282	0.650069	0.34814	1.5383	2.87238	89.710	41.299	131.009	0.95878	1.14220	−48
−47	0.4655	0.4505	0.651216	0.33212	1.5356	3.01095	89.919	41.201	131.120	0.95971	1.14188	−47
−46	0.4894	0.4737	0.652371	0.31698	1.5329	3.15471	90.128	41.103	131.231	0.96063	1.14157	−46

TEMP. °C	PRESSURE kg/cm²	PRESSURE ATM	VOLUME LIQUID l/kg v_f	VOLUME VAPOR m³/kg v_g	DENSITY LIQUID kg/l $1/v_f$	DENSITY VAPOR kg/m³ $1/v_g$	ENTHALPY kcal/kg LIQUID h_f	ENTHALPY kcal/kg LATENT h_{fg}	ENTHALPY kcal/kg VAPOR h_g	ENTROPY kcal/(kg) (°K) LIQUID s_f	ENTROPY kcal/(kg) (°K) VAPOR s_g	TEMP. °C
−45	0.5143	0.4978	0.653534	0.302682	1.5301	3.30379	90.338	41.004	131.342	0.96155	1.14126	−45
−44	0.5402	0.5229	0.654705	0.289157	1.5274	3.45833	90.548	40.905	131.453	0.96246	1.14096	−44
−43	0.5672	0.5489	0.655884	0.276361	1.5247	3.61845	90.758	40.806	131.564	0.96338	1.14067	−43
−42	0.5951	0.5760	0.657070	0.264249	1.5219	3.78430	90.968	40.707	131.675	0.96429	1.14038	−42
−41	0.6242	0.6041	0.658265	0.252779	1.5191	3.95603	91.178	40.607	131.785	0.96519	1.14010	−41
−40	0.6544	0.6333	0.659468	0.241910	1.5164	4.13377	91.389	40.507	131.896	0.96610	1.13982	−40
−39	0.6857	0.6637	0.660680	0.231607	1.5136	4.31766	91.600	40.407	132.006	0.96699	1.13955	−39
−38	0.7182	0.6951	0.661900	0.221835	1.5108	4.50786	91.811	40.306	132.117	0.96789	1.13929	−38
−37	0.7519	0.7277	0.663128	0.212562	1.5080	4.70451	92.022	40.205	132.227	0.96879	1.13903	−37
−36	0.7868	0.7615	0.664366	0.203759	1.5052	4.90777	92.233	40.104	132.337	0.96968	1.13877	−36
−35	0.8230	0.7965	0.665612	0.195398	1.5024	5.11777	92.445	40.002	132.447	0.97056	1.13852	−35
−34	0.8605	0.8328	0.666867	0.187453	1.4995	5.33468	92.656	39.901	132.557	0.97145	1.13828	−34
−33	0.8993	0.8703	0.668131	0.179900	1.4967	5.55866	92.868	39.798	132.667	0.97233	1.13804	−33
−32	0.9394	0.9092	0.669405	0.172716	1.4939	5.78985	93.081	39.696	132.776	0.97321	1.13781	−32
−31	0.9810	0.9494	0.670687	0.165881	1.4910	6.02842	93.293	39.593	132.886	0.97409	1.13758	−31
−30	1.0239	0.9910	0.671979	0.159375	1.4881	6.27453	93.506	39.490	132.995	0.97496	1.13736	−30
−29	1.0683	1.0340	0.673281	0.153178	1.4853	6.52834	93.718	39.386	133.104	0.97583	1.13714	−29
−28	1.1142	1.0784	0.674592	0.147275	1.4824	6.79001	93.931	39.282	133.213	0.97670	1.13692	−28
−27	1.1617	1.1243	0.675913	0.141649	1.4795	7.05973	94.145	39.178	133.322	0.97756	1.13671	−27
−26	1.2107	1.1717	0.677244	0.136284	1.4766	7.33764	94.358	39.073	133.431	0.97842	1.13651	−26
−25	1.2612	1.2207	0.678586	0.131166	1.4737	7.62394	94.572	38.967	133.539	0.97928	1.13631	−25
−24	1.3134	1.2712	0.679937	0.126282	1.4707	7.91878	94.786	38.862	133.648	0.98014	1.13611	−24
−23	1.3673	1.3233	0.681299	0.121620	1.4678	8.22234	95.000	38.756	133.756	0.98100	1.13592	−23
−22	1.4228	1.3770	0.682671	0.117167	1.4648	8.53481	95.215	38.649	133.864	0.98185	1.13573	−22
−21	1.4801	1.4325	0.684054	0.112913	1.4619	8.85636	95.429	38.542	133.972	0.98270	1.13554	−21
−20	1.5391	1.4896	0.685448	0.108847	1.4589	9.18718	95.644	38.435	134.079	0.98354	1.13536	−20
−19	1.5999	1.5485	0.686853	0.104960	1.4559	9.52745	95.859	38.327	134.187	0.98439	1.13518	−19
−18	1.6626	1.6091	0.688269	0.101242	1.4529	9.87735	96.075	38.219	134.294	0.98523	1.13501	−18
−17	1.7271	1.6716	0.689697	0.097684	1.4499	10.23709	96.290	38.110	134.401	0.98607	1.13484	−17
−16	1.7936	1.7359	0.691136	0.094279	1.4469	10.60684	96.506	38.001	134.507	0.98691	1.13468	−16
−15	1.8620	1.8021	0.692586	0.091018	1.4439	10.98681	96.723	37.891	134.614	0.98774	1.13451	−15
−14	1.9323	1.8702	0.694049	0.087895	1.4408	11.37720	96.939	37.781	134.720	0.98857	1.13435	−14
−13	2.0047	1.9402	0.695523	0.084903	1.4378	11.77820	97.156	37.670	134.826	0.98940	1.13420	−13
−12	2.0792	2.0123	0.697010	0.082034	1.4347	12.19002	97.373	37.559	134.932	0.99023	1.13405	−12
−11	2.1557	2.0864	0.698509	0.079284	1.4316	12.61286	97.590	37.447	135.037	0.99106	1.13390	−11
−10	2.2344	2.1625	0.700021	0.076646	1.4285	13.04694	97.808	37.335	135.143	0.99188	1.13375	−10
−9	2.3152	2.2408	0.701545	0.074115	1.4254	13.49246	98.025	37.222	135.248	0.99270	1.13361	−9
−8	2.3983	2.3211	0.703083	0.071686	1.4223	13.94965	98.244	37.109	135.352	0.99352	1.13347	−8
−7	2.4836	2.4037	0.704634	0.069354	1.4192	14.41872	98.462	36.995	135.457	0.99434	1.13333	−7
−6	2.5712	2.4885	0.706198	0.067115	1.4160	14.89989	98.681	36.880	135.561	0.99515	1.13320	−6
−5	2.6611	2.5755	0.707776	0.064963	1.4129	15.39340	98.900	36.765	135.665	0.99597	1.13306	−5
−4	2.7534	2.6648	0.709368	0.062895	1.4097	15.89947	99.119	36.649	135.769	0.99678	1.13294	−4
−3	2.8480	2.7564	0.710974	0.060908	1.4065	16.41832	99.339	36.533	135.872	0.99759	1.13281	−3
−2	2.9452	2.8505	0.712594	0.058996	1.4033	16.95022	99.559	36.416	135.975	0.99839	1.13269	−2
−1	3.0448	2.9469	0.714229	0.057158	1.4001	17.49538	99.779	36.298	136.078	0.99920	1.13257	−1
0	3.1469	3.0457	0.715878	0.055389	1.3969	18.05406	100.000	36.180	136.180	1.00000	1.13245	0
1	3.2517	3.1471	0.717543	0.053687	1.3936	18.62652	100.221	36.061	136.282	1.00080	1.13233	1
2	3.3590	3.2510	0.719223	0.052048	1.3904	19.21299	100.442	35.942	136.384	1.00160	1.13222	2
3	3.4690	3.3574	0.720919	0.050470	1.3871	19.81375	100.664	35.821	136.485	1.00240	1.13211	3
4	3.5816	3.4664	0.722631	0.048950	1.3838	20.42905	100.886	35.700	136.586	1.00319	1.13200	4
5	3.6970	3.5781	0.724359	0.047485	1.3805	21.05916	101.108	35.579	136.687	1.00399	1.13189	5
6	3.8152	3.6925	0.726103	0.046074	1.3772	21.70436	101.331	35.456	136.787	1.00478	1.13179	6
7	3.9362	3.8096	0.727864	0.044713	1.3739	22.36493	101.554	35.333	136.887	1.00557	1.13169	7
8	4.0600	3.9294	0.729643	0.043401	1.3705	23.04114	101.778	35.209	136.987	1.00636	1.13159	8
9	4.1868	4.0521	0.731438	0.042135	1.3672	23.73330	102.002	35.084	137.086	1.00715	1.13149	9

TEMP. °C	PRESSURE kg/cm²	PRESSURE ATM	VOLUME LIQUID l/kg v_f	VOLUME VAPOR m³/kg v_g	DENSITY LIQUID kg/l $1/v_f$	DENSITY VAPOR kg/m³ $1/v_g$	ENTHALPY kcal/kg LIQUID h_f	ENTHALPY kcal/kg LATENT h_{fg}	ENTHALPY kcal/kg VAPOR h_g	ENTROPY kcal/(kg) (°K) LIQUID s_f	ENTROPY kcal/(kg) (°K) VAPOR s_g	TEMP °C
10	4.3164	4.1776	0.733252	0.040914	1.3638	24.44169	102.226	34.959	137.185	1.00793	1.13139	10
11	4.4491	4.3060	0.735083	0.039735	1.3604	25.16662	102.451	34.832	137.284	1.00872	1.13130	11
12	4.5848	4.4374	0.736933	0.038598	1.3570	25.90839	102.677	34.705	137.382	1.00950	1.13120	12
13	4.7236	4.5717	0.738802	0.037499	1.3535	26.66734	102.902	34.577	137.479	1.01028	1.13111	13
14	4.8655	4.7090	0.740690	0.036438	1.3501	27.44375	103.128	34.448	137.577	1.01106	1.13102	14
15	5.0106	4.8494	0.742597	0.035413	1.3466	28.23799	103.355	34.318	137.673	1.01184	1.13094	15
16	5.1588	4.9929	0.744524	0.034423	1.3431	29.05037	103.582	34.188	137.770	1.01262	1.13085	16
17	5.3103	5.1395	0.746472	0.033466	1.3396	29.88124	103.810	34.056	137.866	1.01340	1.13076	17
18	5.4651	5.2894	0.748440	0.032540	1.3361	30.73097	104.038	33.924	137.961	1.01417	1.13068	18
19	5.6232	5.4424	0.750429	0.031646	1.3326	31.59989	104.266	33.790	138.056	1.01495	1.13060	19
20	5.7848	5.5987	0.752440	0.030780	1.3290	32.48840	104.495	33.656	138.151	1.01572	1.13052	20
21	5.9497	5.7584	0.754473	0.029943	1.3254	33.39685	104.725	33.520	138.245	1.01649	1.13044	21
22	6.1181	5.9214	0.756528	0.029133	1.3218	34.32565	104.955	33.383	138.338	1.01726	1.13036	22
23	6.2900	6.0878	0.758606	0.028349	1.3182	35.27519	105.186	33.246	138.431	1.01803	1.13029	23
24	6.4655	6.2576	0.760708	0.027589	1.3146	36.24587	105.417	33.107	138.524	1.01880	1.13021	24
25	6.6446	6.4309	0.762834	0.026854	1.3109	37.23812	105.649	32.967	138.616	1.01957	1.13014	25
26	6.8274	6.6078	0.764984	0.026142	1.3072	38.25237	105.881	32.826	138.707	1.02034	1.13006	26
27	7.0138	6.7883	0.767159	0.025452	1.3035	39.28905	106.114	32.684	138.798	1.02110	1.12999	27
28	7.2040	6.9724	0.769360	0.024784	1.2998	40.34864	106.347	32.541	138.889	1.02187	1.12992	28
29	7.3980	7.1601	0.771587	0.024136	1.2960	41.43158	106.582	32.397	138.978	1.02263	1.12985	29
30	7.5959	7.3516	0.773841	0.023508	1.2923	42.53836	106.816	32.251	139.067	1.02340	1.12978	30
31	7.7976	7.5468	0.776122	0.022899	1.2885	43.66946	107.052	32.104	139.156	1.02416	1.12971	31
32	8.0032	7.7459	0.778431	0.022309	1.2846	44.82543	107.288	31.956	139.244	1.02492	1.12964	32
33	8.2129	7.9488	0.780769	0.021736	1.2808	46.00676	107.525	31.806	139.331	1.02568	1.12957	33
34	8.4266	8.1556	0.783137	0.021180	1.2769	47.21398	107.763	31.655	139.418	1.02645	1.12950	34
35	8.6443	8.3663	0.785534	0.020641	1.2730	48.44770	108.001	31.503	139.504	1.02721	1.12943	35
36	8.8662	8.5811	0.787963	0.020117	1.2691	49.70843	108.240	31.349	139.589	1.02797	1.12937	36
37	9.0922	8.7999	0.790423	0.019609	1.2651	50.99681	108.480	31.194	139.673	1.02873	1.12930	37
38	9.3225	9.0227	0.792916	0.019116	1.2612	52.31345	108.720	31.037	139.757	1.02949	1.12923	38
39	9.5571	9.2497	0.795442	0.018636	1.2572	53.65895	108.962	30.879	139.840	1.03025	1.12917	39
40	9.7960	9.4809	0.798002	0.018171	1.2531	55.03400	109.204	30.719	139.922	1.03101	1.12910	40
41	10.0392	9.7164	0.800597	0.017718	1.2491	56.43926	109.447	30.557	140.004	1.03177	1.12904	41
42	10.2869	9.9561	0.803228	0.017278	1.2450	57.87545	109.691	30.394	140.085	1.03253	1.12897	42
43	10.5390	10.2001	0.805897	0.016851	1.2409	59.34327	109.935	30.229	140.165	1.03329	1.12890	43
44	10.7957	10.4485	0.808603	0.016436	1.2367	60.84349	110.181	30.062	140.244	1.03405	1.12884	44
45	11.0669	10.7014	0.811349	0.016032	1.2325	62.37690	110.428	29.894	140.322	1.03481	1.12877	45
46	11.3228	10.9587	0.814134	0.015639	1.2283	63.94429	110.675	29.724	140.399	1.03557	1.12870	46
47	11.5934	11.2205	0.816961	0.015256	1.2240	65.54653	110.924	29.551	140.475	1.03634	1.12864	47
48	11.8686	11.4870	0.819831	0.014884	1.2198	67.18448	111.173	29.377	140.551	1.03710	1.12857	48
49	12.1487	11.7580	0.822745	0.014522	1.2154	68.85903	111.424	29.201	140.625	1.03786	1.12850	49
50	12.4336	12.0337	0.825703	0.014170	1.2111	70.57118	111.676	29.023	140.699	1.03862	1.12843	50
51	12.7234	12.3142	0.828709	0.013827	1.2067	72.32186	111.928	28.843	140.771	1.03939	1.12836	51
52	13.0181	12.5994	0.831762	0.013493	1.2023	74.11213	112.182	28.660	140.842	1.04015	1.12829	52
53	13.3178	12.8895	0.834805	0.013108	1.1978	75.94304	112.437	28.470	140.913	1.04091	1.12822	53
54	13.6225	13.1844	0.838019	0.012851	1.1933	77.81573	112.693	28.288	140.982	1.04168	1.12815	54
55	13.9324	13.4843	0.841226	0.012542	1.1887	79.73139	112.951	28.099	141.050	1.04245	1.12807	55
56	14.2474	13.7892	0.844488	0.012241	1.1841	81.69118	113.209	27.907	141.116	1.04322	1.12800	56
57	14.5676	14.0991	0.847806	0.011948	1.1795	83.69642	113.469	27.713	141.182	1.04398	1.12792	57
58	14.8931	14.4141	0.851183	0.011662	1.1748	85.74845	113.730	27.516	141.246	1.04475	1.12784	58
59	15.2238	14.7343	0.854620	0.011383	1.1701	87.84864	113.993	27.316	141.309	1.04553	1.12776	59
60	15.5600	15.0596	0.858119	0.011111	1.1653	89.99852	114.257	27.114	141.371	1.04630	1.12768	60
61	15.9016	15.3902	0.861684	0.010846	1.1605	92.19957	114.522	26.909	141.431	1.04707	1.12760	61
62	16.2487	15.7261	0.865316	0.010587	1.1556	94.45348	114.789	26.701	141.490	1.04785	1.12751	62
63	16.6013	16.0674	0.869017	0.010335	1.1507	96.76192	115.057	26.490	141.547	1.04863	1.12743	63
64	16.9595	16.4141	0.872792	0.010088	1.1457	99.12670	115.327	26.275	141.602	1.04941	1.12734	64

TEMP. °C	PRESSURE kg/cm²	PRESSURE ATM	VOLUME LIQUID l/kg v_f	VOLUME VAPOR m³/kg v_g	DENSITY LIQUID kg/l $1/v_f$	DENSITY VAPOR kg/m³ $1/v_g$	ENTHALPY kcal/kg LIQUID h_f	ENTHALPY kcal/kg LATENT h_{fg}	ENTHALPY kcal/kg VAPOR h_g	ENTROPY kcal/(kg) (°K) LIQUID s_f	ENTROPY kcal/(kg) (°K) VAPOR s_g	TEMP. °C
65	17.3234	16.7663	0.876641	0.009847	1.1407	101.54975	115.599	26.058	141.656	1.05019	1.12725	65
66	17.6930	17.1240	0.880570	0.009612	1.1356	104.03300	115.872	25.837	141.709	1.05098	1.12715	66
67	18.0683	17.4873	0.884580	0.009383	1.1305	106.57865	116.147	25.613	141.759	1.05176	1.12706	67
68	18.4495	17.8562	0.888675	0.009158	1.1253	109.18893	116.423	25.385	141.808	1.05255	1.12696	68
69	18.8366	18.2308	0.892859	0.008939	1.1200	111.86621	116.702	25.153	141.855	1.05334	1.12685	69
70	19.2296	18.6112	0.897136	0.008725	1.1147	114.61300	116.982	24.918	141.900	1.05414	1.12675	70
71	19.6286	18.9974	0.901509	0.008516	1.1093	117.43201	117.265	24.678	141.943	1.05493	1.12664	71
72	20.0337	19.3895	0.905984	0.008311	1.1038	120.32611	117.549	24.435	141.984	1.05573	1.12653	72
73	20.4450	19.7875	0.910566	0.008110	1.0982	123.29834	117.835	24.187	142.023	1.05654	1.12641	73
74	20.8624	20.1915	0.915258	0.007914	1.0926	126.35197	118.124	23.935	142.059	1.05734	1.12629	74
75	21.2861	20.6016	0.920068	0.007723	1.0869	129.49045	118.415	23.678	142.093	1.05815	1.12616	75
76	21.7161	21.0177	0.925000	0.007535	1.0811	132.71749	118.708	23.416	142.124	1.05897	1.12603	76
77	22.1524	21.4400	0.930061	0.007351	1.0752	136.03722	119.004	23.150	142.153	1.05979	1.12590	77
78	22.5952	21.8686	0.935259	0.007171	1.0692	139.45386	119.302	22.878	142.179	1.06061	1.12576	78
79	23.0445	22.3035	0.940600	0.006994	1.0632	142.97208	119.602	22.600	142.203	1.06144	1.12561	79
80	23.5004	22.7447	0.946094	0.006821	1.0570	146.59698	119.906	22.317	142.223	1.06227	1.12546	80
81	23.9629	23.1923	0.951748	0.006652	1.0507	150.33395	120.212	22.028	142.240	1.06310	1.12530	81
82	24.4321	23.6464	0.957573	0.006486	1.0443	154.18889	120.521	21.733	142.254	1.06395	1.12514	82
83	24.9081	24.1071	0.963580	0.006322	1.0378	158.16836	120.833	21.430	142.264	1.06479	1.12496	83
84	25.3909	24.5743	0.969779	0.006162	1.0312	162.27935	121.149	21.121	142.270	1.06565	1.12478	84
85	25.8806	25.0483	0.976185	0.006005	1.0244	166.52958	121.467	20.805	142.272	1.06651	1.12459	85
86	26.3772	25.5290	0.982812	0.005850	1.0175	170.92758	121.790	20.481	142.271	1.06737	1.12440	86
87	26.8809	26.0184	0.989674	0.005699	1.0104	175.48283	122.115	20.149	142.264	1.06825	1.12419	87
88	27.3916	26.5108	0.996791	0.005549	1.0032	180.20558	122.445	19.808	142.253	1.06913	1.12397	88
89	27.9096	27.0121	1.004183	0.005402	0.9958	185.10737	122.779	19.458	142.237	1.07002	1.12375	89
90	28.4347	27.5203	1.011870	0.005258	0.9883	190.20135	123.117	19.098	142.216	1.07092	1.12351	90
91	28.9672	28.0357	1.019880	0.005115	0.9805	195.50189	123.460	18.728	142.188	1.07182	1.12325	91
92	29.5071	28.5582	1.028240	0.004974	0.9725	201.02542	123.808	18.347	142.155	1.07274	1.12299	92
93	30.0544	29.0879	1.036983	0.004836	0.9643	206.79062	124.161	17.954	142.114	1.07367	1.12270	93
94	30.6092	29.6248	1.046148	0.004699	0.9559	212.81882	124.519	17.547	142.066	1.07461	1.12240	94
95	31.1716	30.1691	1.055778	0.004563	0.9472	219.13461	124.883	17.127	142.010	1.07556	1.12208	95
96	31.7416	30.7208	1.065924	0.004429	0.9382	225.76660	125.254	16.691	141.945	1.07653	1.12175	96
97	32.3194	31.2801	1.076648	0.004296	0.9288	232.74793	125.632	16.239	141.871	1.07751	1.12138	97
98	32.9050	31.8468	1.088021	0.004165	0.9191	240.11854	126.017	15.768	141.785	1.07851	1.12100	98
99	33.4985	32.4212	1.100131	0.004033	0.9090	247.92562	126.410	15.277	141.688	1.07953	1.12058	99
100	34.0999	33.0033	1.113083	0.003903	0.8984	256.22688	126.813	14.763	141.576	1.08057	1.12013	100
101	34.7094	33.5932	1.127010	0.003772	0.8873	265.09314	127.226	14.223	141.449	1.08163	1.11965	101
102	35.3270	34.1909	1.142078	0.003641	0.8756	274.61372	127.651	13.654	141.304	1.08272	1.11912	102
103	35.9527	34.7966	1.158504	0.003510	0.8632	284.90224	128.088	13.050	141.139	1.08384	1.11853	103
104	36.5868	35.4102	1.176572	0.003377	0.8499	296.10964	128.541	12.407	140.948	1.08500	1.11789	104
105	37.2292	36.0319	1.196671	0.003242	0.8357	308.43978	129.012	11.715	140.727	1.08620	1.11718	105
106	37.8800	36.6618	1.219356	0.003104	0.8201	322.18196	129.505	10.963	140.468	1.08745	1.11636	106
107	38.5393	37.2999	1.245456	0.002961	0.8029	337.76783	130.026	10.135	140.161	1.08877	1.11543	107
108	39.2072	37.9463	1.276302	0.002810	0.7835	355.89504	130.583	9.203	139.787	1.09019	1.11433	108
109	39.8837	38.6011	1.314261	0.002647	0.7609	377.80918	131.193	8.122	139.315	1.09173	1.11298	109
110	40.5690	39.2644	1.364270	0.002462	0.7330	406.20144	131.885	6.794	138.679	1.09348	1.11121	110
111	41.2632	39.9362	1.440385	0.002223	0.6943	449.87416	132.746	4.918	137.664	1.09566	1.10847	111
112.0	41.9662	40.6167	1.791785	0.001792	0.5581	558.08698	135.205		135.205	1.10199	1.10199	112.0

From Published data of E.I. du Pont de Nemours & Co. Inc. Used by Permission.

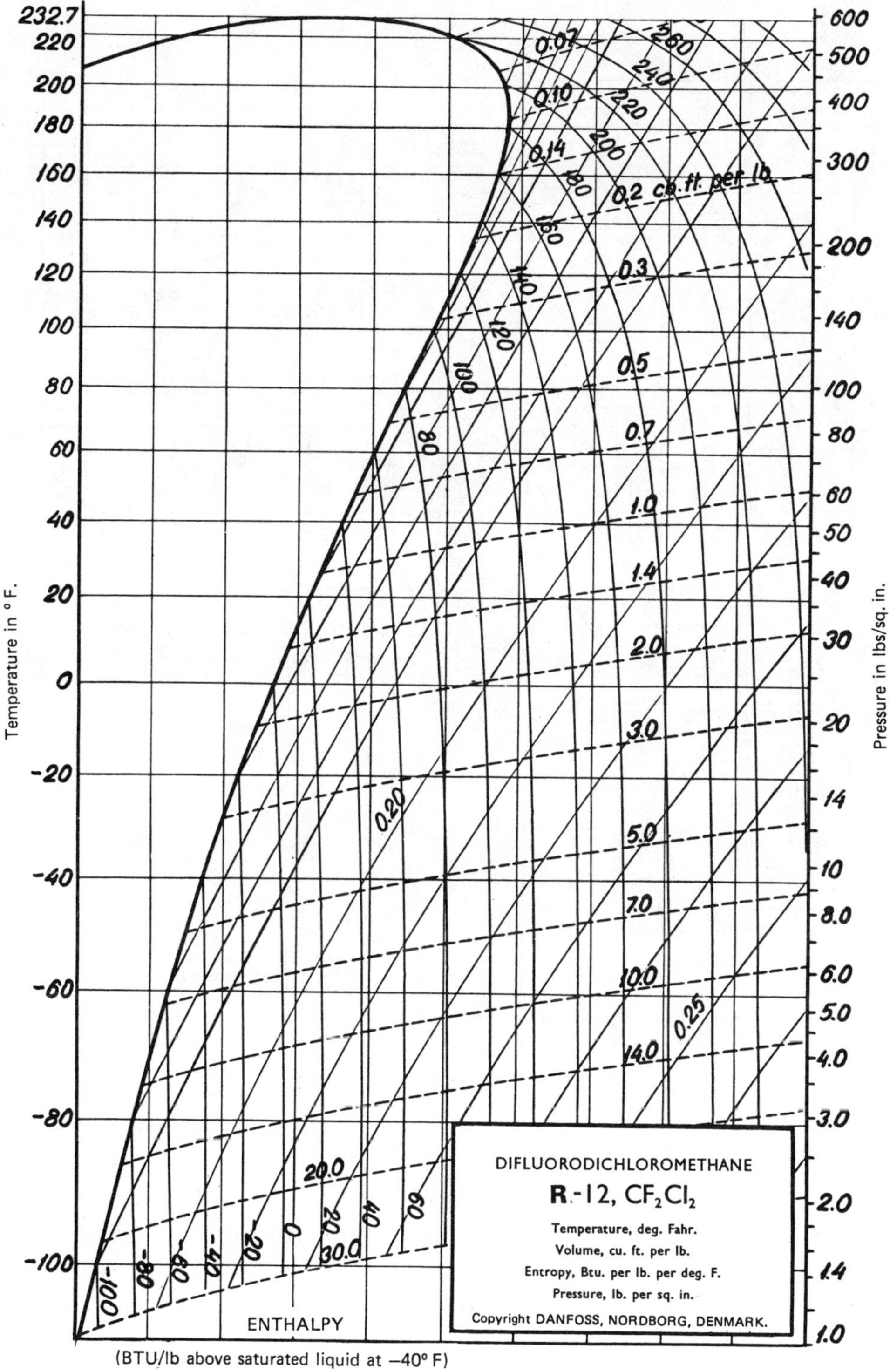

PRESSURE ENTHALPY DIAGRAM FOR R 12 Imperial

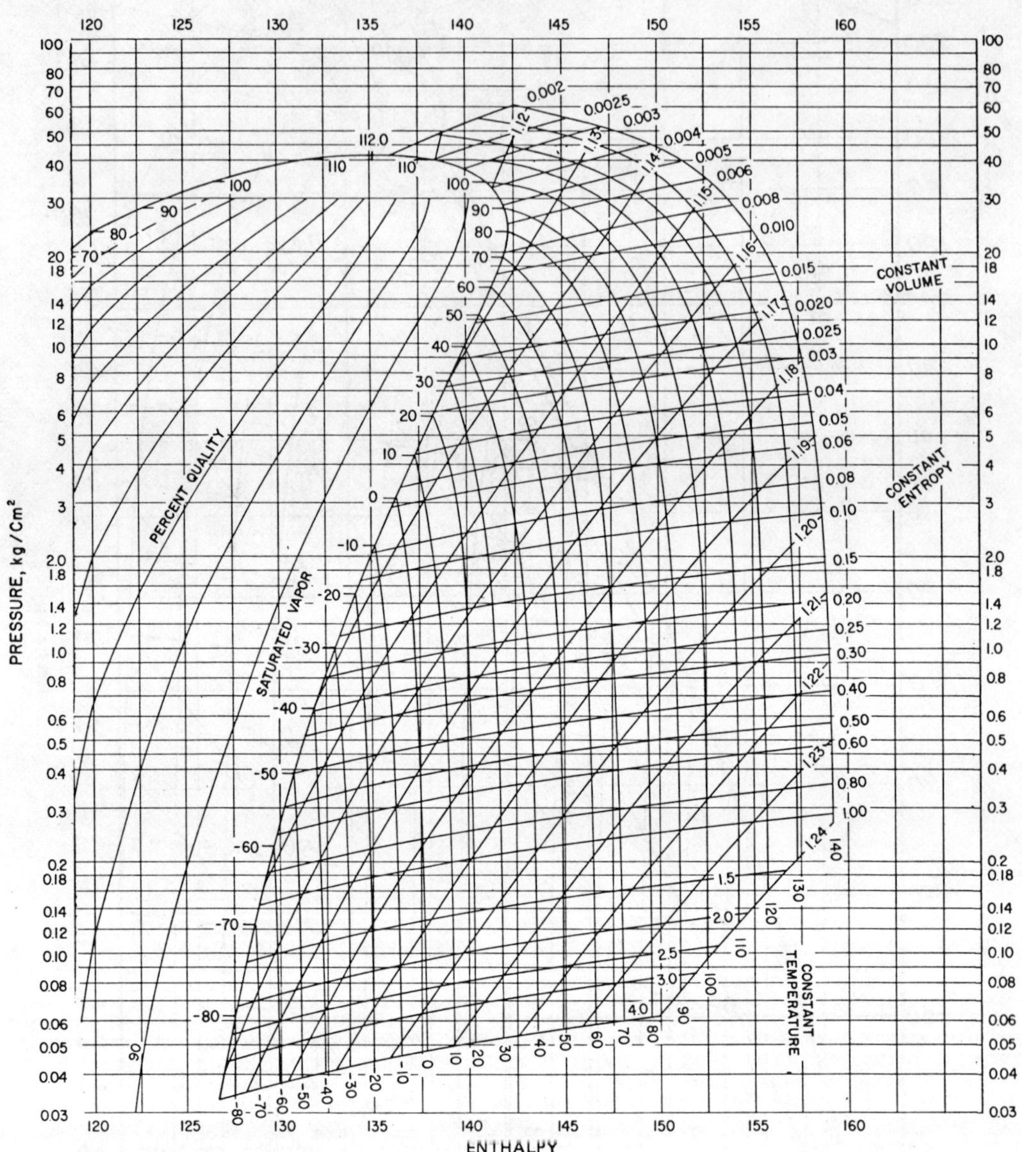

From Published data of E.I. du Pont de Nemours & Co. Inc. Used by Permission.

PRESSURE ENTHALPY DIAGRAM FOR R 12 Metric

DATA SHEET FOR R 13

CF_3Cl	TRIFLUOROMONOCHLOROMETHANE	Molecular Wt. 104.5

Pressure		Temperature				Volume		Density	
39.4	kg/cm²	28.9	°C	83.9	°F	1.72	l/kg	5.79	kg/l.
561	psia	302	°K	544	°R	.0277	ft³/lb	36.1	lbs/ft³
								3.60	Air=1

AT CRITICAL POINT

–40°

–90° C –130° F

Discharge Pressure	
5.98	kg/cm²
88	psia

Inlet Pressure	
.616	kg/cm²
9.05	psia

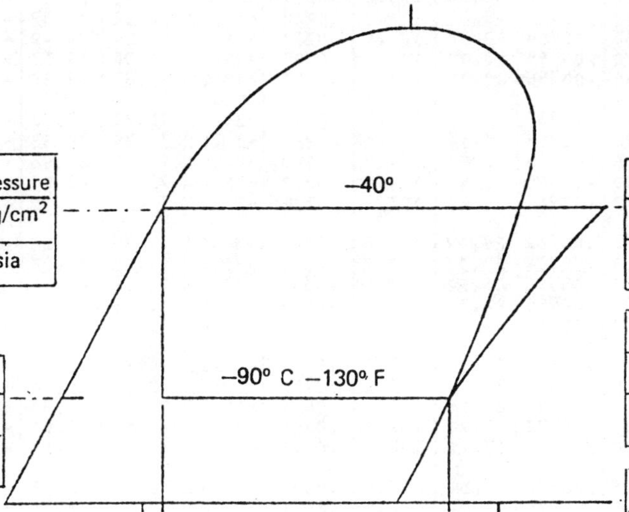

Discharge Temperature			
-19	°C	-2	°F
254	°K	458	°R

Normal Boiling Point			
-81.4	°C	-114.6°	F
192	°K	345	°R

Triple Point			
-181	°C	-294	°F
192	°K	166	°R

Refrigerating Effect	
112.7 kcal/m³	12.6 Btu/ft³

Latent Heat at NBP	35.4 kcal/kg	3700 /kg mol	63.8 Btu/lb	6670 /lb mol

Trouton's No. 19.33	Gas Constant 8.11 kg.m/kg/° K	14.8 ft.lbs/lb/° R

Specific Heat Liquid at 30° C 86° F Fig 14	Gas C_p .158	C_p/C_v 1.145at 25°C

Liquid at 30° C 86° F Density 1.34 kg/l.	84.1 lbs/ft³ Viscosity cp

Reduced Form at NBP Pressure .0262	1 / Temperature 1.505

Reduced value of 1/T	2.05	1.72	1.37
when Reduced Pressure is	a) .001	b) .01	c) .1

TEMP.	PRESSURE		VOLUME cu ft/lb		DENSITY lb/cu ft		ENTHALPY Btu/lb			ENTROPY Btu/(lb)(°R)	
°F	PSIA	PSIG	LIQUID v_f	VAPOR v_g	LIQUID $1/v_f$	VAPOR $1/v_g$	LIQUID h_f	LATENT h_{fg}	VAPOR h_g	LIQUID s_f	VAPOR s_g
-200	.433	-14.26	.00946	61.39	105.6	.0162	-33.83	72.63	38.80	-.0998	.1799
-199	.458	-14.23	.00947	58.23	105.5	.0171	-33.64	72.54	38.89	-.0991	.1792
-198	.484	-14.21	.00948	55.25	105.4	.0181	-33.46	72.46	38.99	-.0984	.1785
-197	.512	-14.18	.00949	52.46	105.2	.0190	-33.28	72.37	39.09	-.0977	.1778
-196	.541	-14.15	.00950	49.82	105.1	.0200	-33.10	72.29	39.19	-.0970	.1771
-195	.572	-14.12	.00951	47.34	105.0	.0211	-32.91	72.21	39.29	-.0963	.1765
-194	.603	-14.09	.00953	45.00	104.9	.0222	-32.73	72.12	39.39	-.0956	.1758
-193	.637	-14.05	.00954	42.80	104.8	.0233	-32.55	72.04	39.49	-.0949	.1752
-192	.672	-14.02	.00955	40.73	104.6	.0245	-32.36	71.95	39.58	-.0942	.1745
-191	.708	-13.98	.00956	38.77	104.5	.0257	-32.18	71.87	39.68	-.0935	.1739
-190	.746	-13.94	.00957	36.92	104.4	.0270	-31.99	71.78	39.78	-.0928	.1733
-189	.786	-13.90	.00958	35.17	104.3	.0284	-31.81	71.70	39.88	-.0921	.1727
-188	.828	-13.86	.00959	33.52	104.2	.0298	-31.62	71.61	39.98	-.0915	.1720
-187	.871	-13.82	.00960	31.96	104.0	.0312	-31.44	71.52	40.08	-.0908	.1714
-186	.916	-13.77	.00961	30.49	103.9	.0328	-31.25	71.44	40.18	-.0901	.1709
-185	.964	-13.73	.00962	29.09	103.8	.0343	-31.07	71.35	40.28	-.0894	.1703
-184	1.013	-13.68	.00964	27.77	103.7	.0360	-30.88	71.26	40.38	-.0888	.1697
-183	1.064	-13.63	.00965	26.52	103.6	.0377	-30.69	71.18	40.48	-.0881	.1691
-182	1.118	-13.57	.00966	25.34	103.4	.0394	-30.51	71.09	40.58	-.0874	.1685
-181	1.174	-13.52	.00967	24.21	103.3	.0412	-30.32	71.00	40.68	-.0867	.1680
-180	1.232	-13.46	.00968	23.15	103.2	.0431	-30.13	70.92	40.78	-.0861	.1674
-179	1.292	-13.40	.00969	22.14	103.1	.0451	-29.94	70.83	40.88	-.0854	.1669
-178	1.355	-13.34	.00970	21.18	103.0	.0471	-29.76	70.74	40.98	-.0847	.1664
-177	1.420	-13.27	.00972	20.28	102.8	.0493	-29.57	70.65	41.08	-.0841	.1658
-176	1.488	-13.20	.00973	19.41	102.7	.0515	-29.38	70.56	41.18	-.0834	.1653
-175	1.559	-13.13	.00974	18.59	102.6	.0537	-29.19	70.48	41.28	-.0827	.1648
-174	1.632	-13.06	.00975	17.81	102.5	.0561	-29.00	70.39	41.38	-.0821	.1643
-173	1.708	-12.98	.00976	17.07	102.3	.0585	-28.81	70.30	41.48	-.0814	.1637
-172	1.788	-12.90	.00977	16.37	102.2	.0610	-28.62	70.21	41.58	-.0807	.1632
-171	1.870	-12.82	.00978	15.70	102.1	.0636	-28.43	70.12	41.68	-.0801	.1627
-170	1.955	-12.74	.00980	15.06	102.0	.0663	-28.24	70.03	41.78	-.0794	.1623
-169	2.043	-12.65	.00981	14.46	101.9	.0691	-28.05	69.94	41.88	-.0788	.1618
-168	2.135	-12.56	.00982	13.88	101.7	.0720	-27.86	69.85	41.98	-.0781	.1613
-167	2.230	-12.46	.00983	13.33	101.6	.0749	-27.67	69.75	42.08	-.0774	.1608
-166	2.328	-12.36	.00984	12.80	101.5	.0780	-27.47	69.66	42.19	-.0768	.1603
-165	2.430	-12.26	.00986	12.30	101.4	.0812	-27.28	69.57	42.29	-.0761	.1599
-164	2.536	-12.15	.00987	11.83	101.2	.0845	-27.09	69.48	42.39	-.0755	.1594
-163	2.645	-12.05	.00988	11.37	101.1	.0878	-26.89	69.39	42.49	-.0748	.1590
-162	2.758	-11.93	.00989	10.94	101.0	.0913	-26.70	69.29	42.59	-.0742	.1585
-161	2.875	-11.82	.00990	10.52	100.9	.0949	-26.51	69.20	42.69	-.0735	.1581
-160	2.996	-11.69	.00992	10.13	100.8	.0986	-26.31	69.11	42.79	-.0729	.1576
-159	3.121	-11.57	.00993	9.75	100.6	.1025	-26.12	69.01	42.89	-.0722	.1572
-158	3.251	-11.44	.00994	9.39	100.5	.1064	-25.92	68.92	42.99	-.0716	.1568
-157	3.384	-11.31	.00995	9.04	100.4	.1105	-25.73	68.82	43.09	-.0709	.1564
-156	3.523	-11.17	.00997	8.71	100.3	.1146	-25.53	68.73	43.19	-.0703	.1559
-155	3.665	-11.03	.00998	8.40	100.1	.1190	-25.34	68.63	43.29	-.0697	.1555
-154	3.813	-10.88	.00999	8.10	100.0	.1234	-25.14	68.54	43.39	-.0690	.1551
-153	3.965	-10.73	.01000	7.81	99.9	.1280	-24.94	68.44	43.49	-.0684	.1547
-152	4.122	-10.57	.01002	7.53	99.8	.1327	-24.74	68.34	43.59	-.0677	.1543
-151	4.284	-10.41	.01003	7.26	99.6	.1375	-24.55	68.24	43.69	-.0671	.1539
-150	4.451	-10.24	.01004	7.01	99.5	.1425	-24.35	68.15	43.79	-.0664	.1535
-149	4.624	-10.07	.01005	6.77	99.4	.1476	-24.15	68.05	43.89	-.0658	.1532
-148	4.801	-9.89	.01007	6.53	99.3	.1529	-23.95	67.95	43.99	-.0652	.1528
-147	4.985	-9.71	.01008	6.31	99.1	.1583	-23.75	67.85	44.09	-.0645	.1524
-146	5.173	-9.52	.01009	6.10	99.0	.1638	-23.55	67.75	44.19	-.0639	.1520

TEMP. °F	PRESSURE PSIA	PRESSURE PSIG	VOLUME cu ft/lb LIQUID v_f	VOLUME cu ft/lb VAPOR v_g	DENSITY lb/cu ft LIQUID $1/v_f$	DENSITY lb/cu ft VAPOR $1/v_g$	ENTHALPY Btu/lb LIQUID h_f	ENTHALPY Btu/lb LATENT h_{fg}	ENTHALPY Btu/lb VAPOR h_g	ENTROPY Btu/(lb)(°R) LIQUID s_f	ENTROPY Btu/(lb)(°R) VAPOR s_g
-145	5.36	-9.32	.01010	5.895	98.92	.1696	-23.35	67.65	44.29	-.0633	.1517
-144	5.56	-9.12	.01012	5.698	98.79	.1754	-23.15	67.55	44.39	-.0626	.1513
-143	5.77	-8.92	.01013	5.509	98.66	.1815	-22.95	67.45	44.49	-.0620	.1509
-142	5.98	-8.70	.01014	5.327	98.54	.1877	-22.75	67.35	44.59	-.0614	.1506
-141	6.20	-8.49	.01016	5.152	98.41	.1940	-22.55	67.24	44.69	-.0607	.1502
-140	6.43	-8.26	.01017	4.984	98.28	.2006	-22.34	67.14	44.79	-.0601	.1499
-139	6.66	-8.03	.01018	4.823	98.16	.2073	-22.14	67.04	44.89	-.0595	.1495
-138	6.90	-7.79	.01020	4.668	98.03	.2141	-21.94	66.93	44.99	-.0588	.1492
-137	7.14	-7.55	.01021	4.519	97.90	.2212	-21.73	66.83	45.09	-.0582	.1488
-136	7.39	-7.29	.01022	4.376	97.77	.2284	-21.53	66.72	45.19	-.0576	.1485
-135	7.65	-7.04	.01024	4.239	97.64	.2359	-21.33	66.62	45.29	-.0569	.1482
-134	7.92	-6.77	.01025	4.106	97.51	.2435	-21.12	66.51	45.39	-.0563	.1479
-133	8.19	-6.50	.01026	3.979	97.39	.2513	-20.92	66.41	45.49	-.0557	.1475
-132	8.47	-6.22	.01028	3.856	97.26	.2593	-20.71	66.30	45.58	-.0550	.1472
-131	8.76	-5.93	.01029	3.738	97.13	.2675	-20.50	66.19	45.68	-.0544	.1469
-130	9.05	-5.63	.01030	3.624	97.00	.2759	-20.30	66.08	45.78	-.0538	.1466
-129	9.36	-5.33	.01032	3.514	96.87	.2845	-20.09	65.97	45.88	-.0532	.1463
-128	9.67	-5.02	.01033	3.409	96.74	.2933	-19.88	65.87	45.98	-.0525	.1460
-127	9.99	-4.70	.01035	3.307	96.61	.3023	-19.67	65.76	46.08	-.0519	.1457
-126	10.32	-4.37	.01036	3.209	96.48	.3115	-19.47	65.65	46.17	-.0513	.1454
-125	10.65	-4.03	.01037	3.114	96.35	.3210	-19.26	65.53	46.27	-.0507	.1451
-124	11.00	-3.69	.01039	3.023	96.22	.3307	-19.05	65.42	46.37	-.0501	.1448
-123	11.35	-3.33	.01040	2.935	96.08	.3406	-18.84	65.31	46.47	-.0494	.1445
-122	11.72	-2.97	.01042	2.851	95.95	.3507	-18.63	65.20	46.56	-.0488	.1442
-121	12.09	-2.60	.01043	2.769	95.82	.3611	-18.42	65.08	46.66	-.0482	.1439
-120	12.47	-2.22	.01045	2.690	95.69	.3717	-18.21	64.97	46.76	-.0476	.1436
-119	12.86	-1.83	.01046	2.613	95.56	.3825	-18.00	64.86	46.86	-.0470	.1433
-118	13.26	-1.43	.01047	2.540	95.43	.3936	-17.78	64.74	46.95	-.0463	.1431
-117	13.67	-1.02	.01049	2.469	95.29	.4050	-17.57	64.63	47.05	-.0457	.1428
-116	14.09	-.60	.01050	2.400	95.16	.4166	-17.36	64.51	47.14	-.0451	.1425
-115	14.52	-.17	.01052	2.334	95.03	.4284	-17.15	64.39	47.24	-.0445	.1423
-114	14.96	.26	.01053	2.269	94.89	.4405	-16.93	64.28	47.34	-.0439	.1420
-113	15.41	.71	.01055	2.207	94.76	.4529	-16.72	64.16	47.43	-.0433	.1417
-112	15.87	1.17	.01056	2.147	94.63	.4655	-16.51	64.04	47.53	-.0426	.1415
-111	16.34	1.64	.01058	2.089	94.49	.4784	-16.29	63.92	47.62	-.0420	.1412
-110	16.82	2.12	.01059	2.033	94.36	.4916	-16.08	63.80	47.72	-.0414	.1410
-109	17.31	2.62	.01061	1.979	94.22	.5051	-15.86	63.68	47.81	-.0408	.1407
-108	17.82	3.12	.01062	1.927	94.09	.5189	-15.64	63.56	47.91	-.0402	.1405
-107	18.33	3.64	.01064	1.876	93.95	.5329	-15.43	63.43	48.00	-.0396	.1402
-106	18.86	4.16	.01065	1.827	93.82	.5473	-15.21	63.31	48.10	-.0390	.1400
-105	19.40	4.70	.01067	1.779	93.68	.5619	-14.99	63.19	48.19	-.0384	.1397
-104	19.95	5.25	.01068	1.733	93.55	.5768	-14.78	63.06	48.28	-.0377	.1395
-103	20.51	5.81	.01070	1.688	93.41	.5921	-14.56	62.94	48.38	-.0371	.1393
-102	21.09	6.39	.01072	1.645	93.27	.6076	-14.34	62.82	48.47	-.0365	.1390
-101	21.67	6.98	.01073	1.603	93.14	.6235	-14.12	62.69	48.56	-.0359	.1388
-100	22.27	7.58	.01075	1.563	93.00	.6397	-13.90	62.56	48.66	-.0353	.1386
-99	22.88	8.19	.01076	1.523	92.86	.6562	-13.68	62.44	48.75	-.0347	.1383
-98	23.51	8.81	.01078	1.485	92.73	.6730	-13.46	62.31	48.84	-.0341	.1381
-97	24.15	9.45	.01080	1.448	92.59	.6902	-13.24	62.18	48.93	-.0335	.1379
-96	24.80	10.10	.01081	1.412	92.45	.7077	-13.02	62.05	49.03	-.0329	.1377
-95	25.46	10.76	.01083	1.378	92.31	.7256	-12.80	61.92	49.12	-.0323	.1374
-94	26.14	11.44	.01084	1.344	92.17	.7438	-12.58	61.79	49.21	-.0317	.1372
-93	26.83	12.13	.01086	1.311	92.03	.7623	-12.35	61.66	49.30	-.0311	.1370
-92	27.54	12.84	.01088	1.279	91.89	.7812	-12.13	61.53	49.39	-.0305	.1368
-91	28.26	13.56	.01089	1.249	91.75	.8005	-11.91	61.39	49.48	-.0299	.1366

TEMP. °F	PRESSURE PSIA	PRESSURE PSIG	VOLUME cu ft/lb LIQUID v_f	VOLUME cu ft/lb VAPOR v_g	DENSITY lb/cu ft LIQUID $1/v_f$	DENSITY lb/cu ft VAPOR $1/v_g$	ENTHALPY Btu/lb LIQUID h_f	ENTHALPY Btu/lb LATENT h_{fg}	ENTHALPY Btu/lb VAPOR h_g	ENTROPY Btu/(lb)(°R) LIQUID s_f	ENTROPY Btu/(lb)(°R) VAPOR s_g
-90	28.99	14.29	.01091	1.219	91.61	.8201	-11.68	61.26	49.57	-.0293	.1364
-89	29.74	15.04	.01093	1.190	91.47	.8401	-11.46	61.13	49.66	-.0287	.1362
-88	30.50	15.81	.01094	1.162	91.33	.8605	-11.24	60.99	49.75	-.0281	.1360
-87	31.28	16.58	.01096	1.134	91.19	.8813	-11.01	60.86	49.84	-.0275	.1357
-86	32.07	17.38	.01098	1.108	91.05	.9024	-10.79	60.72	49.93	-.0269	.1355
-85	32.88	18.18	.01100	1.082	90.91	.924	-10.56	60.58	50.02	-.0263	.1353
-84	33.70	19.01	.01101	1.057	90.77	.945	-10.34	60.44	50.11	-.0257	.1351
-83	34.54	19.85	.01103	1.032	90.62	.968	-10.11	60.31	50.19	-.0251	.1349
-82	35.40	20.70	.01105	1.009	90.48	.991	-9.88	60.17	50.28	-.0245	.1348
-81	36.27	21.57	.01106	.986	90.34	1.014	-9.65	60.03	50.37	-.0239	.1346
-80	37.15	22.46	.01108	.963	90.19	1.037	-9.43	59.89	50.46	-.0233	.1344
-79	38.06	23.36	.01110	.941	90.05	1.061	-9.20	59.75	50.54	-.0227	.1342
-78	38.98	24.28	.01112	.920	89.90	1.086	-8.97	59.60	50.63	-.0221	.1340
-77	39.91	25.22	.01114	.900	89.76	1.111	-8.74	59.46	50.72	-.0215	.1338
-76	40.87	26.17	.01115	.880	89.61	1.136	-8.51	59.32	50.80	-.0209	.1336
-75	41.84	27.14	.01117	.860	89.47	1.162	-8.28	59.17	50.89	-.0203	.1334
-74	42.82	28.13	.01119	.841	89.32	1.188	-8.05	59.03	50.97	-.0197	.1332
-73	43.83	29.13	.01121	.823	89.18	1.214	-7.82	58.88	51.06	-.0191	.1331
-72	44.85	30.16	.01123	.805	89.03	1.241	-7.59	58.74	51.14	-.0185	.1329
-71	45.89	31.20	.01125	.787	88.88	1.269	-7.36	58.59	51.22	-.0180	.1327
-70	46.95	32.25	.01126	.770	88.74	1.297	-7.13	58.44	51.31	-.0174	.1325
-69	48.03	33.33	.01128	.754	88.59	1.326	-6.90	58.29	51.39	-.0168	.1324
-68	49.12	34.43	.01130	.737	88.44	1.355	-6.66	58.14	51.47	-.0162	.1322
-67	50.24	35.54	.01132	.722	88.29	1.384	-6.43	57.99	51.56	-.0156	.1320
-66	51.37	36.67	.01134	.706	88.14	1.414	-6.20	57.84	51.64	-.0150	.1318
-65	52.52	37.82	.01136	.691	87.99	1.445	-5.96	57.69	51.72	-.0144	.1317
-64	53.69	38.99	.01138	.677	87.84	1.476	-5.73	57.54	51.80	-.0138	.1315
-63	54.88	40.18	.01140	.663	87.69	1.507	-5.50	57.38	51.88	-.0133	.1313
-62	56.09	41.39	.01142	.649	87.54	1.539	-5.26	57.23	51.96	-.0127	.1312
-61	57.32	42.62	.01144	.635	87.39	1.572	-5.03	57.08	52.04	-.0121	.1310
-60	58.57	43.87	.01146	.622	87.24	1.605	-4.79	56.92	52.12	-.0115	.1308
-59	59.83	45.14	.01148	.610	87.08	1.639	-4.55	56.76	52.20	-.0109	.1307
-58	61.12	46.43	.01150	.597	86.93	1.673	-4.32	56.61	52.28	-.0103	.1305
-57	62.43	47.74	.01152	.585	86.78	1.708	-4.08	56.45	52.36	-.0098	.1303
-56	63.76	49.07	.01154	.573	86.62	1.743	-3.84	56.29	52.44	-.0092	.1302
-55	65.12	50.42	.01156	.561	86.47	1.779	-3.61	56.13	52.52	-.0086	.1300
-54	66.49	51.79	.01158	.550	86.32	1.816	-3.37	55.97	52.59	-.0080	.1299
-53	67.88	53.19	.01160	.539	86.16	1.853	-3.13	55.81	52.67	-.0074	.1297
-52	69.30	54.60	.01162	.528	86.00	1.890	-2.89	55.64	52.75	-.0069	.1296
-51	70.73	56.04	.01164	.518	85.85	1.929	-2.65	55.48	52.82	-.0063	.1294
-50	72.19	57.50	.01166	.508	85.69	1.968	-2.41	55.32	52.90	-.0057	.1292
-49	73.68	58.98	.01169	.498	85.53	2.007	-2.17	55.15	52.97	-.0051	.1291
-48	75.18	60.48	.01171	.488	85.38	2.047	-1.93	54.99	53.05	-.0045	.1289
-47	76.70	62.01	.01173	.478	85.22	2.088	-1.69	54.82	53.12	-.0040	.1288
-46	78.25	63.56	.01175	.469	85.06	2.130	-1.45	54.65	53.20	-.0034	.1286
-45	79.82	65.13	.01177	.460	84.90	2.172	-1.21	54.48	53.27	-.0028	.1285
-44	81.42	66.72	.01180	.451	84.74	2.215	-.97	54.32	53.34	-.0022	.1283
-43	83.04	68.34	.01182	.442	84.58	2.258	-.73	54.15	53.42	-.0017	.1282
-42	84.68	69.98	.01184	.434	84.42	2.302	-.48	53.97	53.49	-.0011	.1280
-41	86.34	71.65	.01186	.426	84.25	2.347	-.24	53.80	53.56	-.0005	.1279
-40	88.03	73.34	.01189	.417	84.09	2.392	.000	53.63	53.63	.0000	.1278
-39	89.75	75.05	.01191	.409	83.93	2.439	.244	53.46	53.70	.0005	.1276
-38	91.48	76.79	.01193	.402	83.76	2.486	.488	53.28	53.77	.0011	.1275
-37	93.24	78.55	.01196	.394	83.60	2.533	.732	53.11	53.84	.0017	.1273
-36	95.03	80.34	.01198	.387	83.43	2.582	.978	52.93	53.91	.0022	.1272

TEMP.	PRESSURE		VOLUME cu ft/lb		DENSITY lb/cu ft		ENTHALPY Btu/lb			ENTROPY Btu/(lb)(°R)	
°F	PSIA	PSIG	LIQUID v_f	VAPOR v_g	LIQUID $1/v_f$	VAPOR $1/v_g$	LIQUID h_f	LATENT h_{fg}	VAPOR h_g	LIQUID s_f	VAPOR s_g
-35	96.8	82.1	.01200	.380	83.27	2.631	1.223	52.75	53.98	.0028	.1270
-34	98.6	83.9	.01203	.372	83.10	2.681	1.469	52.57	54.04	.0034	.1269
-33	100.5	85.8	.01205	.366	82.93	2.731	1.715	52.39	54.11	.0039	.1268
-32	102.4	87.7	.01208	.359	82.77	2.783	1.962	52.21	54.18	.0045	.1266
-31	104.3	89.6	.01210	.352	82.60	2.835	2.209	52.03	54.24	.0051	.1265
-30	106.2	91.5	.01213	.346	82.43	2.888	2.457	51.85	54.31	.0056	.1263
-29	108.2	93.5	.01215	.339	82.26	2.942	2.705	51.67	54.37	.0062	.1262
-28	110.2	95.5	.01218	.333	82.09	2.996	2.953	51.48	54.44	.0068	.1261
-27	112.2	97.5	.01220	.327	81.92	3.052	3.202	51.30	54.50	.0073	.1259
-26	114.3	99.6	.01223	.321	81.74	3.108	3.451	51.11	54.56	.0079	.1258
-25	116.3	101.6	.01225	.315	81.57	3.165	3.701	50.92	54.63	.0085	.1256
-24	118.4	103.7	.01228	.310	81.40	3.223	3.951	50.74	54.69	.0090	.1255
-23	120.6	105.9	.01231	.304	81.22	3.282	4.201	50.55	54.75	.0096	.1254
-22	122.7	108.0	.01233	.299	81.05	3.342	4.452	50.36	54.81	.0102	.1252
-21	124.9	110.2	.01236	.293	80.87	3.403	4.703	50.17	54.87	.0107	.1251
-20	127.1	112.4	.01239	.288	80.70	3.464	4.955	49.97	54.93	.0113	.1250
-19	129.3	114.6	.01241	.283	80.52	3.527	5.207	49.78	54.99	.0118	.1248
-18	131.6	116.9	.01244	.278	80.34	3.590	5.460	49.58	55.04	.0124	.1247
-17	133.9	119.2	.01247	.273	80.16	3.655	5.713	49.39	55.10	.0130	.1245
-16	136.2	121.5	.01250	.268	79.98	3.720	5.967	49.19	55.16	.0135	.1244
-15	138.6	123.9	.01253	.264	79.80	3.787	6.221	48.99	55.21	.0141	.1243
-14	141.0	126.3	.01256	.259	79.62	3.854	6.475	48.79	55.27	.0146	.1241
-13	143.4	128.7	.01258	.254	79.43	3.923	6.730	48.59	55.32	.0152	.1240
-12	145.8	131.1	.01261	.250	79.25	3.992	6.986	48.39	55.38	.0158	.1239
-11	148.3	133.6	.01264	.246	79.06	4.063	7.242	48.19	55.43	.0163	.1237
-10	150.8	136.1	.01267	.241	78.88	4.134	7.498	47.98	55.48	.0169	.1236
-9	153.3	138.6	.01270	.237	78.69	4.207	7.755	47.78	55.53	.0174	.1235
-8	155.9	141.2	.01273	.233	78.50	4.281	8.013	47.57	55.58	.0180	.1233
-7	158.5	143.8	.01276	.229	78.31	4.356	8.271	47.36	55.63	.0186	.1232
-6	161.1	146.4	.01280	.225	78.12	4.432	8.529	47.15	55.68	.0191	.1231
-5	163.7	149.1	.01283	.221	77.93	4.509	8.788	46.94	55.73	.0197	.1229
-4	166.4	151.7	.01286	.217	77.74	4.588	9.048	46.73	55.78	.0202	.1228
-3	169.2	154.5	.01289	.214	77.55	4.667	9.308	46.52	55.83	.0208	.1227
-2	171.9	157.2	.01292	.210	77.35	4.748	9.568	46.30	55.87	.0213	.1225
-1	174.7	160.0	.01296	.206	77.16	4.830	9.829	46.09	55.92	.0219	.1224
0	177.5	162.8	.01299	.203	76.96	4.914	10.09	45.87	55.96	.0224	.1222
1	180.3	165.7	.01302	.200	76.76	4.999	10.35	45.65	56.00	.0230	.1221
2	183.2	168.5	.01306	.196	76.56	5.085	10.61	45.43	56.05	.0236	.1220
3	186.1	171.4	.01309	.193	76.36	5.172	10.87	45.21	56.09	.0241	.1218
4	189.1	174.4	.01313	.190	76.16	5.261	11.14	44.99	56.13	.0247	.1217
5	192.1	177.4	.01316	.186	75.95	5.351	11.40	44.76	56.17	.0252	.1216
6	195.1	180.4	.01320	.183	75.75	5.442	11.67	44.53	56.21	.0258	.1214
7	198.1	183.4	.01323	.180	75.54	5.535	11.93	44.31	56.24	.0263	.1213
8	201.2	186.5	.01327	.177	75.34	5.630	12.20	44.08	56.28	.0269	.1211
9	204.3	189.6	.01331	.174	75.13	5.725	12.47	43.84	56.32	.0274	.1210
10	207.5	192.8	.01334	.171	74.92	5.823	12.74	43.61	56.35	.0280	.1208
11	210.7	196.0	.01338	.168	74.71	5.922	13.00	43.38	56.38	.0285	.1207
12	213.9	199.2	.01342	.166	74.49	6.022	13.27	43.14	56.42	.0291	.1206
13	217.1	202.4	.01346	.163	74.28	6.125	13.54	42.90	56.45	.0296	.1204
14	220.4	205.7	.01350	.160	74.06	6.229	13.81	42.66	56.48	.0302	.1203
15	223.8	209.1	.01354	.157	73.84	6.334	14.08	42.42	56.51	.0308	.1201
16	227.1	212.4	.01358	.155	73.62	6.441	14.36	42.17	56.53	.0313	.1200
17	230.5	215.8	.01362	.152	73.40	6.550	14.63	41.93	56.56	.0319	.1198
18	234.0	219.3	.01366	.150	73.18	6.661	14.90	41.68	56.59	.0324	.1197
19	237.4	222.7	.01370	.147	72.95	6.774	15.18	41.43	56.61	.0330	.1195

TEMP.	PRESSURE		VOLUME cu ft/lb		DENSITY lb/cu ft		ENTHALPY Btu/lb			ENTROPY Btu/(lb)(°R)	
°F	PSIA	PSIG	LIQUID v_f	VAPOR v_g	LIQUID $1/v_f$	VAPOR $1/v_g$	LIQUID h_f	LATENT h_{fg}	VAPOR h_g	LIQUID s_f	VAPOR s_g
20	241.0	226.3	.0137	.1451	72.72	6.889	15.45	41.18	56.63	.0335	.1194
21	244.5	229.8	.0137	.1427	72.50	7.005	15.73	40.92	56.65	.0341	.1192
22	248.1	233.4	.0138	.1403	72.26	7.124	16.01	40.66	56.67	.0346	.1191
23	251.7	237.0	.0138	.1380	72.03	7.244	16.28	40.41	56.69	.0352	.1189
24	255.4	240.7	.0139	.1357	71.80	7.367	16.56	40.14	56.71	.0358	.1188
25	259.1	244.4	.0139	.1334	71.56	7.492	16.84	39.88	56.73	.0363	.1186
26	262.8	248.1	.0140	.1312	71.32	7.619	17.12	39.61	56.74	.0369	.1185
27	266.6	251.9	.0140	.1290	71.08	7.748	17.40	39.35	56.75	.0374	.1183
28	270.4	255.7	.0141	.1269	70.83	7.880	17.69	39.07	56.77	.0380	.1181
29	274.3	259.6	.0141	.1247	70.59	8.014	17.97	38.80	56.77	.0386	.1180
30	278.2	263.5	.0142	.1226	70.34	8.150	18.26	38.52	56.78	.0391	.1178
31	282.1	267.4	.0142	.1206	70.09	8.289	18.54	38.24	56.79	.0397	.1176
32	286.1	271.4	.0143	.1186	69.83	8.430	18.83	37.96	56.79	.0402	.1175
33	290.1	275.4	.0143	.1166	69.57	8.574	19.12	37.68	56.80	.0408	.1173
34	294.2	279.5	.0144	.1146	69.31	8.721	19.41	37.39	56.80	.0414	.1171
35	298.3	283.6	.0144	.1127	69.05	8.871	19.70	37.09	56.80	.0419	.1169
36	302.4	287.7	.0145	.1108	68.79	9.023	19.99	36.80	56.79	.0425	.1168
37	306.6	291.9	.0145	.1089	68.52	9.179	20.28	36.50	56.79	.0431	.1166
38	310.9	296.2	.0146	.1070	68.24	9.338	20.58	36.20	56.78	.0436	.1164
39	315.1	300.4	.0147	.1052	67.97	9.500	20.87	35.89	56.77	.0442	.1162
40	319.4	304.8	.0147	.1034	67.69	9.66	21.17	35.58	56.76	.0448	.1160
41	323.8	309.1	.0148	.1016	67.41	9.83	21.47	35.27	56.75	.0454	.1158
42	328.2	313.5	.0148	.0999	67.12	10.00	21.77	34.95	56.73	.0459	.1156
43	332.7	318.0	.0149	.0982	66.83	10.18	22.08	34.63	56.71	.0465	.1154
44	337.2	322.5	.0150	.0965	66.54	10.36	22.38	34.31	56.69	.0471	.1152
45	341.7	327.0	.0150	.0948	66.24	10.54	22.69	33.98	56.67	.0477	.1150
46	346.3	331.6	.0151	.0931	65.94	10.73	23.00	33.64	56.64	.0483	.1148
47	350.9	336.2	.0152	.0915	65.63	10.92	23.31	33.30	56.61	.0489	.1146
48	355.6	340.9	.0153	.0899	65.32	11.12	23.62	32.95	56.58	.0494	.1144
49	360.3	345.6	.0153	.0883	65.00	11.32	23.93	32.60	56.54	.0500	.1141
50	365.1	350.4	.0154	.0867	64.68	11.52	24.25	32.25	56.50	.0506	.1139
51	369.9	355.2	.0155	.0851	64.35	11.74	24.57	31.88	56.46	.0512	.1137
52	374.8	360.1	.0156	.0836	64.01	11.95	24.89	31.51	56.41	.0518	.1134
53	379.7	365.0	.0157	.0820	63.68	12.18	25.22	31.14	56.36	.0524	.1132
54	384.7	370.0	.0157	.0805	63.33	12.40	25.55	30.76	56.31	.0531	.1129
55	389.7	375.0	.0158	.0790	62.98	12.64	25.88	30.37	56.25	.0537	.1127
56	394.8	380.1	.0159	.0776	62.62	12.88	26.21	29.97	56.19	.0543	.1124
57	399.9	385.2	.0160	.0761	62.25	13.13	26.55	29.56	56.12	.0549	.1122
58	405.1	390.4	.0161	.0746	61.87	13.38	26.89	29.15	56.05	.0556	.1119
59	410.3	395.6	.0162	.0732	61.49	13.65	27.24	28.73	55.97	.0562	.1116
60	415.6	400.9	.0163	.0718	61.09	13.92	27.59	28.29	55.88	.0568	.1113
61	420.9	406.2	.0164	.0703	60.69	14.20	27.94	27.85	55.80	.0575	.1110
62	426.3	411.6	.0165	.0689	60.28	14.49	28.30	27.39	55.70	.0581	.1107
63	431.7	417.0	.0167	.0675	59.85	14.79	28.67	26.92	55.60	.0580	.1103
64	437.2	422.5	.0168	.0661	59.42	15.11	29.04	26.44	55.48	.0595	.1100
65	442.8	428.1	.0169	.0647	58.97	15.43	29.41	25.95	55.37	.0602	.1096
66	448.4	433.7	.0170	.0634	58.50	15.77	29.80	25.44	55.24	.0609	.1093
67	454.1	439.4	.0172	.0620	58.02	16.12	30.19	24.91	55.10	.0616	.1089
68	459.8	445.1	.0173	.0606	57.52	16.49	30.59	24.36	54.95	.0623	.1085
69	465.6	450.9	.0175	.0592	57.01	16.87	31.00	23.79	54.79	.0630	.1080
70	471.5	456.8	.0177	.0578	56.47	17.28	31.41	23.20	54.62	.0638	.1076
71	477.4	462.7	.0178	.0564	55.91	17.71	31.84	22.58	54.43	.0646	.1071
72	483.4	468.7	.0180	.0550	55.32	18.16	32.29	21.94	54.23	.0654	.1066
73	489.5	474.8	.0182	.0536	54.70	18.64	32.75	21.26	54.01	.0662	.1061
74	495.6	480.9	.0185	.0521	54.05	19.15	33.22	20.53	53.76	.0670	.1055

TEMP. °F	PRESSURE PSIA	PRESSURE PSIG	VOLUME cu ft/lb LIQUID v_f	VOLUME cu ft/lb VAPOR v_g	DENSITY lb/cu ft LIQUID $1/v_f$	DENSITY lb/cu ft VAPOR $1/v_g$	ENTHALPY Btu/lb LIQUID h_f	ENTHALPY Btu/lb LATENT h_{fg}	ENTHALPY Btu/lb VAPOR h_g	ENTROPY Btu/(lb)(°R) LIQUID s_f	ENTROPY Btu/(lb)(°R) VAPOR s_g
75	501.8	487.1	.0187	.0507	53.35	19.71	33.72	19.77	53.49	.0679	.1049
76	508.1	493.4	.0190	.0492	52.61	20.31	34.24	18.94	53.19	.0689	.1042
77	514.5	499.8	.0193	.0476	51.80	20.96	34.79	18.05	52.85	.0698	.1035
78	520.9	506.2	.0196	.0460	50.93	21.69	35.38	17.08	52.47	.0709	.1027
79	527.5	512.8	.0200	.0443	49.95	22.52	36.03	15.99	52.02	.0721	.1017
80	534.1	519.4	.0204	.0425	48.85	23.48	36.74	14.75	51.49	.0733	.1007
81	540.8	526.1	.0210	.0405	47.55	24.63	37.56	13.27	50.84	.0748	.0994
82	547.7	533.0	.0217	.0382	45.94	26.12	38.56	11.42	49.98	.0766	.0977
83	554.6	539.9	.0229	.0352	43.66	28.35	39.93	8.73	48.67	.0791	.0952
83.93	561.3	546.6	.0277	.0277	36.07	36.06	44.27	.000	44.27	.0870	.0870

ENTHALPY (BTU/lb ABOVE SATURATED LIQUID AT –40° F)

From Published data of E.I. du Pont de Nemours & Co. Inc. Used by permission.

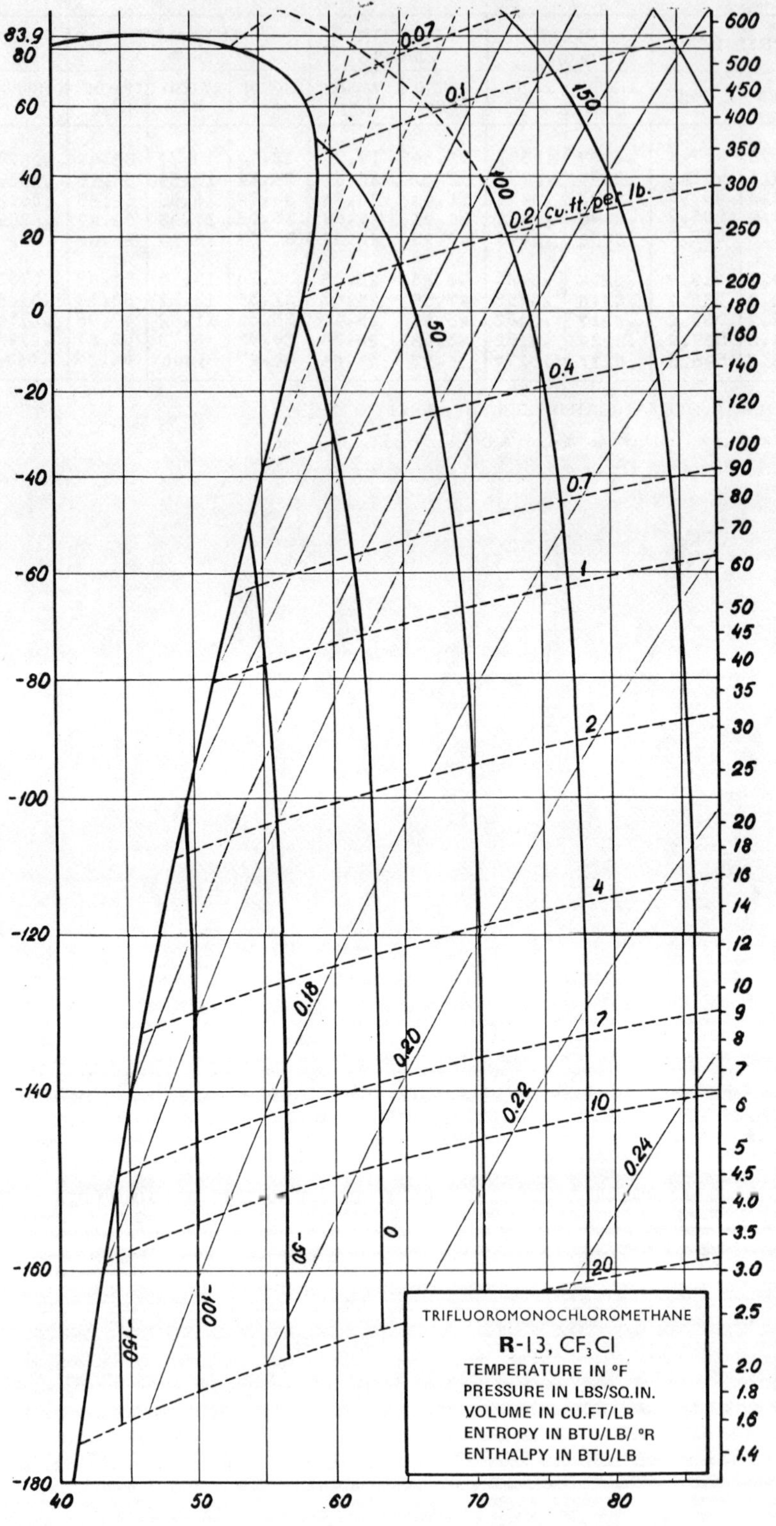
TRIFLUOROMONOCHLOROMETHANE
R-13, CF_3Cl
TEMPERATURE IN °F
PRESSURE IN LBS/SQ.IN.
VOLUME IN CU.FT/LB
ENTROPY IN BTU/LB/ °R
ENTHALPY IN BTU/LB
0.07
0.1
0.2 cu.ft. per lb.
0.4
0.7
1
2
4
7
10
20
0.18
0.20
0.22
0.24
150
100
50
0
-50
-100
-150
83.9
80
60
40
20
0
-20
-40
-60
-80
-100
-120
-140
-160
-180
40
50
60
70
80
600
500
450
400
350
300
250
200
180
160
140
120
100
90
80
70
60
50
45
40
35
30
25
20
18
16
14
12
10
9
8
7
6
5
4.5
4.0
3.5
3.0
2.5
2.0
1.8
1.6
1.4

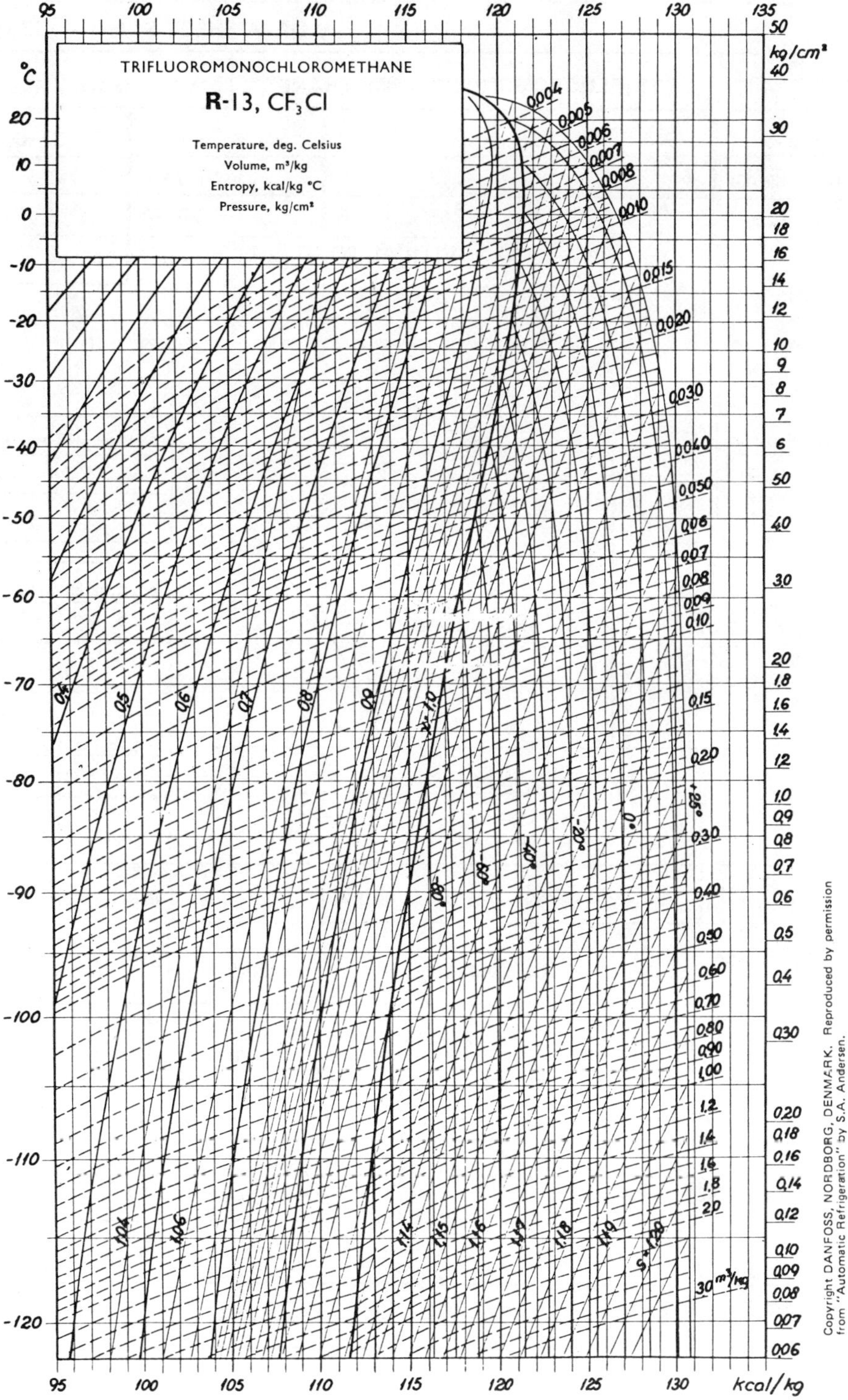

PRESSURE ENTHALPY DIAGRAM FOR R 13 Metric

DATA SHEET FOR R 13B1

$CBrF_3$	TRIFLUOROMONOBROMOMETHANE	Molecular Wt. 149

Pressure		Temperature				Volume		Density	
40.42	kg/cm²	67	°C	153	°F	1.342	l/kg	.745	kg/l.
575	psia	340	°K	613	°R	.0215	ft³/lb	46.5	lbs/ft³
								5.14	Air=1

AT CRITICAL POINT

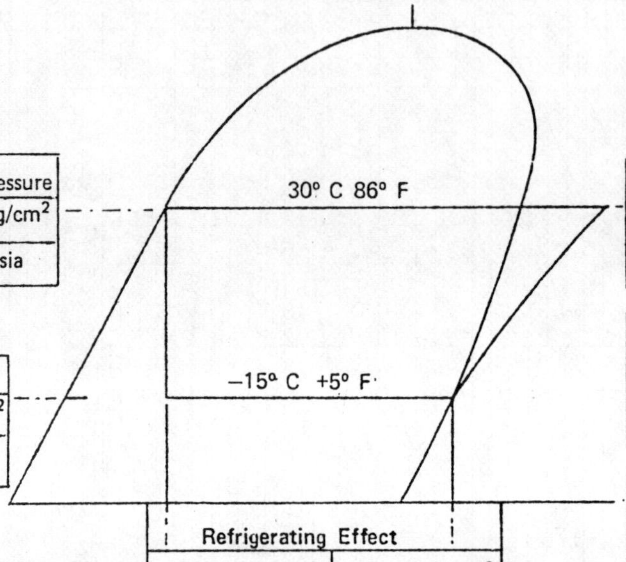

Discharge Pressure	
18.6	kg/cm²
264	psia

Inlet Pressure	
5.5	kg/cm²
77.8	psia

Refrigerating Effect			
685	kcal/m³	77.1	Btu/ft³

Discharge Temperature			
51	°C	124	°F
324	°K	584	°R

Normal Boiling Point			
-58	°C	-72	°F
215	°K	388	°R

Triple Point			
-143	°C	-225	°F
130	°K	235	°R

Latent Heat at NBP	28.38 kcal/kg	4229 /kg mol	51.1 Btu/lb	7614 /lb mol

Trouton's No. 19.7	Gas Constant 5.7 kg.m/kg/° K	10.37 ft.lbs/lb/° R

Specific Heat Liquid at 30° C 86° F .217 Fig 14	Gas C_p .11	C_p/C_v 1.116 at 30°C

Liquid at 30° C 86° F Density 1.5 kg/l.	93.6 lbs/ft³ Viscosity .212 cp

Reduced Form at NBP Pressure .0255	1 / Temperature 1.58

Reduced value of 1/T	2.05	1.72	1.36
when Reduced Pressure is	a) .001	b) .01	c) .1

TEMP. °F	PRESSURE		VOLUME cu ft/lb		DENSITY lb/cu ft		ENTHALPY Btu/lb			ENTROPY Btu/(lb)(°R)		TEMP. °F
	PSIA	PSIG	LIQUID V_f	VAPOR V_g	LIQUID $1/V_f$	VAPOR $1/V_g$	LIQUID h_f	LATENT h_{fg}	VAPOR h_g	LIQUID S_f	VAPOR S_g	
−160	0.6380	28.62*	0.00731	33.748	136.73	0.0296	−18.51	57.11	38.60	−0.0516	0.1389	−160
−159	0.6691	28.56*	0.00732	32.281	136.59	0.0310	−18.36	57.04	38.68	−0.0511	0.1386	−159
−158	0.7015	28.49*	0.00733	30.889	136.46	0.0324	−18.22	56.97	38.76	−0.0507	0.1382	−158
−157	0.7353	28.42*	0.00734	29.567	136.33	0.0338	−18.07	56.91	38.84	−0.0502	0.1378	−157
−156	0.7703	28.35*	0.00734	28.312	136.20	0.0353	−17.93	56.84	38.92	−0.0497	0.1375	−156
−155	0.8067	28.28*	0.00735	27.119	136.07	0.0369	−17.78	56.78	39.00	−0.0492	0.1371	−155
−154	0.8446	28.20*	0.00736	25.985	135.94	0.0385	−17.63	56.71	39.08	−0.0487	0.1368	−154
−153	0.8839	28.12*	0.00736	24.906	135.80	0.0402	−17.49	56.65	39.16	−0.0482	0.1364	−153
−152	0.9248	28.04*	0.00737	23.881	135.67	0.0419	−17.34	56.58	39.24	−0.0478	0.1361	−152
−151	0.9672	27.95*	0.00738	22.904	135.54	0.0437	−17.19	56.52	39.32	−0.0473	0.1358	−151
−150	1.011	27.86*	0.00739	21.975	135.41	0.0455	−17.05	56.45	39.40	−0.0468	0.1355	−150
−149	1.057	27.77*	0.00739	21.090	135.27	0.0474	−16.90	56.38	39.48	−0.0464	0.1351	−149
−148	1.104	27.67*	0.00740	20.247	135.14	0.0494	−16.75	56.32	39.57	−0.0459	0.1348	−148
−147	1.153	27.57*	0.00741	19.444	135.01	0.0514	−16.61	56.25	39.65	−0.0454	0.1345	−147
−146	1.204	27.47*	0.00741	18.678	134.87	0.0535	−16.46	56.19	39.73	−0.0449	0.1342	−146
−145	1.257	27.36*	0.00742	17.948	134.74	0.0557	−16.31	56.12	39.81	−0.0445	0.1339	−145
−144	1.312	27.25*	0.00743	17.252	134.61	0.0580	−16.17	56.06	39.89	−0.0440	0.1336	−144
−143	1.368	27.13*	0.00744	16.587	134.47	0.0603	−16.02	55.99	39.97	−0.0435	0.1333	−143
−142	1.427	27.01*	0.00744	15.953	134.34	0.0627	−15.87	55.93	40.06	−0.0431	0.1330	−142
−141	1.488	26.89*	0.00745	15.348	134.20	0.0652	−15.73	55.86	40.14	−0.0426	0.1327	−141
−140	1.551	26.76*	0.00746	14.769	134.07	0.0677	−15.58	55.80	40.22	−0.0422	0.1324	−140
−139	1.616	26.63*	0.00747	14.217	133.93	0.0703	−15.43	55.73	40.30	−0.0417	0.1321	−139
−138	1.683	26.49*	0.00747	13.689	133.80	0.0731	−15.28	55.67	40.38	−0.0412	0.1318	−138
−137	1.752	26.35*	0.00748	13.184	133.66	0.0758	−15.14	55.60	40.47	−0.0408	0.1315	−137
−136	1.824	26.21*	0.00749	12.701	133.53	0.0787	−14.99	55.53	40.55	−0.0403	0.1312	−136
−135	1.898	26.06*	0.00750	12.239	133.39	0.0817	−14.84	55.47	40.63	−0.0399	0.1310	−135
−134	1.975	25.90*	0.00750	11.797	133.26	0.0848	−14.69	55.40	40.71	−0.0394	0.1307	−134
−133	2.054	25.74*	0.00751	11.375	133.12	0.0879	−14.54	55.34	40.79	−0.0390	0.1304	−133
−132	2.136	25.57*	0.00752	10.970	132.98	0.0912	−14.40	55.27	40.88	−0.0385	0.1302	−132
−131	2.221	25.40*	0.00753	10.582	132.85	0.0945	−14.25	55.21	40.96	−0.0381	0.1299	−131
−130	2.308	25.22*	0.00754	10.211	132.71	0.0979	−14.10	55.14	41.04	−0.0376	0.1296	−130
−129	2.398	25.04*	0.00754	9.8550	132.57	0.1015	−13.95	55.08	41.12	−0.0372	0.1294	−129
−128	2.491	24.85*	0.00755	9.5140	132.44	0.1051	−13.80	55.01	41.21	−0.0367	0.1291	−128
−127	2.586	24.65*	0.00756	9.1870	132.30	0.1088	−13.65	54.94	41.29	−0.0363	0.1289	−127
−126	2.685	24.45*	0.00757	8.8735	132.16	0.1127	−13.50	54.88	41.37	−0.0358	0.1286	−126
−125	2.787	24.25*	0.00757	8.5727	132.02	0.1166	−13.36	54.81	41.46	−0.0354	0.1284	−125
−124	2.892	24.03*	0.00758	8.2841	131.89	0.1207	−13.21	54.75	41.54	−0.0349	0.1282	−124
−123	3.000	23.81*	0.00759	8.0072	131.75	0.1249	−13.06	54.68	41.62	−0.0345	0.1279	−123
−122	3.111	23.59*	0.00760	7.7413	131.61	0.1292	−12.91	54.61	41.70	−0.0340	0.1277	−122
−121	3.226	23.35*	0.00761	7.4859	131.47	0.1336	−12.76	54.55	41.79	−0.0336	0.1274	−121
−120	3.344	23.11*	0.00761	7.2407	131.33	0.1381	−12.61	54.48	41.87	−0.0332	0.1272	−120
−119	3.465	22.86*	0.00762	7.0051	131.19	0.1428	−12.46	54.41	41.95	−0.0327	0.1270	−119
−118	3.591	22.61*	0.00763	6.7787	131.05	0.1475	−12.31	54.35	42.04	−0.0323	0.1268	−118
−117	3.719	22.35*	0.00764	6.5610	130.91	0.1524	−12.16	54.28	42.12	−0.0318	0.1265	−117
−116	3.852	22.08*	0.00765	6.3518	130.77	0.1574	−12.01	54.21	42.20	−0.0314	0.1263	−116
−115	3.988	21.80*	0.00765	6.1505	130.63	0.1626	−11.86	54.14	42.29	−0.0310	0.1261	−115
−114	4.128	21.52*	0.00766	5.9570	130.49	0.1679	−11.71	54.08	42.37	0.0305	0.1259	−114
−113	4.272	21.22*	0.00767	5.7707	130.35	0.1733	−11.56	54.01	42.45	0.0301	0.1257	−113
−112	4.420	20.92*	0.00768	5.5915	130.21	0.1788	−11.41	53.94	42.54	0.0297	0.1255	−112
−111	4.573	20.61*	0.00769	5.4189	130.07	0.1845	−11.25	53.87	42.62	0.0292	0.1253	−111
−110	4.729	20.29*	0.00770	5.2528	129.93	0.1904	−11.10	53.81	42.70	0.0288	0.1251	−110
−109	4.890	19.96*	0.00770	5.0928	129.79	0.1964	−10.95	53.74	42.79	0.0284	0.1249	−109
−108	5.055	19.63*	0.00771	4.9386	129.65	0.2025	−10.80	53.67	42.87	0.0279	0.1247	−108
−107	5.224	19.28*	0.00772	4.7901	129.51	0.2088	−10.65	53.60	42.95	0.0275	0.1245	−107
−106	5.398	18.93*	0.00773	4.6470	129.36	0.2152	−10.50	53.53	43.04	0.0271	0.1243	−106

*Inches of mercury below one atmosphere

TEMP. °F	PRESSURE		VOLUME cu ft/lb		DENSITY lb/cu ft		ENTHALPY Btu/lb			ENTROPY Btu/(lb)(°F)		TEMP. °F
	PSIA	PSIG	LIQUID V_f	VAPOR V_g	LIQUID $1/V_f$	VAPOR $1/V_g$	LIQUID h_f	LATENT h_{fg}	VAPOR h_g	LIQUID S_f	VAPOR S_g	
–105	5.577	18.57*	0.00774	4.5090	129.22	0.2218	–10.34	53.46	43.12	0.0266	0.1241	–105
–104	5.760	18.19*	0.00775	4.3760	129.08	0.2285	–10.19	53.40	43.20	0.0262	0.1239	–104
–103	5.948	17.81*	0.00776	4.2477	128.94	0.2354	–10.04	53.33	43.29	0.0258	0.1237	–103
–102	6.141	17.42*	0.00776	4.1240	128.79	0.2425	– 9.89	53.26	43.37	0.0254	0.1235	–102
–101	6.339	17.01*	0.00777	4.0046	128.65	0.2497	– 9.73	53.19	43.46	0.0249	0.1233	–101
–100	6.542	16.60*	0.00778	3.8894	128.51	0.2571	– 9.58	53.12	43.54	0.0245	0.1232	–100
– 99	6.750	16.18*	0.00779	3.7782	128.36	0.2647	– 9.43	53.05	43.62	0.0241	0.1230	– 99
– 98	6.964	15.74*	0.00780	3.6709	128.22	0.2724	– 9.27	52.98	43.71	0.0237	0.1228	– 98
– 97	7.183	15.30*	0.00781	3.5672	128.07	0.2803	– 9.12	52.91	43.79	0.0232	0.1226	– 97
– 96	7.407	14.84*	0.00782	3.4671	127.93	0.2884	– 8.97	52.84	43.87	0.0228	0.1225	– 96
– 95	7.637	14.37*	0.00783	3.3704	127.78	0.2967	– 8.81	52.77	43.96	0.0224	0.1223	– 95
– 94	7.872	13.89*	0.00783	3.2770	127.64	0.3052	– 8.66	52.70	44.04	0.0220	0.1221	– 94
– 93	8.113	13.40*	0.00784	3.1867	127.49	0.3138	– 8.50	52.62	44.12	0.0216	0.1220	– 93
– 92	8.360	12.90*	0.00785	3.0994	127.35	0.3226	– 8.35	52.55	44.20	0.0211	0.1218	– 92
– 91	8.613	12.39*	0.00786	3.0150	127.20	0.3317	– 8.19	52.48	44.29	0.0207	0.1216	– 91
– 90	8.871	11.86*	0.00787	2.9335	127.06	0.3409	– 8.04	52.41	44.37	0.0203	0.1215	– 90
– 89	9.136	11.32*	0.00788	2.8546	126.91	0.3503	– 7.88	52.34	44.46	0.0199	0.1213	– 89
– 88	9.408	10.77*	0.00789	2.7782	126.76	0.3599	– 7.73	52.27	44.54	0.0195	0.1212	– 88
– 87	9.685	10.20*	0.00790	2.7044	126.61	0.3698	– 7.57	52.19	44.62	0.0190	0.1210	– 87
– 86	9.969	9.62*	0.00791	2.6329	126.47	0.3798	– 7.42	52.12	44.70	0.0186	0.1209	– 86
– 85	10.26	9.03*	0.00792	2.5638	126.32	0.3900	– 7.26	52.05	44.79	0.0182	0.1207	– 85
– 84	10.56	8.43*	0.00793	2.4968	126.17	0.4005	– 7.10	51.97	44.87	0.0178	0.1206	– 84
– 83	10.86	7.81*	0.00794	2.4320	126.02	0.4112	– 6.95	51.90	44.95	0.0174	0.1204	– 83
– 82	11.17	7.18*	0.00794	2.3692	125.88	0.4221	– 6.79	51.83	45.04	0.0170	0.1203	– 82
– 81	11.49	6.53*	0.00795	2.3084	125.73	0.4332	– 6.63	51.75	45.12	0.0165	0.1201	– 81
– 80	11.81	5.87*	0.00796	2.2495	125.58	0.4445	– 6.48	51.68	45.20	0.0161	0.1200	– 80
– 79	12.14	5.19*	0.00797	2.1925	125.43	0.4561	– 6.32	51.61	45.29	0.0157	0.1198	– 79
– 78	12.48	4.50*	0.00798	2.1372	125.28	0.4679	– 6.16	51.53	45.37	0.0153	0.1197	– 78
– 77	12.83	3.80*	0.00799	2.0835	125.13	0.4800	– 6.00	51.45	45.45	0.0149	0.1196	– 77
– 76	13.18	3.08*	0.00800	2.0316	124.98	0.4922	– 5.85	51.38	45.54	0.0145	0.1194	– 76
– 75	13.55	2.34*	0.00801	1.9812	124.83	0.5047	– 5.69	51.31	45.62	0.0141	0.1193	– 75
– 74	13.91	1.59*	0.00802	1.9323	124.68	0.5175	– 5.53	51.23	45.70	0.0137	0.1192	– 74
– 73	14.29	0.82*	0.00803	1.8849	124.53	0.5305	– 5.37	51.16	45.79	0.0133	0.1190	– 73
– 72	14.68	0.04*	0.00804	1.8390	124.37	0.5438	– 5.21	51.08	45.87	0.0128	0.1189	– 72
– 71	15.07	0.37	0.00805	1.7944	124.22	0.5573	– 5.05	51.00	45.95	0.0124	0.1188	– 71
– 70	15.47	0.77	0.00806	1.7511	124.07	0.5711	– 4.89	50.93	46.03	0.0120	0.1187	– 70
– 69	15.88	1.18	0.00807	1.7091	123.92	0.5851	– 4.73	50.85	46.12	0.0116	0.1185	– 69
– 68	16.30	1.60	0.00808	1.6683	123.77	0.5994	– 4.57	50.77	46.20	0.0112	0.1184	– 68
– 67	16.72	2.03	0.00809	1.6287	123.61	0.6140	– 4.41	50.69	46.28	–0.0108	0.1183	– 67
– 66	17.16	2.46	0.00810	1.5903	123.46	0.6288	– 4.25	50.61	46.36	–0.0104	0.1182	– 66
– 65	17.60	2.91	0.00811	1.5529	123.31	0.6439	– 4.09	50.54	46.45	–0.0100	0.1180	– 65
– 64	18.05	3.36	0.00812	1.5167	123.15	0.6593	– 3.93	50.46	46.53	–0.0096	0.1179	– 64
– 63	18.51	3.82	0.00813	1.4814	123.00	0.6750	– 3.77	50.38	46.61	–0.0092	0.1178	– 63
– 62	18.99	4.29	0.00814	1.4472	122.84	0.6910	– 3.61	50.30	46.69	–0.0088	0.1177	– 62
– 61	19.46	4.77	0.00815	1.4139	122.69	0.7072	– 3.45	50.22	46.77	–0.0084	0.1176	– 61
– 60	19.95	5.26	0.00816	1.3816	122.53	0.7238	– 3.29	50.14	46.86	–0.0080	0.1175	– 60
– 59	20.45	5.76	0.00817	1.3502	122.38	0.7406	– 3.12	50.06	46.94	–0.0076	0.1174	– 59
– 58	20.96	6.26	0.00818	1.3196	122.22	0.7578	– 2.96	49.98	47.02	–0.0072	0.1173	– 58
– 57	21.48	6.78	0.00819	1.2899	122.06	0.7752	– 2.80	49.90	47.10	–0.0068	0.1171	– 57
– 56	22.00	7.31	0.00820	1.2611	121.91	0.7930	– 2.64	49.82	47.18	–0.0064	0.1170	– 56
– 55	22.54	7.85	0.00821	1.2330	121.75	0.8111	– 2.47	49.73	47.26	–0.0060	0.1169	– 55
– 54	23.09	8.39	0.00822	1.2056	121.59	0.8294	– 2.31	49.65	47.34	–0.0056	0.1168	– 54
– 53	23.65	8.95	0.00823	1.1790	121.44	0.8482	– 2.15	49.57	47.42	–0.0052	0.1167	– 53
– 52	24.21	9.52	0.00825	1.1532	121.28	0.8672	– 1.98	49.49	47.51	–0.0048	0.1166	– 52
– 51	24.79	10.10	0.00826	1.1280	121.12	0.8865	– 1.82	49.41	47.59	–0.0044	0.1165	– 51

*Inches of mercury below one atmosphere

TEMP. °F	PRESSURE		VOLUME cu ft/lb		DENSITY lb/cu ft		ENTHALPY Btu/lb			ENTROPY Btu/(lb)(°R)		TEMP °F
	PSIA	PSIG	LIQUID V_f	VAPOR V_g	LIQUID $1/V_f$	VAPOR $1/V_g$	LIQUID h_f	LATENT h_{fg}	VAPOR h_g	LIQUID S_f	VAPOR S_g	
– 50	25.38	10.69	0.00827	1.1035	120.96	0.9062	– 1.65	49.32	47.67	–0.0040	0.1164	– 50
– 49	25.98	11.28	0.00828	1.0796	120.80	0.9263	– 1.49	49.24	47.75	–0.0036	0.1163	– 49
– 48	26.59	11.89	0.00829	1.0564	120.64	0.9466	– 1.33	49.16	47.83	–0.0032	0.1162	– 48
– 47	27.21	12.52	0.00830	1.0338	120.48	0.9673	– 1.16	49.07	47.91	–0.0028	0.1161	– 47
– 46	27.84	13.15	0.00831	1.0118	120.32	0.9884	– 1.00	48.99	47.99	–0.0024	0.1160	– 46
– 45	28.49	13.79	0.00832	0.9903	120.16	1.0098	– 0.83	48.90	48.07	–0.0020	0.1159	– 45
– 44	29.14	14.45	0.00833	0.9694	120.00	1.0315	– 0.66	48.81	48.15	–0.0016	0.1158	– 44
– 43	29.81	15.11	0.00834	0.9491	119.84	1.0537	– 0.50	48.73	48.23	–0.0012	0.1158	– 43
– 42	30.49	15.79	0.00836	0.9292	119.68	1.0762	– 0.33	48.64	48.31	–0.0008	0.1157	– 42
– 41	31.17	16.48	0.00837	0.9099	119.51	1.0990	– 0.17	48.56	48.39	–0.0004	0.1156	– 41
– 40	31.88	17.18	0.00838	0.8911	119.35	1.1222	0.00	48.47	48.47	0.0000	0.1155	– 40
– 39	32.59	17.89	0.00839	0.8727	119.19	1.1459	0.17	48.38	48.55	0.0004	0.1154	– 39
– 38	33.31	18.62	0.00840	0.8548	119.02	1.1698	0.33	48.30	48.63	0.0008	0.1153	– 38
– 37	34.05	19.36	0.00841	0.8374	118.86	1.1942	0.50	48.21	48.71	0.0012	0.1152	– 37
– 36	34.80	20.11	0.00842	0.8203	118.70	1.2190	0.67	48.12	48.79	0.0016	0.1151	– 36
– 35	35.56	20.87	0.00844	0.8038	118.53	1.2442	0.84	48.03	48.87	0.0020	0.1151	– 35
– 34	36.34	21.64	0.00845	0.7876	118.37	1.2697	1.00	47.94	48.95	0.0024	0.1150	– 34
– 33	37.13	22.43	0.00846	0.7718	118.20	1.2957	1.17	47.85	49.02	0.0028	0.1149	– 33
– 32	37.93	23.23	0.00847	0.7564	118.04	1.3221	1.34	47.76	49.10	0.0031	0.1148	– 32
– 31	38.74	24.04	0.00848	0.7414	117.87	1.3489	1.51	47.67	49.18	0.0035	0.1147	– 31
– 30	39.57	24.87	0.00850	0.7267	117.70	1.3761	1.68	47.58	49.26	0.0039	0.1147	– 30
– 29	40.41	25.71	0.00851	0.7124	117.54	1.4037	1.85	47.49	49.34	0.0043	0.1146	– 29
– 28	41.26	26.56	0.00852	0.6984	117.37	1.4318	2.02	47.40	49.41	0.0047	0.1145	– 28
– 27	42.13	27.43	0.00853	0.6848	117.20	1.4603	2.19	47.30	49.49	0.0051	0.1144	– 27
– 26	43.01	28.31	0.00854	0.6715	117.03	1.4892	2.36	47.21	49.57	0.0055	0.1143	– 26
– 25	43.90	29.20	0.00856	0.6585	116.86	1.5186	2.53	47.12	49.65	0.0059	0.1143	– 25
– 24	44.81	30.11	0.00857	0.6458	116.69	1.5485	2.70	47.02	49.72	0.0063	0.1142	– 24
– 23	45.73	31.03	0.00858	0.6334	116.52	1.5787	2.87	46.93	49.80	0.0067	0.1141	– 23
– 22	46.67	31.97	0.00859	0.6213	116.35	1.6095	3.04	46.84	49.88	0.0070	0.1140	– 22
– 21	47.62	32.92	0.00861	0.6095	116.18	1.6407	3.21	46.74	49.95	0.0074	0.1140	– 21
– 20	48.58	33.89	0.00862	0.5979	116.01	1.6724	3.38	46.64	50.03	0.0078	0.1139	– 20
– 19	49.56	34.87	0.00863	0.5867	115.84	1.7046	3.56	46.55	50.11	0.0082	0.1138	– 19
– 18	50.56	35.86	0.00865	0.5756	115.67	1.7372	3.73	46.45	50.18	0.0086	0.1138	– 18
– 17	51.57	36.87	0.00866	0.5649	115.49	1.7703	3.90	46.36	50.26	0.0090	0.1137	– 17
– 16	52.59	37.90	0.00867	0.5543	115.32	1.8039	4.07	46.26	50.33	0.0094	0.1136	– 16
– 15	53.63	38.94	0.00868	0.5440	115.15	1.8381	4.25	46.16	50.41	0.0097	0.1135	– 15
– 14	54.69	39.99	0.00870	0.5340	114.97	1.8727	4.42	46.06	50.48	0.0101	0.1135	– 14
– 13	55.76	41.06	0.00871	0.5242	114.80	1.9078	4.59	45.96	50.56	0.0105	0.1134	– 13
– 12	56.85	42.15	0.00872	0.5145	114.62	1.9435	4.77	45.87	50.63	0.0109	0.1133	– 12
– 11	57.95	43.25	0.00874	0.5051	114.45	1.9797	4.94	45.77	50.71	0.0113	0.1133	– 11
– 10	59.07	44.37	0.00875	0.4959	114.27	2.0164	5.12	45.67	50.78	0.0117	0.1132	– 10
– 9	60.20	45.51	0.00876	0.4869	114.10	2.0536	5.29	45.56	50.85	0.0121	0.1132	– 9
– 8	61.35	46.66	0.00878	0.4782	113.92	2.0914	5.47	45.46	50.93	0.0124	0.1131	– 8
– 7	62.52	47.82	0.00879	0.4695	113.74	2.1297	5.64	45.36	51.00	0.0128	0.1130	– 7
– 6	63.70	49.01	0.00881	0.4611	113.56	2.1686	5.81	45.26	51.07	0.0132	0.1130	– 6
– 5	64.90	50.21	0.00882	0.4529	113.38	2.2080	5.99	45.16	51.15	0.0136	0.1129	– 5
– 4	66.12	51.42	0.00883	0.4448	113.20	2.2480	6.17	45.05	51.22	0.0140	0.1128	– 4
– 3	67.35	52.66	0.00885	0.4369	113.02	2.2886	6.34	44.95	51.29	0.0143	0.1128	– 3
– 2	68.60	53.91	0.00886	0.4292	112.84	2.3298	6.52	44.85	51.36	0.0147	0.1127	– 2
– 1	69.87	55.18	0.00888	0.4217	112.66	2.3715	6.70	44.74	51.44	0.0151	0.1126	– 1
0	71.16	56.46	0.00889	0.4143	112.48	2.4139	6.87	44.64	51.51	0.0155	0.1126	0
1	72.46	57.77	0.00890	0.4070	112.30	2.4568	7.05	44.53	51.58	0.0159	0.1125	1
2	73.78	59.09	0.00892	0.3999	112.11	2.5004	7.23	44.42	51.65	0.0163	0.1125	2
3	75.12	60.42	0.00893	0.3930	111.93	2.5445	7.40	44.32	51.72	0.0166	0.1124	3
4	76.48	61.78	0.00895	0.3862	111.75	2.5893	7.58	44.21	51.79	0.0170	0.1124	4

TEMP. °F	PRESSURE		VOLUME cu ft/lb		DENSITY lb/cu ft		ENTHALPY Btu/lb			ENTROPY Btu/(lb)(°R)		TEMP. °F
	PSIA	PSIG	LIQUID V_f	VAPOR V_g	LIQUID $1/V_f$	VAPOR $1/V_g$	LIQUID h_f	LATENT h_{fg}	VAPOR h_g	LIQUID S_f	VAPOR S_g	
5	77.85	63.16	0.00896	0.3795	111.56	2.6347	7.76	44.10	51.86	0.0174	0.1123	5
6	79.25	64.55	0.00898	0.3730	111.38	2.6808	7.94	43.99	51.93	0.0178	0.1122	6
7	80.66	65.96	0.00899	0.3666	111.19	2.7275	8.12	43.88	52.00	0.0181	0.1122	7
8	82.09	67.39	0.00901	0.3604	111.00	2.7749	8.30	43.77	52.07	0.0185	0.1121	8
9	83.54	68.84	0.00902	0.3542	110.82	2.8229	8.48	43.66	52.14	0.0189	0.1121	9
10	85.00	70.31	0.00904	0.3482	110.63	2.8716	8.66	43.55	52.21	0.0193	0.1120	10
11	86.49	71.79	0.00905	0.3424	110.44	2.9210	8.84	43.44	52.28	0.0197	0.1119	11
12	88.00	73.30	0.00907	0.3366	110.25	2.9710	9.02	43.33	52.35	0.0200	0.1119	12
13	89.52	74.83	0.00909	0.3309	110.06	3.0218	9.20	43.22	52.41	0.0204	0.1118	13
14	91.07	76.37	0.00910	0.3254	109.87	3.0732	9.38	43.10	52.48	0.0208	0.1118	14
15	92.63	77.93	0.00912	0.3200	109.68	3.1254	9.56	42.99	52.55	0.0212	0.1117	15
16	94.21	79.52	0.00913	0.3146	109.49	3.1783	9.74	42.87	52.62	0.0215	0.1117	16
17	95.82	81.12	0.00915	0.3094	109.30	3.2319	9.92	42.76	52.68	0.0219	0.1116	17
18	97.44	82.75	0.00917	0.3043	109.10	3.2863	10.11	42.64	52.75	0.0223	0.1116	18
19	99.09	84.39	0.00918	0.2993	108.91	3.3414	10.29	42.53	52.81	0.0227	0.1115	19
20	100.8	86.06	0.00920	0.2944	108.71	3.3973	10.47	42.41	52.88	0.0230	0.1115	20
21	102.4	87.74	0.00921	0.2895	108.52	3.4539	10.65	42.29	52.94	0.0234	0.1114	21
22	104.1	89.45	0.00923	0.2848	108.32	3.5113	10.84	42.17	53.01	0.0238	0.1113	22
23	105.9	91.18	0.00925	0.2801	108.13	3.5695	11.02	42.05	53.07	0.0242	0.1113	23
24	107.6	92.93	0.00927	0.2756	107.93	3.6285	11.20	41.93	53.14	0.0245	0.1112	24
25	109.4	94.69	0.00928	0.2711	107.73	3.6883	11.39	41.81	53.20	0.0249	0.1112	25
26	111.2	96.5	0.00930	0.2667	107.53	3.7490	11.57	41.69	53.27	0.0253	0.1111	26
27	113.0	98.3	0.00932	0.2624	107.33	3.8105	11.76	41.57	53.33	0.0257	0.1111	27
28	114.8	100.1	0.00933	0.2582	107.13	3.8728	11.94	41.45	53.39	0.0260	0.1110	28
29	116.7	102.0	0.00935	0.2541	106.93	3.9359	12.13	41.32	53.45	0.0264	0.1110	29
30	118.6	103.9	0.00937	0.2500	106.73	4.0000	12.32	41.20	53.51	0.0268	0.1109	30
31	120.5	105.8	0.00939	0.2460	106.52	4.0649	12.50	41.07	53.58	0.0272	0.1109	31
32	122.4	107.7	0.00941	0.2421	106.32	4.1307	12.69	40.95	53.64	0.0275	0.1108	32
33	124.3	109.6	0.00942	0.2382	106.11	4.1974	12.88	40.82	53.70	0.0279	0.1108	33
34	126.3	111.6	0.00944	0.2345	105.91	4.2650	13.06	40.69	53.76	0.0283	0.1107	34
35	128.3	113.6	0.00946	0.2308	105.70	4.3335	13.25	40.57	53.82	0.0287	0.1107	35
36	130.3	115.6	0.00948	0.2271	105.49	4.4030	13.44	40.44	53.88	0.0290	0.1106	36
37	132.3	117.6	0.00950	0.2235	105.28	4.4734	13.63	40.31	53.93	0.0294	0.1106	37
38	134.4	119.7	0.00952	0.2200	105.08	4.5448	13.82	40.18	53.99	0.0298	0.1105	38
39	136.5	121.8	0.00954	0.2166	104.86	4.6172	14.01	40.05	54.05	0.0301	0.1104	39
40	138.6	123.9	0.00956	0.2132	104.65	4.6906	14.19	39.91	54.11	0.0305	0.1104	40
41	140.7	126.0	0.00957	0.2099	104.44	4.7649	14.38	39.78	54.16	0.0309	0.1103	41
42	142.9	128.2	0.00959	0.2066	104.23	4.8403	14.57	39.65	54.22	0.0313	0.1103	42
43	145.1	130.4	0.00961	0.2034	104.01	4.9168	14.77	39.51	54.28	0.0316	0.1102	43
44	147.3	132.6	0.00963	0.2002	103.80	4.9943	14.96	39.38	54.33	0.0320	0.1102	44
45	149.5	134.8	0.00965	0.1971	103.58	5.0728	15.15	39.24	54.39	0.0324	0.1101	45
46	151.7	137.1	0.00967	0.1941	103.36	5.1525	15.34	39.10	54.44	0.0327	0.1101	46
47	154.0	139.3	0.00969	0.1911	103.15	5.2332	15.53	38.96	54.50	0.0331	0.1100	47
48	156.3	141.6	0.00972	0.1881	102.93	5.3151	15.72	38.82	54.55	0.0335	0.1100	48
49	158.7	144.0	0.00974	0.1853	102.71	5.3981	15.92	38.68	54.60	0.0339	0.1099	49
50	161.0	146.3	0.00976	0.1824	102.49	5.4822	16.11	38.54	54.65	0.0342	0.1099	50
51	163.4	148.7	0.00978	0.1796	102.26	5.5676	16.31	38.40	54.71	0.0346	0.1098	51
52	165.8	151.1	0.00980	0.1769	102.04	5.6541	16.50	38.26	54.76	0.0350	0.1097	52
53	168.2	153.5	0.00982	0.1742	101.81	5.7418	16.69	38.11	54.81	0.0354	0.1097	53
54	170.7	156.0	0.00984	0.1715	101.59	5.8308	16.89	37.97	54.86	0.0357	0.1096	54
55	173.2	158.5	0.00987	0.1689	101.36	5.9210	17.08	37.82	54.91	0.0361	0.1096	55
56	175.7	161.0	0.00989	0.1663	101.13	6.0124	17.29	37.67	54.96	0.0365	0.1095	56
57	178.2	163.5	0.00991	0.1638	100.90	6.1052	17.48	37.53	55.00	0.0368	0.1095	57
58	180.8	166.1	0.00993	0.1613	100.67	6.1992	17.68	37.38	55.05	0.0372	0.1094	58
59	183.4	168.7	0.00996	0.1589	100.44	6.2946	17.87	37.23	55.10	0.0376	0.1094	59

TEMP. °F	PRESSURE		VOLUME cu ft/lb		DENSITY lb/cu ft		ENTHALPY Btu/lb			ENTROPY Btu/(lb)(°R)		TEMP. °F
	PSIA	PSIG	LIQUID V_f	VAPOR V_g	LIQUID $1/V_f$	VAPOR $1/V_g$	LIQUID h_f	LATENT h_{fg}	VAPOR h_g	LIQUID S_f	VAPOR S_g	
60	186.0	171.3	0.00998	0.1565	100.20	6.3914	18.08	37.07	55.15	0.0380	0.1093	60
61	188.6	173.9	0.01000	0.1541	99.97	6.4895	18.27	36.92	55.19	0.0383	0.1092	61
62	191.3	176.6	0.01003	0.1518	99.73	6.5890	18.47	36.77	55.24	0.0387	0.1092	62
63	194.0	179.3	0.01005	0.1495	99.50	6.6900	18.67	36.61	55.28	0.0391	0.1091	63
64	196.7	182.0	0.01007	0.1472	99.26	6.7924	18.87	36.46	55.32	0.0394	0.1091	64
65	199.5	184.8	0.01010	0.1450	99.02	6.8962	19.07	36.30	55.37	0.0398	0.1090	65
66	202.3	187.6	0.01012	0.1428	98.77	7.0016	19.27	36.14	55.41	0.0402	0.1089	66
67	205.1	190.4	0.01015	0.1407	98.53	7.1085	19.47	35.98	55.45	0.0406	0.1089	67
68	207.9	193.2	0.01017	0.1386	98.29	7.2170	19.67	35.82	55.49	0.0409	0.1088	68
69	210.8	196.1	0.01020	0.1365	98.04	7.3270	19.88	35.66	55.53	0.0413	0.1088	69
70	213.7	199.0	0.01023	0.1344	97.79	7.4387	20.08	35.49	55.57	0.0417	0.1087	70
71	216.6	201.9	0.01025	0.1324	97.54	7.5520	20.28	35.33	55.61	0.0421	0.1086	71
72	219.6	204.9	0.01028	0.1304	97.29	7.6670	20.49	35.16	55.65	0.0424	0.1086	72
73	222.6	207.9	0.01031	0.1285	97.04	7.7837	20.69	34.99	55.69	0.0428	0.1085	73
74	225.6	210.9	0.01033	0.1265	96.78	7.9021	20.90	34.82	55.72	0.0432	0.1084	74
75	228.7	214.0	0.01036	0.1247	96.53	8.0223	21.11	34.65	55.76	0.0436	0.1084	75
76	231.7	217.0	0.01039	0.1228	96.27	8.1443	21.31	34.48	55.79	0.0439	0.1083	76
77	234.8	220.1	0.01042	0.1209	96.01	8.2682	21.52	34.30	55.83	0.0443	0.1082	77
78	238.0	223.3	0.01044	0.1191	95.75	8.3940	21.73	34.13	55.86	0.0447	0.1082	78
79	241.2	226.5	0.01047	0.1173	95.48	8.5216	21.94	33.95	55.89	0.0451	0.1081	79
80	244.4	229.7	0.01050	0.1156	95.22	8.6513	22.15	33.77	55.92	0.0454	0.1080	80
81	247.6	232.9	0.01053	0.1139	94.95	8.7829	22.36	33.59	55.95	0.0458	0.1080	81
82	250.9	236.2	0.01056	0.1121	94.68	8.9166	22.57	33.41	55.98	0.0462	0.1079	82
83	254.2	239.5	0.01059	0.1105	94.41	9.0524	22.78	33.23	56.01	0.0466	0.1078	83
84	257.5	242.8	0.01062	0.1088	94.14	9.1904	23.00	33.04	56.04	0.0470	0.1077	84
85	260.8	246.1	0.01065	0.1072	93.86	9.3305	23.21	32.85	56.06	0.0473	0.1077	85
86	264.2	249.5	0.01069	0.1056	93.58	9.4728	23.42	32.66	56.09	0.0477	0.1076	86
87	267.7	253.0	0.01072	0.1040	93.30	9.6175	23.64	32.47	56.11	0.0481	0.1075	87
88	271.1	256.4	0.01075	0.1024	93.02	9.7645	23.86	32.28	56.14	0.0485	0.1074	88
89	274.6	259.9	0.01078	0.1009	92.74	9.9138	24.07	32.09	56.16	0.0489	0.1073	89
90	278.1	263.4	0.01082	0.0993	92.45	10.0657	24.29	31.89	56.18	0.0493	0.1073	90
91	281.7	267.0	0.01085	0.0978	92.16	10.2200	24.51	31.69	56.20	0.0496	0.1072	91
92	285.3	270.6	0.01089	0.0964	91.87	10.377	24.73	31.49	56.22	0.0500	0.1071	92
93	288.9	274.2	0.01092	0.0949	91.57	10.536	24.95	31.29	56.24	0.0504	0.1070	93
94	292.6	277.9	0.01096	0.0935	91.27	10.699	25.17	31.08	56.25	0.0508	0.1069	94
95	296.3	281.6	0.01099	0.0920	90.97	10.864	25.39	30.87	56.27	0.0512	0.1068	95
96	300.0	285.3	0.01103	0.0906	90.67	11.032	25.62	30.66	56.28	0.0516	0.1068	96
97	303.7	289.0	0.01107	0.0893	90.36	11.202	25.84	30.45	56.29	0.0520	0.1067	97
98	307.5	292.9	0.01110	0.0879	90.05	11.376	26.07	30.24	56.30	0.0524	0.1066	98
99	311.4	296.7	0.01114	0.0866	89.74	11.553	26.29	30.02	56.31	0.0527	0.1065	99
100	315.2	300.6	0.01118	0.0852	89.42	11.733	26.52	29.80	56.32	0.0531	0.1064	100
101	319.2	304.5	0.01122	0.0839	89.10	11.917	26.75	29.58	56.33	0.0535	0.1063	101
102	323.1	308.4	0.01126	0.0826	88.78	12.103	26.98	29.35	56.34	0.0539	0.1062	102
103	327.1	312.4	0.01131	0.0813	88.46	12.293	27.21	29.13	56.34	0.0543	0.1061	103
104	331.1	316.4	0.01135	0.0801	88.13	12.487	27.45	28.90	56.34	0.0547	0.1060	104
105	335.1	320.4	0.01139	0.0788	87.79	12.684	27.68	28.66	56.34	0.0551	0.1059	105
106	339.2	324.5	0.01143	0.0776	87.45	12.886	27.91	28.43	56.34	0.0555	0.1058	106
107	343.3	328.6	0.01148	0.0764	87.11	13.090	28.15	28.19	56.34	0.0559	0.1057	107
108	347.5	332.8	0.01153	0.0752	86.77	13.300	28.39	27.94	56.33	0.0563	0.1056	108
109	351.7	337.0	0.01157	0.0740	86.42	13.513	28.63	27.70	56.33	0.0567	0.1054	109
110	355.9	341.2	0.01162	0.0728	86.06	13.731	28.87	27.45	56.32	0.0571	0.1053	110
111	360.2	345.5	0.01167	0.0717	85.70	13.953	29.11	27.20	56.31	0.0576	0.1052	111
112	364.5	349.8	0.01172	0.0705	85.34	14.179	29.36	26.94	56.30	0.0580	0.1051	112
113	368.9	354.2	0.01177	0.0694	84.97	14.411	29.60	26.68	56.28	0.0584	0.1050	113
114	373.2	358.5	0.01182	0.0683	84.59	14.647	29.85	26.41	56.27	0.0588	0.1048	114

TEMP. °F	PRESSURE		VOLUME cu ft/lb		DENSITY lb/cu ft		ENTHALPY Btu/lb			ENTROPY Btu/(lb)(°R)		TEMP. °F
	PSIA	PSIG	LIQUID V_f	VAPOR V_g	LIQUID $1/V_f$	VAPOR $1/V_g$	LIQUID h_f	LATENT h_{fg}	VAPOR h_g	LIQUID S_f	VAPOR S_g	
115	377.7	363.0	0.01187	0.0672	84.21	14.888	30.10	26.15	56.25	0.0592	0.1047	115
116	382.1	367.4	0.01193	0.0661	83.83	15.135	30.35	25.87	56.23	0.0596	0.1046	116
117	386.6	371.9	0.01199	0.0650	83.43	15.388	30.61	25.60	56.20	0.0601	0.1044	117
118	391.2	376.5	0.01204	0.0639	83.03	15.646	30.86	25.32	56.18	0.0605	0.1043	118
119	395.8	381.1	0.01210	0.0629	82.63	15.910	31.12	25.03	56.15	0.0609	0.1042	119
120	400.4	385.7	0.01216	0.0618	82.22	16.181	31.38	24.74	56.12	0.0613	0.1040	120
121	405.1	390.4	0.01223	0.0608	81.80	16.458	31.64	24.44	56.08	0.0618	0.1039	121
122	409.8	395.1	0.01229	0.0597	81.37	16.742	31.91	24.14	56.05	0.0622	0.1037	122
123	414.5	399.8	0.01236	0.0587	80.93	17.033	32.17	23.83	56.01	0.0627	0.1036	123
124	419.3	404.6	0.01242	0.0577	80.49	17.332	32.44	23.52	55.97	0.0631	0.1034	124
125	424.1	409.4	0.01249	0.0567	80.04	17.639	32.72	23.20	55.92	0.0635	0.1032	125
126	429.0	414.3	0.01257	0.0557	79.57	17.954	32.99	22.88	55.87	0.0640	0.1031	126
127	433.9	419.2	0.01264	0.0547	79.10	18.278	33.27	22.54	55.82	0.0645	0.1029	127
128	438.9	424.2	0.01272	0.0537	78.62	18.611	33.55	22.20	55.76	0.0649	0.1027	128
129	443.9	429.2	0.01280	0.0528	78.12	18.954	33.84	21.86	55.70	0.0654	0.1025	129
130	448.9	434.2	0.01288	0.0518	77.61	19.307	34.13	21.50	55.63	0.0659	0.1023	130
131	454.0	439.3	0.01297	0.0508	77.09	19.672	34.42	21.14	55.56	0.0663	0.1021	131
132	459.1	444.4	0.01306	0.0499	76.56	20.048	34.72	20.77	55.49	0.0668	0.1019	132
133	464.3	449.6	0.01316	0.0489	76.01	20.437	35.02	20.38	55.41	0.0673	0.1017	133
134	469.5	454.8	0.01326	0.0480	75.44	20.840	35.33	19.99	55.32	0.0678	0.1015	134
135	474.8	460.1	0.01336	0.0470	74.85	21.257	35.64	19.59	55.23	0.0683	0.1012	135
136	480.1	465.4	0.01347	0.0461	74.24	21.690	35.96	19.17	55.13	0.0688	0.1010	136
137	485.4	470.7	0.01358	0.0452	73.61	22.140	36.28	18.74	55.02	0.0693	0.1007	137
138	490.8	476.1	0.01371	0.0442	72.96	22.608	36.62	18.30	54.91	0.0699	0.1005	138
139	496.3	481.6	0.01384	0.0433	72.28	23.097	36.95	17.84	54.79	0.0704	0.1002	139
140	501.8	487.1	0.01397	0.0424	71.57	23.608	37.30	17.36	54.66	0.0710	0.0999	140
141	507.3	492.6	0.01412	0.0414	70.82	24.144	37.66	16.86	54.52	0.0715	0.0996	141
142	512.9	498.2	0.01428	0.0405	70.03	24.707	38.03	16.34	54.37	0.0721	0.0993	142
143	518.5	503.8	0.01445	0.0395	69.20	25.301	38.41	15.80	54.21	0.0727	0.0989	143
144	524.2	509.5	0.01464	0.0386	68.31	25.931	38.80	15.23	54.03	0.0734	0.0986	144
145	529.9	515.2	0.01485	0.0376	67.36	26.601	39.22	14.62	53.84	0.0740	0.0982	145
146	535.7	521.0	0.01508	0.0366	66.32	27.318	39.65	13.98	53.63	0.0747	0.0978	146
147	541.5	526.8	0.01534	0.0356	65.18	28.091	40.11	13.29	53.40	0.0754	0.0973	147
148	547.4	532.7	0.01565	0.0346	63.91	28.931	40.61	12.53	53.14	0.0762	0.0969	148
149	553.3	538.6	0.01601	0.0335	62.45	29.857	41.15	11.70	52.85	0.0771	0.0963	149
150	559.2	544.5	0.01647	0.0324	60.71	30.891	41.77	10.74	52.51	0.0781	0.0957	150
151	565.2	550.6	0.01711	0.0312	58.46	32.07	42.52	9.61	52.12	0.0793	0.0950	151
152	571.3	556.6	0.01819	0.0299	54.96	33.48	43.59	8.05	51.65	0.0810	0.0942	152
152.6 (Critical)	574.9	560.2	0.0215	0.0215	46.5	46.5						

ENTHALPY (Btu/lb above saturated liquid at −40° F)

From Published data of E.I. du Pont de Nemours & Co. Inc. Used by Permission.

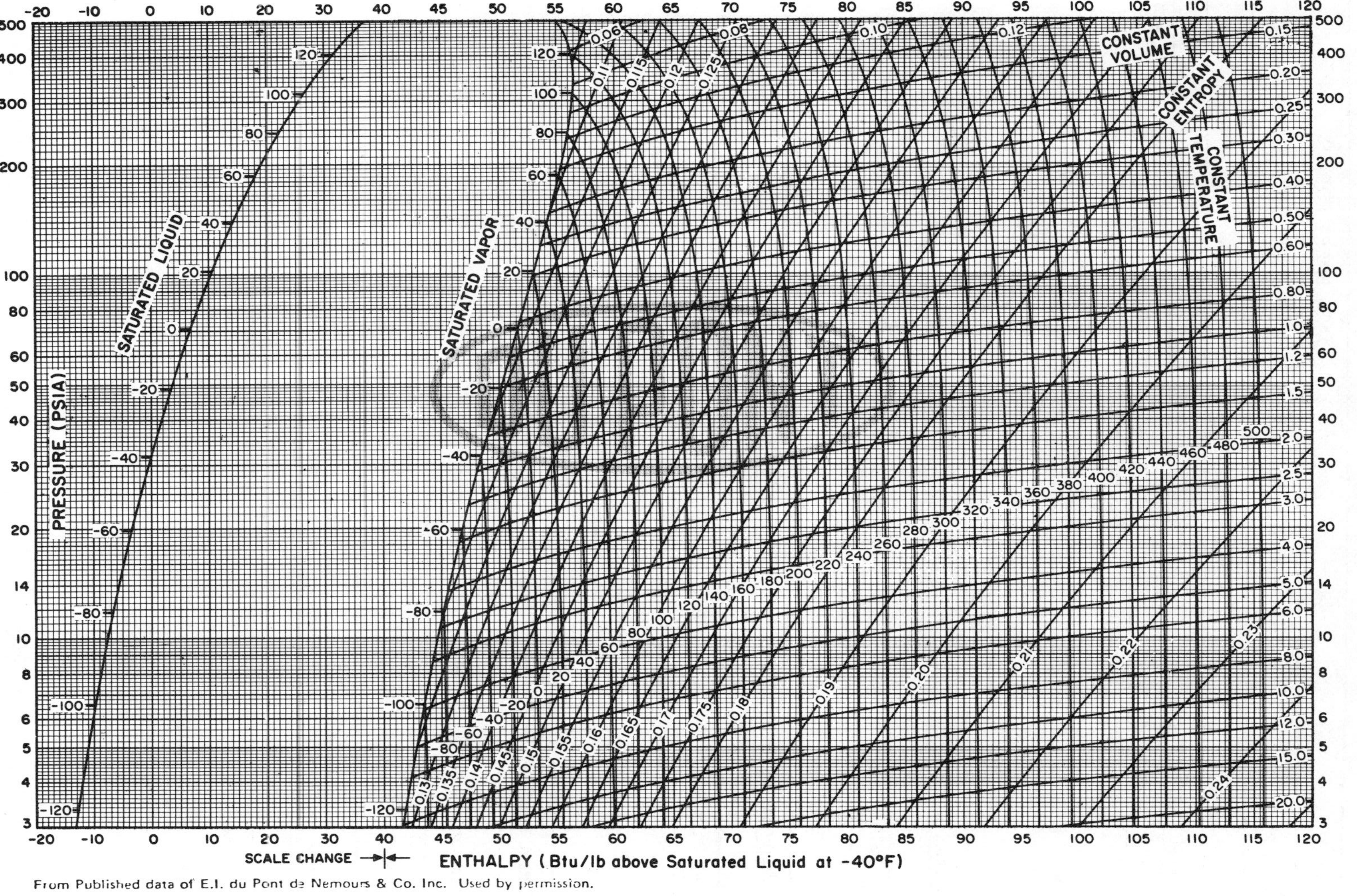

From Published data of E.I. du Pont de Nemours & Co. Inc. Used by permission.

DATA SHEET FOR R 14

CF_4	CARBON TETRAFLUORIDE	Molecular Wt. 88

Pressure		Temperature				Volume		Density	
38.1	kg/cm²	-45.7	°C	-50.2	°F	1.60	l/kg	.626	kg/l.
543	psia	227	°K	410	°R	.0256	ft³/lb	39.1	lbs/ft³
								3.03	Air=1

AT CRITICAL POINT

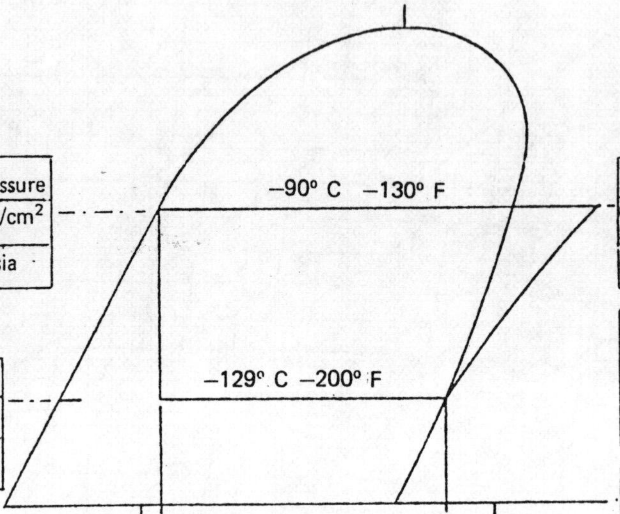

Discharge Pressure	
8.22	kg/cm²
117	psia

−90° C −130° F

Discharge Temperature			
-23	°C	-10	°F
250	°K	450	°R

Inlet Pressure	
.97	kg/cm²
13.75	psia

−129° C −200° F

Normal Boiling Point			
-128	°C	-198	°F
145	°K	262	°R

Triple Point			
-184	°C	-299	°F
89	°K	161	°R

Refrigerating Effect	
164 kcal/m³	18.4 Btu/ft³

Latent Heat at NBP	32.5 kcal/kg	28.60 /kg mol	58.5 Btu/lb	5148 /lb mol
Trouton's No. 19.65	Gas Constant 9.63 kg.m/kg/° K	17.55 ft.lbs/lb/° R		
Specific Heat Liquid at 30° C 86° F Fig 14	Gas C_p .169	C_p/C_v 1.159 at 25°C		
Liquid at 30° C 86° F Density 1.39 kg/l.	87.1 lbs/ft³ Viscosity cp			
Reduced Form at NBP Pressure .027	1 / Temperature 1.565			

Reduced value of 1/T when Reduced Pressure is	a) .001	b) .01	c) .1
	2.05	1.72	1.36

TEMP.	PRESSURE		VOLUME cu ft/lb		DENSITY lb/cu ft		ENTHALPY Btu/lb			ENTROPY Btu/(lb)(°R)		TEMP.
°F	PSIA	PSIG	LIQUID v_f	VAPOR v_g	LIQUID $1/v_f$	VAPOR $1/v_g$	LIQUID h_f	LATENT h_{fg}	VAPOR h_g	LIQUID s_f	VAPOR s_g	°F
−230	3.4630	22.8694*	0.009278	8.0017	107.780	0.12497	42.084	62.082	104.166	0.3552	0.6255	−230
−229	3.6497	22.4892*	0.009293	7.6218	107.603	0.13120	42.284	61.980	104.264	0.3561	0.6248	−229
−228	3.8445	22.0926*	0.009309	7.2635	107.426	0.13767	42.484	61.879	104.363	0.3569	0.6240	−228
−227	4.0477	21.6789*	0.009324	6.9253	107.249	0.14440	42.685	61.776	104.462	0.3578	0.6233	−227
−226	4.2595	21.2477*	0.009340	6.6060	107.071	0.15138	42.887	61.673	104.560	0.3587	0.6226	−226
−225	4.4802	20.7984*	0.009355	6.3041	106.893	0.15863	43.089	61.569	104.658	0.3595	0.6219	−225
−224	4.7100	20.3305*	0.009371	6.0188	106.715	0.16615	43.293	61.464	104.757	0.3604	0.6212	−224
−223	4.9493	19.8433*	0.009387	5.7490	106.536	0.17394	43.496	61.359	104.855	0.3613	0.6205	−223
−222	5.1982	19.3364*	0.009402	5.4937	106.356	0.18203	43.701	61.253	104.953	0.3621	0.6198	−222
−221	5.4572	18.8092*	0.009418	5.2519	106.176	0.19041	43.906	61.146	105.052	0.3630	0.6192	−221
−220	5.7264	18.2610*	0.009434	5.0230	105.996	0.19909	44.112	61.038	105.150	0.3638	0.6185	−220
−219	6.0062	17.6913*	0.009450	4.8060	105.815	0.20807	44.319	60.929	105.248	0.3647	0.6178	−219
−218	6.2969	17.0995*	0.009467	4.6003	105.634	0.21738	44.526	60.820	105.346	0.3656	0.6172	−218
−217	6.5988	16.4849*	0.009483	4.4051	105.452	0.22701	44.734	60.709	105.443	0.3664	0.6166	−217
−216	6.9121	15.8470*	0.009499	4.2199	105.270	0.23697	44.943	60.598	105.541	0.3673	0.6159	−216
−215	7.2372	15.1851*	0.009516	4.0441	105.087	0.24727	45.153	60.486	105.639	0.3681	0.6153	−215
−214	7.5744	14.4986*	0.009533	3.8772	104.904	0.25792	45.364	60.373	105.736	0.3690	0.6147	−214
−213	7.9239	13.7869*	0.009549	3.7185	104.720	0.26892	45.575	60.259	105.833	0.3698	0.6141	−213
−212	8.2862	13.0492*	0.009566	3.5676	104.536	0.28030	45.788	60.143	105.931	0.3707	0.6135	−212
−211	8.6616	12.2850*	0.009583	3.4242	104.351	0.29204	46.000	60.028	106.028	0.3716	0.6129	−211
−210	9.0503	11.4935*	0.009600	3.2877	104.166	0.30416	46.214	59.910	106.124	0.3724	0.6123	−210
−209	9.4528	10.6742*	0.009617	3.1579	103.980	0.31667	46.428	59.793	106.221	0.3733	0.6118	−209
−208	9.8692	9.8262*	0.009634	3.0342	103.794	0.32958	46.643	59.675	106.318	0.3741	0.6112	−208
−207	10.300	8.949*	0.009652	2.9163	103.608	0.34289	46.859	59.555	106.414	0.3750	0.6107	−207
−206	10.746	8.042*	0.009669	2.8041	103.420	0.35662	47.076	59.434	106.510	0.3758	0.6101	−206
−205	11.206	7.104*	0.009687	2.6971	103.233	0.37077	47.294	59.313	106.606	0.3767	0.6096	−205
−204	11.682	6.135*	0.009705	2.5950	103.044	0.38536	47.512	59.190	106.702	0.3775	0.6090	−204
−203	12.174	5.133*	0.009722	2.4976	102.856	0.40039	47.731	59.066	106.798	0.3784	0.6085	−203
−202	12.682	4.099*	0.009740	2.4046	102.666	0.41587	47.952	58.941	106.893	0.3792	0.6080	−202
−201	13.207	3.032*	0.009758	2.3158	102.476	0.43181	48.173	58.816	106.988	0.3801	0.6074	−201
−200	13.748	1.930*	0.009777	2.2311	102.286	0.44822	48.394	58.689	107.083	0.3809	0.6069	−200
−199	14.306	0.793*	0.009795	2.1501	102.095	0.46510	48.617	58.561	107.178	0.3818	0.6064	−199
−198	14.882	0.186	0.009813	2.0726	101.903	0.4825	48.840	58.432	107.272	0.3826	0.6059	−198
−197	15.476	0.780	0.009832	1.9986	101.711	0.5004	49.065	58.301	107.366	0.3835	0.6054	−197
−196	16.088	1.392	0.009850	1.9277	101.519	0.5187	49.291	58.169	107.460	0.3843	0.6049	−196
−195	16.719	2.023	0.009869	1.8599	101.325	0.5377	49.517	58.036	107.554	0.3852	0.6045	−195
−194	17.370	2.674	0.009888	1.7950	101.131	0.5571	49.745	57.902	107.647	0.3861	0.6040	−194
−193	18.039	3.343	0.009907	1.7330	100.937	0.5770	49.970	57.770	107.740	0.3869	0.6035	−193
−192	18.728	4.032	0.009926	1.6735	100.742	0.5975	50.199	57.634	107.833	0.3877	0.6030	−192
−191	19.438	4.742	0.009946	1.6165	100.546	0.6186	50.428	57.497	107.925	0.3886	0.6026	−191
−190	20.168	5.472	0.009965	1.5620	100.350	0.6402	50.655	57.363	108.018	0.3894	0.6021	−190
−189	20.919	6.223	0.009985	1.5095	100.153	0.6625	50.889	57.220	108.109	0.3903	0.6017	−189
−188	21.692	6.996	0.010004	1.4592	99.955	0.6853	51.123	57.078	108.201	0.3912	0.6012	−188
−187	22.486	7.790	0.010024	1.4111	99.757	0.7087	51.354	56.938	108.292	0.3920	0.6008	−187
−186	23.303	8.607	0.010044	1.3649	99.558	0.7326	51.584	56.799	108.383	0.3928	0.6004	−186
−185	24.142	9.446	0.010065	1.3204	99.359	0.7574	51.823	56.650	108.473	0.3937	0.5999	−185
−184	25.004	10.308	0.010085	1.2778	99.159	0.7826	52.057	56.506	108.563	0.3945	0.5995	−184
−183	25.889	11.193	0.010105	1.2368	98.958	0.8085	52.293	56.360	108.653	0.3954	0.5991	−183
−182	26.798	12.102	0.010126	1.1975	98.757	0.8351	52.530	56.213	108.742	0.3962	0.5987	−182
−181	27.732	13.036	0.010147	1.1596	98.554	0.8624	52.771	56.059	108.831	0.3971	0.5983	−181
−180	28.690	13.994	0.010168	1.1232	98.352	0.8903	53.007	55.912	108.920	0.3979	0.5978	−180
−179	29.673	14.977	0.010189	1.0883	98.148	0.9188	53.244	55.765	109.008	0.3988	0.5974	−179
−178	30.682	15.986	0.010210	1.0546	97.944	0.9482	53.486	55.609	109.096	0.3996	0.5970	−178
−177	31.716	17.020	0.010231	1.0224	97.739	0.9781	53.723	55.460	109.183	0.4005	0.5967	−177
−176	32.777	18.081	0.010253	0.9911	97.533	1.0090	53.969	55.300	109.270	0.4013	0.5963	−176
−175	33.864	19.168	0.010275	0.9611	97.327	1.0405	54.214	55.142	109.356	0.4022	0.5959	−175
−174	34.979	20.283	0.010297	0.9322	97.120	1.0727	54.455	54.987	109.442	0.4030	0.5955	−174
−173	36.121	21.425	0.010319	0.9044	96.912	1.1057	54.700	54.827	109.527	0.4039	0.5951	−173
−172	37.290	22.594	0.010341	0.8776	96.703	1.1395	54.945	54.668	109.613	0.4047	0.5947	−172
−171	38.489	23.793	0.010363	0.8517	96.493	1.1741	55.192	54.505	109.697	0.4056	0.5944	−171
−170	39.716	25.020	0.010386	0.8268	96.283	1.2095	55.440	54.341	109.781	0.4064	0.5940	−170

*Inches of mercury below one atmosphere.

TEMP. °F	PRESSURE		VOLUME cu ft/lb		DENSITY lb/cu ft		ENTHALPY Btu/lb			ENTROPY Btu/(lb)(°R)		TEMP °F
	PSIA	PSIG	LIQUID v_f	VAPOR v_g	LIQUID $1/v_f$	VAPOR $1/v_g$	LIQUID h_f	LATENT h_{fg}	VAPOR h_g	LIQUID s_f	VAPOR s_g	
−170	39.716	25.020	0.010386	0.8268	96.283	1.2095	55.440	54.341	109.781	0.4064	0.5940	−170
−169	40.972	26.276	0.010409	0.8028	96.072	1.2457	55.688	54.177	109.864	0.4073	0.5936	−169
−168	42.258	27.562	0.010432	0.7796	95.860	1.2827	55.936	54.012	109.947	0.4081	0.5933	−168
−167	43.574	28.878	0.010455	0.7573	95.648	1.3205	56.181	53.850	110.030	0.4089	0.5929	−167
−166	44.921	30.225	0.010478	0.7357	95.434	1.3593	56.434	53.678	110.112	0.4098	0.5926	−166
−165	46.298	31.602	0.010502	0.7148	95.220	1.3990	56.687	53.506	110.193	0.4106	0.5922	−165
−164	47.707	33.011	0.010526	0.6947	95.005	1.4395	56.939	53.335	110.274	0.4115	0.5919	−164
−163	49.148	34.452	0.010550	0.6733	94.788	1.4809	57.192	53.163	110.354	0.4123	0.5915	−163
−162	50.621	35.925	0.010574	0.6564	94.572	1.5234	57.450	52.983	110.433	0.4132	0.5912	−162
−161	52.126	37.430	0.010598	0.6384	94.554	1.5665	57.698	52.815	110.513	0.4140	0.5908	−161
−160	53.665	38.969	0.010623	0.6209	94.135	1.6106	57.952	52.639	110.592	0.4148	0.5905	−160
−159	55.237	40.541	0.010648	0.6039	93.915	1.6560	58.212	52.457	110.669	0.4157	0.5902	−159
−158	56.843	42.147	0.010678	0.5875	93.695	1.7020	58.466	52.281	110.747	0.4165	0.5898	−158
−157	58.484	43.788	0 010698	0.5717	93.473	1.7493	58.726	52.097	110.823	0.4174	0.5895	−157
−156	60.160	45.464	0.010724	0.5564	93.251	1.7974	58.984	51.915	110.899	0.4182	0.5892	−156
−155	61.870	47.174	0.010750	0.5415	93.027	1.8466	59.242	51.732	110.974	0.4191	0.5888	−155
−154	63.617	48.921	0.010773	0.5272	92.803	1.8969	59.502	51.546	111.049	0.4199	0.5885	−154
−153	65.399	50.703	0.010802	0.5133	92.578	1.9482	59.763	51.359	111.122	0.4207	0.5882	−153
−152	67.219	52.523	0.010828	0.4998	92.351	2.0006	60.025	51.171	111.196	0.4216	0.5879	−152
−151	69.075	54.379	0.010855	0.4868	92.124	2.0542	60.288	50.980	111.268	0.4224	0.5876	−151
−150	70.969	56.273	0.010882	0.4742	91.895	2.1089	60.553	50.786	111.339	0.4233	0.5873	−150
−149	72.901	58.203	0.010909	0.4619	91.666	2.1650	60.823	50.587	111.409	0.4241	0.5869	−149
−148	74.871	60.175	0.010937	0.4501	91.435	2.2215	61.078	50.403	111.481	0.4249	0.5866	−148
−147	76.880	62.184	0.010963	0.4386	91.203	2.2801	61.353	50.196	111.549	0.4258	0.5863	−147
−146	78.929	64.233	0.010993	0.4275	90.971	2.3393	61.615	50.003	111.618	0.4266	0.5860	−146
−145	81.017	66.321	0.011021	0.4167	90.737	2.3998	61.880	49.807	111.687	0.4275	0.5857	−145
−144	83.145	68.445	0.011050	0.4063	90.501	2.4615	62.146	49.608	111.754	0.4283	0.5854	−144
−143	85.314	70.618	0.011078	0.3961	90.265	2.5244	62.411	49.410	111.821	0.4291	0.5851	−143
−142	87.525	72.829	0.011108	0.3862	90.028	2.5891	62.686	49.199	111.885	0.4300	0.5848	−142
−141	89.776	75.080	0.011137	0.3767	89.789	2.6547	62.955	48.995	111.950	0.4308	0.5845	−141
−140	92.070	77.374	0.011167	0.3674	89.549	2.7216	63.224	48.790	112.014	0.4316	0.5842	−140
−139	94.406	79.710	0.011197	0.3584	89.308	2.7899	63.493	48.584	112.077	0.4324	0.5839	−139
−138	96.786	82.090	0.011228	0.3497	89.065	2.8600	63.771	48.367	112.138	0.4333	0.5836	−138
−137	99.208	84.512	0.011239	0.3412	88.821	2.9313	64.045	48.154	112.199	0.4341	0.5834	−137
−136	101.675	86.979	0.011290	0.3329	88.576	3.0036	64.315	47.945	112.260	0.4349	0.5831	−136
−135	104.186	89.490	0.011321	0.3250	88.330	3.0774	64.585	47.735	112.320	0.4358	0.5828	−135
−134	106.741	92.043	0.011353	0.3171	88.082	3.1532	64.863	47.515	112.377	0.4366	0.5825	−134
−133	109.342	94.646	0.011385	0.3096	87.832	3.2302	65.137	47.298	112.435	0.4374	0.5822	−133
−132	111.989	97.293	0.011418	0.3022	87.582	3.3094	65.421	47.069	112.489	0.4383	0.5819	−132
−131	114.681	99.985	0.011451	0.2951	87.329	3.3893	65.693	46.852	112.546	0.4391	0.5816	−131
−130	117.42	102.72	0.011484	0.2881	87.076	3.4714	65.975	46.624	112.599	0.4399	0.5813	−130
−129	120.21	105.51	0.011518	0.2813	86.820	3.5548	66.252	46.400	112.652	0.4407	0.5811	−129
−128	123.04	108.35	0.011552	0.2747	86.564	3.6401	66.533	46.171	112.704	0.4416	0.5808	−128
−127	125.92	111.23	0.011587	0.2683	86.305	3.7272	66.815	45.938	112.754	0.4424	0.5805	−127
−126	128.86	114.16	0.011622	0.2621	86.045	3.8158	67.096	45.708	112.804	0.4432	0.5802	−126
−125	131.84	117.14	0.011657	0.2560	85.784	3.9064	67.380	45.472	112.852	0.4441	0.5799	−125
−124	134.87	120.17	0.011693	0.2501	85.520	3.9981	67.657	45.243	112.900	0.4449	0.5796	−124
−123	137.94	123.25	0.011729	0.2444	85.255	4.0923	67.941	45.004	112.946	0.4457	0.5794	−123
−122	141.08	126.38	0.011766	0.2387	84.988	4.1888	68.230	44.759	112.990	0.4465	0.5791	−122
−121	144.26	129.56	0.011804	0.2333	84.720	4.2871	68.517	44.515	113.033	0.4474	0.5788	−121
−120	147.49	132.00	0.011841	0.2280	84.449	4.3867	68.801	44.275	113.075	0.4482	0.5785	−120
−119	150.77	136.00	0.011860	0.2228	84.177	4.4887	69.087	44.029	113.116	0.4490	0.5782	−119
−118	154.11	139.42	0.011919	0.2177	83.903	4.5931	69.377	43.778	113.155	0.4498	0.5779	−118
−117	157.50	142.81	0.011958	0.2128	83.626	4.6992	69.665	43.529	113.193	0.4506	0.5777	−117
−116	160.95	146.25	0.011998	0.2080	83.348	4.8079	69.956	43.273	113.229	0.4515	0.5774	−116
−115	164.45	149.75	0.012038	0.2033	83.067	4.9184	70.244	43.021	113.265	0.4523	0.5771	−115
−114	168.00	153.30	0.012079	0.1988	82.785	5.0309	70.533	42.766	113.299	0.4531	0.5768	−114
−113	171.61	156.91	0.012121	0.1943	82.500	5.1464	70.826	42.505	113.331	0.4539	0.5765	−113
−112	175.27	160.57	0.012163	0.1900	82.213	5.2640	71.118	42.244	113.362	0.4547	0.5762	−112
−111	178.99	164.30	0.012206	0.1857	81.924	5.3844	71.414	41.976	113.390	0.4556	0.5759	−111
−110	182.77	168.05	0.012250	0.1816	81.632	5.5069	71.707	41.711	113.418	0.4564	0.5757	−110

TEMP.	PRESSURE		VOLUME cu ft/lb		DENSITY lb/cu ft		ENTHALPY Btu/lb			ENTROPY Btu/(lb)(°R)		TEMP.
°F	PSIA	PSIG	LIQUID v_f	VAPOR v_g	LIQUID $1/v_f$	VAPOR $1/v_g$	LIQUID h_f	LATENT h_{fg}	VAPOR h_g	LIQUID s_f	VAPOR s_g	°F
−110	182.77	168.08	0.012250	0.1816	81.632	5.5069	71.707	41.711	113.418	0.4564	0.5757	−110
−109	186.61	171.91	0.012294	0.1776	81.338	5.6322	72.003	41.440	113.444	0.4572	0.5754	−109
−108	190.50	175.60	0.012339	0.1736	81.042	5.7600	72.299	41.169	113.468	0.4580	0.5751	−108
−107	194.45	179.76	0.012385	0.1698	80.742	5.8903	72.594	40.896	113.490	0.4588	0.5748	−107
−106	198.46	183.77	0.012432	0.1660	80.441	6.0237	72.892	40.618	113.511	0.4596	0.5745	−106
−105	202.53	187.84	0.012479	0.1623	80.136	6.1610	73.197	40.331	113.528	0.4605	0.5742	−105
−104	206.67	191.97	0.012527	0.1587	79.829	6.3008	73.500	40.044	113.544	0.4613	0.5739	−104
−103	210.66	196.16	0.012576	0.1552	79.518	6.4437	73.804	39.754	113.557	0.4621	0.5736	−103
−102	215.11	200.42	0.012625	0.1518	79.205	6.5894	74.107	39.463	113.570	0.4630	0.5733	−102
−101	219.43	204.73	0.012676	0.1484	78.889	6.7372	74.406	39.176	113.582	0.4638	0.5730	−101
−100	223.81	209.11	0.012726	0.1451	78.569	6.8904	74.715	38.873	113.589	0.4646	0.5727	−100
− 99	228.25	213.55	0.012780	0.1419	78.247	7.0457	75.020	38.575	113.595	0.4654	0.5724	− 99
− 98	232.76	218.06	0.012834	0.1388	77.921	7.2051	75.329	38.270	113.599	0.4662	0.5720	− 98
− 97	237.33	222.63	0.012888	0.1357	77.591	7.3670	75.633	37.969	113.601	0.4670	0.5717	− 97
− 96	241.97	227.27	0.012944	0.1327	77.258	7.5347	75.947	37.652	113.599	0.4679	0.5714	− 96
− 95	246.67	231.97	0.013000	0.1298	76.921	7.7053	76.258	37.337	113.595	0.4687	0.5711	− 95
− 94	251.44	236.75	0.013058	0.1269	76.580	7.8802	76.571	37.018	113.589	0.4695	0.5708	− 94
− 93	256.28	241.59	0.013117	0.1241	76.235	8.0593	76.886	36.694	113.580	0.4704	0.5704	− 93
− 92	261.19	246.49	0.013178	0.1213	75.886	8.2427	77.202	36.367	113.569	0.4712	0.5701	− 92
− 91	266.17	251.47	0.013239	0.1186	75.532	8.4303	77.518	36.037	113.555	0.4720	0.5697	− 91
− 90	271.21	256.52	0.013302	0.1160	75.174	8.6229	77.837	35.700	113.537	0.4728	0.5694	− 90
− 89	276.33	261.64	0.013367	0.1134	74.812	8.8215	78.161	35.354	113.515	0.4737	0.5691	− 89
− 88	281.52	266.83	0.013433	0.1108	74.444	9.0245	78.484	35.007	113.491	0.4745	0.5687	− 88
− 87	286.79	272.09	0.013501	0.1083	74.071	9.2328	78.810	34.654	113.464	0.4754	0.5683	− 87
− 86	292.12	277.43	0.013570	0.1059	73.693	9.4464	79.136	34.297	113.433	0.4762	0.5680	− 86
− 85	297.53	282.84	0.013641	0.1035	73.309	9.6656	79.464	33.936	113.399	0.4770	0.5676	− 85
− 84	303.02	288.32	0.013714	0.1011	72.919	9.8922	79.798	33.563	113.360	0.4779	0.5672	− 84
− 83	308.58	293.89	0.013789	0.0988	72.524	10.1245	80.132	33.186	113.318	0.4787	0.5668	− 83
− 82	314.23	299.53	0.013866	0.0965	72.121	10.3616	80.463	32.810	113.274	0.4796	0.5664	− 82
− 81	319.95	305.25	0.013945	0.0943	71.712	10.6090	80.806	32.415	113.221	0.4804	0.5660	− 81
− 80	325.74	311.05	0.014026	0.0920	71.296	10.8648	81.154	32.009	113.163	0.4813	0.5656	− 80
− 79	331.62	316.93	0.014110	0.0899	70.872	11.1247	81.496	31.609	113.105	0.4822	0.5652	− 79
− 78	337.59	322.89	0.014197	0.0878	70.440	11.3960	81.848	31.190	113.038	0.4831	0.5648	− 78
− 77	343.63	328.94	0.014286	0.0857	69.999	11.6741	82.199	30.769	112.968	0.4839	0.5643	− 77
− 76	349.76	335.07	0.014378	0.0836	69.550	11.9632	82.557	30.333	112.890	0.4848	0.5639	− 76
− 75	355.98	341.28	0.014474	0.0816	69.091	12.2609	82.917	29.891	112.808	0.4857	0.5634	− 75
− 74	362.28	347.58	0.014573	0.0796	68.621	12.5707	83.284	29.434	112.718	0.4866	0.5629	− 74
− 73	368.67	353.98	0.014675	0.0776	68.141	12.8916	83.655	28.966	112.621	0.4876	0.5625	− 73
− 72	375.15	360.46	0.014782	0.0756	67.649	13.2240	84.030	28.488	112.518	0.4885	0.5620	− 72
− 71	381.73	367.03	0.014893	0.0737	67.144	13.5704	84.412	27.994	112.406	0.4894	0.5614	− 71
− 70	388.39	373.70	0.015009	0.0718	66.625	13.9295	84.799	27.489	112.287	0.4904	0.5609	− 70
− 69	395.15	380.46	0.015130	0.0699	66.092	14.3062	85.195	26.963	112.157	0.4913	0.5603	− 69
− 68	402.01	387.51	0.015257	0.0680	65.542	14.7016	85.601	26.415	112.016	0.4923	0.5598	− 68
− 67	408.97	394.27	0.015391	0.0662	64.975	15.1103	86.009	25.860	111.868	0.4933	0.5592	− 67
− 66	416.02	401.33	0.015531	0.0643	64.388	15.5448	86.433	25.271	111.704	0.4943	0.5585	− 66
− 65	423.18	408.49	0.015679	0.0625	63.780	15.9949	86.858	24.674	111.532	0.4954	0.5579	− 65
− 64	430.45	415.75	0.015836	0.0607	63.147	16.4772	87.305	24.036	111.341	0.4964	0.5572	− 64
− 63	437.82	423.13	0.016003	0.0589	62.488	16.9837	87.760	23.377	111.136	0.4975	0.5565	− 63
− 62	445.30	430.61	0.016182	0.0571	61.798	17.5223	88.230	22.683	110.913	0.4987	0.5557	− 62
− 61	452.90	438.20	0.016374	0.0552	61.072	18.1028	88.723	21.943	110.666	0.4998	0.5549	− 61
− 60	460.61	445.91	0.016582	0.0534	60.307	18.7266	89.236	21.158	110.394	0.5011	0.5540	− 60
− 59	468.44	453.74	0.016809	0.0516	59.494	19.3949	89.769	20.330	110.099	0.5023	0.5531	− 59
− 58	476.39	461.69	0.017058	0.0497	58.623	20.1207	90.329	19.444	109.772	0.5037	0.5521	− 58
− 57	484.47	469.77	0.017336	0.0478	57.683	20.9293	90.930	18.460	109.399	0.5051	0.5510	− 57
− 56	492.67	477.98	0.017650	0.0458	56.656	21.8198	91.569	17.414	108.984	0.5066	0.5498	− 56
− 55	501.01	486.32	0.018013	0.0438	55.514	22.8519	92.278	16.212	108.490	0.5083	0.5484	− 55
− 54	509.49	494.79	0.018445	0.0416	54.216	24.0558	93.064	14.842	107.907	0.5102	0.5468	− 54
− 53	518.11	503.42	0.018981	0.0392	52.683	25.4972	93.958	13.244	107.202	0.5123	0.5449	− 53
− 52	526.88	512.19	0.019703	0.0365	50.755	27.3673	95.042	11.239	106.280	0.5149	0.5424	− 52
− 51	535.81	521.11	0.020862	0.0330	47.935	30.3214	96.562	8.254	104.816	0.5185	0.5387	− 51
− 50.19	543.16	528.46	0.025602	0.0256	39.060	39.0600	100.636	0.000	100.636	0.5284	0.5284	− 50.19

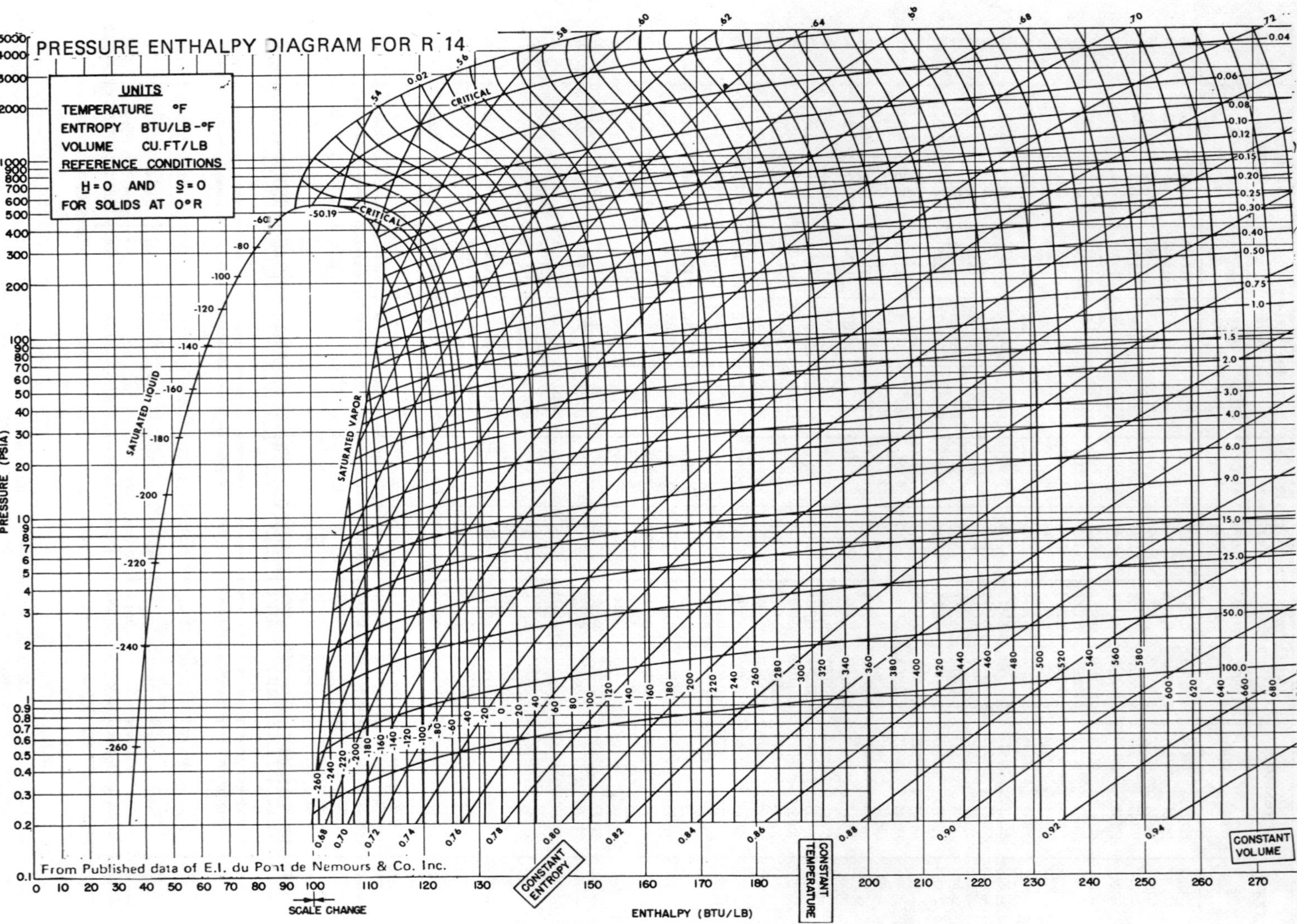

PRESSURE ENTHALPY DIAGRAM FOR R 14
UNITS
TEMPERATURE °F
ENTROPY BTU/LB-°F
VOLUME CU.FT/LB
REFERENCE CONDITIONS
H=0 AND S=0
FOR SOLIDS AT 0°R
CRITICAL
-50.19
SATURATED LIQUID
SATURATED VAPOR
CONSTANT ENTROPY
CONSTANT TEMPERATURE
CONSTANT VOLUME
PRESSURE (PSIA)
ENTHALPY (BTU/LB)
SCALE CHANGE
From Published data of E.I. du Pont de Nemours & Co. Inc.

DATA SHEET FOR R 21

$CHCl_2F$	DICHLOROMONOFLUOROMETHANE	Molecular Wt. 103

Pressure		Temperature				Volume		Density	
52.7	kg/cm²	178.5	°C	353.3	°F	1.916	l/kg	.522	kg/l.
750	psia	451.5	°K	813	°R	.0307	ft³/lb	32.6	lbs/ft³
								3.55	Air=1

AT CRITICAL POINT

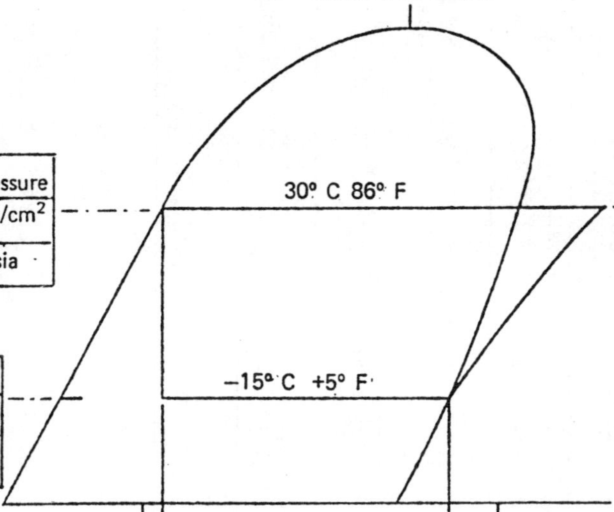

30° C 86° F

−15° C +5° F

Discharge Pressure	
2.2	kg/cm²
31.2	psia

Inlet Pressure	
.37	kg/cm²
5.3	psia

Discharge Temperature			
61	°C	142	°F
334	°K	602	°R

Normal Boiling Point			
8.9	°C	48	°F
282	°K	507	°R

Triple Point			
-135	°C	-211	°F
138	°K	249	°R

Refrigerating Effect	
87 kcal/m³	9.79 Btu/ft³

Latent Heat at NBP	57.9 kcal/kg	5964 /kg mol	104.2 Btu/lb	10733 /lb mol

Trouton's No.	21.17	Gas Constant	8.24 kg.m/kg/° K	15.01 ft.lbs/lb/° R

Specific Heat Liquid at 30° C 86° F	.256 Fig 13	Gas C_p .14	C_p/C_v 1.175 at 25°C

Liquid at 30° C 86° F Density	1.35 kg/l.	84.5 lbs/ft³	Viscosity .33 cp

Reduced Form at NBP Pressure	.0196	1 / Temperature	1.6

Reduced value of 1/T when Reduced Pressure is	2.02	1.69	1.37
	a) .001	b) .01	c) .1

TEMP.	PRESSURE		VOLUME cu ft/lb		DENSITY lb/cu ft		ENTHALPY Btu/lb			ENTROPY Btu/(lb)(°R)		TEMP
°F	PSIA	PSIG	LIQUID v_f	VAPOR v_g	LIQUID $1/v_f$	VAPOR $1/v_g$	LIQUID h_f	LATENT h_{fg}	VAPOR h_g	LIQUID s_f	VAPOR s_g	°F
-40	1.358	27.16*	0.01058	32.09	94.52	0.0311	0.00	114.56	114.56	0.0000	0.2730	-40
-38	1.453	26.96*	.01060	30.13	94.38	.0331	0.47	114.33	114.80	.0011	.2723	-38
-36	1.553	26.76*	.01061	28.32	94.23	.0353	0.94	114.10	115.04	.0022	.2716	-36
-34	1.659	26.54*	.01063	26.63	94.08	.0375	1.42	113.86	115.28	.0033	.2709	-34
-32	1.770	26.32*	.01065	25.07	93.93	.0398	1.89	113.63	115.52	.0044	.2702	-32
-30	1.888	26.08*	0.01066	23.61	93.79	0.0423	2.36	113.40	115.76	0.0055	0.2695	-30
-28	2.012	25.82*	.01068	22.25	93.64	.0449	2.83	113.17	116.00	.0066	.2688	-28
-26	2.143	25.56*	.01070	20.98	93.49	.0476	3.30	112.94	116.24	.0077	.2682	-26
-24	2.281	25.28*	.01071	19.80	93.34	.0505	3.77	112.71	116.48	.0088	.2675	-24
-22	2.426	24.98*	.01073	18.69	93.19	.0535	4.24	112.48	116.72	.0099	.2669	-22
-20	2.578	24.67*	0.01075	17.66	93.04	0.0566	4.71	112.25	116.96	0.0109	0.2663	-20
-18	2.739	24.34*	.01077	16.70	92.89	.0599	5.19	112.02	117.21	.0120	.2657	-18
-16	2.907	24.00*	.01078	15.80	92.74	.0633	5.66	111.79	117.45	.0131	.2651	-16
-14	3.083	23.64*	.01080	14.96	92.59	.0668	6.13	111.56	117.69	.0141	.2645	-14
-12	3.269	23.26*	.01082	14.17	92.44	.0706	6.60	111.33	117.93	.0152	.2639	-12
-10	3.463	22.87*	0.01084	13.43	92.28	0.0744	7.07	111.10	118.17	0.0162	0.2633	-10
- 8	3.667	22.45*	.01085	12.73	92.13	.0785	7.55	110.86	118.41	.0173	.2627	- 8
- 6	3.881	22.02*	.01087	12.08	91.98	.0828	8.02	110.63	118 65	.0183	.2622	- 6
- 4	4.104	21.56*	.01089	11.47	91.83	.0872	8.49	110.40	118.89	.0194	.2617	- 4
- 2	4.338	21.09*	.01091	10.89	91.67	.0918	8.97	110.16	119.13	.0204	.2611	- 2
0	4.582	20.59*	0.01093	10.35	91.52	0.0966	9.44	109.93	119.37	0.0214	0.2606	0
2	4.838	20.07*	.01095	9.840	91.36	.1016	9.92	109.69	119.61	.0225	.2601	2
4	5.105	19.53*	.01096	9.361	91.21	.1068	10.39	109.46	119.85	.0235	.2596	4
5†	**5.243**	**19.25***	**.01097**	**9.132**	**91.13**	**.1095**	**10.63**	**109.34**	**119.97**	**.0240**	**.2593**	**5†**
6	5.384	18.96*	.01098	8.910	91.05	.1122	10.86	109.23	120.09	.0245	.2591	6
8	5.674	18.37*	.01100	8.486	90.90	.1178	11.33	109.00	120.33	.0255	.2586	8
10	5.978	17.75*	0.01102	8.085	90.74	0.1237	11.81	108.76	120.57	0.0265	0.2581	10
12	6.294	17.11*	.01104	7.707	90.59	.1298	12.29	108.52	120.81	.0275	.2577	12
14	6.625	16.43*	.01106	7.349	90.43	.1361	12.77	108.28	121.05	.0286	.2572	14
16	6.968	15.73*	.01108	7.012	90.27	.1426	13.25	108.05	121.30	.0296	.2567	16
18	7.325	15.01*	.01110	6.694	90.11	.1494	13.73	107.81	121.54	.0306	.2563	18
20	7.699	14.25*	0.01112	6.392	89.96	0.1565	14.21	107.57	121.78	0.0316	0.2559	20
22	8.087	13.46*	.01114	6.107	89.80	.1638	14.68	107.34	122.02	.0326	.2555	22
24	8.488	12.64*	.01116	5.838	89.64	.1713	15.16	107.10	122.26	.0336	.2550	24
26	8.906	11.79*	.01118	5.584	89.48	.1791	15.64	106.86	122.50	.0346	.2546	26
28	9.341	10.90*	.01120	5.342	89.32	.1872	16.12	106.62	122.74	.0356	.2542	28
30	9.793	9.98*	0.01122	5.112	89.16	0.1956	16.61	106.37	122.98	0.0365	0.2538	30
32	10.26	9.03*	.01124	4.894	89.00	.2043	17.09	106.13	123.22	.0375	.2534	32
34	10.75	8.03*	.01126	4.688	88.84	.2133	17.58	105.88	123.46	.0385	.2530	34
36	11.26	7.00*	.01128	4.492	88.68	.2226	18.07	105.64	123.71	.0395	.2526	36
38	11.78	5.94*	.01130	4.306	88.52	.2322	18.55	105.40	123.95	.0405	.2523	38
40	12.32	4.84*	0.01132	4.130	88.35	0.2421	19.04	105.15	124.19	0.0414	0.2519	40
42	12.88	3.70*	.01134	3.963	88.19	.2523	19.53	104.90	124.43	.0424	.2516	42
44	13.46	2.52*	.01136	3.804	88.03	.2629	20.02	104.65	124.67	.0434	.2512	44
46	14.07	1.28*	.01138	3.651	87.86	.2739	20.51	104.40	124.91	.0444	.2509	46
48	14.69	0.01*	.01140	3.507	87.70	.2852	21.00	104.15	125.15	.0453	.2505	48
50	15.33	0.63	0.01142	3.370	87.54	0.2968	21.49	103.90	125.39	0.0463	0.2502	50
52	16.00	1.30	.01145	3.239	87.37	.3088	21.99	103.64	125.63	.0473	.2499	52
54	16.69	1.99	.01147	3.114	87.21	.3212	22.48	103.39	125.87	.0482	.2495	54
56	17.40	2.70	.01149	2 995	87.04	.3339	22.98	103.13	126.11	.0492	.2492	56
58	18.14	3.44	.01151	2.881	86.88	.3471	23.48	102.88	126.36	.0502	.2489	58
60	18.90	4.20	0.01152	2.773	86.71	0.3606	[illegible]	[illegible]	[illegible]	0.0511	[illegible]	60
62	19.69	4.99	.01155	2 669	86.54	.3747	24.49	102.35	126.84	.0521	.2483	62
64	20.50	5.80	.01158	2.570	86.38	.3891	24.99	102.09	127.08	.0530	.2480	64
66	21.34	6.64	.01160	2.476	86.21	.4039	25.49	101.83	127.32	.0540	.2477	66
68	22.20	7.50	.01162	2.386	86.04	.4191	25.99	101.57	127.56	.0549	.2474	68
70	23.08	8.38	0.01164	2.300	85.87	0.4346	26.49	101.30	127.79	0.0559	0.2471	70
72	24 00	9.30	.01167	2 218	85.71	.4508	27.00	101.03	128.03	.0568	.2469	72
74	24.95	10.25	.01169	2 139	85.54	.4675	27.51	100.76	128.27	.0577	.2466	74
76	25.92	11.22	.01171	2.064	85.37	.4846	28.02	100.49	128.51	.0587	.2463	76
78	26.93	12.23	.01174	1.992	85.20	.5021	28.52	100.22	128.74	.0596	.2460	78

*Inches of mercury below one atmosphere

TEMP.	PRESSURE		VOLUME cu ft/lb		DENSITY lb/cu ft		ENTHALPY Btu/lb			ENTROPY Btu/(lb)(°R)		TEMP
°F	PSIA	PSIG	LIQUID v_f	VAPOR v_g	LIQUID $1/v_f$	VAPOR $1/v_g$	LIQUID h_f	LATENT h_{fg}	VAPOR h_g	LIQUID s_f	VAPOR s_g	°F
80	27.96	13.26	0.01176	1.923	85.03	0.5201	29.03	99.95	128.98	0.0606	0.2458	80
82	29.02	14.32	.01178	1.857	84.86	.5386	29.54	99.67	129.21	.0615	.2455	82
84	30.11	15.41	.01181	1.794	84.69	.5575	30.05	99.39	129.44	.0624	.2453	84
86†	**31.23**	**16.53**	**.01183**	**1.733**	**84.52**	**.5770**	**30.56**	**99.12**	**129.68**	**.0634**	**.2450**	**86†**
88	32.39	17.69	.01186	1.675	84.34	.5971	31.07	98.84	129.91	.0643	.2448	88
90	33.58	18.88	0.01188	1.619	84.17	0.6177	31.59	98.55	130.14	0.0652	0.2446	90
92	34.80	20.10	.01190	1.565	84.00	.6388	32.11	98.27	130.38	.0662	.2443	92
94	36.06	21.36	.01193	1.513	83.83	.6606	32.63	97.98	130.61	.0671	.2441	94
96	37.35	22.65	.01195	1.464	83.65	.6829	33.15	97.69	130.84	.0680	.2439	96
98	38.68	23.98	.01198	1.417	83.48	.7058	33.67	97.40	131.07	.0690	.2437	98
100	40.04	25.34	0.01200	1.371	83.31	0.7292	34.18	97.11	131.29	0.0699	0.2434	100
102	41.44	26.74	.01203	1.327	83.13	.7533	34.70	96.82	131.52	.0708	.2432	102
104	42.88	28.18	.01205	1.285	82.96	.7779	35.23	96.52	131.75	.0717	.2430	104
106	44.35	29.65	.01208	1.245	82.78	.8033	35.75	96.22	131.97	.0727	.2428	106
108	45.85	31.15	.01211	1.206	82.61	.8291	36.27	95.93	132.20	.0736	.2426	108
110	47.40	32.70	0.01213	1.169	82.43	0.8557	36.79	95.63	132.42	0.0745	0.2424	110
112	48.98	34.28	.01216	1.133	82.25	.8827	37.32	95.32	132.64	.0754	.2422	112
114	50.61	35.91	.01218	1.098	82.08	.9107	37.86	95.01	132.87	.0763	.2420	114
116	52.28	37.58	.01221	1.064	81.90	.9393	38.39	94.70	133.09	.0773	.2418	116
118	54.00	39.30	.01224	1.032	81.72	.9688	38.92	94.39	133.31	.0782	.2416	118
120	55.75	41.05	0.01226	1.001	81.54	0.9986	39.46	94.08	133.53	0.0791	0.2414	120
122	57.54	42.84	.01229	0.971	81.36	1.029	39.99	93.76	133.75	.0800	.2412	122
124	59.38	44.68	.01232	.942	81.19	1.061	40.53	93.44	133.97	.0809	.2410	124
126	61.26	46.56	.01235	.914	81.01	1.093	41.06	93.12	134.18	.0818	.2409	126
128	63.18	48.48	.01237	.888	80.83	1.126	41.60	92.80	134.40	.0827	.2407	128
130	65.15	50.45	0.01240	0.862	80.65	1.160	42.13	92.48	134.61	0.0837	0.2405	130
132	67.17	52.47	.01243	.837	80.47	1.195	42.67	92.15	134.82	.0846	.2403	132
134	69.23	54.53	.01246	.813	80.28	1.230	43.22	91.82	135.03	.0855	.2402	134
136	71.34	56.64	.01248	.790	80.10	1.266	43.76	91.48	135.24	.0864	.2400	136
138	73.51	58.81	.01251	.767	79.92	1.303	44.31	91.14	135.45	.0873	.2398	138
140	75.72	61.02	0.01254	0.745	79.74	1.341	44.86	90.80	135.66	0.0882	0.2396	140
142	77.97	63.27	.01257	.724	79.56	1.380	45.41	90.46	135.87	.0891	.2395	142
144	80.29	65.59	.01260	.704	79.37	1.420	45.96	90.11	136.07	.0900	.2393	144
146	82.64	67.94	.01263	.684	79.19	1.461	46.51	89.77	136.28	.0909	.2392	146
148	85.05	70.35	.01266	.665	79.01	1.502	47.06	89.42	136.48	.0918	.2390	148
150	87.51	72.81	0.01269	0.647	78.82	1.544	47.62	89.06	136.68	0.0927	0.2388	150
152	90.01	75.31	.01272	.630	78.64	1.587	48.18	88.71	136.89	.0936	.2387	152
154	92.58	77.88	.01275	.612	78.45	1.631	48.74	88.35	137.09	.0945	.2385	154
156	95.20	80.50	.01278	.596	78.27	1.677	49.30	87.99	137.29	.0954	.2384	156
158	97.88	83.18	.01281	.580	78.08	1.724	49.87	87.62	137.49	.0963	.2382	158
160	100.6	85.91	0.01284	0.564	77.90	1.771	50.43	87.26	137.69	0.0972	0.2381	160

ENTHALPY (BTU/lb ABOVE SATURATED LIQUID AT –40° F

From Published data of E.I. du Pont de Nemours & Co. Inc. Used by Permission.

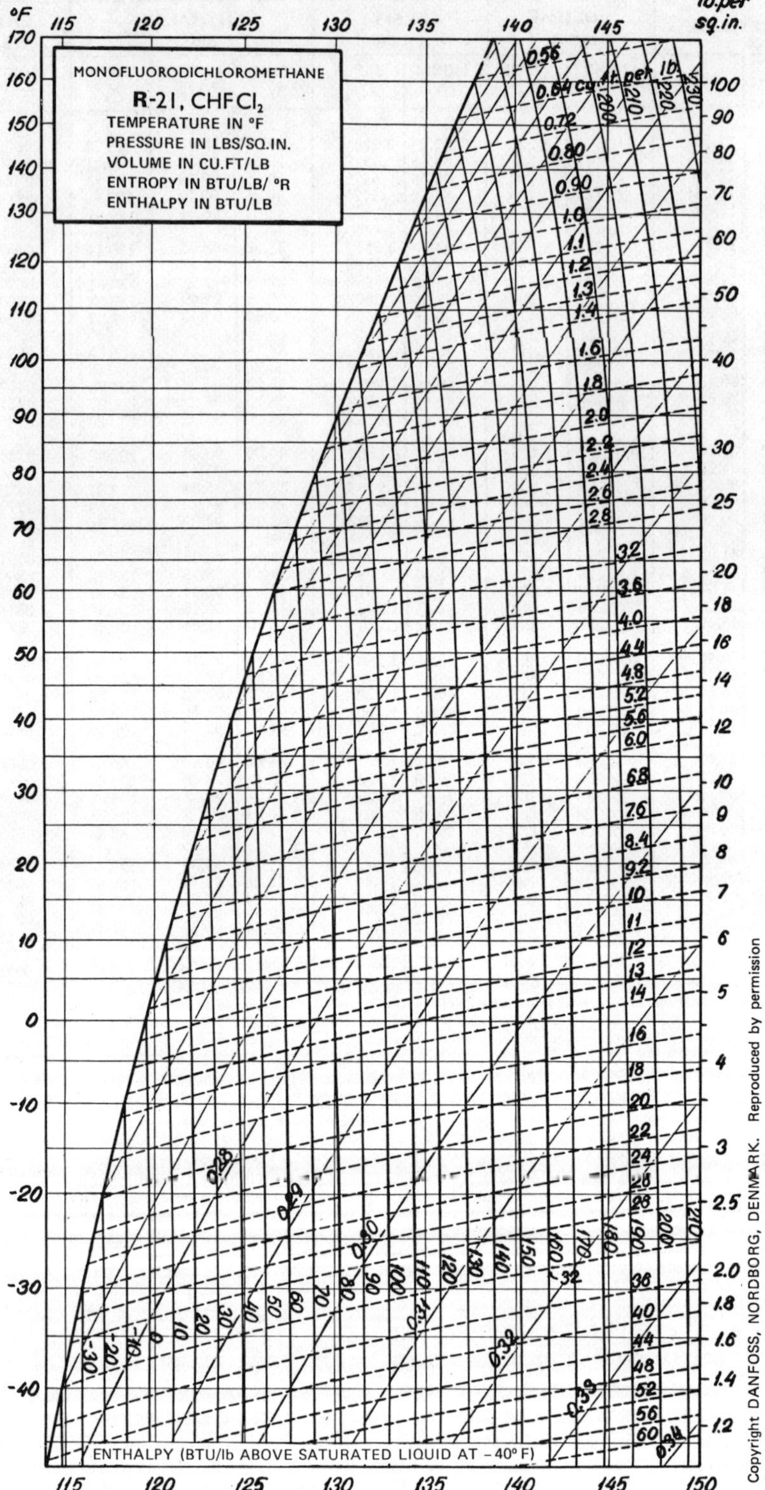
MONOFLUORODICHLOROMETHANE
R-21, $CHFCl_2$
TEMPERATURE IN °F
PRESSURE IN LBS/SQ.IN.
VOLUME IN CU.FT/LB
ENTROPY IN BTU/LB/ °R
ENTHALPY IN BTU/LB
°F
lb.per sq.in.
cu.ft.per lb.
ENTHALPY (BTU/lb ABOVE SATURATED LIQUID AT −40° F)

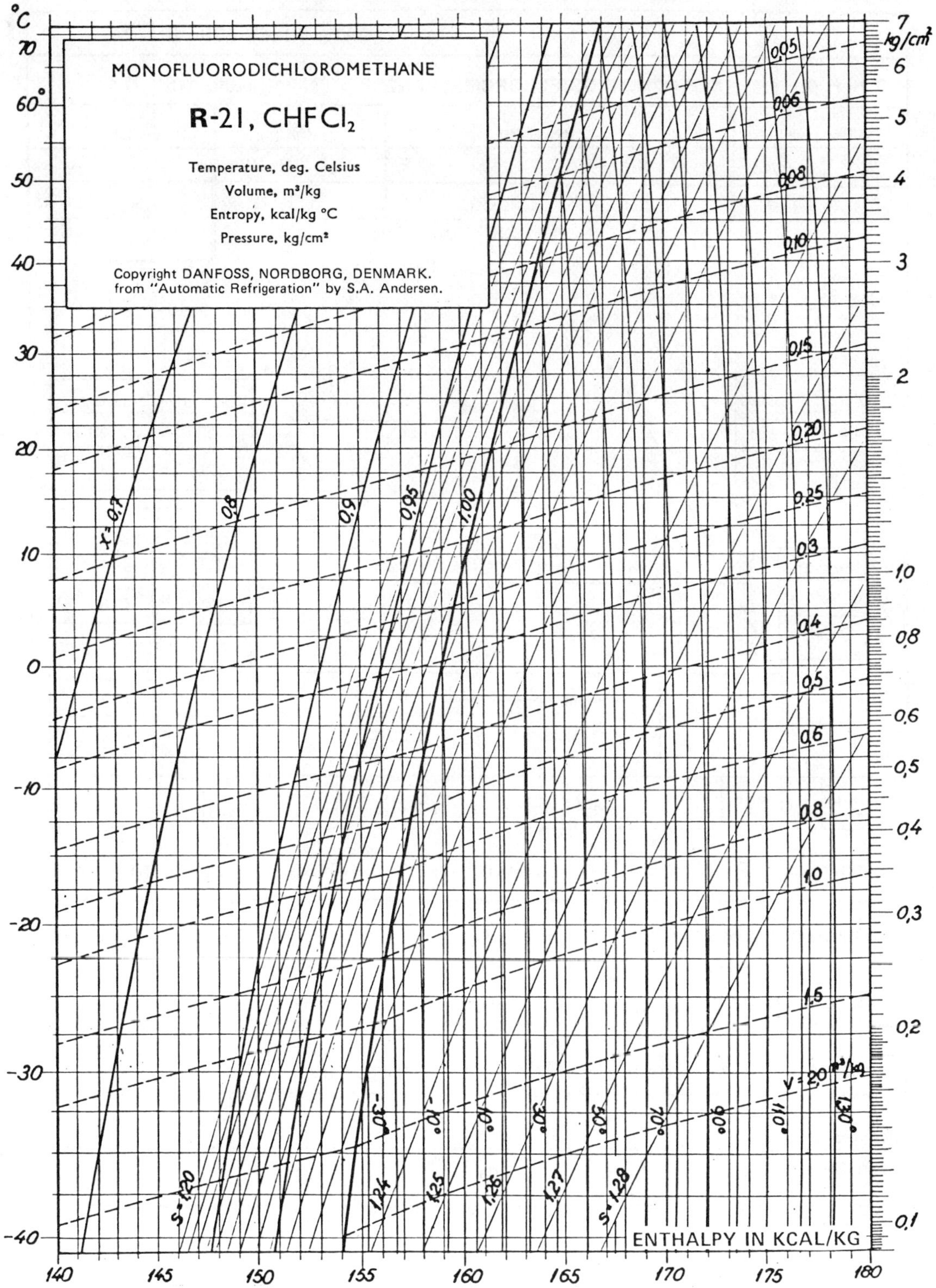
MONOFLUORODICHLOROMETHANE
R-21, CHFCl₂
Temperature, deg. Celsius
Volume, m³/kg
Entropy, kcal/kg °C
Pressure, kg/cm²
Copyright DANFOSS, NORDBORG, DENMARK.
from "Automatic Refrigeration" by S.A. Andersen.
°C
70
60°
50
40
30
20
10
0
-10
-20
-30
-40
7
kg/cm²
6
5
4
3
2
1,0
0,8
0,6
0,5
0,4
0,3
0,2
0,1
0,05
0,06
0,08
0,10
0,15
0,20
0,25
0,3
0,4
0,5
0,6
0,8
1,0
1,5
v = 20 m³/kg
x = 0,7
0,8
0,9
0,95
1,00
-30°
-10°
10°
30°
50°
70°
90°
110°
130°
s = 1,20
1,24
1,25
1,26
1,27
s = 1,28
ENTHALPY IN KCAL/KG
140
145
150
155
160
165
170
175
180

DATA SHEET FOR R 22

CHF_2Cl	MONOCHLORODIFLUOROMETHANE	Molecular Wt. 86.5

Pressure		Temperature				Volume		Density	
50.3	kg/cm²	96	°C	205	°F	1.904	l/kg	.525	kg/l.
716	psia	369	°K	655	°R	.0305	ft³/lb	32.8	lbs/ft³
								2.98	Air=1

AT CRITICAL POINT

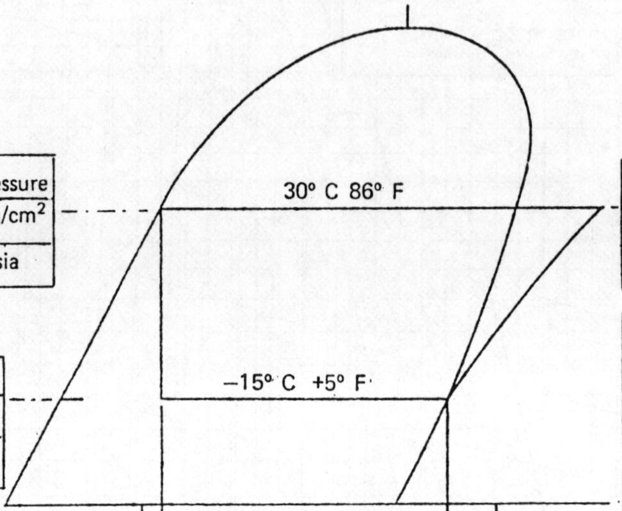

Discharge Pressure	
12.2	kg/cm²
173	psia

Inlet Pressure	
3.02	kg/cm²
42.9	psia

Discharge Temperature			
53	°C	128	°F
326	°K	588	°R

Normal Boiling Point			
-40.8	°C	-41.4	°F
232	°K	418	°R

Triple Point			
-160	°C	-256	°F
113	°K	204	°R

Refrigerating Effect			
501	kcal/m³	56.4	Btu/ft³

Latent Heat at NBP	56 kcal/kg	4844 /kg mol	100.7 Btu/lb	8710 /lb mol

Trouton's No.	20.85	Gas Constant 9.80 kg.m/kg/° K	17.78 ft.lbs/lb/° R

Specific Heat Liquid at 30° C 86° F	Fig 13	Gas C_p .157	C_p/C_v 1.184 at 25°C

Liquid at 30° C 86° F Density .117 kg/l.	73.3 lbs/ft³ Viscosity .18 cp

Reduced Form at NBP Pressure .0205	1 / Temperature 1.629

Reduced value of 1/T	2.02	1.169	1.37
when Reduced Pressure is	a) .001	b) .01	c) .1

TEMP.	PRESSURE		VOLUME cu ft/lb		DENSITY lb/cu ft		ENTHALPY Btu/lb			ENTROPY Btu/(lb)(°R)		TEMP.
°F	PSIA	PSIG	LIQUID v_f	VAPOR v_g	LIQUID $1/v_f$	VAPOR $1/v_g$	LIQUID h_f	LATENT h_{fg}	VAPOR h_g	LIQUID s_f	VAPOR s_g	°F
−155	0.208	29.49 *	0.01013	180.8	98.66	0.00553	−27.095	114.0	86.94	−0.07512	0.2991	−155
−154	0.220	29.47 *	0.01014	171.9	98.58	0.00581	−26.870	113.9	87.06	−0.07438	0.2983	−154
−153	0.232	29.44 *	0.01015	163.6	98.49	0.00611	−26.646	113.8	87.17	−0.07365	0.2974	−153
−152	0.244	29.42 *	0.01016	155.7	98.40	0.00642	−26.422	113.7	87.29	−0.07292	0.2966	−152
−151	0.257	29.39 *	0.01017	148.2	98.32	0.00674	−26.198	113.6	87.40	−0.07219	0.2958	−151
−150	0.271	29.36 *	0.01018	141.2	98.23	0.00708	−25.974	113.4	87.52	−0.07147	0.2950	−150
−149	0.285	29.33 *	0.01018	134.5	98.14	0.00743	−25.749	113.3	87.63	−0.07074	0.2942	−149
−148	0.300	29.30 *	0.01019	128.2	98.06	0.00779	−25.525	113.2	87.75	−0.07002	0.2934	−148
−147	0.316	29.27 *	0.01020	122.3	97.97	0.00817	−25.300	113.1	87.86	−0.06930	0.2926	−147
−146	0.332	29.24 *	0.01021	116.6	97.88	0.00857	−25.075	113.0	87.98	−0.06859	0.2918	−146
−145	0.349	29.20 *	0.01022	111.3	97.80	0.00898	−24.851	112.9	88.10	−0.06787	0.2910	−145
−144	0.367	29.17 *	0.01023	106.2	97.71	0.00940	−24.626	112.8	88.21	−0.06716	0.2902	−144
−143	0.386	29.13 *	0.01024	101.4	97.62	0.00985	−24.401	112.7	88.33	−0.06645	0.2895	−143
−142	0.405	29.09 *	0.01025	96.9	97.53	0.0103	−24.175	112.6	88.44	−0.06574	0.2887	−142
−141	0.425	29.05 *	0.01026	92.6	97.45	0.0107	−23.950	112.5	88.56	−0.06503	0.2880	−141
−140	0.446	29.01 *	0.01027	88.5	97.36	0.0112	−23.725	112.4	88.68	−0.06432	0.2872	−140
−139	0.468	28.96 *	0.01028	84.6	97.27	0.0118	−23.499	112.2	88.79	−0.06362	0.2865	−139
−138	0.491	28.92 *	0.01028	80.9	97.18	0.0123	−23.274	112.1	88.91	−0.06291	0.2858	−138
−137	0.515	28.87 *	0.01029	77.4	97.09	0.0129	−23.048	112.0	89.03	−0.06221	0.2851	−137
−136	0.540	28.82 *	0.01030	74.1	97.01	0.0134	−22.822	111.9	89.14	−0.06152	0.2844	−136
−135	0.565	28.76 *	0.01031	70.9	96.92	0.0140	−22.596	111.8	89.26	−0.06082	0.2836	−135
−134	0.592	28.71 *	0.01032	67.9	96.83	0.0147	−22.369	111.7	89.38	−0.06012	0.2829	−134
−133	0.620	28.65 *	0.01033	65.1	96.74	0.0153	−22.143	111.6	89.49	−0.05943	0.2823	−133
−132	0.649	28.59 *	0.01034	62.4	96.65	0.0160	−21.917	111.5	89.61	−0.05874	0.2816	−132
−131	0.679	28.53 *	0.01035	59.8	96.56	0.0167	−21.690	111.4	89.73	−0.05805	0.2809	−131
−130	0.710	28.47 *	0.01036	57.3	96.48	0.0174	−21.463	111.31	89.84	−0.05736	0.2802	−130
−129	0.743	28.40 *	0.01037	55.0	96.39	0.0181	−21.236	111.20	89.96	−0.05667	0.2796	−129
−128	0.776	28.33 *	0.01038	52.7	96.30	0.0189	−21.009	111.09	90.08	−0.05598	0.2789	−128
−127	0.811	28.26 *	0.01039	50.6	96.21	0.0197	−20.781	110.98	90.19	−0.05530	0.2782	−127
−126	0.847	28.19	0.01040	48.6	96.12	0.0205	−20.554	110.87	90.31	−0.05462	0.2776	−126
−125	0.885	28.11 *	0.01041	46.6	96.03	0.0214	−20.326	110.7	90.43	−0.05394	0.2770	−125
−124	0.924	28.03 *	0.01042	44.8	95.94	0.0222	−20.098	110.6	90.55	−0.05326	0.2763	−124
−123	0.964	27.95 *	0.01043	43.0	95.85	0.0232	−19.870	110.5	90.68	−0.05258	0.2757	−123
−122	1.00	27.87 *	0.01044	41.4	95.76	0.0241	−19.642	110.4	90.78	−0.05190	0.2751	−122
−121	1.05	27.78 *	0.01045	39.8	95.67	0.0251	−19.413	110.3	90.90	−0.05123	0.2744	−121
−120	1.09	27.69 *	0.01046	38.2	95.58	0.0261	−19.185	110.2	91.02	−0.05055	0.2738	−120
−119	1.14	27.59 *	0.01047	36.8	95.49	0.0271	−18.956	110.0	91.13	−0.04988	0.2732	−119
−118	1.19	27.49	0.01048	35.4	95.40	0.0282	−18.727	109.9	91.25	−0.04921	0.2726	−118
−117	1.24	27.39 *	0.01049	34.0	95.31	0.0293	−18.497	109.8	91.37	−0.04854	0.2720	−117
−116	1.29	27.29 *	0.01050	32.8	95.22	0.0304	−18.268	109.7	91.49	−0.04787	0.2714	−116
−115	1.34	27.18 *	0.01051	31.5	95.13	0.0316	−18.038	109.6	91.60	−0.04720	0.2709	−115
−114	1.40	27.06 *	0.01052	30.4	95.04	0.0328	−17.808	109.5	91.72	−0.04654	0.2703	−114
−113	1.45	26.95 *	0.01053	29.3	94.95	0.0341	−17.578	109.4	91.84	−0.04587	0.2697	−113
−112	1.51	26.83 *	0.01054	28.2	94.86	0.0354	−17.348	109.3	91.96	−0.04521	0.2691	−112
−111	1.57	26.70 *	0.01055	27.2	94.77	0.0367	−17.117	109.1	92.07	−0.04455	0.2686	−111
−110	1.64	26.57 *	0.01056	26.2	94.68	0.0381	−16.886	109.0	92.19	−0.04389	0.2680	−110
−109	1.70	26.44 *	0.01057	25.3	94.59	0.0395	−16.655	108.9	92.31	−0.04323	0.2675	−109
−108	1.77	26.30 *	0.01058	24.4	94.50	0.0409	−16.424	108.8	92.43	−0.04257	0.2669	−108
−107	1.84	26.16 *	0.01059	23.5	94.41	0.0424	−16.192	108.7	92.54	−0.04191	0.2664	−107
−106	1.91	26.02 *	0.01060	22.7	94.32	0.0440	−15.960	108.6	92.66	−0.04125	0.2658	−106
−105	1.99	25.86 *	0.01061	21.9	94.22	0.0455	−15.728	108.5	92.78	−0.04060	0.2653	−105
−104	2.06	25.71 *	0.01062	21.1	94.13	0.0472	−15.496	108.3	92.90	−0.03995	0.2648	−104
−103	2.14	25.55 *	0.01063	20.4	94.04	0.0489	−15.263	108.2	93.01	−0.03929	0.2642	−103
−102	2.22	25.38 *	0.01064	19.7	93.95	0.0506	−15.031	108.1	93.13	−0.03864	0.2637	−102
−101	2.31	25.21 *	0.01065	19.0	93.86	0.0524	−14.798	108.0	93.25	−0.03799	0.2632	−101

*Inches of mercury below one atmosphere

TEMP.	PRESSURE		VOLUME cu ft/lb		DENSITY lb/cu ft		ENTHALPY Btu/lb			ENTROPY Btu/(lb)(°R)		TEMP.
°F	PSIA	PSIG	LIQUID v_f	VAPOR v_g	LIQUID $1/v_f$	VAPOR $1/v_g$	LIQUID h_f	LATENT h_{fg}	VAPOR h_g	LIQUID s_f	VAPOR s_g	°F
−100	2.39	25.03 *	0.01066	18.43	93.77	0.05425	−14.56	107.9	93.37	−0.03734	0.2627	−100
− 99	2.48	24.85 *	0.01067	17.81	93.67	0.05613	−14.33	107.8	93.48	−0.03670	0.2622	− 99
− 98	2.57	24.66 *	0.01068	17.22	93.58	0.05806	−14.09	107.7	93.60	−0.03605	0.2617	− 98
− 97	2.67	24.47 *	0.01069	16.65	93.49	0.06005	−13.86	107.5	93.72	−0.03540	0.2612	− 97
− 96	2.77	24.27 *	0.01070	16.10	93.40	0.06209	−13.62	107.4	93.84	−0.03476	0.2607	− 96
− 95	2.87	24.07 *	0.01071	15.57	93.30	0.06419	−13.39	107.3	93.95	−0.03411	0.2602	− 95
− 94	2.97	23.86 *	0.01072	15.07	93.21	0.06634	−13.15	107.2	94.07	−0.03347	0.2597	− 94
− 93	3.08	23.64 *	0.01073	14.58	93.12	0.06855	−12.92	107.1	94.19	−0.03283	0.2592	− 93
− 92	3.19	23.41 *	0.01074	14.11	93.03	0.07083	−12.68	106.9	94.30	−0.03219	0.2588	− 92
− 91	3.30	23.18 *	0.01076	13.66	92.93	0.07316	−12.45	106.8	94.42	−0.03155	0.2583	− 91
− 90	3.42	22.95 *	0.01077	13.23	92.84	0.07555	−12.21	106.7	94.54	−0.03091	0.2578	− 90
− 89	3.54	22.70 *	0.01078	12.81	92.75	0.07801	−11.97	106.6	94.66	−0.03027	0.2574	− 89
− 88	3.66	22.45 *	0.01079	12.41	92.65	0.08053	−11.74	106.5	94.77	−0.02963	0.2569	− 88
− 87	3.79	22.20 *	0.01080	12.03	92.56	0.08312	−11.50	106.4	94.89	−0.02900	0.2564	− 87
− 86	3.92	21.93 *	0.01081	11.65	92.46	0.08577	−11.26	106.2	95.01	−0.02836	0.2560	− 86
− 85	4.05	21.66 *	0.01082	11.30	92.37	0.08849	−11.03	106.1	95.12	−0.02773	0.2556	− 85
− 84	4.19	21.38 *	0.01083	10.95	92.28	0.09128	−10.79	106.0	95.24	−0.02709	0.2551	− 84
− 83	4.33	21.09 *	0.01084	10.62	92.18	0.09413	−10.55	105.9	95.36	−0.02646	0.2547	− 83
− 82	4.48	20.79 *	0.01085	10.30	92.09	0.09706	−10.31	105.7	95.47	−0.02583	0.2542	− 82
− 81	4.62	20.49 *	0.01087	9.992	91.99	0.1000	−10.07	105.6	95.59	−0.02520	0.2538	− 81
− 80	4.78	20.18 *	0.01088	9.694	91.90	0.1031	− 9.838	105.5	95.71	−0.02457	0.2534	− 80
− 79	4.93	19.86 *	0.01089	9.407	91.81	0.1063	− 9.599	105.4	95.82	−0.02394	0.2529	− 79
− 78	5.10	19.53 *	0.01090	9.130	91.71	0.1095	− 9.359	105.3	95.94	−0.02331	0.2525	− 78
− 77	5.26	19.19 *	0.01091	8.862	91.62	0.1128	− 9.119	105.1	96.05	−0.02269	0.2521	− 77
− 76	5.43	18.85 *	0.01092	8.604	91.52	0.1162	− 8.878	105.0	96.17	−0.02206	0.2517	− 76
− 75	5.61	18.49 *	0.01093	8.355	91.43	0.1196	− 8.638	104.9	96.29	−0.02143	0.2513	− 75
− 74	5.78	18.13 *	0.01094	8.114	91.33	0.1232	− 8.397	104.8	96.405	−0.02081	0.2509	− 74
− 73	5.97	17.75 *	0.01096	7.882	91.24	0.1268	− 8.155	104.6	96.52	−0.02019	0.2505	− 73
− 72	6.16	17.37 *	0.01097	7.657	91.14	0.1305	− 7.914	104.5	96.63	−0.01956	0.2501	− 72
− 71	6.35	16.98 *	0.01098	7.441	91.04	0.1343	− 7.672	104.4	96.75	−0.01894	0.2497	− 71
− 70	6.55	16.58 *	0.01099	7.231	90.95	0.1382	− 7.429	104.2	96.86	−0.01832	0.2493	− 70
− 69	6.75	16.16 *	0.01100	7.029	90.85	0.1422	− 7.187	104.1	96.98	−0.01770	0.2489	− 69
− 68	6.96	15.74 *	0.01101	6.833	90.76	0.1463	− 6.944	104.0	97.09	−0.01708	0.2485	− 68
− 67	7.17	15.31 *	0.01103	6.644	90.66	0.1504	− 6.700	103.9	97.21	−0.01646	0.2481	− 67
− 66	7.39	14.86 *	0.01104	6.462	90.56	0.1547	− 6.457	103.7	97.32	−0.01584	0.2477	− 66
− 65	7.61	14.41 *	0.01105	6.285	90.47	0.1591	− 6.213	103.6	97.44	−0.01522	0.2474	− 65
− 64	7.84	13.94 *	0.01106	6.114	90.37	0.1635	− 5.968	103.5	97.55	−0.01460	0.2470	− 64
− 63	8.08	13.46 *	0.01107	5.949	90.27	0.1680	− 5.724	103.3	97.67	−0.01399	0.2466	− 63
− 62	8.32	12.98 *	0.01108	5.789	90.18	0.1727	− 5.479	103.2	97.78	−0.01337	0.2462	− 62
− 61	8.56	12.47 *	0.01110	5.634	90.08	0.1774	− 5.233	103.1	97.90	−0.01276	0.2459	− 61
− 60	8.81	11.96 *	0.01111	5.484	89.98	0.1823	− 4.987	103.0	98.01	−0.01214	0.2455	− 60
− 59	9.07	11.44 *	0.01112	5.339	89.88	0.1872	− 4.741	102.8	98.12	−0.01153	0.2452	− 59
− 58	9.33	10.90 *	0.01113	5.198	89.79	0.1923	− 4.495	102.7	98.24	−0.01092	0.2448	− 58
− 57	9.60	10.35 *	0.01114	5.062	89.69	0.1975	− 4.248	102.6	98.35	−0.01030	0.2444	− 57
− 56	9.88	9.79 *	0.01116	4.931	89.59	0.2027	− 4.001	102.4	98.46	−0.00969	0.2441	− 56
− 55	10.1	9.22 *	0.01117	4.803	89.49	0.2081	− 3.754	102.3	98.58	−0.00908	0.2437	− 55
− 54	10.4	8.63 *	0.01118	4.679	89.39	0.2136	− 3.506	102.2	98.69	−0.00847	0.2434	− 54
− 53	10.7	8.03 *	0.01119	4.560	89.30	0.2192	− 3.258	102.0	98.80	−0.00786	0.2431	− 53
− 52	11.0	7.42 *	0.01121	4.444	89.20	0.2250	− 3.009	101.9	98.92	−0.00725	0.2427	− 52
− 51	11.3	6.79 *	0.01122	4.331	89.10	0.2308	− 2.761	101.7	99.03	−0.00665	0.2424	− 51
− 50	11.6	6.15 *	0.01123	4.222	89.00	0.2368	− 2.511	101.6	99.14	−0.00604	0.2420	− 50
− 49	11.9	5.49 *	0.01124	4.116	88.90	0.2429	− 2.262	101.5	99.25	−0.00543	0.2417	− 49
− 48	12.3	4.82 *	0.01126	4.014	88.80	0.2491	− 2.012	101.3	99.36	−0.00483	0.2414	− 48
− 47	12.6	4.14 *	0.01127	3.914	88.70	0.2554	− 1.762	101.2	99.48	−0.00422	0.2411	− 47
− 46	13.0	3.44 *	0.01128	3.817	88.60	0.2619	− 1.511	101.1	99.59	−0.00361	0.2407	− 46

* Inches of mercury below one atmosphere

TEMP.	PRESSURE		VOLUME cu ft/lb		DENSITY lb/cu ft		ENTHALPY Btu/lb			ENTROPY Btu/(lb)(°R)		TEMP.
°F	PSIA	PSIG	LIQUID v_f	VAPOR v_g	LIQUID $1/v_f$	VAPOR $1/v_g$	LIQUID h_f	LATENT h_{fg}	VAPOR h_g	LIQUID s_f	VAPOR s_g	°F
−45	13.35	2.732*	0.01129	3.724	88.50	0.2685	−1.260	100.9	99.70	−0.00301	0.2404	−45
−44	13.71	2.002*	0.01131	3.633	88.40	0.2752	−1.009	100.8	99.81	−0.00241	0.2401	−44
−43	14.07	1.258*	0.01132	3.545	88.30	0.2820	−0.757	100.6	99.92	−0.00181	0.2398	−43
−42	14.45	0.498*	0.01133	3.459	88.20	0.2890	−0.505	100.5	100.0	−0.00120	0.2395	−42
−41	14.83	0.137	0.01135	3.376	88.10	0.2961	−0.253	100.3	100.1	−0.00060	0.2391	−41
−40	15.22	0.526	0.01136	3.295	88.00	0.3034	0.000	100.2	100.2	0.00000	0.2388	−40
−39	15.61	0.923	0.01137	3.217	87.90	0.3108	0.253	100.1	100.3	0.00060	0.2385	−39
−38	16.02	1.32	0.01138	3.141	87.80	0.3183	0.506	99.9	100.4	0.00120	0.2382	−38
−37	16.43	1.74	0.01140	3.067	87.70	0.3260	0.760	99.8	100.5	0.00180	0.2379	−37
−36	16.85	2.16	0.01141	2.995	87.60	0.3338	1.014	99.6	100.6	0.00240	0.2376	−36
−35	17.29	2.59	0.01142	2.925	87.50	0.3418	1.269	99.5	100.8	0.00300	0.2373	−35
−34	17.72	3.03	0.01144	2.857	87.39	0.3499	1.524	99.3	100.9	0.00359	0.2370	−34
−33	18.17	3.48	0.01145	2.791	87.29	0.3581	1.779	99.2	101.0	0.00419	0.2367	−33
−32	18.63	3.93	0.01146	2.727	87.19	0.3666	2.035	99.0	101.1	0.00479	0.2364	−32
−31	19.09	4.40	0.01148	2.665	87.09	0.3751	2.291	98.9	101.2	0.00538	0.2362	−31
−30	19.57	4.87	0.01149	2.604	86.99	0.3838	2.547	98.8	101.3	0.00598	0.2359	−30
−29	20.05	5.36	0.01150	2.546	86.88	0.3927	2.804	98.6	101.4	0.00657	0.2356	−29
−28	20.54	5.85	0.01152	2.488	86.78	0.4018	3.061	98.5	101.5	0.00716	0.2353	−28
−27	21.05	6.53	0.01153	2.432	86.68	0.4110	3.318	98.3	101.6	0.00776	0.2350	−27
−26	21.56	6.86	0.01155	2.378	86.57	0.4204	3.576	98.2	101.7	0.00835	0.2347	−26
−25	22.08	7.39	0.01156	2.326	86.47	0.4299	3.834	98.0	101.8	0.00894	0.2345	−25
−24	22.61	7.92	0.01157	2.274	86.37	0.4396	4.093	97.89	101.9	0.00953	0.2342	−24
−23	23.15	8.46	0.01159	2.224	86.26	0.4495	4.352	97.74	102.0	0.01013	0.2339	−23
−22	23.71	9.01	0.01160	2.176	86.16	0.4595	4.611	97.59	102.2	0.01072	0.2336	−22
−21	24.27	9.57	0.01162	2.128	86.06	0.4697	4.871	97.43	102.3	0.01131	0.2334	−21
−20	24.84	10.1	0.01163	2.082	85.95	0.4801	5.131	97.28	102.4	0.01189	0.2331	−20
−19	25.42	10.7	0.01164	2.037	85.85	0.4907	5.391	97.12	102.5	0.01248	0.2328	−19
−18	26.02	11.3	0.01166	1.994	85.74	0.5015	5.652	96.97	102.6	0.01307	0.2326	−18
−17	26.62	11.9	0.01167	1.951	85.64	0.5124	5.913	96.81	102.7	0.01366	0.2323	−17
−16	27.23	12.5	0.01169	1.909	85.53	0.5235	6.175	96.66	102.8	0.01425	0.2321	−16
−15	27.86	13.1	0.01170	1.869	85.43	0.5348	6.436	96.50	102.9	0.01483	0.2318	−15
−14	28.50	13.8	0.01172	1.830	85.32	0.5464	6.699	96.34	103.0	0.01542	0.2315	−14
−13	29.14	14.4	0.01173	1.791	85.22	0.5581	6.961	96.18	103.1	0.01600	0.2313	−13
−12	29.80	15.1	0.01174	1.754	85.11	0.5699	7.224	96.02	103.2	0.01659	0.2310	−12
−11	30.48	15.7	0.01176	1.718	85.00	0.5820	7.488	95.86	103.3	0.01717	0.2308	−11
−10	31.16	16.4	0.01177	1.682	84.90	0.5943	7.751	95.70	103.4	0.01776	0.2305	−10
− 9	31.85	17.1	0.01179	1.647	84.79	0.6068	8.015	95.54	103.5	0.01834	0.2303	− 9
− 8	32.56	17.8	0.01180	1.614	84.68	0.6195	8.280	95.38	103.6	0.01892	0.2300	− 8
− 7	33.28	18.5	0.01182	1.581	84.58	0.6324	8.545	95.21	103.7	0.01950	0.2298	− 7
− 6	34.01	19.3	0.01183	1.549	84.47	0.6455	8.810	95.05	103.8	0.02009	0.2296	− 6
− 5	34.75	20.0	0.01185	1.517	84.36	0.6588	9.075	94.88	103.9	0.02067	0.2293	− 5
− 4	35.50	20.8	0.01186	1.487	84.25	0.6724	9.341	94.72	104.0	0.02125	0.2291	− 4
− 3	36.27	21.5	0.01188	1.457	84.15	0.6861	9.608	94.55	104.1	0.02183	0.2288	− 3
− 2	37.05	22.3	0.01189	1.428	84.04	0.7001	9.874	94.39	104.2	0.02241	0.2286	− 2
1	37.85	23.1	0.01191	1.400	83.93	0.7143	10.14	94.22	104.3	0.02299	0.2284	− 1
0	38.65	23.9	0.01193	1.372	83.82	0.7287	10.40	94.05	104.4	0.02357	0.2281	0
1	39.47	24.7	0.01194	1.345	83.71	0.7433	10.67	93.88	104.5	0.02414	0.2279	1
2	40.30	25.6	0.01196	1.318	83.60	0.7582	10.94	93.71	104.6	0.02472	0.2277	2
3	41.15	26.4	0.01197	1.293	83.49	0.7733	11.21	93.54	104.7	0.02530	0.2274	3
4	42.01	27.3	0.01199	1.268	83.38	0.7886	11.48	93.37	104.8	0.02587	0.2272	4
5	42.88	28.1	0.01200	1.243	83.27	0.8042	11.75	93.20	104.9	0.02645	0.2270	5
6	43.77	29.0	0.01202	1.219	83.16	0.8200	12.02	93.03	105.0	0.02703	0.2268	6
7	44.67	29.9	0.01204	1.196	83.05	0.8360	12.29	92.86	105.1	0.02760	0.2265	7
8	45.59	30.8	0.01205	1.173	82.94	0.8523	12.56	92.68	105.2	0.02818	0.2263	8
9	46.52	31.8	0.01207	1.150	82.83	0.8689	12.83	92.51	105.3	0.02875	0.2261	9

*Inches of mercury below one atmosphere

TEMP. °F	PRESSURE		VOLUME cu ft/lb		DENSITY lb/cu ft		ENTHALPY Btu/lb			ENTROPY Btu/(lb)(°R)		TEMP °F
	PSIA	PSIG	LIQUID v_f	VAPOR v_g	LIQUID $1/v_f$	VAPOR $1/v_g$	LIQUID h_f	LATENT h_{fg}	VAPOR h_g	LIQUID s_f	VAPOR s_g	
10	47.46	32.76	0.01208	1.129	82.72	0.8857	13.10	92.33	105.4	0.02932	0.2259	10
11	48.42	33.72	0.01210	1.107	82.61	0.9027	13.37	92.16	105.5	0.02990	0.2257	11
12	49.39	34.70	0.01212	1.086	82.50	0.9200	13.64	91.98	105.6	0.03047	0.2254	12
13	50.38	35.68	0.01213	1.066	82.38	0.9376	13.92	91.80	105.7	0.03104	0.2252	13
14	51.38	36.69	0.01215	1.046	82.27	0.9554	14.19	91.63	105.8	0.03161	0.2250	14
15	52.40	37.70	0.01217	1.027	82.16	0.9735	14.46	91.45	105.9	0.03218	0.2248	15
16	53.43	38.74	0.01218	1.008	82.05	0.9918	14.73	91.27	106.0	0.03275	0.2246	16
17	54.48	39.79	0.01220	0.9896	81.93	1.010	15.01	91.09	106.1	0.03332	0.2244	17
18	55.55	40.85	0.01222	0.9714	81.82	1.029	15.28	90.91	106.1	0.03389	0.2242	18
19	56.63	41.93	0.01223	0.9536	81.71	1.048	15.56	90.72	106.2	0.03446	0.2240	19
20	57.72	43.03	0.01225	0.9363	81.59	1.068	15.83	90.54	106.3	0.03503	0.2237	20
21	58.83	44.14	0.01227	0.9193	81.48	1.087	16.11	90.36	106.4	0.03560	0.2235	21
22	59.96	45.27	0.01229	0.9027	81.36	1.107	16.38	90.17	106.5	0.03617	0.2233	22
23	61.11	46.41	0.01230	0.8864	81.25	1.128	16.66	89.99	106.6	0.03674	0.2231	23
24	62.27	47.57	0.01232	0.8705	81.13	1.148	16.94	89.80	106.7	0.03730	0.2229	24
25	63.45	48.75	0.01234	0.8550	81.02	1.169	17.21	89.62	106.8	0.03787	0.2227	25
26	64.64	49.94	0.01236	0.8397	80.90	1.190	17.49	89.43	106.9	0.03844	0.2225	26
27	65.85	51.15	0.01237	0.8248	80.79	1.212	17.77	89.24	107.0	0.03900	0.2223	27
28	67.08	52.38	0.01239	0.8103	80.67	1.234	18.05	89.05	107.1	0.03958	0.2221	28
29	68.32	53.63	0.01241	0.7960	80.55	1.256	18.33	88.86	107.1	0.04013	0.2219	29
30	69.59	54.89	0.01243	0.7820	80.44	1.278	18.60	88.67	107.2	0.04070	0.2217	30
31	70.87	56.17	0.01245	0.7684	80.32	1.301	18.88	88.48	107.3	0.04126	0.2215	31
32	72.16	57.47	0.01246	0.7550	80.20	1.324	19.16	88.29	107.4	0.04182	0.2213	32
33	73.48	58.78	0.01248	0.7419	80.08	1.347	19.44	88.09	107.5	0.04239	0.2211	33
34	74.81	60.12	0.01250	0.7291	79.97	1.371	19.72	87.90	107.6	0.04295	0.2210	34
35	76.17	61.47	0.01252	0.7165	79.85	1.395	20.01	87.70	107.7	0.04351	0.2208	35
36	77.54	62.84	0.01254	0.7042	79.73	1.419	20.29	87.51	107.8	0.04407	0.2206	36
37	78.92	64.23	0.01256	0.6922	79.61	1.444	20.57	87.31	107.8	0.04464	0.2204	37
38	80.33	65.64	0.01257	0.6804	79.49	1.469	20.85	87.11	107.9	0.04520	0.2202	38
39	81.76	67.06	0.01259	0.6688	79.37	1.495	21.13	86.92	108.0	0.04576	0.2200	39
40	83.20	68.51	0.01261	0.6575	79.25	1.520	21.42	86.72	108.1	0.04632	0.2198	40
41	84.67	69.97	0.01263	0.6464	79.13	1.546	21.70	86.52	108.2	0.04688	0.2196	41
42	86.15	71.45	0.01265	0.6355	79.01	1.573	21.98	86.31	108.3	0.04744	0.2194	42
43	87.65	72.95	0.01267	0.6249	78.89	1.600	22.27	86.11	108.3	0.04800	0.2193	43
44	89.17	74.48	0.01269	0.6144	78.77	1.627	22.55	85.91	108.4	0.04855	0.2191	44
45	90.71	76.02	0.01271	0.6042	78.64	1.654	22.84	85.71	108.5	0.04911	0.2189	45
46	92.28	77.58	0.01273	0.5942	78.52	1.682	23.12	85.50	108.6	0.04967	0.2187	46
47	93.86	79.16	0.01275	0.5844	78.40	1.711	23.41	85.30	108.7	0.05023	0.2185	47
48	95.46	80.76	0.01277	0.5747	78.28	1.739	23.70	85.09	108.7	0.05079	0.2183	48
49	97.08	82.38	0.01279	0.5653	78.15	1.768	23.98	84.88	108.8	0.05134	0.2182	49
50	98.72	84.03	0.01281	0.5560	78.03	1.798	24.27	84.67	108.9	0.05190	0.2180	50
51	100.3	85.63	0.01283	0.5469	77.90	1.828	24.56	84.46	109.0	0.05245	0.2178	51
52	102.0	87.38	0.01285	0.5380	77.78	1.858	24.85	84.25	109.1	0.05301	0.2176	52
53	103.7	89.08	0.01287	0.5293	77.65	1.889	25.13	84.04	109.1	0.05357	0.2175	53
54	105.5	90.81	0.01289	0.5207	77.53	1.920	25.42	83.83	109.2	0.05412	0.2173	54
55	107.2	92.56	0.01291	0.5123	77.40	1.951	25.71	83.62	109.3	0.05468	0.2171	55
56	109.0	94.32	0.01294	0.5041	77.28	1.983	26.00	83.40	109.4	0.05523	0.2169	56
57	110.8	96.11	0.01296	0.4960	77.15	2.015	26.29	83.19	109.4	0.05579	0.2167	57
58	112.6	97.93	0.01298	0.4881	77.02	2.048	26.58	82.97	109.5	0.05634	0.2166	58
59	114.4	99.76	0.01300	0.4803	76.90	2.081	26.88	82.75	109.6	0.05689	0.2164	59
60	116.3	101.6	0.01302	0.4727	76.77	2.115	27.17	82.54	109.7	0.05745	0.2162	60
61	118.1	103.4	0.01304	0.4652	76.64	2.149	27.46	82.32	109.7	0.05800	0.2161	61
62	120.0	105.3	0.01306	0.4578	76.51	2.184	27.75	82.10	109.8	0.05855	0.2159	62
63	122.0	107.3	0.01309	0.4506	76.38	2.219	28.05	81.87	109.9	0.05910	0.2157	63
64	123.9	109.2	0.01311	0.4435	76.25	2.254	28.34	81.65	110.0	0.05966	0.2155	64

TEMP. °F	PRESSURE PSIA	PSIG	VOLUME cu ft/lb LIQUID v_f	VAPOR v_g	DENSITY lb/cu ft LIQUID $1/v_f$	VAPOR $1/v_g$	ENTHALPY Btu/lb LIQUID h_f	LATENT h_{fg}	VAPOR h_g	ENTROPY Btu/(lb)(°R) LIQUID s_f	VAPOR s_g	TEMP. °F
65	125.9	111.2	0.01313	0.4366	76.12	2.290	28.63	81.43	110.0	0.06021	0.2154	65
66	127.9	113.2	0.01315	0.4298	75.99	2.326	28.93	81.20	110.1	0.06076	0.2152	66
67	129.9	115.2	0.01318	0.4231	75.86	2.363	29.22	80.98	110.2	0.06131	0.2150	67
68	131.9	117.2	0.01320	0.4165	75.73	2.400	29.52	80.75	110.2	0.06186	0.2149	68
69	134.0	119.3	0.01322	0.4100	75.60	2.438	29.81	80.52	110.3	0.06241	0.2147	69
70	136.1	121.4	0.01325	0.4037	75.46	2.476	30.11	80.29	110.4	0.06296	0.2145	70
71	138.2	123.5	0.01327	0.3975	75.33	2.515	30.41	80.06	110.4	0.06351	0.2143	71
72	140.3	125.6	0.01329	0.3913	75.20	2.555	30.71	79.83	110.5	0.06406	0.2142	72
73	142.5	127.8	0.01332	0.3853	75.06	2.594	31.00	79.60	110.6	0.06461	0.2140	73
74	144.7	130.0	0.01334	0.3794	74.93	2.635	31.30	79.37	110.6	0.06516	0.2138	74
75	146.9	132.2	0.01336	0.3736	74.79	2.676	31.60	79.13	110.7	0.06571	0.2137	75
76	149.1	134.4	0.01339	0.3680	74.66	2.717	31.90	78.89	110.8	0.06626	0.2135	76
77	151.4	136.7	0.01341	0.3624	74.52	2.759	32.20	78.66	110.8	0.06681	0.2133	77
78	153.6	138.9	0.01344	0.3569	74.39	2.801	32.50	78.42	110.9	0.06736	0.2132	78
79	155.9	141.3	0.01346	0.3515	74.25	2.844	32.80	78.18	110.9	0.06791	0.2130	79
80	158.3	143.6	0.01349	0.3462	74.11	2.888	33.10	77.94	111.0	0.06846	0.2128	80
81	160.6	145.9	0.01351	0.3409	73.97	2.932	33.41	77.70	111.1	0.06901	0.2127	81
82	163.0	148.3	0.01354	0.3358	73.83	2.977	33.71	77.45	111.1	0.06956	0.2125	82
83	165.4	150.7	0.01356	0.3308	73.70	3.022	34.01	77.21	111.2	0.07011	0.2123	83
84	167.9	153.2	0.01359	0.3258	73.56	3.068	34.32	76.96	111.2	0.07065	0.2122	84
85	170.3	155.6	0.01362	0.3210	73.42	3.115	34.62	76.71	111.3	0.07120	0.2120	85
86	172.8	158.1	0.01364	0.3162	73.27	3.162	34.93	76.47	111.4	0.07175	0.2118	86
87	175.3	160.6	0.01367	0.3115	73.13	3.210	35.23	76.22	111.4	0.07230	0.2117	87
88	177.9	163.2	0.01370	0.3069	72.99	3.258	35.54	75.96	111.5	0.07285	0.2115	88
89	180.5	165.8	0.01372	0.3023	72.85	3.307	35.85	75.71	111.5	0.07339	0.2113	89
90	183.0	168.4	0.01375	0.2978	72.70	3.357	36.15	75.46	111.6	0.07394	0.2112	90
91	185.7	171.0	0.01378	0.2934	72.56	3.407	36.46	75.20	111.6	0.07449	0.2110	91
92	188.3	173.6	0.01380	0.2891	72.41	3.458	36.77	74.94	111.7	0.07504	0.2108	92
93	191.0	176.3	0.01383	0.2849	72.27	3.509	37.08	74.69	111.7	0.07559	0.2107	93
94	193.7	179.0	0.01386	0.2807	72.12	3.562	37.39	74.43	111.8	0.07613	0.2105	94
95	196.5	181.8	0.01389	0.2766	71.98	3.615	37.70	74.16	111.8	0.07668	0.2103	95
96	199.2	184.5	0.01392	0.2725	71.83	3.668	38.01	73.90	111.9	0.07723	0.2102	96
97	202.0	187.3	0.01395	0.2685	71.68	3.723	38.32	73.64	111.9	0.07778	0.2100	97
98	204.8	190.1	0.01397	0.2646	71.53	3.778	38.64	73.37	112.0	0.07832	0.2098	98
99	207.7	193.0	0.01400	0.2608	71.38	3.834	38.95	73.10	112.0	0.07887	0.2097	99
100	210.6	195.9	0.01403	0.2570	71.23	3.890	39.26	72.83	112.1	0.07942	0.2095	100
101	213.5	198.8	0.01406	0.2532	71.08	3.948	39.58	72.56	112.1	0.07997	0.2093	101
102	216.4	201.7	0.01409	0.2496	70.93	4.006	39.89	72.29	112.1	0.08052	0.2092	102
103	219.4	204.7	0.01412	0.2460	70.78	4.065	40.21	72.02	112.2	0.08107	0.2090	103
104	222.4	207.7	0.01415	0.2424	70.62	4.124	40.53	71.74	112.2	0.08161	0.2088	104
105	225.4	210.7	0.01419	0.2389	70.47	4.185	40.84	71.46	112.3	0.08216	0.2087	105
106	228.5	213.8	0.01422	0.2354	70.31	4.246	41.16	71.18	112.3	0.08271	0.2085	106
107	231.5	216.9	0.01425	0.2320	70.16	4.308	41.48	70.90	112.3	0.08326	0.2083	107
108	234.7	220.0	0.01428	0.2287	70.00	4.371	41.80	70.62	112.4	0.08381	0.2082	108
109	237.8	223.1	0.01431	0.2254	69.84	4.435	42.12	70.33	112.4	0.08436	0.2080	109
110	241.0	226.3	0.01435	0.2222	69.68	4.500	42.44	70.05	112.4	0.08491	0.2078	110
111	244.2	229.5	0.01438	0.2190	69.52	4.565	42.76	69.76	112.5	0.08546	0.2077	111
112	247.5	232.8	0.01441	0.2158	69.36	4.632	43.09	69.47	112.5	0.08601	0.2075	112
113	250.7	236.0	0.01444	0.2127	69.20	4.699	43.41	69.18	112.5	0.08656	0.2073	113
114	254.0	239.3	0.01448	0.2097	69.04	4.767	43.73	68.88	112.6	0.08711	0.2071	114
115	257.4	242.7	0.01451	0.2067	68.88	4.837	44.06	68.59	112.6	0.08766	0.2070	115
116	260.7	246.1	0.01455	0.2037	68.71	4.907	44.39	68.29	112.6	0.08821	0.2068	116
117	264.2	249.5	0.01458	0.2008	68.55	4.978	44.71	67.99	112.7	0.08876	0.2066	117
118	267.6	252.9	0.01462	0.1980	68.38	5.050	45.04	67.68	112.7	0.08932	0.2064	118
119	271.1	256.4	0.01465	0.1951	68.22	5.123	45.37	67.38	112.7	0.08987	0.2063	119

TEMP. °F	PRESSURE PSIA	PRESSURE PSIG	VOLUME cu ft/lb LIQUID v_f	VOLUME cu ft/lb VAPOR v_g	DENSITY lb/cu ft LIQUID $1/v_f$	DENSITY lb/cu ft VAPOR $1/v_g$	ENTHALPY Btu/lb LIQUID h_f	ENTHALPY Btu/lb LATENT h_{fg}	ENTHALPY Btu/lb VAPOR h_g	ENTROPY Btu/(lb)(°R) LIQUID s_f	ENTROPY Btu/(lb)(°R) VAPOR s_g	TEMP. °F
120	274.6	259.9	0.01469	0.1923	68.05	5.198	45.70	67.07	112.7	0.09042	0.2061	120
121	278.1	263.4	0.01473	0.1896	67.88	5.273	46.03	66.76	112.8	0.09098	0.2059	121
122	281.7	267.0	0.01476	0.1869	67.71	5.349	46.36	66.45	112.8	0.09153	0.2057	122
123	285.3	270.6	0.01480	0.1842	67.54	5.427	46.70	66.14	112.8	0.09208	0.2056	123
124	288.9	274.2	0.01484	0.1816	67.37	5.505	47.03	65.82	112.8	0.09264	0.2054	124
125	292.6	277.9	0.01488	0.1790	67.19	5.585	47.36	65.50	112.8	0.09320	0.2052	125
126	296.3	281.6	0.01492	0.1764	67.02	5.666	47.70	65.18	112.8	0.09375	0.2050	126
127	300.0	285.3	0.01496	0.1739	66.84	5.748	48.04	64.86	112.9	0.09431	0.2048	127
128	303.8	289.1	0.01499	0.1714	66.67	5.831	48.38	64.53	112.9	0.09487	0.2046	128
129	307.6	292.9	0.01503	0.1690	66.49	5.916	48.71	64.20	112.9	0.09543	0.2044	129
130	311.5	296.8	0.01508	0.1666	66.31	6.002	49.05	63.87	112.9	0.09598	0.2043	130
131	315.3	300.6	0.01512	0.1642	66.13	6.089	49.40	63.54	112.9	0.09654	0.2041	131
132	319.2	304.6	0.01516	0.1618	65.94	6.177	49.74	63.20	112.9	0.09711	0.2039	132
133	323.2	308.5	0.01520	0.1595	65.76	6.267	50.08	62.86	112.9	0.09767	0.2037	133
134	327.2	312.5	0.01524	0.1572	65.58	6.358	50.43	62.52	112.9	0.09823	0.2035	134
135	331.2	316.5	0.01529	0.1550	65.39	6.451	50.77	62.17	112.9	0.09879	0.2033	135
136	335.3	320.6	0.01533	0.1527	65.20	6.545	51.12	61.82	112.9	0.09936	0.2031	136
137	339.4	324.7	0.01538	0.1505	65.01	6.640	51.47	61.47	112.9	0.09992	0.2029	137
138	343.5	328.8	0.01542	0.1484	64.82	6.737	51.82	61.12	112.9	0.1004	0.2027	138
139	347.7	333.0	0.01547	0.1462	64.63	6.835	52.17	60.76	112.9	0.1010	0.2025	139
140	351.9	337.2	0.01551	0.1441	64.44	6.936	52.52	60.40	112.9	0.1016	0.2023	140
141	356.1	341.5	0.01556	0.1420	64.24	7.037	52.88	60.03	112.9	0.1022	0.2021	141
142	360.4	345.7	0.01561	0.1400	64.04	7.141	53.23	59.67	112.9	0.1027	0.2019	142
143	364.8	350.1	0.01566	0.1380	63.84	7.246	53.59	59.29	112.8	0.1033	0.2017	143
144	369.1	354.4	0.01571	0.1360	63.64	7.352	53.95	58.92	112.8	0.1039	0.2015	144
145	373.5	358.8	0.01576	0.1340	63.44	7.461	54.31	58.54	112.8	0.1044	0.2013	145
146	378.0	363.3	0.01581	0.1320	63.24	7.571	54.67	58.15	112.8	0.1050	0.2010	146
147	382.5	367.8	0.01586	0.1301	63.03	7.684	55.04	57.77	112.8	0.1056	0.2008	147
148	387.0	372.3	0.01591	0.1282	62.82	7.798	55.40	57.38	112.7	0.1062	0.2006	148
149	391.5	376.8	0.01597	0.1263	62.61	7.914	55.77	56.98	112.7	0.1068	0.2004	149
150	396.1	381.5	0.01602	0.1244	62.40	8.033	56.14	56.58	112.7	0.1073	0.2002	150
151	400.8	386.1	0.01608	0.1226	62.18	8.153	56.51	56.18	112.6	0.1079	0.1999	151
152	405.5	390.8	0.01613	0.1208	61.97	8.276	56.88	55.77	112.6	0.1085	0.1997	152
153	410.2	395.5	0.01619	0.1190	61.75	8.401	57.26	55.35	112.6	0.1091	0.1995	153
154	415.0	400.3	0.01625	0.1172	61.52	8.528	57.63	54.93	112.5	0.1097	0.1992	154
155	419.8	405.1	0.01631	0.1155	61.30	8.658	58.01	54.51	112.5	0.1103	0.1990	155
156	424.6	409.9	0.01637	0.1137	61.07	8.790	58.39	54.08	112.4	0.1109	0.1987	156
157	429.5	414.8	0.01643	0.1120	60.84	8.924	58.78	53.65	112.4	0.1115	0.1985	157
158	434.5	419.8	0.01649	0.1103	60.61	9.062	59.16	53.21	112.3	0.1121	0.1982	158
159	439.5	424.8	0.01656	0.1086	60.38	9.202	59.55	52.76	112.3	0.1127	0.1980	159
160	444.5	429.8	0.01662	0.1070	60.14	9.344	59.94	52.31	112.2	0.1133	0.1977	160
161	449.5	434.9	0.01669	0.1053	59.90	9.490	60.34	51.85	112.1	0.1139	0.1975	161
162	454.7	440.0	0.01676	0.1037	59.66	9.639	60.73	51.39	112.1	0.1146	0.1972	162
163	459.8	445.1	0.01683	0.1021	59.41	9.791	61.13	50.92	112.0	0.1151	0.1969	163
164	465.0	450.3	0.01690	0.1005	59.16	9.946	61.53	50.44	111.9	0.1158	0.1966	164
165	470.3	455.6	0.01697	0.09895	58.90	10.10	61.94	49.96	111.9	0.1164	0.1964	165
166	475.6	460.9	0.01705	0.09739	58.65	10.26	62.35	49.46	111.8	0.1170	0.1961	166
167	480.9	466.2	0.01712	0.09584	58.39	10.43	62.76	48.96	111.7	0.1176	0.1958	167
168	486.3	471.6	0.01720	0.09430	58.12	10.60	63.17	48.46	111.6	0.1183	0.1955	168
169	491.7	477.0	0.01728	0.09278	57.85	10.77	63.59	47.94	111.5	0.1189	0.1952	169
170	497.2	482.5	0.01736	0.09127	57.58	10.95	64.01	47.41	111.4	0.1195	0.1949	170
171	502.7	488.0	0.01745	0.08978	57.30	11.13	64.44	46.88	111.3	0.1202	0.1945	171
172	508.3	493.6	0.01753	0.08829	57.01	11.32	64.87	46.34	111.2	0.1208	0.1942	172
173	513.9	499.3	0.01762	0.08682	56.73	11.51	65.31	45.78	111.0	0.1215	0.1939	173
174	519.6	504.9	0.01771	0.08536	56.43	11.71	65.75	45.22	110.9	0.1222	0.1935	174

TEMP. °F	PRESSURE		VOLUME cu ft/lb		DENSITY lb/cu ft		ENTHALPY Btu/lb			ENTROPY Btu/(lb)(°R)		TEMP. °F
	PSIA	PSIG	LIQUID v_f	VAPOR v_g	LIQUID $1/v_f$	VAPOR $1/v_g$	LIQUID h_f	LATENT h_{fg}	VAPOR h_g	LIQUID s_f	VAPOR s_g	
175	525.3	510.7	0.01781	0.08391	56.13	11.91	66.19	44.64	110.8	0.1228	0.1932	175
176	531.1	516.4	0.01791	0.08247	55.83	12.12	66.64	44.05	110.6	0.1235	0.1928	176
177	537.0	522.3	0.01801	0.08104	55.52	12.34	67.09	43.45	110.5	0.1242	0.1925	177
178	542.8	528.1	0.01811	0.07961	55.20	12.56	67.55	42.84	110.4	0.1249	0.1921	178
179	548.8	534.1	0.01822	0.07820	54.88	12.78	68.02	42.21	110.2	0.1256	0.1917	179
180	554.7	540.0	0.01833	0.07679	54.54	13.02	68.49	41.57	110.0	0.1263	0.1913	180
181	560.8	546.1	0.01844	0.07538	54.20	13.26	68.97	40.91	109.8	0.1270	0.1909	181
182	566.9	552.2	0.01856	0.07398	53.86	13.51	69.46	40.23	109.7	0.1277	0.1905	182
183	573.0	558.3	0.01869	0.07259	53.50	13.77	69.96	39.54	109.5	0.1285	0.1900	183
184	579.2	564.5	0.01882	0.07120	53.13	14.04	70.46	38.82	109.2	0.1292	0.1896	184
185	585.4	570.7	0.01895	0.06981	52.75	14.32	70.97	38.09	109.0	0.1300	0.1891	185
186	591.8	577.1	0.01909	0.06842	52.37	14.61	71.50	37.33	108.8	0.1308	0.1886	186
187	598.1	583.4	0.01924	0.06703	51.96	14.91	72.03	36.54	108.5	0.1316	0.1881	187
188	604.5	589.8	0.01939	0.06563	51.55	15.23	72.57	35.73	108.3	0.1324	0.1876	188
189	611.0	596.3	0.01956	0.06424	51.12	15.56	73.13	34.89	108.0	0.1332	0.1870	189
190	617.5	602.8	0.01973	0.06283	50.67	15.91	73.71	34.02	107.7	0.1340	0.1864	190
191	624.1	609.4	0.01991	0.06142	50.21	16.28	74.30	33.11	107.4	0.1349	0.1858	191
192	630.8	616.1	0.02011	0.05999	49.72	16.66	74.90	32.16	107.0	0.1358	0.1852	192
193	637.5	622.8	0.02031	0.05855	49.21	17.07	75.53	31.16	106.6	0.1367	0.1845	193
194	644.3	629.6	0.02054	0.05709	48.68	17.51	76.18	30.11	106.2	0.1377	0.1838	194
195	651.1	636.4	0.02078	0.05560	48.11	17.98	76.86	28.99	105.8	0.1387	0.1830	195
196	658.0	643.3	0.02104	0.05408	47.51	18.48	77.56	27.81	105.3	0.1397	0.1821	196
197	665.0	650.3	0.02133	0.05252	46.87	19.03	78.31	26.54	104.8	0.1408	0.1812	197
198	672.0	657.3	0.02165	0.05091	46.17	19.64	79.10	25.16	104.2	0.1420	0.1802	198
199	679.1	664.4	0.02201	0.04922	45.41	20.31	79.94	23.66	103.6	0.1432	0.1791	199
200	686.3	671.6	0.02243	0.04743	44.57	21.08	80.86	21.99	102.8	0.1446	0.1779	200
201	693.6	678.9	0.02293	0.04551	43.60	21.97	81.87	20.08	101.9	0.1460	0.1764	201
202	700.9	686.2	0.02354	0.04337	42.47	23.05	83.03	17.84	100.8	0.1477	0.1747	202
203	708.2	693.5	0.02435	0.04086	41.05	24.46	84.41	15.02	99.4	0.1498	0.1725	203
204	715.7	701.0	0.02564	0.03754	38.99	26.63	86.30	10.95	97.2	0.1526	0.1691	204
204.81	721.9	707.2	0.03052	0.03052	32.76	32.76	91.32	0.000	91.32	0.1601	0.1601	204.81

ENTHALPY (BTU/lb ABOVE SATURATED LIQUID AT –40° F)

From Published data of E.I. du Pont de Nemours & Co. Inc. Used by permission.

TEMP. °C	PRESSURE kg/cm²	VOLUME LIQUID l/kg v_f	VOLUME VAPOR m³/kg v_g	DENSITY LIQUID kg/l $1/v_f$	DENSITY VAPOR kg/m³ $1/v_g$	ENTHALPY kcal/kg LIQUID h_f	ENTHALPY kcal/kg LATENT h_{fg}	ENTHALPY kcal/kg VAPOR h_g	ENTROPY kcal/(kg) (°K) LIQUID s_f	ENTROPY kcal/(kg) (°K) VAPOR s_g	TEMP °C
−100	0.0212	0.6366	8.0089	1.5708	0.1249	75.154	62.973	138.127	0.88808	1.25175	−100
−99	0.0232	0.6376	7.3535	1.5683	0.1360	75.379	62.864	138.243	0.88937	1.25033	−99
−98	0.0253	0.6386	6.7595	1.5658	0.1479	75.604	62.755	138.359	0.89066	1.24894	−98
−97	0.0277	0.6397	6.2204	1.5633	0.1608	75.829	62.646	138.475	0.89194	1.24757	−97
−96	0.0302	0.6407	5.7306	1.5608	0.1745	76.054	62.537	138.591	0.89322	1.24622	−96
−95	0.0330	0.6418	5.2851	1.5582	0.1892	76.280	62.428	138.708	0.89449	1.24489	−95
−94	0.0359	0.6428	4.8793	1.5557	0.2050	76.506	62.319	138.824	0.89575	1.24359	−94
−93	0.0390	0.6439	4.5093	1.5532	0.2218	76.732	62.209	138.941	0.89701	1.24231	−93
−92	0.0424	0.6449	4.1717	1.5506	0.2397	76.958	62.100	139.058	0.89826	1.24105	−92
−91	0.0461	0.6460	3.8632	1.5481	0.2589	77.185	61.990	139.175	0.89951	1.23982	−91
−90	0.0500	0.6470	3.5810	1.5455	0.2793	77.412	61.880	139.292	0.90075	1.23860	−90
−89	0.0541	0.6481	3.3226	1.5429	0.3010	77.639	61.770	139.409	0.90199	1.23741	−89
−88	0.0586	0.6492	3.0857	1.5404	0.3241	77.867	61.660	139.526	0.90322	1.23623	−88
−87	0.0633	0.6503	2.8684	1.5378	0.3486	78.095	61.549	139.644	0.90445	1.23507	−87
−86	0.0684	0.6514	2.6689	1.5352	0.3747	78.323	61.438	139.761	0.90567	1.23394	−86
−85	0.0738	0.6525	2.4854	1.5326	0.4024	78.551	61.327	139.878	0.90689	1.23282	−85
−84	0.0796	0.6536	2.3166	1.5301	0.4317	78.780	61.216	139.996	0.90810	1.23172	−84
−83	0.0858	0.6547	2.1611	1.5275	0.4627	79.010	61.104	140.113	0.90931	1.23064	−83
−82	0.0923	0.6558	2.0178	1.5249	0.4956	79.239	60.992	140.231	0.91052	1.22958	−82
−81	0.0993	0.6569	1.8855	1.5223	0.5304	79.469	60.879	140.349	0.91171	1.22853	−81
−80	0.1067	0.6581	1.7633	1.5196	0.5671	79.700	60.767	140.466	0.91291	1.22750	−80
−79	0.1145	0.6592	1.6504	1.5170	0.6059	79.931	60.653	140.584	0.91410	1.22649	−79
−78	0.1228	0.6603	1.5459	1.5144	0.6469	80.162	60.540	140.702	0.91529	1.22550	−78
−77	0.1316	0.6615	1.4492	1.5118	0.6901	80.394	60.426	140.819	0.91647	1.22452	−77
−76	0.1410	0.6626	1.3595	1.5091	0.7356	80.626	60.311	140.937	0.91765	1.22355	−76
−75	0.1509	0.6638	1.2764	1.5065	0.7835	80.858	60.196	141.055	0.91883	1.22260	−75
−74	0.1613	0.6650	1.1992	1.5039	0.8339	81.092	60.081	141.172	0.92000	1.22167	−74
−73	0.1723	0.6661	1.1275	1.5012	0.8870	81.325	59.965	141.290	0.92117	1.22075	−73
−72	0.1840	0.6673	1.0608	1.4985	0.9427	81.559	59.848	141.407	0.92234	1.21985	−72
−71	0.1963	0.6685	0.9988	1.4959	1.0012	81.794	59.731	141.525	0.92350	1.21896	−71
−70	0.2093	0.6697	0.9410	1.4932	1.0627	82.029	59.613	141.642	0.92466	1.21809	−70
−69	0.2229	0.6709	0.8872	1.4905	1.1272	82.264	59.495	141.759	0.92581	1.21723	−69
−68	0.2373	0.6721	0.8370	1.4879	1.1948	82.500	59.376	141.876	0.92696	1.21638	−68
−67	0.2525	0.6733	0.7901	1.4852	1.2656	82.737	59.257	141.993	0.92811	1.21554	−67
−66	0.2684	0.6746	0.7464	1.4825	1.3398	82.974	59.136	142.110	0.92926	1.21472	−66
−65	0.2852	0.6758	0.7055	1.4798	1.4174	83.212	59.016	142.227	0.93040	1.21391	−65
−64	0.3027	0.6770	0.6673	1.4770	1.4986	83.450	58.894	142.344	0.93154	1.21312	−64
−63	0.3212	0.6783	0.6315	1.4743	1.5835	83.688	58.772	142.460	0.93268	1.21233	−63
−62	0.3406	0.6795	0.5980	1.4716	1.6722	83.928	58.649	142.577	0.93381	1.21156	−62
−61	0.3609	0.6808	0.5666	1.4689	1.7649	84.168	58.525	142.693	0.93495	1.21080	−61
−60	0.3822	0.6821	0.5372	1.4661	1.8616	84.408	58.401	142.809	0.93608	1.21005	−60
−59	0.4045	0.6833	0.5096	1.4634	1.9625	84.649	58.276	142.925	0.93720	1.20932	−59
50	0.4270	0.0040	0.4830	1.4606	2.0677	84.891	58.150	143.041	0.93833	1.20859	−58
−57	0.4523	0.6859	0.4593	1.4579	2.1773	85.133	58.023	143.156	0.93945	1.20787	−57
−56	0.4778	0.6872	0.4364	1.4551	2.2916	85.376	57.896	143.272	0.94057	1.20717	−56
−55	0.5045	0.6885	0.4149	1.4523	2.4105	85.619	57.768	143.387	0.94168	1.20648	−55
−54	0.5324	0.6899	0.3946	1.4496	2.5343	85.863	57.638	143.501	0.94279	1.20579	−54
−53	0.5615	0.6912	0.3755	1.4468	2.6631	86.107	57.508	143.616	0.94391	1.20512	−53
−52	0.5918	0.6925	0.3575	1.4440	2.7971	86.353	57.378	143.730	0.94501	1.20445	−52
−51	0.6235	0.6939	0.3406	1.4412	2.9363	86.598	57.246	143.844	0.94612	1.20380	−51
−50	0.6565	0.6952	0.3246	1.4384	3.0810	86.845	57.113	143.958	0.94723	1.20316	−50
−49	0.6910	0.6966	0.3095	1.4355	3.2312	87.092	56.980	144.072	0.94833	1.20252	−49
−48	0.7268	0.6980	0.2952	1.4327	3.3872	87.340	56.845	144.185	0.94943	1.20189	−48
−47	0.7641	0.6994	0.2818	1.4299	3.5490	87.588	56.710	144.298	0.95053	1.20128	−47
−46	0.8029	0.7008	0.2690	1.4270	3.7169	87.837	56.573	144.410	0.95162	1.20067	−46

TEMP. °C	PRESSURE kg/cm²	VOLUME LIQUID l/kg v_f	VOLUME VAPOR m³/kg v_g	DENSITY LIQUID kg/l $1/v_f$	DENSITY VAPOR kg/m³ $1/v_g$	ENTHALPY kcal/kg LIQUID h_f	ENTHALPY kcal/kg LATENT h_{fg}	ENTHALPY kcal/kg VAPOR h_g	ENTROPY kcal/(kg) (°K) LIQUID s_f	ENTROPY kcal/(kg) (°K) VAPOR s_g	TEMP °C
−45	0.8433	0.7022	0.2570	1.4242	3.8910	88.086	56.436	144.523	0.95271	1.20007	−45
−44	0.8853	0.7036	0.2456	1.4213	4.0715	88.337	56.298	144.634	0.95380	1.19948	−44
−43	0.9289	0.7050	0.2348	1.4184	4.2584	88.587	56.159	144.746	0.95489	1.19889	−43
−42	0.9743	0.7064	0.2246	1.4156	4.4521	88.839	56.018	144.857	0.95598	1.19832	−42
−41	1.0213	0.7079	0.2149	1.4127	4.6526	89.091	55.877	144.968	0.95707	1.19775	−41
−40	1.0701	0.7093	0.2058	1.4098	4.8601	89.344	55.735	145.079	0.95815	1.19719	−40
−39	1.1208	0.7108	0.1971	1.4069	5.0749	89.597	55.592	145.189	0.95923	1.19664	−39
−38	1.1733	0.7128	0.1888	1.4039	5.2970	89.851	55.447	145.299	0.96031	1.19609	−38
−37	1.2278	0.7138	0.1809	1.4010	5.5267	90.106	55.302	145.408	0.96139	1.19556	−37
−36	1.2842	0.7153	0.1735	1.3981	5.7640	90.361	55.156	145.517	0.96246	1.19503	−36
−35	1.3487	0.7168	0.1664	1.3951	6.0093	90.617	55.008	145.625	0.96353	1.19450	−35
−34	1.4032	0.7183	0.1597	1.3922	6.2627	90.874	54.859	145.733	0.96461	1.19399	−34
−33	1.4658	0.7198	0.1533	1.3892	6.5244	91.131	54.710	145.841	0.96567	1.19348	−33
−32	1.5306	0.7214	0.1472	1.3862	6.7946	91.389	54.559	145.948	0.96674	1.19298	−32
−31	1.5976	0.7229	0.1414	1.3833	7.0734	91.648	54.407	146.055	0.96781	1.19248	−31
−30	1.6669	0.7245	0.1359	1.3803	7.3611	91.907	54.254	146.161	0.96887	1.19199	−30
−29	1.7386	0.7261	0.1306	1.3773	7.6578	92.167	54.100	146.267	0.96993	1.19151	−29
−28	1.8126	0.7277	0.1256	1.3742	7.9638	92.428	53.944	146.372	0.97099	1.19103	−28
−27	1.8890	0.7293	0.1208	1.3712	8.2793	92.689	53.788	146.477	0.97205	1.19056	−27
−26	1.9679	0.7309	0.1162	1.3682	8.6044	92.951	53.630	146.581	0.97311	1.19009	−26
−25	2.0493	0.7325	0.1118	1.3651	8.9394	93.214	53.471	146.685	0.97416	1.18963	−25
−24	2.1333	0.7342	0.1077	1.3621	9.2846	93.477	53.311	146.788	0.97522	1.18918	−24
−23	2.2200	0.7358	0.1037	1.3590	9.6400	93.741	53.150	146.891	0.97627	1.18873	−23
−22	2.3094	0.7375	0.0999	1.3559	10.0060	94.006	52.987	146.993	0.97732	1.18829	−22
−21	2.4015	0.7392	0.0963	1.3528	10.3827	94.271	52.824	147.095	0.97836	1.18785	−21
−20	2.4964	0.7409	0.0928	1.3497	10.7704	94.537	52.659	147.196	0.97941	1.18742	−20
−19	2.5942	0.7426	0.0895	1.3466	11.1693	94.804	52.493	147.296	0.98046	1.18699	−19
−18	2.6949	0.7443	0.0864	1.3435	11.5797	95.071	52.325	147.396	0.98150	1.18657	−18
−17	2.7986	0.7461	0.0833	1.3403	12.0018	95.339	52.157	147.495	0.98254	1.18615	−17
−16	2.9053	0.7478	0.0804	1.3372	12.4357	95.608	51.987	147.594	0.98358	1.18574	−16
−15	3.0152	0.7496	0.0776	1.3340	12.8819	95.877	51.815	147.692	0.98462	1.18533	−15
−14	3.1281	0.7514	0.0750	1.3308	13.3404	96.147	51.643	147.790	0.98565	1.18492	−14
−13	3.2443	0.7532	0.0724	1.3276	13.8117	96.418	51.469	147.886	0.98669	1.18452	−13
−12	3.3638	0.7550	0.0700	1.3244	14.2958	96.689	51.294	147.983	0.98772	1.18413	−12
−11	3.4865	0.7569	0.0676	1.3212	14.7932	96.961	51.117	148.078	0.98875	1.18374	−11
−10	3.6127	0.7587	0.0653	1.3180	15.3040	97.234	50.939	148.173	0.98978	1.18335	−10
−9	3.7423	0.7606	0.0632	1.3147	15.8286	97.507	50.760	148.267	0.99081	1.18297	−9
−8	3.8754	0.7625	0.0611	1.3115	16.3671	97.781	50.579	148.361	0.99184	1.18259	−8
−7	4.0121	0.7644	0.0591	1.3082	16.9200	98.056	50.397	148.453	0.99286	1.18221	−7
−6	4.1524	0.7663	0.0572	1.3049	17.4874	98.332	50.214	148.546	0.99389	1.18184	−6
−5	4.2964	0.7683	0.0553	1.3016	18.0697	98.608	50.029	148.637	0.99491	1.18147	−5
−4	4.4441	0.7703	0.0536	1.2083	18.6671	98.885	49.842	148.728	0.99593	1.18111	−4
−3	4.5957	0.7722	0.0519	1.2949	19.2800	99.163	49.655	148.817	0.99695	1.18075	−3
−2	4.7511	0.7742	0.0502	1.2916	19.9087	99.441	49.466	148.907	0.99797	1.18039	−2
−1	4.9104	0.7763	0.0487	1.2882	20.5535	99.720	49.275	148.995	0.99898	1.18004	−1
0	5.0738	0.7783	0.0471	1.2848	21.2147	100.000	49.083	149.083	1.00000	1.17968	0
1	5.2412	0.7804	0.0457	1.2814	21.8927	100.281	48.889	149.169	1.00101	1.17934	1
2	5.4127	0.7825	0.0443	1.2780	22.5877	100.562	48.694	149.255	1.00203	1.17899	2
3	5.5884	0.7846	0.0429	1.2746	23.3002	100.844	48.497	149.341	1.00304	1.17865	3
4	5.7684	0.7867	0.0416	1.2711	24.0305	101.126	48.298	149.425	1.00405	1.17831	4
5	5.9527	0.7889	0.0404	1.2676	24.7788	101.410	48.098	149.508	1.00506	1.17797	5
6	6.1413	0.7910	0.0391	1.2642	25.5457	101.694	47.897	149.591	1.00606	1.17764	6
7	6.3344	0.7932	0.0380	1.2606	26.3315	101.979	47.694	149.673	1.00707	1.17731	7
8	6.5320	0.7955	0.0369	1.2571	27.1366	102.265	47.489	149.754	1.00807	1.17698	8
9	6.7342	0.7977	0.0358	1.2536	27.9613	102.551	47.282	149.834	1.00908	1.17665	9

TEMP. °C	PRES-SURE kg/cm²	VOLUME LIQUID l/kg v_f	VOLUME VAPOR m³/kg v_g	DENSITY LIQUID kg/l $1/v_f$	DENSITY VAPOR kg/m³ $1/v_g$	ENTHALPY kcal/kg LIQUID h_f	ENTHALPY kcal/kg LATENT h_{fg}	ENTHALPY kcal/kg VAPOR h_g	ENTROPY kcal/(kg) (°K) LIQUID s_f	ENTROPY kcal/(kg) (°K) VAPOR s_g	TEMP. °C
10	6.9410	0.8000	0.0347	1.2500	28.8061	102.839	47.074	149.913	1.01008	1.17633	10
11	7.1525	0.8023	0.0337	1.2464	29.6714	103.127	46.864	149.991	1.01108	1.17601	11
12	7.3687	0.8046	0.0327	1.2428	30.5576	103.416	46.653	150.068	1.01208	1.17569	12
13	7.5898	0.8070	0.0318	1.2392	31.4651	103.705	46.439	150.145	1.01308	1.17537	13
14	7.8158	0.8094	0.0309	1.2355	32.3945	103.996	46.224	150.220	1.01408	1.17505	14
15	8.0468	0.8118	0.0300	1.2319	33.3461	104.287	46.007	150.294	1.01508	1.17474	15
16	8.2828	0.8142	0.0291	1.2282	34.3204	104.579	45.788	150.367	1.01607	1.17442	16
17	8.5239	0.8167	0.0283	1.2245	35.3180	104.872	45.568	150.440	1.01707	1.17411	17
18	8.7701	0.8192	0.0275	1.2207	36.3392	105.166	45.345	150.511	1.01806	1.17380	18
19	9.0216	0.8217	0.0267	1.2170	37.3848	105.460	45.120	150.581	1.01906	1.17350	19
20	9.2784	0.8243	0.0260	1.2132	38.4552	105.756	44.894	150.650	1.02005	1.17319	20
21	9.5406	0.8269	0.0253	1.2094	39.5508	106.052	44.665	150.718	1.02104	1.17288	21
22	9.8082	0.8295	0.0246	1.2055	40.6724	106.350	44.435	150.785	1.02203	1.17258	22
23	10.081	0.8322	0.0239	1.2017	41.8204	106.648	44.202	150.850	1.02302	1.17228	23
24	10.360	0.8349	0.0233	1.1978	42.9955	106.947	43.968	150.915	1.02401	1.17197	24
25	10.644	0.8376	0.0226	1.1939	44.1983	107.247	43.731	150.978	1.02500	1.17167	25
26	10.935	0.8404	0.0220	1.1899	45.4295	107.548	43.492	151.040	1.02599	1.17137	26
27	11.230	0.8432	0.0214	1.1859	46.6896	107.850	43.250	151.100	1.02698	1.17107	27
28	11.532	0.8461	0.0208	1.1819	47.9795	108.153	43.007	151.160	1.02797	1.17077	28
29	11.840	0.8490	0.0203	1.1779	49.2998	108.457	42.761	151.218	1.02896	1.17047	29
30	12.153	0.8519	0.0197	1.1738	50.6513	108.762	42.513	151.275	1.02994	1.17018	30
31	12.473	0.8549	0.0192	1.1698	52.0347	109.068	42.262	151.330	1.03093	1.16988	31
32	12.799	0.8579	0.0187	1.1656	53.4508	109.375	42.009	151.384	1.03192	1.16958	32
33	13.131	0.8610	0.0182	1.1615	54.9005	109.683	41.753	151.436	1.03290	1.16928	33
34	13.470	0.8641	0.0177	1.1573	56.3846	109.993	41.495	151.487	1.03389	1.16898	34
35	13.815	0.8673	0.0173	1.1531	57.9039	110.303	41.234	151.537	1.03488	1.16868	35
36	14.166	0.8705	0.0168	1.1488	59.4596	110.615	40.970	151.585	1.03586	1.16838	36
37	14.524	0.8738	0.0164	1.1445	61.0524	110.927	40.704	151.631	1.03685	1.16808	37
38	14.888	0.8771	0.0160	1.1402	62.6834	111.241	40.435	151.676	1.03783	1.16778	38
39	15.259	0.8805	0.0155	1.1358	64.3537	111.556	40.163	151.719	1.03882	1.16748	39
40	15.637	0.8839	0.0151	1.1314	66.0644	111.873	39.888	151.761	1.03981	1.16718	40
41	16.022	0.8874	0.0147	1.1269	67.8166	112.191	39.610	151.800	1.04080	1.16688	41
42	16.413	0.8909	0.0144	1.1224	69.6115	112.510	39.328	151.838	1.04178	1.16657	42
43	16.812	0.8946	0.0140	1.1179	71.4504	112.830	39.044	151.874	1.04277	1.16627	43
44	17.218	0.8983	0.0136	1.1133	73.3345	113.152	38.756	151.908	1.04376	1.16596	44
45	17.631	0.9020	0.0133	1.1086	75.2653	113.475	38.465	151.940	1.04475	1.16565	45
46	18.051	0.9058	0.0129	1.1040	77.2443	113.800	38.170	151.970	1.04574	1.16534	46
47	18.478	0.9097	0.0126	1.0992	79.2728	114.126	37.872	151.998	1.04674	1.16503	47
48	18.913	0.9137	0.0123	1.0944	81.3526	114.454	37.570	152.024	1.04773	1.16471	48
49	19.356	0.9178	0.0120	1.0896	83.4853	114.784	37.264	152.047	1.04872	1.16439	49
50	19.806	0.9219	0.0117	1.0847	85.6727	115.115	36.954	152.068	1.04972	1.16407	50
51	20.263	0.9261	0.0114	1.0798	87.9166	115.447	36.640	152.087	1.05072	1.16375	51
52	20.729	0.9304	0.0111	1.0748	90.2191	115.782	36.322	152.104	1.05172	1.16342	52
53	21.202	0.9348	0.0108	1.0697	92.5821	116.118	35.999	152.117	1.05272	1.16309	53
54	21.683	0.9394	0.0105	1.0646	95.0079	116.457	35.672	152.129	1.05372	1.16276	54
55	22.173	0.9440	0.0103	1.0594	97.4989	116.797	35.340	152.137	1.05473	1.16242	55
56	22.670	0.9487	0.0100	1.0541	100.0575	117.139	35.004	152.143	1.05573	1.16208	56
57	23.176	0.9535	0.0097	1.0487	102.6863	117.483	34.662	152.146	1.05674	1.16173	57
58	23.690	0.9585	0.0095	1.0433	105.3881	117.830	34.316	152.145	1.05776	1.16138	58
59	24.212	0.9635	0.0092	1.0378	108.1659	118.178	33.963	152.142	1.05877	1.16102	59
60	24.743	0.9687	0.0090	1.0323	111.1002	118.545	33.580	152.125	1.05984	1.16063	60
61	25.283	0.9741	0.0088	1.0266	114.0513	118.900	33.213	152.113	1.06087	1.16026	61
62	25.832	0.9796	0.0085	1.0209	117.0904	119.258	32.840	152.098	1.06190	1.15988	62
63	26.389	0.9852	0.0083	1.0150	120.2217	119.619	32.460	152.079	1.06294	1.15950	63
64	26.955	0.9910	0.0081	1.0091	123.4497	119.983	32.074	152.056	1.06398	1.15911	64

TEMP. °C	PRES-SURE kg/cm²	VOLUME		DENSITY		ENTHALPY kcal/kg			ENTROPY kcal/(kg) (°K)		TEMP. °C
		LIQUID l/kg v_f	VAPOR m³/kg v_g	LIQUID kg/l $1/v_f$	VAPOR kg/m³ $1/v_g$	LIQUID h_f	LATENT h_{fg}	VAPOR h_g	LIQUID s_f	VAPOR s_g	
65	27.531	0.9970	0.0079	1.0030	126.7793	120.349	31.679	152.029	1.06502	1.15871	65
66	28.116	1.0031	0.0077	0.9969	130.2157	120.720	31.277	151.997	1.06608	1.15830	66
67	28.710	1.0095	0.0075	0.9906	133.7649	121.093	30.867	151.960	1.06714	1.15788	67
68	29.313	1.0160	0.0073	0.9842	137.4330	121.470	30.448	151.919	1.06820	1.15745	68
69	29.926	1.0228	0.0071	0.9777	141.2272	121.851	30.020	151.872	1.06927	1.15701	69
70	30.549	1.0298	0.0069	0.9710	145.1548	122.237	29.582	151.819	1.07035	1.15656	70
71	31.182	1.0371	0.0067	0.9642	149.2245	122.626	29.134	151.760	1.07144	1.15610	71
72	31.824	1.0446	0.0065	0.9573	153.4454	123.021	28.675	151.695	1.07254	1.15562	72
73	32.477	1.0525	0.0063	0.9501	157.8279	123.420	28.204	151.624	1.07365	1.15513	73
74	33.140	1.0606	0.0061	0.9428	162.3836	123.824	27.720	151.545	1.07477	1.15462	74
75	33.814	1.0691	0.0060	0.9354	167.1255	124.235	27.223	151.458	1.07590	1.15409	75
76	34.498	1.0780	0.0058	0.9277	172.0683	124.652	26.712	151.363	1.07705	1.15355	76
77	35.193	1.0873	0.0056	0.9197	177.2286	125.075	26.184	151.259	1.07821	1.15299	77
78	35.899	1.0970	0.0055	0.9116	182.6254	125.506	25.640	151.145	1.07939	1.15240	78
79	36.616	1.1073	0.0053	0.9031	188.2807	125.944	25.076	151.021	1.08058	1.15179	79
80	37.344	1.1181	0.0051	0.8944	194.2200	126.392	24.492	150.884	1.08180	1.15115	80
81	38.084	1.1295	0.0050	0.8854	200.4731	126.849	23.886	150.735	1.08303	1.15048	82
82	38.836	1.1416	0.0048	0.8760	207.0752	127.317	23.254	150.572	1.08430	1.14977	82
83	39.599	1.1545	0.0047	0.8662	214.0686	127.798	22.595	150.392	1.08559	1.14903	83
84	40.375	1.1683	0.0045	0.8559	221.5046	128.292	21.903	150.195	1.08692	1.14824	84
85	41.163	1.1832	0.0044	0.8451	229.4462	128.801	21.176	149.977	1.08828	1.14740	85
86	41.964	1.1993	0.0042	0.8338	237.9727	129.329	20.407	149.736	1.08969	1.14651	86
87	42.778	1.2169	0.0040	0.8217	247.1853	129.877	19.590	149.467	1.09115	1.14554	87
88	43.605	1.2363	0.0039	0.8089	257.2175	130.451	18.715	149.166	1.09267	1.14449	88
89	44.446	1.2579	0.0037	0.7950	268.2499	131.055	17.771	148.825	1.09427	1.14334	89
90	45.300	1.2822	0.0036	0.7799	280.5378	131.696	16.740	148.436	1.09597	1.14207	90
91	46.169	1.3103	0.0034	0.7632	294.4605	132.385	15.598	147.983	1.09780	1.14063	91
92	47.053	1.3435	0.0032	0.7443	310.6234	133.139	14.307	147.445	1.09979	1.13897	92
93	47.952	1.3844	0.0030	0.7223	330.0958	133.985	12.798	146.783	1.10202	1.13697	93
94	48.866	1.4384	0.0028	0.6952	355.0986	134.980	10.936	145.917	1.10465	1.13444	94
95	49.797	1.5205	0.0026	0.6577	391.8991	136.272	8.355	144.627	1.10808	1.13077	95
96.01	50.750	1.9056	0.0019	0.5248	524.7653	140.150	0.	140.150	1.11850	1.11850	96.01

From Published data of E.I. du Pont de Nemours & Co. Inc. Used by permission.

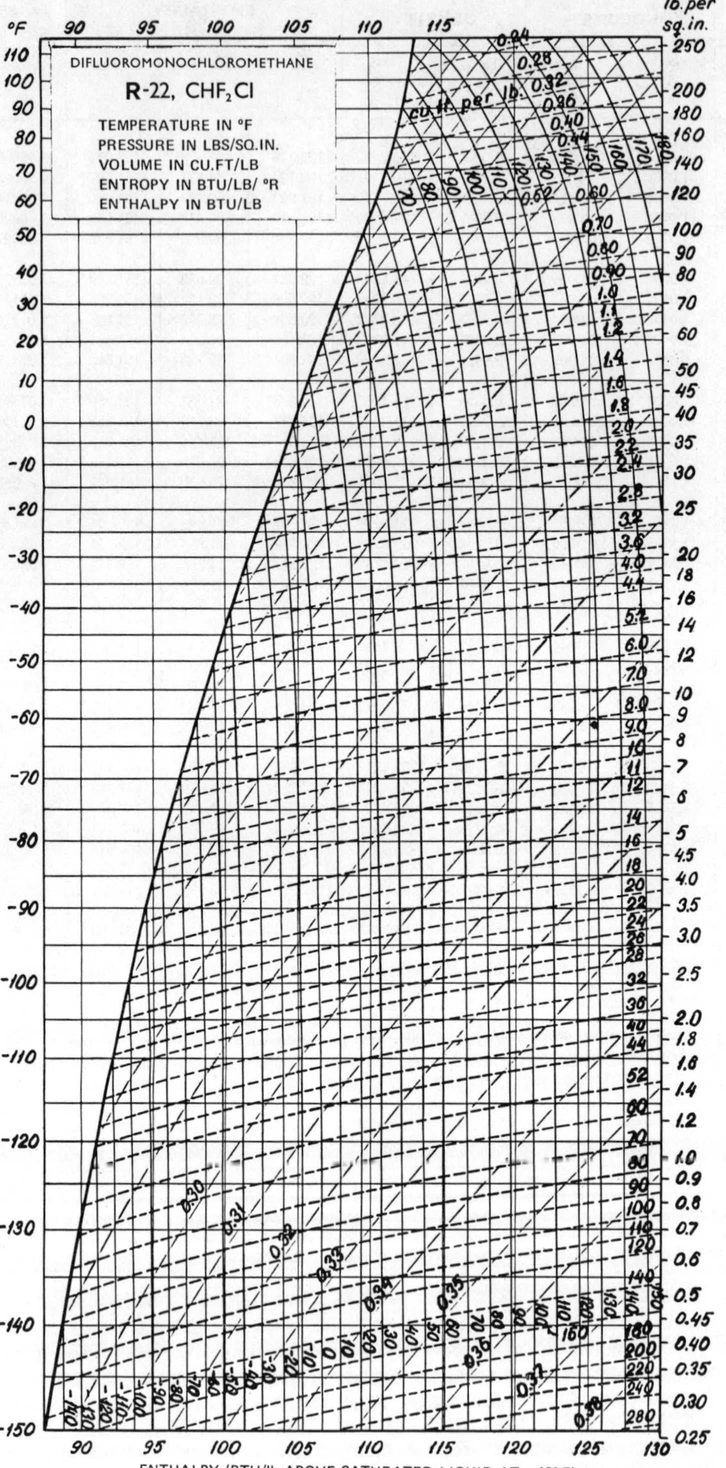
DIFLUOROMONOCHLOROMETHANE
R-22, CHF_2Cl
TEMPERATURE IN °F
PRESSURE IN LBS/SQ.IN.
VOLUME IN CU.FT/LB
ENTROPY IN BTU/LB/ °R
ENTHALPY IN BTU/LB
°F
lb.per sq.in.
cu.ft. per lb.
ENTHALPY (BTU/lb ABOVE SATURATED LIQUID AT −40° F)

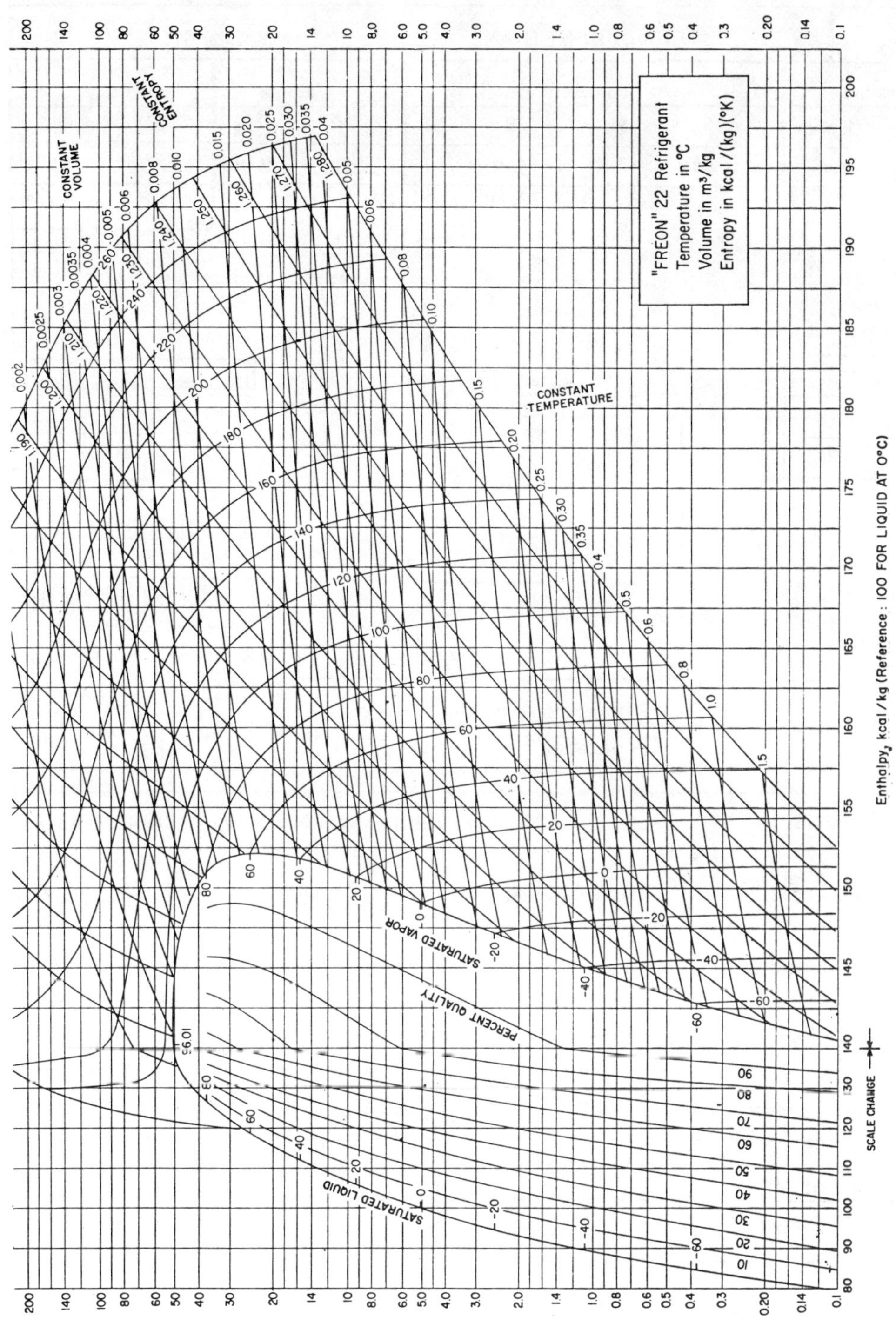

From Published data of E.I. du Pont de Nemours & Co. Inc. Used by permission.

DATA SHEET FOR R 23

CHF_3	TRIFLUOROMETHANE	Molecular Wt. 70

Pressure		Temperature				Volume		Density	
49.3	kg/cm^2	26	°C	79	°F	1.904	l/kg	.525	kg/l.
701	psia	299	°K	539	°R	.0305	ft^3/lb	32.8	lbs/ft^3
								2.41	Air=1

AT CRITICAL POINT

Discharge Pressure	
8.86	kg/cm^2
126	psia

−34° C −30° F

Discharge Temperature			
4.4	°C	40	°F
277	°K	500	°R

Normal Boiling Point			
-82	°C	-116	°F
191	°K	344	°R

Inlet Pressure	
1.6	kg/cm^2
23.7	psia

−73° C −100° F

Triple Point			
-155	°C	-247	°F
118	°K	213	°R

Refrigerating Effect			
238	kcal/m^3	26.8	Btu/ft^3

Latent Heat at NBP	57.2 kcal/kg	4004 /kg mol	103 Btu/lb	7222 /lb mol

Trouton's No. 21.0	Gas Constant 12.11 kg.m/kg/° K	22.1 ft.lbs/lb/° R

Specific Heat Liquid at 30° C 86° F Fig 14	Gas C_p .176	C_p/C_v 1.91 at 25°C

Liquid at 30° C 86° F Density 12.4 kg/l.	77.8 lbs/ft^3 Viscosity .23 cp

Reduced Form at NBP Pressure .021	1 / Temperature 1.565

Reduced value of 1/T	2.02	1.69	1.37
when Reduced Pressure is	a) .001	b) .01	c) .1

TEMP. °F	PRESSURE		VOLUME cu ft/lb	DENSITY lb/cu ft	ENTHALPY Btu/lb		ENTROPY Btu/(lb)(°R)	
	psia	psig	Vapor	Liquid	Liquid	Vapor	Liquid	Vapor
−190	0.622	28.7*	66.18	97.06	−44.44	71.64	−0.1301	0.3004
−180	1.068	27.7*	39.87	96.22	−41.28	72.83	−0.1185	0.2895
−170	1.755	26.3*	25.06	95.34	−38.28	74.01	−0.1080	0.2796
−160	2.774	24.3*	16.34	94.42	−35.39	75.17	−0.0982	0.2707
−150	4.237	21.3*	11.00	93.47	−32.58	76.31	−0.0890	0.2626
−148	4.595	20.6*	10.20	93.27	−32.03	76.54	−0.0872	0.2611
−146	4.976	19.8*	9.467	93.07	−31.48	76.76	−0.0854	0.2596
−144	5.384	19.0*	8.797	92.87	−30.93	76.98	−0.0837	0.2581
−142	5.818	18.1*	8.183	92.67	−30.37	77.20	−0.0820	0.2567
−140	6.280	17.1*	7.619	92.46	−29.82	77.42	−0.0802	0.2552
−138	6.771	16.1*	7.101	92.26	−29.28	77.64	−0.0785	0.2538
−136	7.293	15.1*	6.625	92.05	−28.73	77.85	−0.0768	0.2524
−134	7.847	13.9*	6.187	91.84	−28.18	78.07	−0.0751	0.2511
−132	8.435	12.7*	5.783	91.63	−27.63	78.28	−0.0735	0.2497
−130	9.058	11.5*	5.410	91.42	−27.08	78.49	−0.0718	0.2484
−128	9.717	10.1*	5.066	91.20	−26.53	78.70	−0.0701	0.2471
−126	10.41	8.72*	4.748	90.99	−25.98	78.91	−0.0685	0.2458
−124	11.15	7.22*	4.454	90.77	−25.42	79.12	−0.0668	0.2446
−122	11.93	5.63*	4.181	90.55	−24.87	79.32	−0.0652	0.2434
−120	12.75	3.96*	3.928	90.32	−24.32	79.53	−0.0636	0.2421
−118	13.62	2.20*	3.694	90.10	−23.76	79.73	−0.0619	0.2409
−116	14.53	0.340*	3.476	89.87	−23.20	79.93	−0.0603	0.2398
−114	15.49	0.793	3.273	89.64	−22.64	80.12	−0.0587	0.2386
−112	16.50	1.80	3.084	89.41	−22.07	80.32	−0.0571	0.2374
−110	17.56	2.86	2.909	89.18	−21.51	80.51	−0.0554	0.2363
−108	18.68	3.98	2.745	88.94	−20.94	80.70	−0.0538	0.2352
−106	19.85	5.15	2.592	88.71	−20.37	80.89	−0.0522	0.2341
−104	21.07	6.38	2.450	88.47	−19.80	81.08	−0.0506	0.2330
−102	22.36	7.66	2.317	88.23	−19.23	81.27	−0.0490	0.2319
−100	23.71	9.01	2.192	87.98	−18.65	81.45	−0.0474	0.2309
− 98	25.12	10.4	2.076	87.74	−18.08	81.54	−0.0458	0.2298
− 96	26.59	11.9	1.967	87.49	−17.49	81.81	−0.0442	0.2288
− 94	28.14	13.4	1.865	87.24	−16.91	81.99	−0.0426	0.2278
− 92	29.75	15.1	1.769	86.99	−16.32	82.16	−0.0411	0.2268
− 90	31.43	16.7	1.679	86.73	−15.73	82.33	−0.0395	0.2258
− 88	33.19	18.5	1.594	86.47	−15.14	82.50	−0.0379	0.2248
− 86	35.02	20.3	1.515	86.21	−14.55	82.67	−0.0363	0.2239
− 84	36.93	22.2	1.440	85.95	−13.95	82.84	−0.0347	0.2229
− 82	38.92	24.2	1.370	85.69	−13.35	83.00	−0.0331	0.2220
− 80	40.99	26.3	1.300	85.42	−12.74	83.16	−0.0315	0.2210
−78	43.14	28.4	1.242	85.15	−12.14	83.32	−0.0300	0.2201
−76	45.38	30.7	1.183	84.88	−11.52	83.47	−0.0284	0.2192
−74	47.71	33.0	1.128	84.60	−10.91	83.63	−0.0268	0.2183
−72	50.13	35.4	1.075	84.32	−10.30	83.78	−0.0252	0.2174
−70	52.64	37.9	1.026	84.04	− 9.67	83.92	−0.0236	0.2166
−68	55.25	40.6	0.9793	83.76	− 9.04	84.07	−0.0220	0.2157
−66	57.95	43.3	0.9352	83.47	− 8.42	84.21	−0.0204	0.2148
−64	60.75	46.1	0.8934	83.18	− 7.79	84.35	−0.0189	0.2140
−62	63.66	49.0	0.8539	82.89	− 7.15	84.48	−0.0173	0.2131
−60	66.67	52.0	0.8165	82.60	− 6.52	84.62	−0.0157	0.2123
−58	69.79	55.1	0.7810	82.30	− 5.88	84.75	−0.0141	0.2115
−56	73.01	58.3	0.7474	82.00	− 5.24	84.88	−0.0126	0.2107
−54	76.35	61.7	0.7155	81.69	− 4.59	85.00	−0.0110	0.2099
−52	79.80	65.1	0.6852	81.39	− 3.95	85.12	−0.0094	0.2091
−50	83.37	68.7	0.6564	81.07	− 3.30	85.24	−0.0078	0.2083
−48	87.05	72.4	0.6291	80.76	− 2.65	85.35	−0.0063	0.2075
−46	90.86	76.2	0.6031	80.44	− 1.99	85.47	−0.0047	0.2067
−44	94.79	80.1	0.5784	80.12	− 1.33	85.58	−0.0031	0.2059
−42	98.84	84.1	0.5549	79.80	− 0.66	85.68	−0.0016	0.2052
−40	103.0	88.3	0.5325	79.47	− 0.00	85.78	−0.0000	0.2044
−38	107.3	92.6	0.5112	79.14	0.67	85.88	0.0016	0.2036
−36	111.8	97.1	0.4908	78.81	1.34	85.98	0.0031	0.2029
−34	116.4	102	0.4715	78.47	2.01	86.07	0.0047	0.2021
−32	121.1	106	0.4530	78.12	2.68	86.15	0.0062	0.2014
−30	126.0	111	0.4354	77.78	3.36	86.24	0.0078	0.2007

*Inches of mercury below one atmosphere

TEMP. °F	PRESSURE		VOLUME cu ft/lb	DENSITY lb/cu ft	ENTHALPY Btu/lb		ENTROPY Btu/(lb)(°R)	
	psia	psig	Vapor	Liquid	Liquid	Vapor	Liquid	Vapor
−28	131.0	116	0.4185	77.43	4.04	86.32	0.0093	0.1999
−26	136.1	121	0.4024	77.07	4.72	86.39	0.0109	0.1992
−24	141.4	127	0.3871	76.71	5.40	86.47	0.0124	0.1985
−22	146.9	132	0.3724	76.35	6.08	86.53	0.0140	0.1978
−20	152.5	138	0.3583	75.98	6.78	86.60	0.0155	0.1970
−10	182.9	168	0.2965	74.07	10.25	86.84	0.0232	0.1935
0	217.6	203	0.2463	72.03	13.80	86.95	0.0307	0.1899
10	257.0	242	0.2052	69.84	17.41	86.90	0.0383	0.1863
20	301.6	287	0.1710	67.46	21.13	86.65	0.0459	0.1825
30	351.7	337	0.1422	64.85	25.00	86.14	0.0539	0.1784
40	408.2	393	0.1176	61.93	29.12	85.25	0.0616	0.1739
50	471.6	457	0.0962	58.59	33.64	83.82	0.0702	0.1686
60	543.0	528	0.0768	54.55	38.89	81.48	0.0799	0.1618
70	623.4	609	0.0581	49.05	45.63	77.18	0.0922	0.1518
78.66	701.4	687	0.0305	32.78	60.77	60.77	0.1198	0.1198

Based on 0 for the saturated liquid at −40 F. From unpublished data of Allied Chemical Corp. (1970)

DATA SHEET FOR R 30

CH_2Cl_2	METHYLENE CHLORIDE	Molecular Wt.	84.9

Pressure		Temperature				Volume		Density	
61	kg/cm²	235	°C	455	°F	2.0	l/kg	.5	kg/l.
867	psia	508	°K	915	°R	.032	ft³/lb	31.25	lbs/ft³
								2.93	Air=1

AT CRITICAL POINT

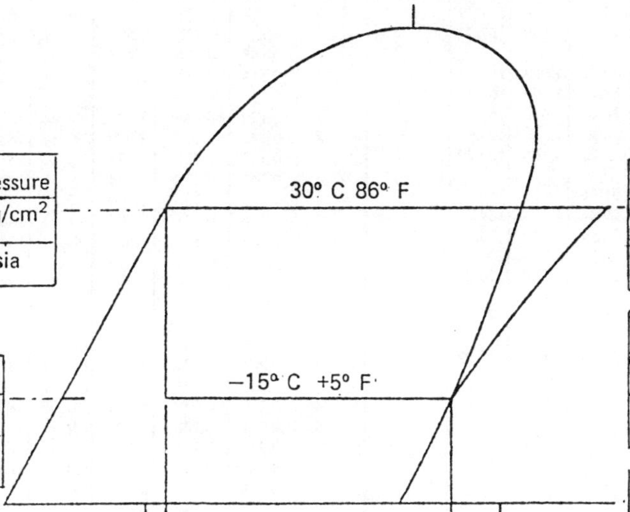

30° C 86° F

−15° C +5° F

Discharge Pressure	
.703	kg/cm²
10.4	psia

Inlet Pressure	
.086	kg/cm²
1.23	psia

Discharge Temperature			
96	°C	205	°F
369	°K	665	°R

Normal Boiling Point			
40	°C	104	°F
313	°K	564	°R

Triple Point			
-97	°C	-142	°F
167	°K	318	°R

Refrigerating Effect			
23.9	kcal/m³	2.69	Btu/ft³

Latent Heat at NBP	78.9 kcal/kg	6670 /kg mol	142 Btu/lb	12006 /lb mol

Trouton's No. 21.87	Gas Constant 10 kg.m/kg/° K	18.2 ft.lbs/lb/° R

Specific Heat Liquid at 30° C 86° F	Gas C_p	C_p/C_v 1.124 at 0°C

Liquid at 30° C 86° F Density 1.3 kg/l.	81.3 lbs/ft³ Viscosity cp

Reduced Form at NBP Pressure .0169	1 / Temperature 1.62

Reduced value of 1/T	1.99	1.69	
when Reduced Pressure is	a) .001	b) .01	c) .1

TEMP. °F	PRESS psia	VOLUME cu ft/lb		ENTHALPY Btu/lb	
		Liquid	Vapor	Liquid	Vapor
− 4	0.92	0.01147	61.66	9.76	165.58
− 2	1.00	0.01149	58.06	10.30	165.91
0	1.05	0.01150	54.71	10.84	166.27
+ 2	1.12	0.01151	51.58	11.38	166.61
+ 4	1.19	0.01152	48.65	11.90	166.93
+ 5	1.24	0.01153	47.24	12.17	167.11
+ 6	1.28	0.01155	45.88	12.44	167.29
+ 8	1.35	0.01157	43.33	12.98	167.62
+ 10	1.44	0.01158	40.95	13.52	167.96
+ 12	1.52	0.01160	38.70	14.06	168.12
+ 14	1.62	0.01161	36.61	14.60	168.62
+ 16	1.72	0.01163	34.65	15.14	168.95
+ 18	1.82	0.01164	32.81	15.68	169.29
+ 20	1.93	0.01166	31.08	16.22	169.58
+ 22	2.05	0.01168	29.44	16.76	169.90
+ 24	2.18	0.01170	27.94	17.30	170.23
+ 26	2.29	0.01172	26.35	17.84	170.55
+ 28	2.42	0.01174	25.15	18.36	170.86
+ 30	2.56	0.01175	23.89	18.90	171.18
+ 32	2.70	0.01177	22.68	19.44	171.49
+ 34	2.84	0.01179	21.56	19.98	171.79
+ 36	3.00	0.01180	20.51	20.52	172.08
+ 38	3.17	0.01182	19.51	21.06	172.37
+ 40	3.34	0.01184	18.58	21.60	172.67
+ 42	3.53	0.01186	17.69	22.14	173.00
+ 44	3.71	0.01187	16.85	22.68	173.29
+ 46	3.91	0.01189	16.07	23.22	173.57
+ 48	4.12	0.01191	15.33	23.74	173.86
+ 50	4.32	0.01193	14.61	27.28	174.15
+ 52	4.55	0.01195	13.97	24.82	174.46
+ 54	4.78	0.01196	13.33	25.36	174.74
+ 56	5.02	0.01198	12.72	25.90	175.00
+ 58	5.26	0.01200	12.14	26.44	175.27
+ 60	5.55	0.01202	11.61	26.98	175.55
+ 62	5.82	0.01204	11.10	27.34	175.82
+ 64	6.07	0.01206	10.62	28.06	176.11
+ 66	6.38	0.01208	10.17	28.60	176.38
+ 68	6.68	0.01209	9.74	29.14	176.62
+ 70	7.00	0.01211	9.32	29.66	176.87
+ 72	7.34	0.01212	8.92	30.20	177.14
+ 74	7.66	0.01214	8.55	30.74	177.41
+ 76	8.02	0.01216	8.21	31.28	177.64
+ 78	8.39	0.01218	7.87	31.82	177.89
+ 80	8.77	0.01220	7.56	32.36	178.15
+ 82	9.17	0.01222	7.26	32.90	178.38
+ 84	9.57	0.01224	6.97	33.44	178.65
+ 86	10.00	0.01227	6.68	33.98	178.88
+ 88	10.42	0.01229	6.42	34.52	179.10
+ 90	10.88	0.01231	6.17	35.06	179.37
+ 92	11.35	0.01232	5.94	35.60	179.60
+ 94	11.83	0.01234	5.72	36.14	179.82
+ 96	12.34	0.01236	5.49	36.68	180.05
+ 98	12.85	0.01238	5.30	37.22	180.27
+100	13.40	0.01240	5.06	37.76	180.50
+102	13.94	0.01241	4.90	38.30	180.72
+104	14.50	0.01243	4.73	38.84	180.90

TEMP. °C	PRESS kg/cm²	VOLUME l/kg	m³/kg	ENTHALPY kcal/kg	
		Liquid	Vapor	Liquid	Vapor
− 20	0.065	0.716	3.849	94.62	181.19
− 19	0.069	0.717	3.645	94.89	181.36
− 18	0.073	0.718	3.455	95.16	181.53
− 17	0.078	0.719	3.275	95.43	181.70
− 16	0.082	0.719	3.107	95.69	181.87
− 15	0.087	0.720	2.949	95.96	182.04
− 14	0.092	0.721	2.797	96.23	182.21
− 13	0.097	0.722	2.659	96.50	182.38
− 12	0.102	0.723	2.527	96.77	182.55
− 11	0.108	0.724	2.402	97.04	182.71
− 10	0.114	0.725	2.285	97.31	182.88
− 9	0.120	0.726	2.174	97.58	183.04
− 8	0.127	0.727	2.070	97.85	183.21
− 7	0.133	0.728	1.971	98.12	183.37
− 6	0.141	0.729	1.878	98.39	183.52
− 5	0.148	0.730	1.790	98.65	183.68
− 4	0.156	0.731	1.707	98.92	183.84
− 3	0.164	0.732	1.628	99.19	184.00
− 2	0.172	0.733	1.553	99.46	184.16
− 1	0.181	0.734	1.483	99.73	184.32
0	0.190	0.735	1.416	100.00	184.47
+ 1	0.199	0.736	1.353	100.27	184.62
+ 2	0.209	0.737	1.293	100.52	184.77
+ 3	0.219	0.738	1.236	100.81	184.91
+ 4	0.230	0.739	1.183	101.08	185.07
+ 5	0.241	0.740	1.131	101.35	185.23
+ 6	0.253	0.741	1.083	101.62	185.37
+ 7	0.265	0.742	1.037	101.88	185.52
+ 8	0.278	0.743	0.993	102.15	185.66
+ 9	0.291	0.744	0.952	102.42	185.81
+ 10	0.304	0.745	0.912	102.69	185.95
+ 11	0.318	0.746	0.874	102.96	186.10
+ 12	0.333	0.747	0.839	103.23	186.25
+ 13	0.348	0.748	0.805	103.50	186.38
+ 14	0.363	0.749	0.772	103.77	186.51
+ 15	0.380	0.750	0.741	104.04	186.65
+ 16	0.397	0.751	0.712	104.31	186.79
+ 17	0.414	0.752	0.684	104.58	186.93
+ 18	0.432	0.753	0.657	104.85	187.07
+ 19	0.451	0.754	0.632	105.12	187.20
+ 20	0.470	0.755	0.608	105.39	187.32
+ 21	0.490	0.756	0.584	105.65	187.45
+ 22	0.511	0.757	0.562	105.92	187.58
+ 23	0.532	0.758	0.541	106.19	187.72
+ 24	0.554	0.759	0.521	106.46	187.84
+ 25	0.577	0.760	0.502	106.73	187.96
+ 26	0.601	0.761	0.483	107.00	188.09
+ 27	0.625	0.762	0.466	107.27	188.21
+ 28	0.650	0.763	0.449	107.54	188.33
+ 29	0.676	0.765	0.433	107.81	188.46
+ 30	0.703	0.766	0.417	108.08	188.58

SATURATION PROPERTIES – R 30

Imperial Metric

DATA SHEET FOR R 40

CH_3Cl	METHYL CHLORIDE	Molecular Wt. 50.5

Pressure		Temperature				Volume		Density	
68.1	kg/cm²	143	°C	289.6	°F	2.83	l/kg	.353	kg/l.
968	psia	416	°K	758.6	°R	.0454	ft³/lb	22.02	lbs/ft³
								1.74	Air=1

AT CRITICAL POINT

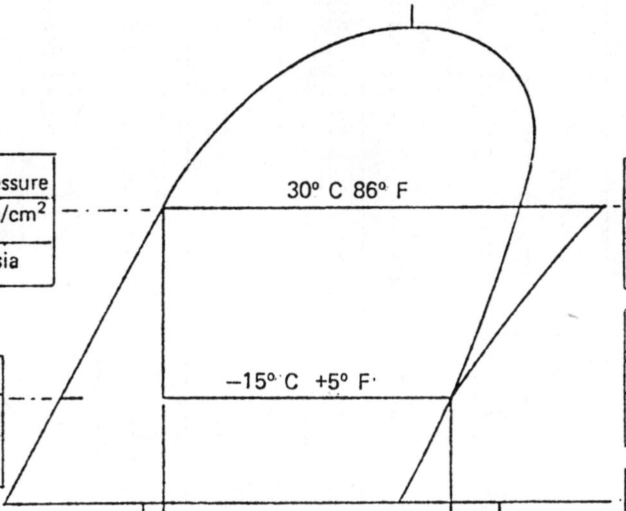

Discharge Pressure	
6.66	kg/cm²
94.7	psia

30° C 86° F

Discharge Temperature			
85	°C	185	°F
358	°K	645	°R

Normal Boiling Point			
-24	°C	-11	°F
249	°K	449	°R

Inlet Pressure	
1.49	kg/cm²
21.2	psia

−15° C +5° F

Triple Point			
-97.5	°C	-144	°F
175	°K	316	°R

Refrigerating Effect	
kcal/m³	Btu/ft³

Latent Heat at NBP	102.3 kcal/kg	5168 /kg mol	184.2 Btu/lb	9302 /lb mol

Trouton's No. 20.7	Gas Constant 16.8 kg.m/kg/° K	30.6 ft.lbs/lb/° R

Specific Heat Liquid at 30° C 86° F	Gas C_p	C_p/C_v 1.27 at 0°C

Liquid at 30° C 86° F Density .904 kg/l.	56.2 lbs/ft³ Viscosity cp

Reduced Form at NBP Pressure .0152	1 / Temperature 1.67

Reduced value of 1/T when Reduced Pressure is		1.76	1.139
	a) .001	b) .01	c) .1

TEMP. °F	PRESSURE		VOLUME cu ft/lb		DENSITY lb/cu ft		ENTHALPY Btu/lb			ENTROPY Btu/(lb) (°R)	
	PSIA	PSIG	LIQUID v_f	VAPOR v_g	LIQUID $1/v_f$	VAPOR $1/v_g$	LIQUID h_f	LATENT h_{fg}	VAPOR h_g	LIQUID s_f	VAPOR s_g
−80	1·95	25·94*	0·01493	41·08	66·98	0·0243	−13·88	198·64	184·75	−0·0351	0·4882
−70	2·75	24·32*	0·01508	29·84	66·31	0·0335	−10·52	196·77	186·25	−0·0261	0·4790
−60	3·79	22·19*	0·01523	22·09	65·66	0·0452	− 7·03	194·78	187·74	−0·0172	0·4703
−50	5·15	19·43*	0·01538	16·64	65·02	0·0601	− 3·53	192·72	189·19	−0·0085	0·4620
−40	6·87	15·92*	0·01553	12·72	64·39	0·0786	0·00	190·66	190·66	0·0000	0·4544
−38	7·27	15·12*	0·01556	12·08	64·27	0·0827	0·71	190·23	190·95	0·0017	0·4529
−36	7·68	14·28*	0·01559	11·48	64·14	0·0871	1·42	189·81	191·23	0·0034	0·4515
−34	8·11	13·40*	0·01562	10·91	64·02	0·0916	2·13	189·38	191·51	0·0051	0·4500
−32	8·56	12·48*	0·01565	10·38	63·90	0·0963	2·85	188·95	191·80	0·0067	0·4486
−30	9·03	11·52*	0·01568	9·873	63·78	0·1013	3·56	188·52	192·08	0·0084	0·4472
−28	9·52	10·53*	0·01571	9·399	63·65	0·1064	4·27	188·09	192·37	0·0100	0·4458
−26	10·04	9·490*	0·01574	8·953	63·53	0·1117	4·99	187·65	192·65	0·0117	0·4445
−24	10·57	8·399*	0·01577	8·533	63·41	0·1172	5·71	187·22	192·93	0·0133	0·4431
−22	11·13	7·267*	0·01580	8·136	63·29	0·1229	6·42	186·78	193·21	0·0150	0·4418
−20	11·71	6·090*	0·01583	7·761	63·17	0·1289	7·14	186·34	193·49	0·0166	0·4405
−18	12·31	4·866*	0·01586	7·408	63·05	0·1350	7·86	185·90	193·76	0·0183	0·4393
−16	12·93	3·594*	0·01589	7·074	62·93	0·1414	8·58	185·46	194·04	0·0199	0·4380
−14	13·58	2·268*	0·01592	6·758	62·81	0·1480	9·30	185·01	194·32	0·0215	0·4367
−12	14·26	0·890*	0·01595	6·459	62·70	0·1548	10·03	184·56	194·59	0·0232	0·4355
−10	14·96	0·266	0·01598	6·176	62·58	0·1619	10·75	184·11	194·87	0·0247	0·4343
− 8	15·69	0·996	0·01601	5·908	62·46	0·1693	11·48	183·66	195·14	0·0263	0·4331
− 6	16·45	1·754	0·01604	5·654	62·34	0·1769	12·20	183·21	195·42	0·0279	0·4319
− 4	17·24	2·540	0·01607	5·413	62·23	0·1847	12·93	182·76	195·69	0·0295	0·4307
− 2	18·05	3·356	0·01610	5·185	62·11	0·1929	13·66	182·30	195·96	0·0311	0·4296
0	18·90	4·201	0·01613	4·969	62·00	0·2013	14·39	181·85	196·23	0·0327	0·4284
2	19·77	5·077	0·01616	4·763	61·88	0·2100	15·12	181·39	196·51	0·0343	0·4273
4	20·68	5·985	0·01619	4·568	61·77	0·2189	15·85	180·93	196·78	0·0359	0·4262
5	21·15	6·455	0·01622	4·471	61·65	0·2237	16·21	180·70	196·92	0·0367	0·4257
6	21·62	6·924	0·01625	4·379	61·54	0·2284	16·58	180·47	197·05	0·0375	0·4251
8	22·59	7·896	0·01628	4·206	61·43	0·2378	17·31	180·01	197·31	0·0390	0·4240
10	23·60	8·903	0·01631	4·038	61·31	0·2477	18·04	179·53	197·58	0·0406	0·4229
12	24·64	9·943	0·01634	3·878	61·20	0·2579	18·77	179·06	197·83	0·0422	0·4218
14	25·72	11·02	0·01637	3·726	61·09	0·2684	19·51	178·58	198·09	0·0437	0·4208
16	26·83	12·13	0·01640	3·581	60·98	0·2792	20·25	178·10	189·34	0·0453	0·4198
18	27·97	13·28	0·01644	3·443	60·83	0·2904	20·98	177·61	198·59	0·0468	0·4187
20	29·16	14·46	0·01647	3·312	60·72	0·3019	21·73	177·11	198·84	0·0484	0·4177
22	30·38	15·69	0·01650	3·186	60·61	0·3138	22·47	176·61	199·08	0·0499	0·4166
24	31·64	16·95	0·01654	3·067	60·46	0·3261	23·21	176·11	199·32	0·0514	0·4156
26	32·95	18·25	0·01658	2·952	60·31	0·3388	23·95	175·61	199·56	0·0530	0·4146
28	34·29	19·60	0·01662	2·843	60·17	0·3517	24·70	175·10	199·79	0·0545	0·4136
30	35·68	20·98	0·01665	2·739	60·06	0·3650	25·44	174·59	200·03	0·0560	0·4126
32	37·11	22·41	0·01669	2·640	59·92	0·3787	26·18	174·08	200·26	0·0575	0·4117
34	38·58	23·88	0·01673	2·546	59·77	0·3928	26·93	173·56	200·49	0·0590	0·4107
36	40·09	25·39	0·01677	2·455	59·63	0·4073	27·67	173·05	200·72	0·0605	0·4098
38	41·65	26·95	0·01681	2·369	59·49	0·4222	28·42	172·53	200·95	0·0621	0·4088
40	43·25	28·56	0·01684	2·286	59·38	0·4375	29·17	172·00	201·17	0·0636	0·4079
42	44·91	30·21	0·01688	2·206	59·24	0·4532	29·92	171·48	201·40	0·0651	0·4070
44	46·61	31·91	0·01692	2·130	59·10	0·4694	30·67	170·95	201·62	0·0665	0·4061
46	48·35	33·66	0·01696	2·057	58·96	0·4861	31·42	170·42	201·84	0·0680	0·4052
48	50·15	35·45	0·01700	1·987	58·82	0·5033	32·17	169·89	202·06	0·0695	0·4043
50	51·99	37·29	0·01704	1·920	58·69	0·5208	32·93	169·35	202·28	0·0710	0·4034
52	53·88	39·18	0·01708	1·856	58·55	0·5388	33·68	168·81	202·49	0·0725	0·4025
54	55·83	41·13	0·01712	1·794	58·41	0·5573	34·44	168·27	202·71	0·0740	0·4017
56	57·83	43·13	0·01716	1·735	58·28	0·5763	35·19	167·72	202·91	0·0754	0·4008
58	59·88	45·19	0·01720	1·679	58·14	0·5958	35·95	167·18	203·13	0·0769	0·3999
60	62·00	47·30	0·01724	1·624	58·00	0·6158	36·71	166·62	203·33	0·0784	0·3991
62	64·17	49·47	0·01728	1·572	57·87	0·6362	37·47	166·07	203·54	0·0798	0·3983
64	66·39	51·70	0·01732	1·522	57·74	0·6572	38·23	165·51	203·74	0·0813	0·3974
66	68·67	53·98	0·01736	1·473	57·60	0·6788	39·00	164·95	203·95	0·0827	0·3966
68	71·01	56·32	0·01740	1·427	57·47	0·7008	39·76	164·39	204·15	0·0842	0·3958

*Inches of mercury below one atmosphere

TEMP. °F	PRESSURE		VOLUME cu ft/lb		DENSITY lb/cu ft		ENTHALPY Btu/lb			ENTROPY Btu/(lb) (°R)	
	PSIA	PSIG	LIQUID v_f	VAPOR v_g	LIQUID $1/v_f$	VAPOR $1/v_g$	LIQUID h_f	LATENT h_{fg}	VAPOR h_g	LIQUID s_f	VAPOR s_g
70	73·41	58·71	0·01744	1·382	57·34	0·7234	40·52	163·82	204·34	0·0856	0·3950
72	75·86	61·17	0·01748	1·339	57·21	0·7467	41·29	163·24	204·53	0·0870	0·3941
74	78·37	63·68	0·01752	1·298	57·08	0·7704	42·06	162·66	204·72	0·0885	0·3933
76	80·94	66·25	0·01756	1·258	56·95	0·7948	42·82	162·08	204·90	0·0899	0·3925
78	83·57	68·87	0·01760	1·220	56·82	0·8196	43·59	161·50	205·09	0·0913	0·3918
80	86·26	71·56	0·01764	1·183	56·69	0·8451	44·36	160·91	205·27	0·0928	0·3910
82	89·01	74·31	0·01768	1·148	56·56	0·8710	45·13	160·32	205·45	0·0942	0·3902
84	91·82	77·13	0·01773	1·114	56·40	0·8979	45·90	159·72	205·62	0·0956	0·3894
86	94·70	80·00	0·01778	1·081	56·24	0·9253	46·67	159·13	205·80	0·0970	0·3887
88	97·64	82·94	0·01782	1·049	56·12	0·9531	47·44	158·52	205·96	0·0984	0·3879
90	100·6	85·95	0·01786	1·018	55·99	0·9819	48·21	157·92	206·13	0·0998	0·3872
92	103·7	89·02	0·01791	0·9889	55·83	1·011	48·99	157·31	206·30	0·1012	0·3865
94	106·9	92·16	0·01796	0·9603	55·68	1·041	49·77	156·69	206·46	0·1026	0·3857
96	110·1	95·37	0·01800	0·9333	55·56	1·072	50·54	156·08	206·62	0·1041	0·3850
98	113·4	98·65	0·01804	0·9069	55·43	1·103	51·32	155·46	206·78	0·1055	0·3843
100	116·7	102·0	0·01808	0·8814	55·31	1·135	52·09	154·85	206·94	0·1069	0·3836
102	120·1	105·4	0·01813	0·8568	55·15	1·167	52·87	154·22	207·09	0·1082	0·3828
104	123·6	108·9	0·01818	0·8331	55·01	1·200	53·65	153·60	207·25	0·1096	0·3822
106	127·2	112·5	0·01823	0·8105	54·85	1·234	54·43	152·97	207·40	0·1110	0·3815
108	130·8	116·1	0·01828	0·7884	54·70	1·268	55·22	152·33	207·55	0·1124	0·3808
110	134·5	119·8	0·01833	0·7672	54·55	1·303	56·00	151·70	207·70	0·1138	0·3801
112	138·3	123·6	0·01838	0·7466	54·41	1·339	56·78	151·06	207·84	0·1151	0·3794
114	142·2	127·5	0·01843	0·7268	54·26	1·376	57·57	150·41	207·98	0·1165	0·3787
116	146·1	131·4	0·01848	0·7075	54·11	1·414	58·36	149·77	208·13	0·1179	0·3781
118	150·1	135·4	0·01853	0·6889	53·97	1·452	59·15	149·11	208·26	0·1193	0·3774
120	154·2	139·5	0·01859	0·6710	53·79	1·490	59·93	148·46	208·39	0·1206	0·3768
122	158·4	143·7	0·01865	0·6534	53·62	1·530	60·73	147·80	208·53	0·1220	0·3762
124	162·6	147·9	0·01870	0·6367	53·48	1·571	61·51	147·14	208·65	0·1234	0·3755
126	167·0	152·3	0·01875	0·6201	53·33	1·613	62·31	146·47	208·78	0·1247	0·3749
128	171·4	156·7	0·01881	0·6043	53·16	1·655	63·10	145·80	208·90	0·1261	0·3742
130	175·9	161·1	0·01887	0·5889	52·99	1·698	63·89	145·13	209·02	0·1274	0·3736
132	180·4	165·7	0·01893	0·5741	52·83	1·742	64·69	144·45	209·14	0·1288	0·3730
134	185·1	170·4	0·01898	0·5596	52·69	1·787	65·48	143·77	209·25	0·1301	0·3723
136	189·8	175·1	0·01904	0·5455	52·52	1·833	66·28	143·09	209·37	0·1314	0·3717
138	194·7	180·0	0·01909	0·5320	52·38	1·880	67·08	142·40	209·48	0·1328	0·3711
140	199·6	184·9	0·01915	0·5189	52·22	1·927	67·87	141·71	209·58	0·1341	0·3705
150	225·4	210·7	0·01945	0·4586	51·41	2·181	71·87	138·23	210·10	0·1407	0·3674
160	253·5	238·8	0·01978	0·4070	50·56	2·457	75·90	134·66	210·56	0·1473	0·3646
170	283·9	269·2	0·02015	0·3613	49·63	2·768	79·97	130·96	210·93	0·1538	0·3618

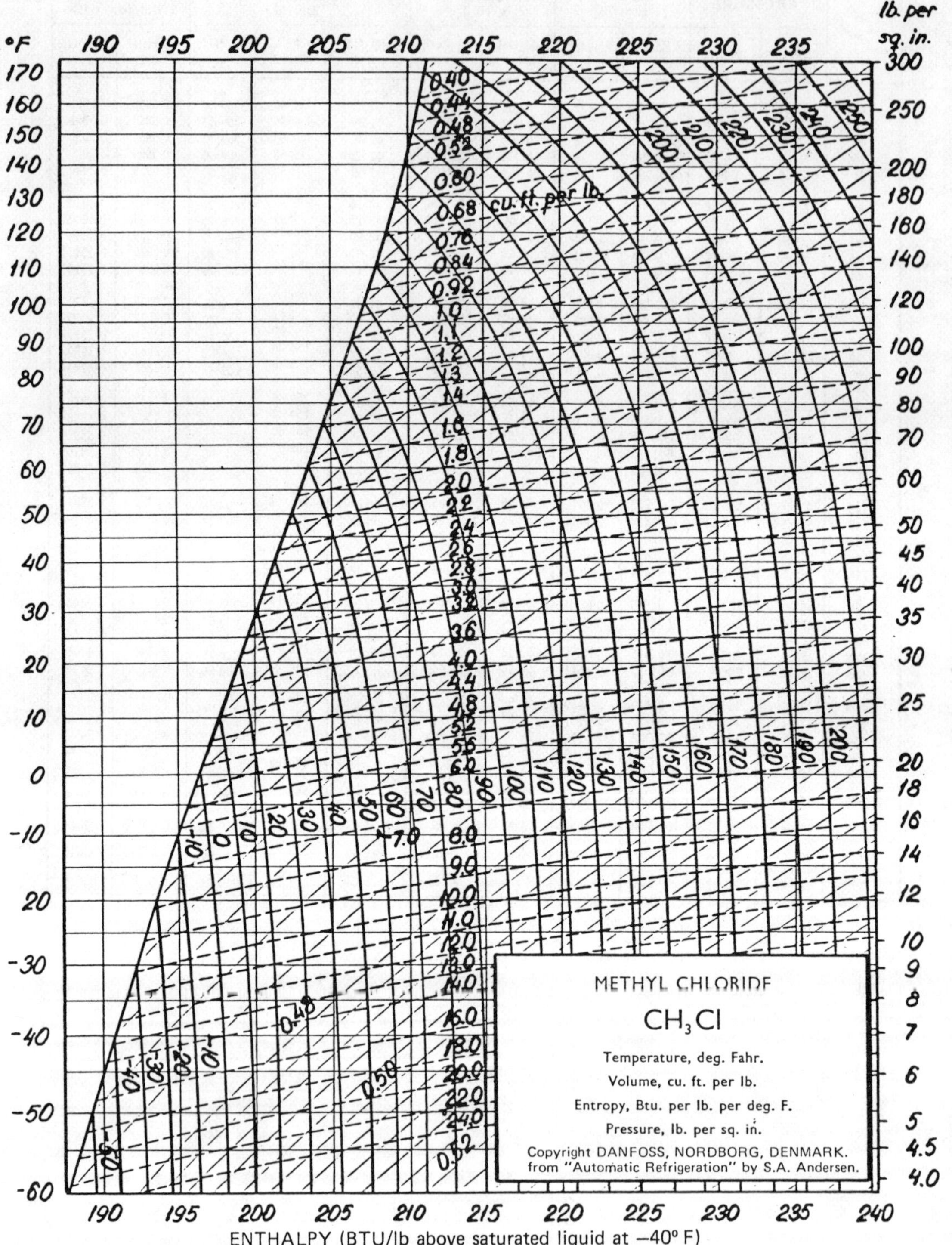
°F
lb. per sq. in.
cu. ft. per lb.
METHYL CHLORIDE
CH3Cl
Temperature, deg. Fahr.
Volume, cu. ft. per lb.
Entropy, Btu. per lb. per deg. F.
Pressure, lb. per sq. in.
Copyright DANFOSS, NORDBORG, DENMARK.
from "Automatic Refrigeration" by S.A. Andersen.
ENTHALPY (BTU/lb above saturated liquid at −40° F)

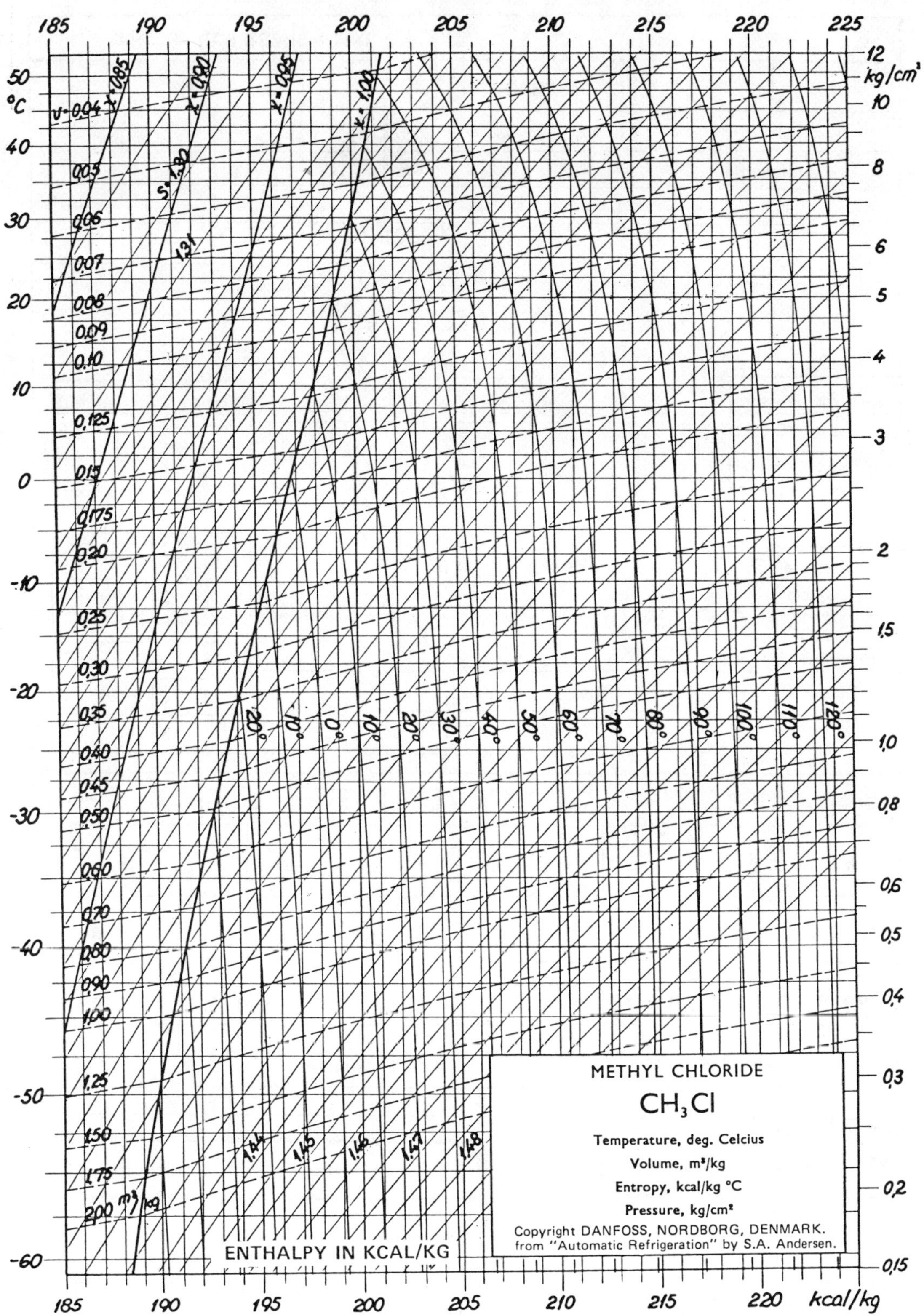
METHYL CHLORIDE
CH_3Cl
Temperature, deg. Celcius
Volume, m³/kg
Entropy, kcal/kg °C
Pressure, kg/cm²
Copyright DANFOSS, NORDBORG, DENMARK.
from "Automatic Refrigeration" by S.A. Andersen.
ENTHALPY IN KCAL/KG
kcal/kg
kg/cm²
°C

DATA SHEET FOR R 50

CH_4	METHANE	Molecular Wt.	16

Pressure		Temperature				Volume		Density	
47.3	kg/cm²	-82	°C	-116	°F	6.2	l/kg	.162	kg/l.
673	psia	191	°K	344	°R	.099	ft³/lb	10.1	lbs/ft³
								.55	Air=1

AT CRITICAL POINT

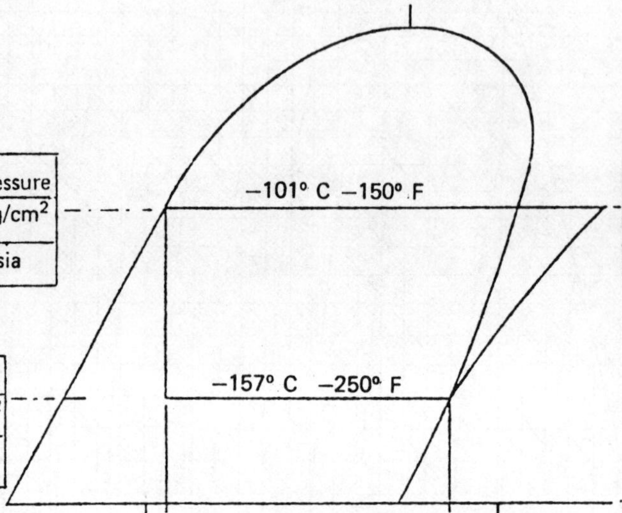

−101° C −150° F

−157° C −250° F

Discharge Pressure	
25.6	kg/cm²
364	psia

Discharge Temperature			
-29	°C	-20	°F
244	°K	440	°R

Inlet Pressure	
1.53	kg/cm²
21.7	psia

Normal Boiling Point			
-161	°C	-259	°F
112	°K	201	°R

Triple Point			
-182	°C	-296	°F
91	°K	164	°R

Refrigerating Effect			
176	kcal/m³	19.8	Btu/ft³

Latent Heat at NBP	122 kcal/kg	1956 /kg mol	220 Btu/lb	3521 /lb mol
Trouton's No. 17.5	Gas Constant 53 kg.m/kg/° K	96.56 ft.lbs/lb/° R		
Specific Heat Liquid at 30° C 86° F 1.02 Fig 19	Gas C_p	C_p C_v at		
Liquid at 30° C 86° F Density .308 kg/l.	19.2 lbs/ft³	Viscosity cp		
Reduced Form at NBP Pressure .0218	1 / Temperature 1.70			

Reduced value of 1/T when Reduced Pressure is	a) .001	b) .01	c) .1
		1.84	1.42

TEMP. °F	PRESSURE		VOLUME cu ft/lb	DENSITY lb/cu ft	ENTHALPY Btu/lb		ENTROPY Btu/(lb)(°R)	
	psia	psig	Vapor	Liquid	Liquid	Vapor	Liquid	Vapor
−280	4.90	19.94*	24.04	27.51	−1934.	−1705	1.087	2.357
−275	6.49	16.71*	18.68	27.28	−1929	−1703	1.108	2.333
−270	8.44	12.74*	14.61	27.04	−1925.	−1701	1.129	2.311
−265	10.82	7.90*	11.64	26.80	−1921.	−1699	1.150	2.290
−260	13.80	1.82*	9.31	26.55	−1917.	−1697	1.169	2.270
−258.68	14.696	0	8.71	26.46	−1917.	−1697	1.174	2.263
−255	17.40	2.70	7.48	26.32	−1912.	−1696	1.190	2.247
−250	21.71	7.01	6.11	26.04	−1908.	−1694	1.208	2.230
−245	26.60	11.90	5.07	25.77	−1904.	−1692	1.227	2.214
−240	32.4	17.70	4.22	25.51	−1900.	−1690	1.244	2.198
−235	39.0	24.3	3.56	25.25	−1896.	−1689	1.262	2.184
−230	46.4	31.7	3.03	25.00	−1892.	−1687	1.282	2.171
−225	54.8	40.1	2.59	24.69	−1888.	−1686	1.300	2.157
−220	64.5	49.8	2.23	24.45	−1883	−1685	1.317	2.144
−215	75.2	60.5	1.92	24.15	−1879	−1684	1.334	2.132
−210	87.6	72.9	1.66	23.87	−1875	−1683	1.351	2.120
−205	101.0	86.3	1.45	23.53	−1870.	−1682.	1.368	2.108
−200	115.7	101.0	1.27	23.20	−1865.	−1681.	1.387	2.098
−195	132.0	117.3	1.12	22.88	−1861.	−1680.	1.404	2.087
−190	150.0	135.3	0.984	22.57	−1856.	−1679.	1.423	2.077
−185	169.7	155.0	0.867	22.22	−1851.	−1679.	1.442	2.067
−180	191.5	176.8	0.768	21.83	−1846.	−1678.	1.459	2.057
−175	214.9	200.3	.682	21.46	−1841.	−1678.	1.477	2.048
−170	240.0	225.3	.607	21.05	−1836.	−1678.	1.494	2.038
−165	267.3	252.6	.541	20.66	−1831.	−1678.	1.510	2.029
−160	297.0	282.3	.481	20.24	−1826.	−1679.	1.527	2.018
−155	329.0	314.3	0.428	19.76	−1820.	−1680.	1.545	2.006
−150	364.0	349.3	.380	19.23	−1815.	−1681.	1.562	1.994
−145	401.0	386.3	.338	18.69	−1809.	−1683.	1.583	1.983
−140	440.0	425.3	.301	18.12	−1802.	−1684.	1.601	1.970
−135	482.0	467.3	.267	17.42	−1795.	−1686.	1.623	1.957
−130	527.0	512.3	0.232	16.67	−1787.	−1689.	1.646	1.943
−125	575.0	560.3	.199	15.72	−1776.	−1692.	1.677	1.926
−120	627.0	612.3	.161	14.37	−1757.	−1698.	1.730	1.905
−116.5	673.1	658.4	.099	10.10	−1730.	−1730.	1.810	1.810

*Inches of mercury below one atmosphere

From *Fluid Thermodynamic Properties for Light Petroleum Systems* by Kenneth E. Starling.

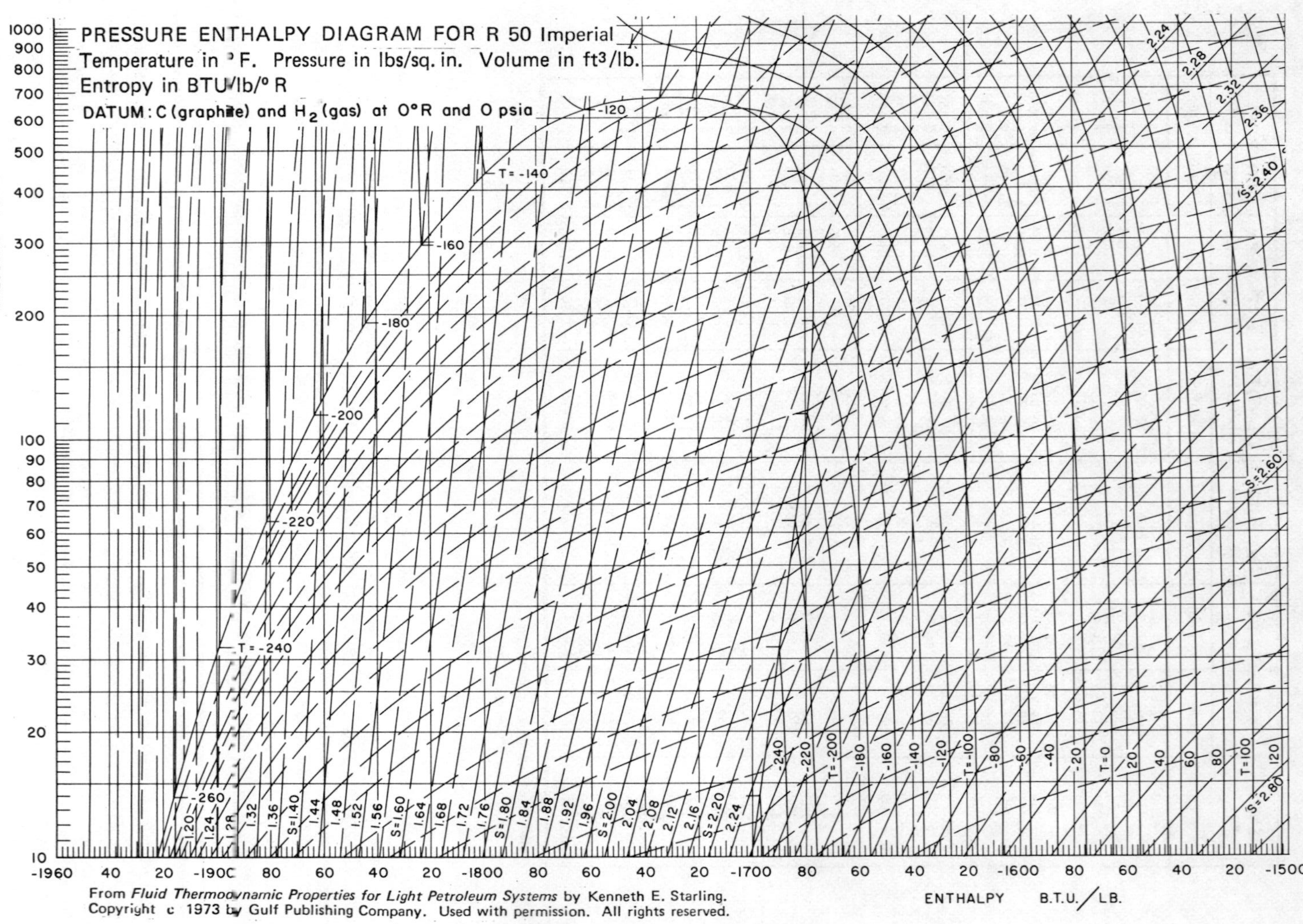

From *Fluid Thermodynamic Properties for Light Petroleum Systems* by Kenneth E. Starling.

DATA SHEET FOR R 113

$C_2Cl_3F_3$	TRICHLOROTRIFLUOROETHANE	Molecular Wt.	187.4

Pressure		Temperature				Volume		Density	
34.8	kg/cm²	214	°C	417	°F	1.736	l/kg	.576	kg/l.
4.95	psia	487	°K	877	°R	.0278	ft³/lb	36.0	lbs/ft³
								6.46	Air=1

AT CRITICAL POINT

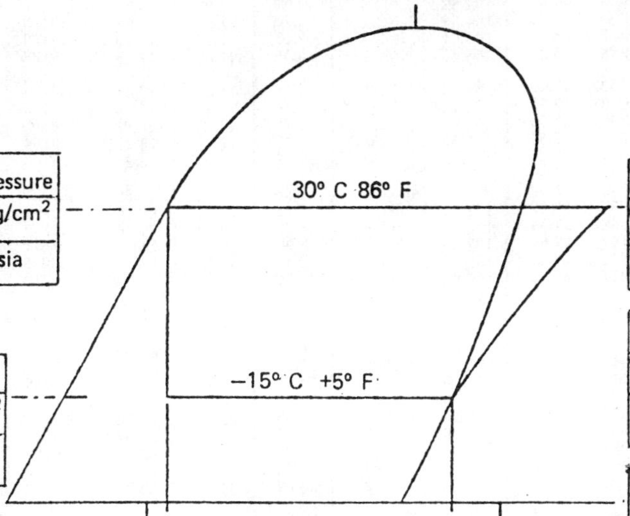

30° C 86° F

−15° C +5° F

Discharge Pressure	
.553	kg/cm²
7.86	psia

Inlet Pressure	
.068	kg/cm²
.98	psia

Discharge Temperature			
30	°C	86	°F
303	°K	546	°R

Normal Boiling Point			
47.6	°C	117.6	°F
321	°K	578	°R

Triple Point			
-35	°C	-31	°F
238	°K	429	°R

Refrigerating Effect	
17.5 kcal/m³	1.97 Btu/ft³

Latent Heat at NBP	35.1 kcal/kg	6578 /kg mol	63.12 Btu/lb	11829 /lb mol

Trouton's No. 20.46	Gas Constant 4.52 kg.m/kg/° K	8.25 ft.lbs/lb/° R

Specific Heat Liquid at 30° C 86° F .218	Gas C_p .161	C_p/C_v 1.08 at 60°C

Liquid at 30° C 86° F Density 1.55 kg/l.	97 lbs/ft³ Viscosity .619 cp

Reduced Form at NBP Pressure .0297	1 / Temperature 1.52

Reduced value of 1/T	1.98	1.67	1.34
when Reduced Pressure is	a) .001	b) .01	c) .1

TEMP.	PRESSURE		VOLUME cu ft/lb		DENSITY lb/cu ft		ENTHALPY Btu/lb			ENTROPY Btu/(lb)(°R)		TEMP.
°F	PSIA	PSIG	LIQUID v_f	VAPOR v_g	LIQUID $1/v_f$	VAPOR $1/v_g$	LIQUID h_f	LATENT h_{fg}	VAPOR h_g	LIQUID s_f	VAPOR s_g	°F
−30	0.298	29.31*	0.00947	82.26	105.64	0.0121	1.97	72.68	74.65	0.0047	0.1738	−30
−28	.321	29.27*	.00948	76.81	105.50	.0130	2.36	72.57	74.93	.0056	.1737	−28
−26	.345	29.22*	.00949	71.71	105.37	.0139	2.76	72.45	75.21	.0065	.1736	−26
−24	.371	29.16*	.00950	66.99	105.23	.0149	3.16	72.33	75.49	.0074	.1735	−24
−22	.399	29.11*	.00952	62.63	105.09	.0159	3.56	72.21	75.77	.0083	.1733	−22
−20	0.428	29.05*	0.00953	58.61	104.96	0.0170	3.96	72.09	76.05	0.0092	0.1732	−20
−18	.460	28.98*	.00954	54.88	104.82	.0182	4.36	71.98	76.34	.0101	.1731	−18
−16	.493	28.92*	.00955	51.42	104.68	.0194	4.76	71.86	76.62	.0110	.1730	−16
−14	.528	28.85*	.00957	48.23	104.54	.0207	5.16	71.74	76.90	.0119	.1729	−14
−12	.565	28.77*	.00958	45.25	104.40	.0221	5.56	71.62	77.18	.0128	.1729	−12
−10	0.604	28.69*	0.00959	42.48	104.26	0.0235	5.96	71.51	77.47	0.0137	0.1728	−10
− 8	.646	28.60*	.00960	39.92	104.12	.0250	6.36	71.39	77.75	.0146	.1727	− 8
− 6	.690	28.51*	.00962	37.54	103.98	.0266	6.76	71.27	78.03	.0155	.1726	− 6
− 4	.736	28.42*	.00963	35.31	103.84	.0283	7.17	71.15	78.32	.0164	.1726	− 4
− 2	.786	28.32*	.00964	33.24	103.70	.0300	7.57	71.03	78.60	.0173	.1725	− 2
0	0.837	28.21*	0.00966	31.31	103.56	0.0319	7.98	70.92	78.89	0.0182	0.1725	0
2	.892	28.10*	.00967	29.52	103.41	.0338	8.38	70.80	79.18	.0190	.1724	2
4	.950	27.99*	.00968	27.84	103.27	.0359	8.78	70.68	79.46	.0199	.1724	4
5†	**0.980**	**27.92***	**.00969**	**27.04**	**103.20**	**.0369**	**8.98**	**70.62**	**79.60**	**.0203**	**.1723**	**5†**
6	1.011	27.86*	.00970	26.27	103.13	.0380	9.19	70.56	79.75	.0208	.1723	6
8	1.075	27.73*	.00971	24.81	102.98	.0403	9.59	70.44	80.03	.0216	.1723	8
10	1.142	27.60*	0.00972	23.45	102.84	0.0426	10.00	70.32	80.32	0.0225	0.1723	10
12	1.213	27.45*	.00974	22.17	102.69	.0451	10.41	70.20	80.61	.0234	.1722	12
14	1.288	27.30*	.00975	20.97	102.55	.0476	10.81	70.08	80.89	.0242	.1722	14
16	1.366	27.14*	.00977	19.84	102.40	.0504	11.22	69.96	81.18	.0251	.1722	16
18	1.448	26.97*	.00978	18.79	102.25	.0532	11.62	69.84	81.46	.0259	.1722	18
20	1.534	26.80*	0.00979	17.81	102.10	0.0561	12.03	69.72	81.75	0.0268	0.1722	20
22	1.624	26.61*	.00981	16.89	101.96	.0592	12.44	69.60	82.04	.0276	.1721	22
24	1.719	26.42*	.00982	16.02	101.81	.0624	12.85	69.48	82.33	.0285	.1721	24
26	1.818	26.22*	.00984	15.20	101.66	.0657	13.26	69.36	82.62	.0293	.1722	26
28	1.922	26.01*	.00985	14.43	101.51	.0692	13.67	69.24	82.91	.0302	.1722	28
30	2.031	25.79*	0.00987	13.71	101.36	0.0729	14.08	69.12	83.20	0.0310	0.1722	30
32	2.145	25.55*	.00988	13.03	101.21	.0767	14.49	69.00	83.49	.0318	.1722	32
34	2.264	25.31*	.00990	12.39	101.06	.0807	14.91	68.87	83.78	.0327	.1722	34
36	2.388	25.06*	.00991	11.79	100.91	.0848	15.32	68.75	84.07	.0335	.1722	36
38	2.519	24.79*	.00993	11.22	100.76	.0891	15.74	68.62	84.36	.0343	.1722	38
40	2.655	24.52*	0.00994	10.68	100.60	0.0936	16.16	68.50	84.65	0.0352	0.1723	40
42	2.797	24.23*	.00996	10.18	100.45	.0982	16.57	68.37	84.94	.0360	.1723	42
44	2.944	23.93*	.00997	9.703	100.30	.1031	16.99	68.25	85.24	.0368	.1723	44
46	3.098	23.61*	.00999	9.253	100.14	.1081	17.41	68.12	85.53	.0377	.1724	46
48	3.258	23.29*	.01000	8.830	99.99	.1133	17.82	68.00	85.82	.0385	.1724	48
50	3.427	22.94*	0.01002	8.426	99.83	0.1187	18.24	67.87	86.11	0.0393	0.1725	50
52	3.602	22.59*	.01003	8.044	99.68	.1243	18.66	67.74	86.40	.0401	.1726	52
54	3.784	22.22*	.01005	7.682	99.52	.1302	19.08	67.61	86.69	.0410	.1726	54
56	3.973	21.83*	.01006	7.342	99.37	.1362	19.50	67.48	86.98	.0418	.1727	56
58	4.170	21.43*	.01008	7.018	99.21	.1425	19.93	67.35	87.28	.0426	.1727	58
60	4.374	21.02*	0.01010	6.713	99.05	0.1490	20.35	67.22	87.57	0.0434	0.1728	60
62	4.586	20.59*	.01011	6.424	98.89	.1557	20.77	67.09	87.86	.0442	.1729	62
64	4.807	20.14*	.01013	6.149	98.73	.1626	21.19	66.96	88.15	.0450	.1729	64
66	5.036	19.67*	.01015	5.889	98.58	.1698	21.62	66.83	88.45	.0459	.1730	66
68	5.275	19.18*	.01016	5.640	98.42	.1773	22.05	66.69	88.74	.0467	.1731	68
70	5.523	18.68*	0.01018	5.404	98.26	0.1851	22.48	66.56	89.04	0.0475	0.1731	70
72	5.780	18.16*	.01019	5.180	98.10	.1931	22.90	66.43	89.33	.0483	.1732	72
74	6.042	17.62*	.01021	4.971	97.93	.2012	23.33	66.29	89.62	.0491	.1733	74
76	6.320	17.06*	.01023	4.769	97.77	.2097	23.76	66.16	89.92	.0499	.1734	76
78	6.607	16.47*	.01025	4.574	97.61	.2186	24.19	66.02	90.21	.0507	.1735	78
80	6.902	15.87*	0.01026	4.392	97.45	0.2277	24.63	65.88	90.51	0.0515	0.1736	80
82	7.208	15.25*	.01028	4.218	97.28	.2371	25.06	65.74	90.80	.0523	.1737	82
84	7.527	14.60*	.01030	4.051	97.12	.2468	25.49	65.60	91.09	.0531	.1738	84
86†	**7.856**	**13.93***	**.01031**	**3.893**	**96.96**	**.2569**	**25.93**	**65.46**	**91.39**	**.0539**	**.1739**	**86†**
88	8.194	13.24*	.01033	3.742	96.79	.2672	26.36	65.32	91.68	.0547	.1740	88

*Inches of mercury below one atmosphere

TEMP.	PRESSURE		VOLUME cu ft/lb		DENSITY lb/cu ft		ENTHALPY Btu/lb			ENTROPY Btu/(lb)(°R)		TEMP.
°F	PSIA	PSIG	LIQUID v_f	VAPOR v_g	LIQUID $1/v_f$	VAPOR $1/v_g$	LIQUID h_f	LATENT h_{fg}	VAPOR h_g	LIQUID s_f	VAPOR s_g	°F
90	8.545	12.53*	0.01035	3.600	96.63	0.2778	26.80	65.18	91.98	0.0555	0.1741	90
92	8.908	11.79*	.01037	3.463	96.46	.2888	27.24	65.04	92.28	.0563	.1742	92
94	9.281	11.03*	.01039	3.333	96.30	.3001	27.67	64.90	92.57	.0571	.1743	94
96	9.668	10.24*	.01040	3.208	96.13	.3117	28.11	64.75	92.86	.0578	.1744	96
98	10.07	9.42*	.01042	3.089	95.96	.3237	28.55	64.60	93.15	.0586	.1745	98
100	10.48	8.59*	0.01044	2.976	95.79	0.3360	28.99	64.46	93.45	0.0594	0.1746	100
102	10.91	7.71*	.01046	2.867	95.63	.3488	29.44	64.31	93.75	.0602	.1747	102
104	11.35	6.82*	.01048	2.762	95.46	.3620	29.89	64.16	94.05	.0610	.1748	104
106	11.81	5.88*	.01050	2.662	95.29	.3756	30.33	64.01	94.34	.0618	.1750	106
108	12.28	4.93*	.01051	2.567	95.12	.3896	30.78	63.86	94.64	.0626	.1751	108
110	12.76	3.95*	0.01053	2.477	94.95	0.4038	31.22	63.71	94.93	0.0634	0.1752	110
112	13.25	2.95*	.01055	2.391	94.78	.4182	31.67	63.56	95.23	.0641	.1753	112
114	13.76	1.91*	.01057	2.308	94.61	.4333	32.12	63.40	95.52	.0649	.1755	114
116	14.29	0.83*	.01059	2.228	94.43	.4489	32.57	63.25	95.82	.0657	.1756	116
118	14.84	0.14	.01061	2.151	94.26	.4649	33.03	63.09	96.12	.0665	.1757	118
120	15.40	0.70	0.01063	2.078	94.09	0.4813	33.48	62.93	96.41	0.0673	0.1758	120
122	15.97	1.27	.01065	2.008	93.92	.4981	33.93	62.78	96.71	.0680	.1760	122
124	16.56	1.86	.01067	1.941	93.74	.5153	34.38	62.62	97.00	.0688	.1761	124
126	17.17	2.47	.01069	1.876	93.57	.5330	34.83	62.46	97.29	.0696	.1763	126
128	17.80	3.10	.01071	1.814	93.39	.5514	35.29	62.30	97.59	.0704	.1764	128
130	18.45	3.74	0.01073	1.754	93.22	0.5702	35.75	62.14	97.89	0.0712	0.1765	130
132	19.11	4.41	.01075	1.697	93.04	.5894	36.21	61.97	98.18	.0719	.1767	132
134	19.79	5.09	.01077	1.642	92.86	.6091	36.67	61.80	98.47	.0727	.1768	134
136	20.48	5.78	.01079	1.590	92.69	.6290	37.13	61.64	98.77	.0735	.1770	136
138	21.19	6.49	.01081	1.540	92.51	.6494	37.59	61.48	99.06	.0742	.1771	138
140	21.93	7.23	0.01083	1.491	92.33	0.6707	38.05	61.31	99.36	0.0750	0.1773	140
142	22.69	7.99	.01085	1.444	92.15	.6926	38.52	61.13	99.65	.0758	.1774	142
144	23.47	8.77	.01087	1.399	91.98	.7150	38.98	60.96	99.94	.0765	.1775	144
146	24.27	9.57	.01089	1.355	91.80	.7379	39.45	60.79	100.24	.0773	.1777	146
148	25.09	10.39	.01092	1.313	91.62	.7615	39.92	60.61	100.53	.0781	.1778	148
150	25.93	11.23	0.01094	1.273	91.44	0.7856	40.38	60.44	100.82	0.0789	0.1780	150
152	26.79	12.09	.01096	1.234	91.25	.8102	40.85	60.27	101.11	.0796	.1782	152
154	27.67	12.97	.01098	1.197	91.07	.8353	41.32	60.09	101.41	.0804	.1783	154
156	28.56	13.86	.01100	1.162	90.89	.8608	41.79	59.91	101.70	.0812	.1785	156
158	29.48	14.78	.01102	1.128	90.71	.8869	42.26	59.73	101.99	.0819	.1786	158
160	30.44	15.74	0.01105	1.094	90.53	0.9141	42.74	59.55	102.29	0.0827	0.1788	160
170	35.53	20.83	.01116	0.9442	89.60	1.059	45.12	58.62	103.74	.0865	.1796	170
180	41.22	26.52	.01128	.8193	88.67	1.221	47.53	57.66	105.19	.0903	.1804	180
190	47.60	32.90	.01140	.7134	87.72	1.402	49.97	56.66	106.63	.0940	.1813	190
200	54.66	39.96	.01153	.6241	86.76	1.602	52.45	55.62	108.07	.0978	.1821	200
210	62.50	47.80	.01166	.5477	85.79	1.826	54.96	54.54	109.50	.1015	.1830	210
220	71.07	56.37	0.01179	0.4827	84.80	2.072	57.49	53.43	110.92	0.1052	0.1839	220

*Inches of mercury below one atmosphere (760 mm Hg) ENTHALPY (BTU/lb ABOVE SATURATED LIQUID AT –40° F

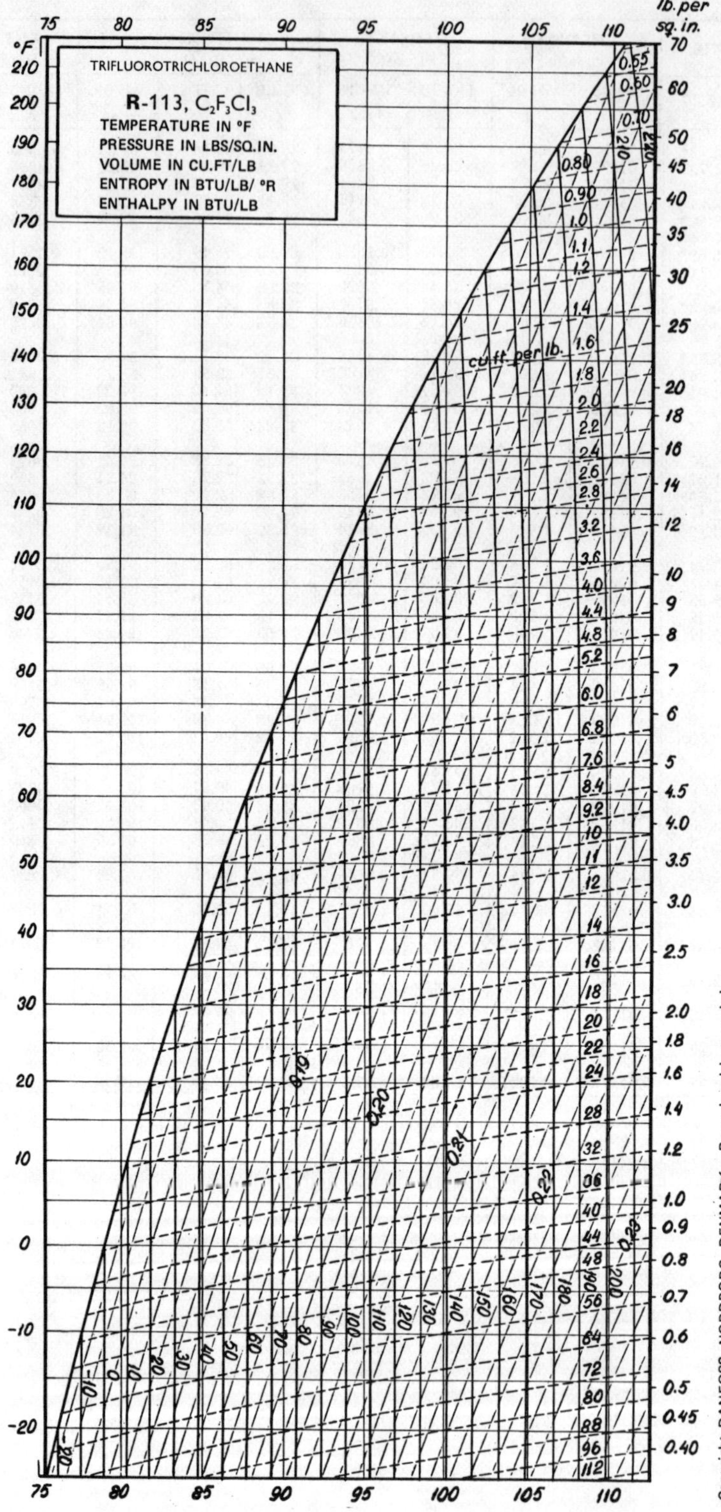
TRIFLUOROTRICHLOROETHANE
R-113, $C_2F_3Cl_3$
TEMPERATURE IN °F
PRESSURE IN LBS/SQ.IN.
VOLUME IN CU.FT/LB
ENTROPY IN BTU/LB/ °R
ENTHALPY IN BTU/LB
°F
lb.per sq.in.
cu.ft. per lb.

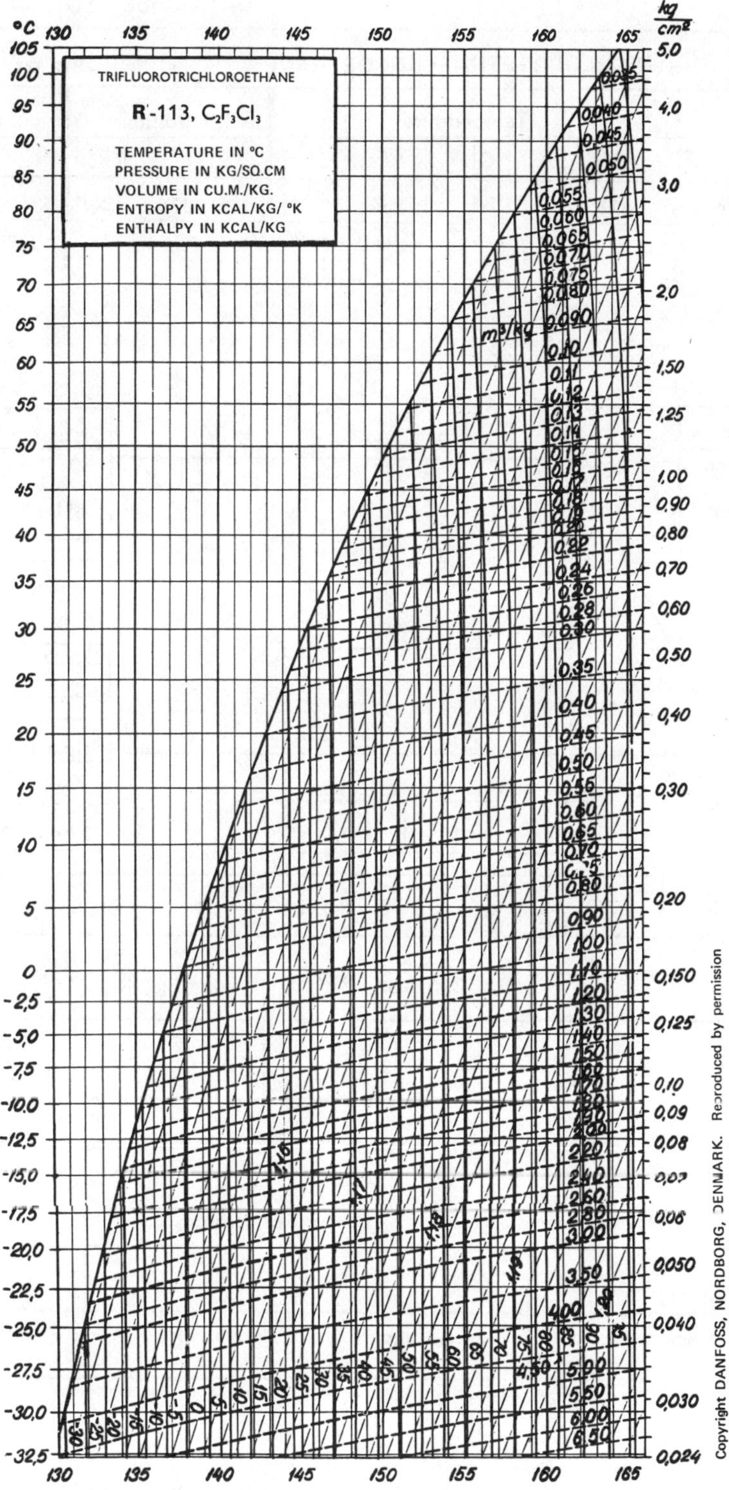
TRIFLUOROTRICHLOROETHANE
R-113, $C_2F_3Cl_3$
TEMPERATURE IN °C
PRESSURE IN KG/SQ.CM
VOLUME IN CU.M./KG.
ENTROPY IN KCAL/KG/ °K
ENTHALPY IN KCAL/KG
°C
kg/cm²
m³/kg

DATA SHEET FOR R 114

$C_2Cl_2F_4$	DICHLOROTETRAFLUOROETHANE	Molecular Wt.	171

Pressure		Temperature				Volume		Density	
33.25	kg/cm²	146	°C	294	°F	1.72	l/kg	.582	kg/l.
473	psia	419	°K	754	°R	.0275	ft³/lb	36.3	lbs/ft³
								5.89	Air=1

AT CRITICAL POINT

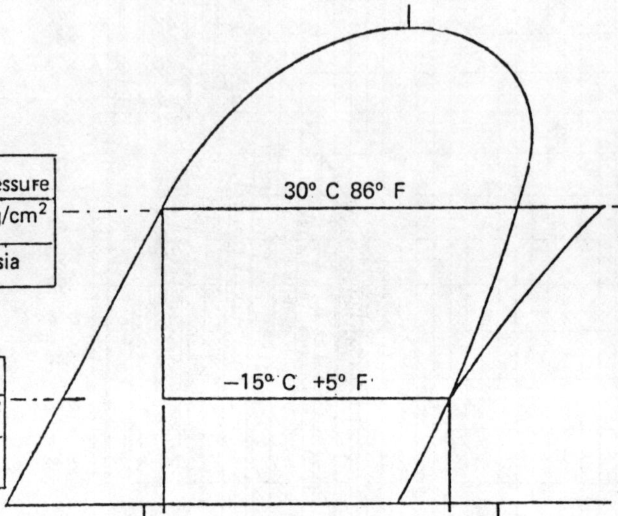

30° C 86° F

−15° C +5° F

Discharge Pressure	
2.58	kg/cm²
36.7	psia

Inlet Pressure	
.473	kg/cm²
6.73	psia

Discharge Temperature			
30	°C	86	°F
303	°K	546	°R

Normal Boiling Point			
3.5	°C	39	°F
276.5	°K	499	°R

Triple Point			
-94	°C	-137	°F
179	°K	323	°R

Refrigerating Effect			
88	kcal/m³	9.9	Btu/ft³

Latent Heat at NBP	33 kcal/kg	5605 /kg mol	59 Btu/lb	10089 /lb mol

Trouton's No.	20.3	Gas Constant	4.95 kg.m/kg/°K	9.04 ft.lbs/lb/°R

Specific Heat Liquid at 30° C 86° F	.238 Fig 15	Gas C_p .170	C_p/C_v 1.084 at 25°C

Liquid at 30° C 86° F Density	1.44 kg/l.	89.9 lbs/ft³	Viscosity .36 cp

Reduced Form at NBP	Pressure .0311	1 / Temperature 1.51

Reduced value of 1/T			
when Reduced Pressure is	a) .001	b) .01	c) .1

TEMP. °F	PRESSURE PSIA	PRESSURE PSIG	VOLUME cu ft/lb LIQUID v_f	VOLUME cu ft/lb VAPOR v_g	DENSITY lb/cu ft LIQUID $1/v_f$	DENSITY lb/cu ft VAPOR $1/v_g$	ENTHALPY Btu/lb LIQUID h_f	ENTHALPY Btu/lb LATENT h_{fg}	ENTHALPY Btu/lb VAPOR h_g	ENTROPY Btu/(lb)(°R) LIQUID s_f	ENTROPY Btu/(lb)(°R) VAPOR s_g	TEMP °F
−40	1.866	26.12*	0.00978	14.02	102.25	0.07132	0.00	65.91	65.91	0.0000	0.1571	−40
−38	1.990	25.87*	.00980	13.20	102.08	.07574	0.45	65.74	66.19	.0011	.1570	−38
−36	2.121	25.60*	.00981	12.44	101.90	.08038	0.91	65.56	66.47	.0021	.1569	−36
−34	2.259	25.32*	.00983	11.73	101.72	.08524	1.36	65.38	66.74	.0032	.1568	−34
−32	2.404	25.03*	.00985	11.07	101.55	.09034	1.81	65.21	67.02	.0042	.1567	−32
−30	2.557	24.72*	0.00987	10.45	101.37	0.09568	2.27	65.03	67.30	0.0053	0.1567	−30
−28	2.718	24.39*	.00988	9.877	101.19	.1013	2.72	64.86	67.58	.0063	.1567	−28
−26	2.887	24.04*	.00990	9.338	101.01	.1071	3.17	64.68	67.86	.0074	.1566	−26
−24	3.064	23.68*	.00992	8.833	100.83	.1132	3.63	64.51	68.14	.0084	.1565	−24
−22	3.249	23.31*	.00994	8.362	100.65	.1196	4.08	64.34	68.42	.0095	.1565	−22
−20	3.444	22.91*	0.00995	7.921	100.47	0.1263	4.54	64.16	68.70	0.0105	0.1565	−20
−18	3.648	22.49*	.00997	7.508	100.29	.1332	4.99	63.99	68.98	.0116	.1565	−18
−16	3.862	22.06*	.00999	7.121	100.11	.1404	5.44	63.81	69.26	.0126	.1564	−16
−14	4.085	21.61*	.01001	6.757	99.92	.1480	5.90	63.64	69.54	.0136	.1564	−14
−12	4.319	21.13*	.01003	6.416	99.74	.1559	6.35	63.46	69.82	.0146	.1564	−12
−10	4.564	20.63*	0.01005	6.095	99.56	0.1641	6.81	63.29	70.10	0.0157	0.1564	−10
− 8	4.819	20.11*	.01006	5.794	99.37	.1726	7.26	63.11	70.38	.0167	.1564	− 8
− 6	5.086	19.57*	.01008	5.510	99.19	.1815	7.72	62.94	70.66	.0177	.1564	− 6
− 4	5.365	19.00*	.01010	5.244	99.00	.1907	8.18	62.77	70.94	.0187	.1564	− 4
− 2	5.655	18.41*	.01012	4.992	98.81	.2003	8.63	62.59	71.22	.0197	.1565	− 2
0	5.958	17.79*	0.01014	4.756	98.62	0.2103	9.09	62.42	71.50	0.0207	0.1565	0
2	6.274	17.15*	.01016	4.533	98.44	.2206	9.54	62.24	71.78	.0217	.1565	2
4	6.603	16.48*	.01018	4.322	98.25	.2314	10.00	62.07	72.07	.0227	.1565	4
5†	6.772	16.14*	.01019	4.221	98.15	.2369	10.23	61.98	72.21	.0232	.1565	5†
6	6.945	15.78*	.01020	4.123	98.06	.2425	10.46	61.89	72.35	.0236	.1566	6
8	7.301	15.06*	.01022	3.935	97.87	.2541	10.91	61.71	72.63	.0246	.1566	8
10	7.671	14.31*	0.01024	3.758	97.68	0.2661	11.37	61.54	72.91	0.0256	0.1566	10
12	8.057	13.52*	.01026	3.591	97.48	.2785	11.83	61.36	73.19	.0266	.1567	12
14	8.457	12.71*	.01028	3.432	97.29	.2914	12.29	61.19	73.47	.0275	.1567	14
16	8.873	11.86*	.01030	3.282	97.10	.3047	12.75	61.01	73.75	.0285	.1568	16
18	9.305	10.98*	.01032	3.140	96.90	.3185	13.20	60.83	74.04	.0295	.1568	18
20	9.753	10.07*	0.01034	3.005	96.71	0.3328	13.66	60.65	74.32	0.0304	0.1569	20
22	10.22	9.12*	.01036	2.877	96.51	.3476	14.12	60.48	74.60	.0314	.1569	22
24	10.70	8.14*	.01038	2.756	96.32	.3629	14.58	60.30	74.88	.0323	.1570	24
26	11.20	7.12*	.01040	2.641	96.12	.3786	15.05	60.12	75.17	.0333	.1571	26
28	11.72	6.07*	.01043	2.532	95.92	.3949	15.51	59.94	75.45	.0342	.1571	28
30	12.25	4.99*	0.01045	2.429	95.73	0.4118	15.97	59.76	75.73	0.0352	0.1572	30
32	12.81	3.85*	.01047	2.330	95.53	.4292	16.43	59.58	76.01	.0361	.1573	32
34	13.38	2.69*	.01049	2.236	95.33	.4472	16.89	59.40	76.29	.0370	.1574	34
36	13.98	1.47*	.01051	2.147	95.13	.4658	17.36	59.22	76.58	.0380	.1575	36
38	14.59	0.22*	.01053	2.062	94.93	.4849	17.82	59.04	76.86	.0389	.1575	38
40	15.22	0.52	0.01056	1.982	94.73	0.5047	18.28	58.86	77.14	0.0398	0.1576	40
42	15.88	1.18	.01058	1.905	94.52	.5250	18.75	58.67	77.42	.0408	.1577	42
44	16.56	1.86	.01060	1.832	94.32	.5460	19.21	58.49	77.70	.0417	.1578	44
46	17.26	2.56	.01063	1.762	94.12	.5676	19.68	58.31	77.99	.0426	.1579	46
48	17.98	3.28	.01065	1.695	93.91	.5899	20.14	58.12	78.27	.0435	.1580	48
50	18.73	4.03	0.01067	1.632	93.71	0.6129	20.61	57.94	78.55	0.0444	0.1581	50
52	19.50	4.80	.01070	1.571	93.50	.6365	21.08	57.75	78.83	.0453	.1582	52
54	20.29	5.59	.01072	1.513	93.30	.6609	21.54	57.56	79.11	.0463	.1583	54
56	21.11	6.41	.01074	1.458	93.09	.6859	22.01	57.38	79.39	.0472	.1584	56
58	21.96	7.26	.01077	1.405	92.88	.7117	22.48	57.19	79.67	.0481	.1585	58
60	22.83	8.13	0.01079	1.354	92.68	0.7383	22.95	57.00	79.95	0.0490	0.1587	60
62	23.72	9.02	.01082	1.306	92.47	.7655	23.42	56.81	80.23	.0499	.1588	62
64	24.64	9.94	.01084	1.260	92.26	.7936	23.89	56.62	80.51	.0508	.1589	64
66	25.59	10.89	.01086	1.216	92.05	.8225	24.36	56.43	80.79	.0517	.1590	66
68	26.57	11.87	.01089	1.174	91.84	.8521	24.83	56.24	81.07	.0526	.1591	68
70	27.57	12.87	0.01091	1.133	91.63	0.8826	25.30	56.04	81.35	0.0534	0.1593	70
72	28.61	13.91	.01094	1.094	91.41	.9140	25.78	55.85	81.62	.0543	.1594	72
74	29.67	14.97	.01097	1.057	91.20	.9462	26.25	55.65	81.90	.0552	.1595	74
76	30.76	16.06	.01099	1.021	90.99	0.9793	26.73	55.45	82.18	.0561	.1596	76
78	31.88	17 18	.01102	0.9869	90.77	1.013	27.20	55.26	82.46	.0570	.1597	78
80	33.04	18.34	0.01104	0.9541	90.56	1.048	27.68	55.06	82.73	0.0579	0.1599	80
82	34.22	19.52	.01107	.9226	90.34	1.084	28.15	54.86	83.01	.0587	.1600	82
84	35.44	20.74	.01110	.8923	90.13	1.121	28.03	54.00	83.29	.0590	.1001	84
86†	36.69	21.99	.01112	.8632	89.91	1.159	29.11	54.46	83.56	.0605	.1603	86†
88	37.97	23.27	.01115	.8353	89.69	1.197	29.58	54.25	83.84	.0613	.1604	88
90	39.29	24.59	0.01118	0.8084	89.47	1.237	30.06	54.05	84.11	0.0622	0.1605	90
92	40.64	25.94	.01120	.7827	89.25	1.278	30.54	53.84	84.39	.0631	.1607	92
94	42.02	27.32	.01123	.7579	89.03	1.320	31.02	53.64	84.66	.0639	.1608	94
96	43.44	28.74	.01126	.7340	88.81	1.362	31.50	53.43	84.93	.0648	.1609	96
98	44.89	30.19	.01129	.7111	88.59	1.406	31.99	53.22	85.21	.0656	.1611	98
100	46.39	31.69	0.01132	0.6890	88.37	1.452	32.47	53.01	85.48	0.0665	0.1612	100
110	54.41	39.71	0.01146	0.5901	87.25	1.695	34.89	51.94	86.83	0.0708	0.1619	110
120	63.44	48.74	0.01162	0.5077	86.08	1.970	37.33	50.83	88.16	0.0750	0.1627	120
130	73.54	58.84	0.01178	0.4387	84.89	2.280	39.80	49.67	89.47	0.0792	0.1634	130
140	84.79	70.09	0.01195	0.3803	83.66	2.629	42.29	48.47	90.76	0.0833	0.1641	140

*Inches of mercury below one atmosphere

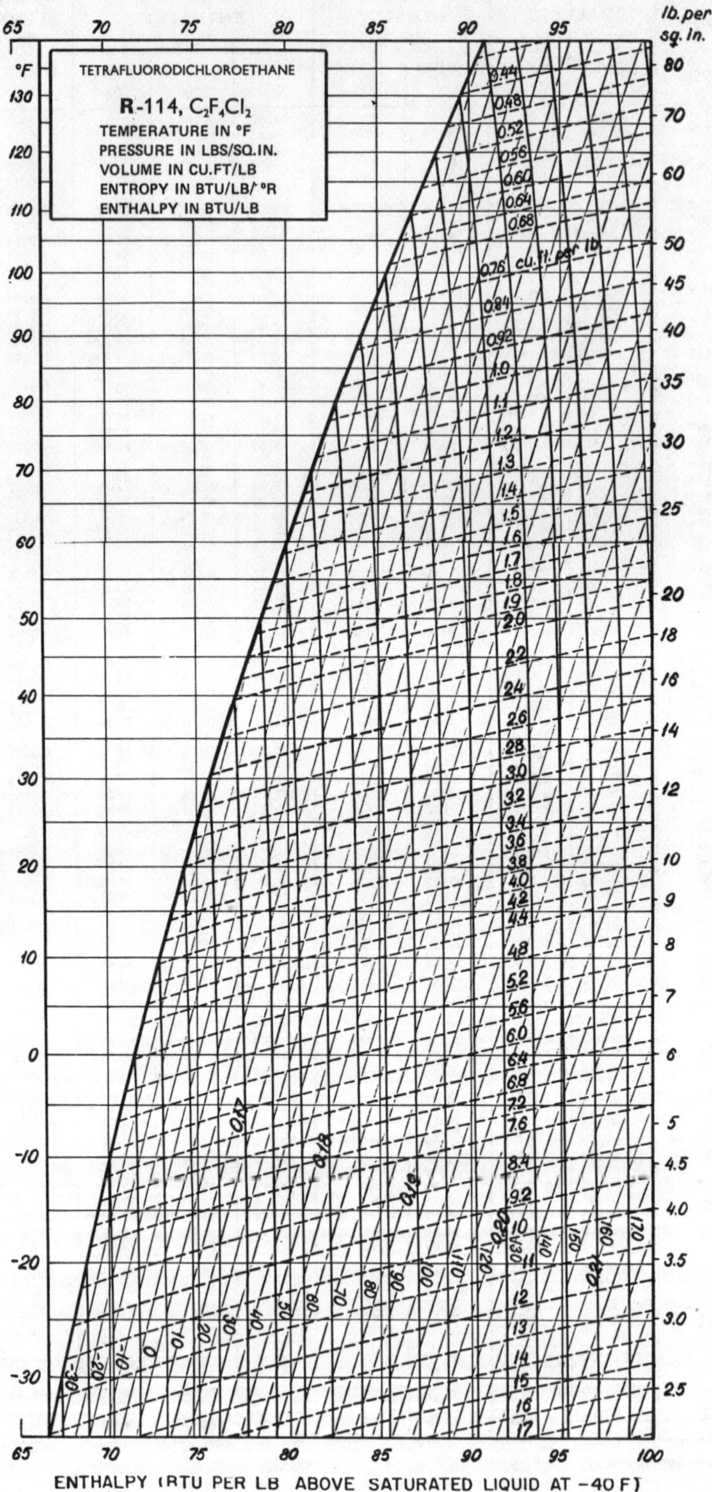
TETRAFLUORODICHLOROETHANE
R-114, $C_2F_4Cl_2$
TEMPERATURE IN °F
PRESSURE IN LBS/SQ.IN.
VOLUME IN CU.FT/LB
ENTROPY IN BTU/LB/°R
ENTHALPY IN BTU/LB
°F
lb. per sq. in.
cu. ft. per lb
ENTHALPY (BTU PER LB ABOVE SATURATED LIQUID AT -40 F)

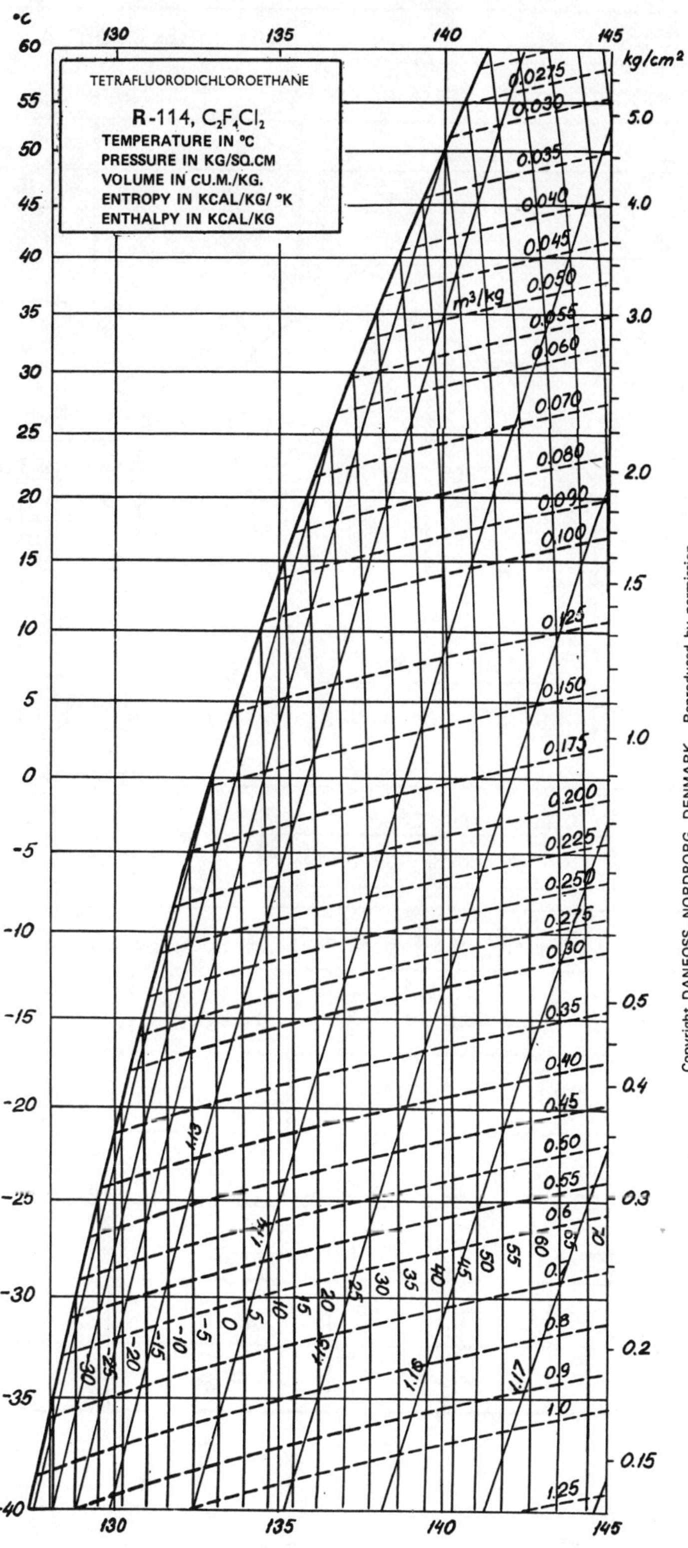
TETRAFLUORODICHLOROETHANE
R-114, $C_2F_4Cl_2$
TEMPERATURE IN °C
PRESSURE IN KG/SQ.CM
VOLUME IN CU.M./KG.
ENTROPY IN KCAL/KG/ °K
ENTHALPY IN KCAL/KG
°C
kg/cm²
m³/kg
130
135
140
145

DATA SHEET FOR R 115

C_2ClF_5	MONOCHLOROPENTAFLUOROETHANE	Molecular Wt. 154.5

Pressure		Temperature				Volume		Density	
31.8	kg/cm²	80	°C	176	°F	1.68	l/kg	.596	kg/l.
453	psia	353	°K	636	°R	.0269	ft³/lb	37.2	lbs/ft³
								5.33	Air=1

AT CRITICAL POINT

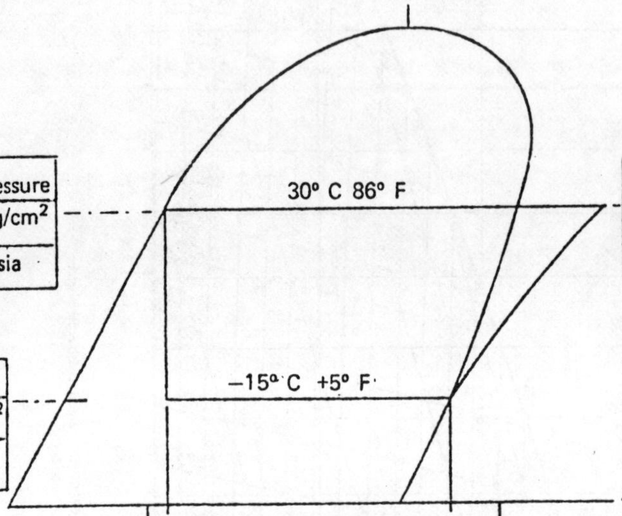

30° C 86° F

−15° C +5° F

Discharge Pressure	
10.47	kg/cm²
149	psia

Inlet Pressure	
2.66	kg/cm²
37.8	psia

Discharge Temperature			
30	°C	86	°F
303	°K	546	°R

Normal Boiling Point			
-39	°C	-38	°F
234	°K	422	°R

Triple Point			
-106	°C	-159	°F
167	°K	301	°R

Refrigerating Effect			
336	kcal/m³	37.8	Btu/ft³

Latent Heat at NBP	30	kcal/kg	4652	/kg mol	54.2	Btu/lb	8374	/lb mol

Trouton's No.	19.84	Gas Constant	5.49	kg.m/kg/° K	10	ft.lbs/lb/° R

Specific Heat Liquid at 30° C 86° F	Fig 15	Gas C_p		C_p/C_v 1.091 at 0°C

Liquid at 30° C 86° F Density	1.286	kg/l.	80	lbs/ft³	Viscosity		cp

Reduced Form at NBP	Pressure	.0324	1 / Temperature	1.51

Reduced value of 1/T	1.99	1.67	1.34
when Reduced Pressure is	a) .001	b) .01	c) .1

TEMP. °F	PRESSURE	VOLUME cu ft/lb		ENTHALPY Btu/lb			ENTROPY Btu/(lb)(°R)
	PSIA	LIQUID	VAPOR	LIQUID	LATENT	VAPOR	VAPOR
-140	0.44	0.00926	49.40	-21.16	61.45	40.29	0.1348
-130	0.70	0.00935	32.36	-19.12	60.76	41.64	0.1331
-120	1.07	0.00944	21.83	-17.08	60.09	43.00	0.1318
-110	1.59	0.00954	15.12	-15.03	59.42	44.39	0.1308
-100	2.30	0.00964	10.73	-12.96	58.75	45.78	0.1300
- 90	3.25	0.00974	7.784	-10.86	58.07	47.20	0.1295
- 80	4.50	0.00985	5.756	- 8.75	57.37	48.62	0.1292
- 70	6.12	0.00996	4.332	- 6.610	56.66	50.05	0.1291
- 60	8.17	0.01008	3.313	- 4.438	55.93	51.49	0.1291
- 50	10.73	0.01019	2.571	- 2.236	55.18	52.94	0.1293
- 40	13.88	0.01032	2.021	0.000	54.39	54.39	0.1296
- 30	17.73	0.01045	1.608	2.271	53.57	55.84	0.1300
- 20	22.35	0.01058	1.293	4.577	52.71	57.28	0.1304
- 10	27.86	0.01072	1.050	6.922	51.80	58.73	0.1310
0	34.37	0.01087	0.860	9.308	50.85	60.16	0.1317
10	41.98	0.0110	0.710	11.73	49.85	61.58	0.1324
20	50.81	0.0112	0.590	14.20	48.78	62.98	0.1331
30	60.98	0.0113	0.494	16.71	47.66	64.37	0.1338
40	72.63	0.0115	0.416	19.27	46.46	65.73	0.1346
50	85.87	0.0117	0.351	21.88	45.18	67.06	0.1354
60	100.8	0.0119	0.298	24.54	43.82	68.36	0.1362
70	117.7	0.0122	0.254	27.26	42.35	69.62	0.1369
80	136.6	0.0124	0.217	30.04	40.78	70.82	0.1377
90	157.6	0.0127	0.185	32.88	39.08	71.96	0.1383
100	181.0	0.0131	0.159	35.80	37.22	73.03	0.1389
110	206.9	0.0134	0.136	38.81	35.19	74.01	0.1394
120	235.4	0.0139	0.116	41.92	32.95	74.88	0.1397

TEMP. °F	PRESSURE	VOLUME cu ft/lb		ENTHALPY Btu/lb			ENTROPY Btu/(lb)(°R)
	PSIA	LIQUID	VAPOR	LIQUID	LATENT	VAPOR	VAPOR
130	266.8	0.0144	0.0992	45.15	30.44	75.60	0.1400
135	283.6	0.0147	0.0914	46.83	29.06	75.89	0.1399
140	301.2	0.0150	0.0840	48.54	27.58	76.13	0.1398
145	319.6	0.0154	0.0770	50.31	25.98	76.30	0.1396
150	338.9	0.0158	0.0704	52.14	24.24	76.39	0.1394
155	359.0	0.0163	0.0640	54.05	22.32	76.37	0.1389
160	380.0	0.0170	0.0579	56.06	20.16	76.23	0.1383
165	401.9	0.0178	0.0519	58.24	17.65	75.89	0.1374
170	424.8	0.0191	0.0459	60.72	14.56	75.28	0.1361
175.89	453.0	0.0268	0.0268	68.16		68.16	0.1245

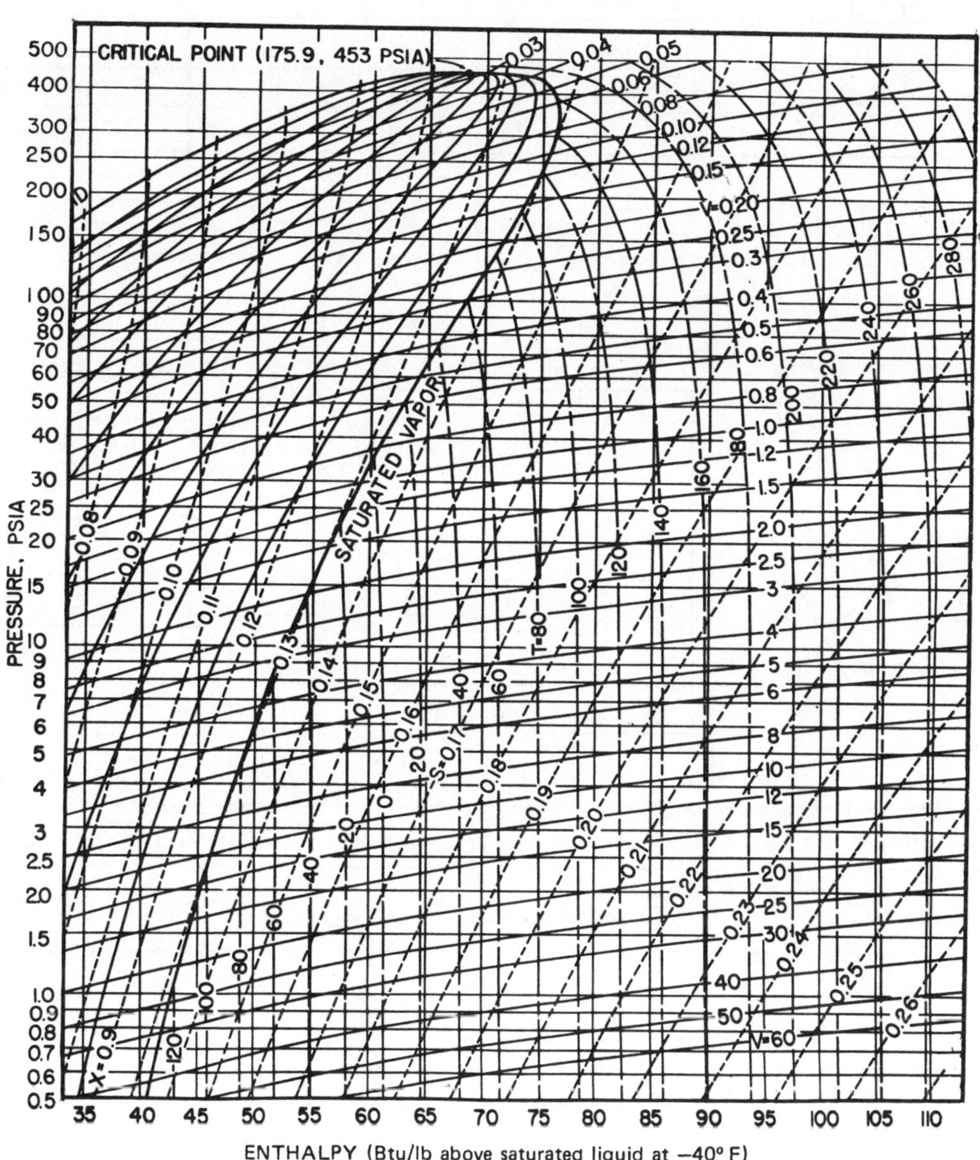

Temperature in °F. Pressure in lbs/sq.in. Volume in ft^3/lb.
Entropy in BTU/lb/ °R.

From Published data of E.I. du Pont de Nemours & Co. Inc. Used by permission.

DATA SHEET FOR R 142b

$C_2H_3F_2Cl$	DIFLUOROMONOCHLOROETHANE	Molecular Wt.	100.5

Pressure		Temperature				Volume		Density	
42.3	kg/cm²	137	°C	279	°F	2.3	l/kg	.435	kg/l.
602	psia	410	°K	739	°R	.0368	ft³/lb	27.2	lbs/ft³
AT CRITICAL POINT								3.46	Air=1

30° C 86° F

−15° C +5° F

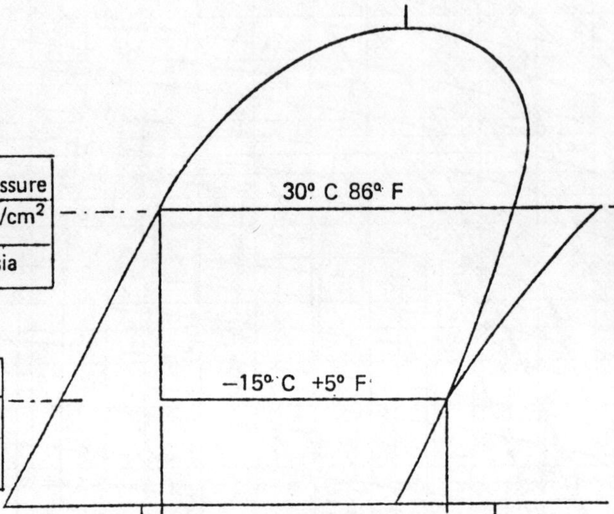

Discharge Pressure	
4.0	kg/cm²
57	psia

Inlet Pressure	
.81	kg/cm²
11.5	psia

Discharge Temperature			
37	°C	100	°F
310	°K	560	°R

Normal Boiling Point			
-9	°C	14	°F
264	°K	474	°R

Triple Point			
-131	°C	-203	°F
142	°K	257	°R

Refrigerating Effect			
159	kcal/m³	17.9	Btu/ft³

Latent Heat at NBP	51	kcal/kg	5137	/kg mol	92	Btu/lb	9246	/lb mol

Trouton's No.	19.6	Gas Constant	8.45	kg.m/kg/° K	15.4	ft.lbs/lb/° R

Specific Heat Liquid at 30° C 86° F	.311 Fig 16	Gas C_p	C_p/C_v at

Liquid at 30° C 86° F Density	1.10	kg/l.	68.6	lbs/ft³	Viscosity	cp

Reduced Form at NBP	Pressure	.0244	1 / Temperature	1.56

Reduced value of 1/T when Reduced Pressure is	a) .001	b) .01	c) .1
	2.04	1.69	1.34

TEMP. °F	PRESSURE		VOLUME cu ft/lb	DENSITY lb/cu ft	ENTHALPY Btu/lb		ENTROPY Btu/(lb)(°R)	
	psia	psig	Vapor	Liquid	Liquid	Vapor	Liquid	Vapor
-100	0.545	28.8*	70.24	83.19	-10.82	85.97	-0.0284	0.2407
-95	0.657	28.6*	59.05	82.83	-10.10	86.73	- .0263	.2392
-90	0.788	28.3*	49.87	82.47	-9.05	87.50	- .0234	.2378
-85	0.941	28.0*	42.30	82.10	-8.21	88.28	- .0211	.2364
-80	1.118	27.6*	36.05	81.74	-7.44	89.06	- .0190	.2351
-75	1.323	27.2*	30.84	81.37	-6.52	89.85	-0.0166	0.2339
-70	1.559	26.7*	26.49	81.00	-5.61	90.63	- .0142	.2328
-65	1.828	26.2*	22.86	80.64	-4.70	91.43	- .0118	.2317
-60	2.136	25.6*	19.79	80.27	-3.74	92.22	- .0093	.2308
-55	2.486	24.9*	17.19	79.90	-2.84	93.02	- .0070	.2298
-50	2.883	24.1*	14.99	79.52	-1.89	93.82	-0.0048	0.2290
-45	3.331	23.1*	13.11	79.15	-0.97	94.62	- .0024	.2281
-40	3.835	22.1*	11.50	78.78	0.00	95.43	.0000	.2274
-35	4.402	21.0*	10.12	78.40	1.02	96.24	.0025	.2267
-30	5.036	19.7*	8.937	78.02	2.01	97.04	.0048	.2260
-25	5.743	18.2*	7.911	77.65	3.04	97.85	0.0073	0.2254
-20	6.530	16.6*	7.022	77.26	4.08	98.66	.0097	.2248
-15	7.403	14.8*	6.250	76.88	5.12	99.47	.0121	.2243
-10	8.370	12.9*	5.575	76.50	6.20	100.28	.0146	.2238
-5	9.438	10.7*	4.986	76.11	7.30	101.09	.0171	.2234
0	10.61	8.31*	4.469	75.72	8.40	101.89	0.0195	0.2229
5	11.91	5.68*	4.015	75.33	9.53	102.70	.0220	.2225
10	13.32	2.79*	3.614	74.93	10.68	103.50	.0246	.2222
15	14.87	0.18	3.261	74.54	11.83	104.30	.0271	.2219
20	16.57	1.87	2.947	74.14	13.03	105.09	.0296	.2216
25	18.41	3.72	2.669	73.73	14.23	105.89	0.0322	0.2213
30	20.42	5.72	2.421	73.33	15.47	106.67	.0348	.2210
35	22.60	7.90	2.200	72.92	16.71	107.46	.0373	.2208
40	24.96	10.3	2.003	72.50	17.96	108.23	.0399	.2206
45	27.51	12.8	1.826	72.08	19.24	109.00	.0425	.2204
50	30.26	15.6	1.668	71.66	20.55	109.77	0.0451	0.2202
55	33.23	18.5	1.525	71.24	21.86	110.52	.0477	.2200
60	36.42	21.7	1.397	70.81	23.19	111.27	.0504	.2199
65	39.85	25.2	1.281	70.37	24.54	112.01	.0530	.2197
70	43.53	28.8	1.177	69.93	25.92	112.74	.0557	.2196
75	47.47	32.8	1.082	69.49	27.32	113.46	0.0583	0.2194
80	51.69	37.0	0.9964	69.03	28.72	114.17	.0610	.2193
85	56.19	41.5	.9184	68.58	30.15	114.87	.0636	.2192
90	60.99	46.3	.8475	68.11	31.58	115.56	.0663	.2191
95	66.11	51.4	.7829	67.64	33.04	116.23	.0690	.2190
100	71.54	56.8	0.7240	67.16	34.52	116.89	0.0717	0.2188
105	77.32	62.6	.6702	66.68	36.01	117.53	.0743	.2187
110	83.45	68.8	.6210	66.18	37.51	118.16	.0770	.2186
115	89.95	75.3	.5758	65.68	39.03	118.77	.0797	.2185
120	96.83	82.1	.5345	65.17	40.57	119.36	.0824	.2183
125	104.1	89.4	0.4965	64.65	42.12	119.94	0.0851	0.2182
130	111.8	97.1	.4616	64.11	43.68	120.49	.0878	.2180
135	119.9	105	.4294	63.57	45.25	121.03	.0904	.2179
140	128.4	114	.3998	63.01	46.83	121.55	.0931	.2177
145	137.4	123	.3725	62.45	48.43	122.04	.0958	.2175
150	146.9	132	0.3472	61.86	50.04	122.51	0.0984	0.2173
155	156.9	142	.3239	61.27	51.65	122.95	.1011	.2171
160	167.3	153	.3023	60.65	53.28	123.38	.1037	.2168
165	178.2	164	.2822	60.02	54.92	123.77	.1063	.2165
170	189.7	175	.2637	59.38	56.57	124.14	.1090	.2163
175	201.7	187	0.2465	58.71	58.23	124.48	0.1116	0.2160
180	214.3	200	.2304	58.02	59.89	124.79	.1142	.2156
185	227.4	213	.2155	57.30	61.57	125.07	.1168	.2153
190	241.1	226	.2017	56.57	63.26	125.31	.1194	.2149
195	255.5	241	.1887	55.80	64.96	125.53	.1219	.2145
200	270.4	256	0.1767	55.00	66.68	125.71	0.1245	0.2140
205	286.0	271	.1654	54.17	68.42	125.85	.1271	.2135
210	302.2	288	.1548	53.29	70.18	125.96	.1297	.2130
215	319.1	304	.1449	52.38	71.96	126.02	.1323	.2125
220	336.6	322	.1356	51.42	73.77	126.04	.1350	.2119
230	373.8	359	0.1186	49.32	77.54	125.94	0.1404	0.2105
240	413.8	399	.1032	46.92	81.61	125.58	.1461	.2089
250	456.8	442	.0887	44.10	86.21	124.83	.1525	.2069
260	502.9	488	.0736	40.63	92.07	123.14	.1605	.2036
265	527.1	512	.0274	38.48	117.08	117.99	.1951	1964

*Inches of mercury below one atmosphere

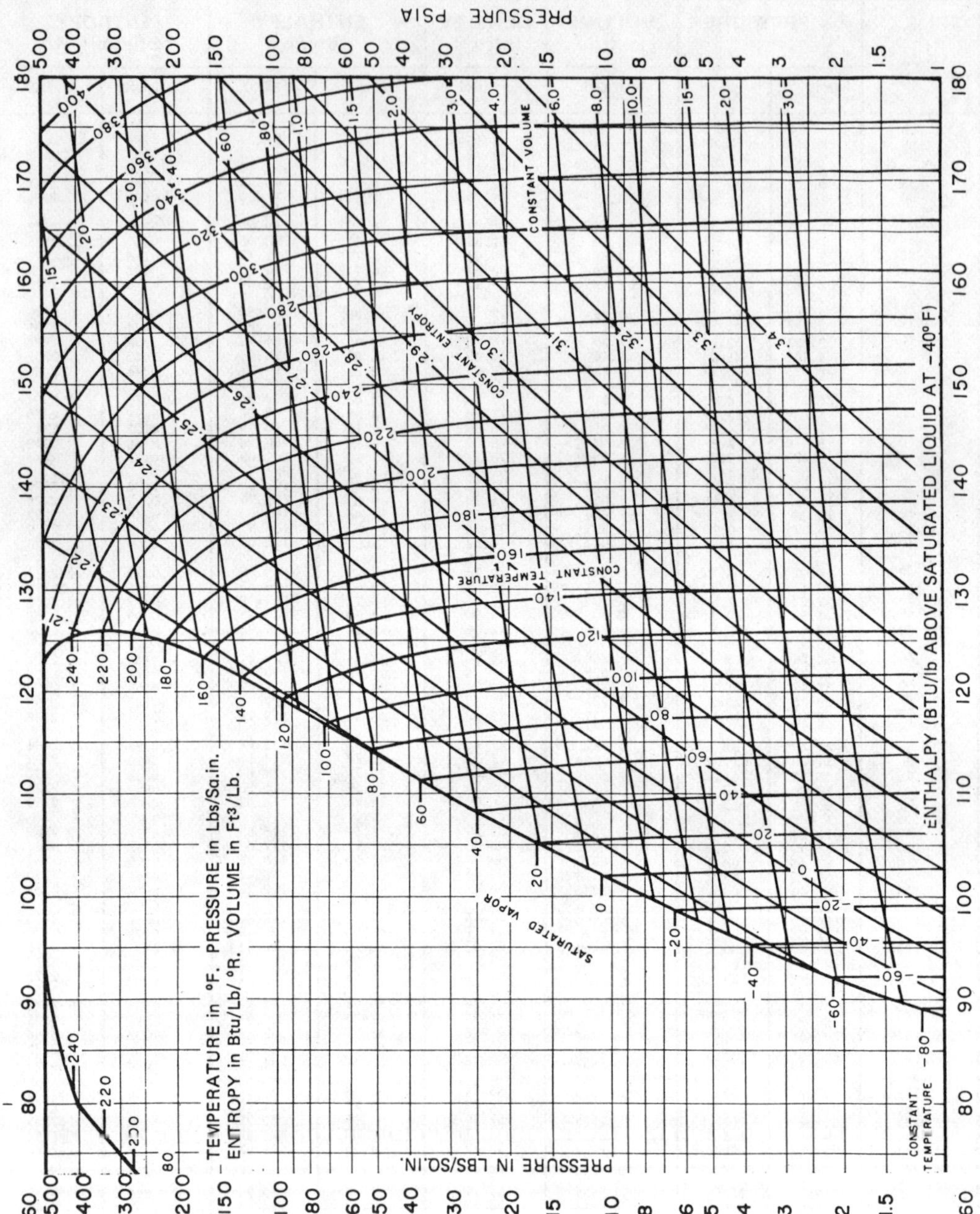

PRESSURE ENTHALPY DIAGRAM FOR R 142b Imperial

DATA SHEET FOR R 152a

$C_2H_4F_2$	DIFLUOROETHANE	Molecular Wt. 66

Pressure		Temperature				Volume		Density	
45.8	kg/cm^2	84	°C	236	°F	2.74	l/kg	.364	kg/l.
652	psia	387	°K	696	°R	.044	ft^3/lb	22.7	lbs/ft^3
								2.28	Air=1

AT CRITICAL POINT

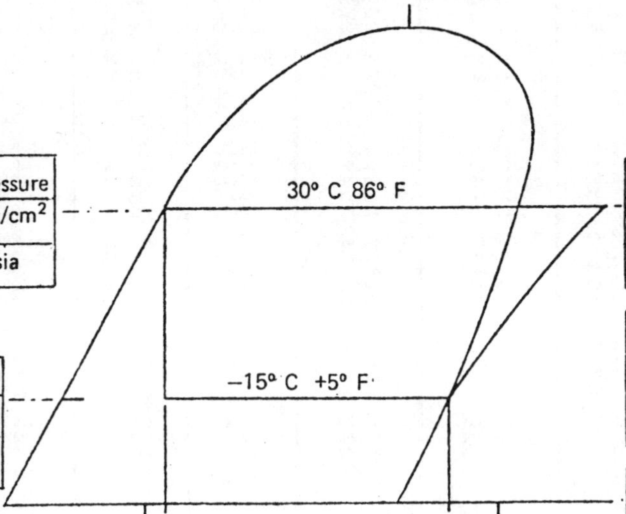

30° C 86° F

−15° C +5° F

Discharge Pressure	
6.97	kg/cm^2
99.2	psia

Discharge Temperature			
49	°C	120	°F
322	°K	580	°R

Inlet Pressure	
1.56	kg/cm^2
22.2	psia

Normal Boiling Point			
-25	°C	-13	°F
248	°K	447	°R

Refrigerating Effect			
289	kcal/m^3	32.5	Btu/ft^3

Triple Point			
-116	°C	-178	°F
157	°K	282	°R

Latent Heat at NBP	76 kcal/kg	5023 /kg mol	137 Btu/lb	9042 /lb mol

Trouton's No. 20.25	Gas Constant 12.85 kg.m/kg/° K	23.4 ft.lbs/lb/° R

Specific Heat Liquid at 30° C 86° F	Gas C_p	C_p C_v at

Liquid at 30° C 86° F Density .893 kg/l.	55.4 lbs/ft^3 Viscosity cp

Reduced Form at NBP Pressure .0225	1 / Temperature 1.56

Reduced value of 1/T	2.02	1.68	1.338
when Reduced Pressure is	a) .001	b) .01	c) .1

TEMP. °F	PRESSURE		VOLUME cu ft/lb	DENSITY lb/cu ft	ENTHALPY Btu/lb		ENTROPY Btu/(lb)(°R)	
	psia	psig	Vapor	Liquid	Liquid	Vapor	Liquid	Vapor
-150	0.1360	29.6*	369.7	71.69	-24.05	117.92	-0.0593	0.3992
-145	.1730	29.6*	295.2	71.40	-23.90	118.83	- .0592	.3944
-140	.2190	29.5*	236.9	71.11	-21.82	119.74	- .0530	.3898
-135	.2750	29.4*	191.6	70.82	-20.92	120.67	- .0507	.3854
-130	.3430	29.2*	155.9	70.53	-20.14	121.59	- .0487	.3812
-125	0.4250	29.1*	127.7	70.24	-19.84	122.53	-0.0481	0.3772
-120	.5240	28.9*	105.1	69.94	-18.54	123.47	- .0447	.3734
-115	.6410	28.6*	87.12	69.64	-17.29	124.41	- .0414	.3697
-110	.7800	28.3*	72.59	69.34	-16.51	125.36	- .0394	.3663
-105	.9440	28.0*	60.81	69.04	-15.36	126.32	- .0365	.3630
-100	1.136	27.6*	51.21	68.74	-14.16	127.28	-0.0334	0.3598
-95	1.360	27.2*	43.34	68.43	-13.29	128.25	- .0313	.3568
-90	1.621	26.6*	36.83	68.13	-12.10	129.22	- .0284	.3339
-85	1.922	26.0*	31.45	67.82	-10.88	130.20	- .0253	.3512
-80	2.270	25.3*	26.96	67.51	-9.91	131.17	- .0230	.3485
-75	2.669	24.5*	23.21	67.20	-8.71	132.16	-0.0201	0.3460
-70	3.125	23.6*	20.05	66.88	-7.49	133.14	- .0172	.3436
-65	3.644	22.5*	17.39	66.56	-6.29	134.13	- .0144	.3414
-60	4.233	21.3*	15.14	66.24	-5.10	135.12	- .0116	.3392
-55	4.899	19.9*	13.23	65.92	-3.91	136.11	- .0089	.3371
-50	5.650	18.4*	11.59	65.60	-2.62	137.11	-0.0059	0.3351
-45	6.494	16.7*	10.19	65.27	-1.34	138.10	- .0030	.3332
-40	7.439	14.8*	8.982	64.94	0.00	139.09	.0000	.3314
-35	8.494	12.6*	7.942	64.61	1.32	140.08	.0029	.3297
-30	9.669	10.2*	7.042	64.27	2.71	141.08	.0060	.3280
-25	10.97	7.58*	6.261	63.94	4.08	142.07	0.0090	0.3265
-20	12.42	4.64*	5.581	63.59	5.46	143.05	.0120	.3249
-15	14.02	1.39*	4.987	63.25	6.89	144.04	.0151	.3235
-10	15.77	1.08	4.466	62.90	8.37	145.02	.0182	.3221
-5	17.71	3.01	4.009	62.55	9.89	146.00	.0215	.3208
0	19.83	5.13	3.607	62.20	11.41	146.97	0.0246	0.3195
5	22.15	7.45	3.251	61.84	12.96	147.93	.0278	.3183
10	24.68	9.99	2.937	61.47	14.55	148.89	.0311	.3171
15	27.45	12.8	2.658	61.11	16.16	149.84	.0344	.3160
20	30.45	15.8	2.410	60.74	17.80	150.78	.0377	.3149
25	33.71	19.0	2.189	60.36	19.49	151.72	0.0411	0.3139
30	37.24	22.5	1.992	59.98	21.18	152.64	.0445	.3129
35	41.06	26.4	1.815	59.60	22.92	153.55	.0479	.3119
40	45.18	30.5	1.657	59.21	24.70	154.45	.0513	.3110
45	49.62	34.9	1.514	58.81	26.49	155.33	.0548	.3101
50	54.40	39.7	1.386	58.41	28.32	156.21	0.0583	0.3092
55	59.53	44.8	1.271	58.00	30.19	157.06	.0618	.3084
60	65.03	50.3	1.166	57.59	32.10	157.90	.0652	.3075
65	70.92	56.2	1.072	57.17	34.02	158.72	.0690	.3067
70	77.22	62.5	0.9859	56.75	35.99	159.53	.0726	.3059
75	83.94	69.2	0.9080	56.31	37.98	160.31	0.0763	0.3051
80	91.11	76.4	.8373	55.87	40.00	161.07	.0799	.3043
85	98.74	84.0	.7728	55.43	42.05	161.80	.0836	.3035
90	106.9	92.2	.7139	54.97	44.15	162.52	.0874	.3027
95	115.5	101	.6601	54.50	46.28	163.20	.0911	.3019
100	124.6	110	0.6109	54.03	48.43	163.86	0.0949	0.3011
105	134.3	120	.5658	53.54	50.61	164.48	.0987	.3003
110	144.5	130	.5244	53.05	52.83	165.08	.1025	.2995
115	155.3	141	.4863	52.54	55.08	165.64	.1063	.2987
120	166.8	152	.4512	52.02	57.38	166.16	.1102	.2979
125	178.8	164	0.4189	51.49	59.70	166.65	0.1141	0.2970
130	191.5	177	.3891	50.95	62.05	167.09	.1180	.2961
135	204.9	190	.3616	50.38	64.44	167.49	.1219	.2952
140	218.9	204	.3361	49.81	66.87	167.83	.1259	.2942
145	233.7	219	.3124	49.21	69.34	168.13	.1299	.2933
150	249.1	234	0.2905	48.60	71.84	168.37	0.1339	0.2922
155	265.4	251	.2700	47.96	74.40	168.55	.1379	.2911
160	282.4	268	.2510	47.30	77.00	168.67	.1420	.2900
165	300.1	285	.2332	46.62	79.66	168.69	.1462	.2887
170	318.7	304	.2165	45.90	82.37	166.64	.1504	.2874
175	338.2	323	0.2008	45.15	85.16	168.48	0.1547	0.2860
180	358.4	344	.1860	44.37	88.03	168.22	.1590	.2844
190	401.6	387	.1586	42.67	94.12	167.24	.1682	.2807
200	448.5	434	.1326	40.73	100.98	165.29	.1784	.2758
210	499.2	484	.0996	38.45	111.16	159.15	.1933	.2650

* Inches of mercury below one standard atmosphere.
Based on 0 for the saturated liquid at −40 F.

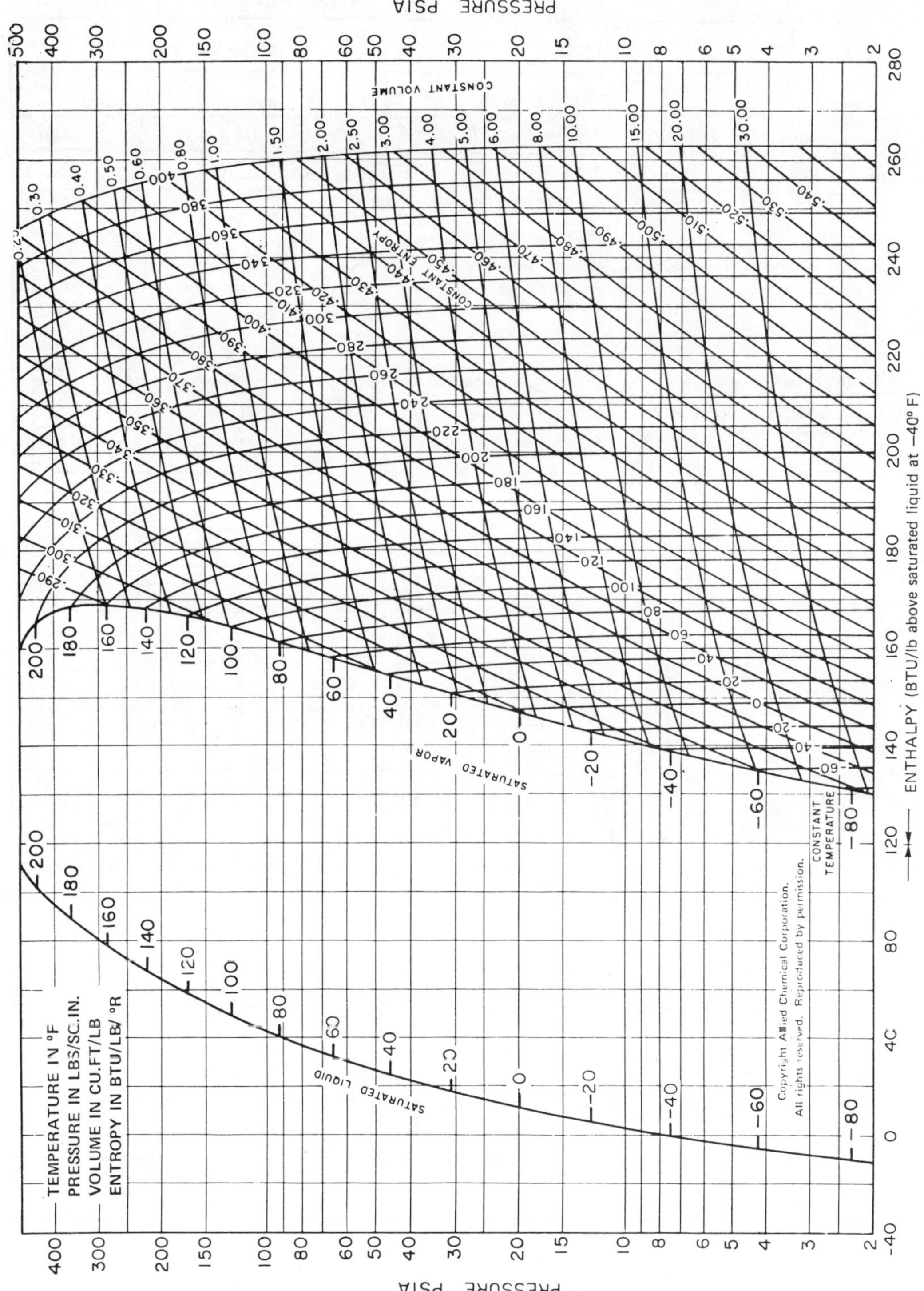
PRESSURE PSIA
CONSTANT VOLUME
CONSTANT ENTROPY
SATURATED VAPOR
SATURATED LIQUID
CONSTANT TEMPERATURE
TEMPERATURE IN °F
PRESSURE IN LBS/SQ.IN.
VOLUME IN CU.FT/LB
ENTROPY IN BTU/LB/ °R
Copyright Allied Chemical Corporation.
All rights reserved. Reproduced by permission.
ENTHALPY (BTU/lb above saturated liquid at −40° F)

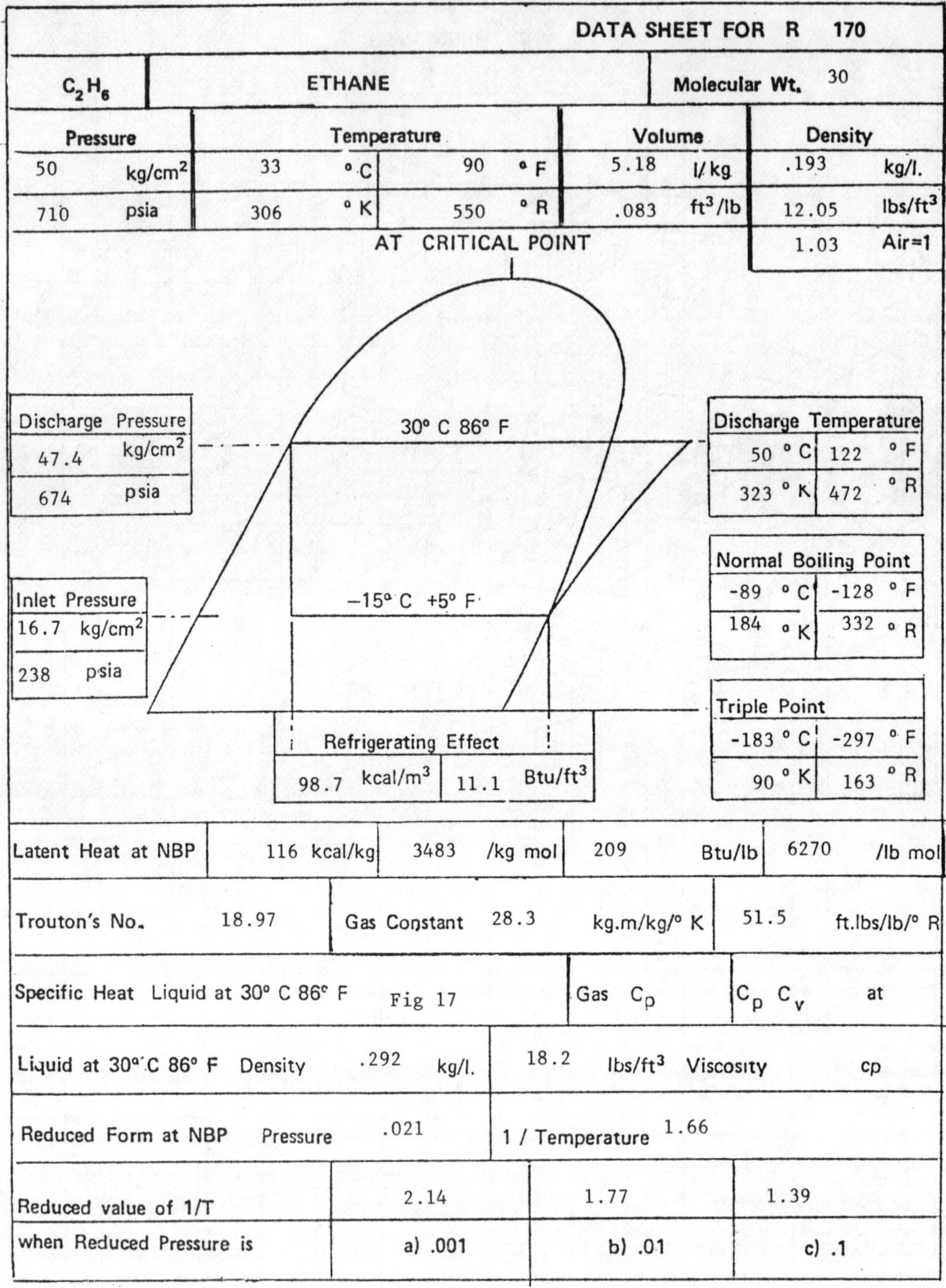

DATA SHEET FOR R 170

C_2H_6	ETHANE	Molecular Wt. 30

Pressure		Temperature				Volume		Density	
50	kg/cm²	33	°C	90	°F	5.18	l/kg	.193	kg/l.
710	psia	306	°K	550	°R	.083	ft³/lb	12.05	lbs/ft³
								1.03	Air=1

AT CRITICAL POINT

Discharge Pressure	
47.4	kg/cm²
674	psia

Inlet Pressure	
16.7	kg/cm²
238	psia

Refrigerating Effect			
98.7	kcal/m³	11.1	Btu/ft³

Discharge Temperature			
50	°C	122	°F
323	°K	472	°R

Normal Boiling Point			
-89	°C	-128	°F
184	°K	332	°R

Triple Point			
-183	°C	-297	°F
90	°K	163	°R

Latent Heat at NBP	116 kcal/kg	3483 /kg mol	209 Btu/lb	6270 /lb mol

Trouton's No. 18.97	Gas Constant 28.3 kg.m/kg/° K	51.5 ft.lbs/lb/° R

Specific Heat Liquid at 30° C 86° F Fig 17	Gas C_p	C_p C_v at

Liquid at 30° C 86° F Density .292 kg/l.	18.2 lbs/ft³ Viscosity cp

Reduced Form at NBP Pressure .021	1 / Temperature 1.66

Reduced value of 1/T	2.14	1.77	1.39
when Reduced Pressure is	a) .001	b) .01	c) .1

TEMP. °F	PRESSURE		VOLUME cu ft/lb	DENSITY lb/cu ft	ENTHALPY Btu/lb		ENTROPY Btu/(lb)(°R)	
	psia	psig	Vapor	Liquid	Liquid	Vapor	Liquid	Vapor
−220	0.27	29.41*	310.50	37.47	−1157.4	−921.1	0.824	1.810
−215	0.37	29.21*	235.62	37.29	−1154.7	−919.6	.835	1.795
−210	0.50	28.95*	179.20	37.12	−1151.9	−918.2	.846	1.782
−205	0.65	28.65*	138.48	36.94	−1149.1	−916.7	.857	1.769
−200	0.85	28.25*	107.80	36.76	−1146.3	−915.3	.868	1.757
−195	1.10	27.74*	85.704	36.59	−1143.5	−913.8	.878	1.746
−190	1.40	27.13*	68.430	36.42	−1140.7	−912.4	0.889	1.735
−185	1.76	26.41*	55.323	36.23	−1137.9	−911.0	.901	1.727
−180	2.20	25.52*	44.900	36.05	−1135.1	−909.5	.912	1.719
−175	2.73	24.46*	36.803	35.87	−1132.3	−908.1	.922	1.710
−170	3.36	23.18*	30.320	35.69	−1129.5	−906.7	.932	1.701
−165	4.10	21.70*	25.238	35.51	−1126.7	−905.2	.941	1.693
−160	4.97	19.93*	21.110	35.32	−1123.9	−903.8	0.951	1.685
−155	5.99	17.87*	17.818	35.14	−1121.0	−902.5	.960	1.678
−150	7.14	15.56*	15.100	34.95	−1118.2	−901.1	0.970	1.671
−145	8.45	12.90*	12.885	34.76	−1115.5	−899.8	0.979	1.664
−140	9.97	9.84*	11.050	34.57	−1112.7	−898.5	0.988	1.658
−135	11.76	6.28*	9.548	34.39	−1109.7	−897.5	0.996	1.651
−130	13.80	2.16*	8.282	34.21	−1106.7	−896.5	1.005	1.645
−127.85	14.696	0	7.762	34.12	−1105.4	−896.0	1.009	1.639
−125	16.00	1.30	7.274	34.00	−1103.8	−895.3	1.014	1.636
−120	18.48	3.78	6.274	33.82	−1100.9	−894.1	1.023	1.631
−115	20.13	5.48	5.474	33.62	−1097.9	−892.9	1.032	1.626
−110	24.35	9.65	4.854	33.43	−1094.9	−891.7	1.041	1.622
−105	27.80	13.10	4.312	33.22	−1091.9	−890.5	1.049	1.617
−100	31.55	16.85	3.812	33.01	−1088.9	−889.4	1.058	1.613
− 95	35.80	21.10	3.397	32.81	−1085.9	−888.2	1.067	1.609
− 90	40.27	25.57	3.032	32.61	−1082.9	−887.1	1.076	1.605
− 85	45.30	30.60	2.719	32.39	−1079.8	−886.0	1.084	1.601
− 80	50.66	35.96	2.444	32.18	−1076.8	−884.9	1.092	1.597
− 75	56.40	41.70	2.202	31.95	−1073.8	−883.8	1.100	1.593
− 70	62.97	48.27	1.989	31.73	−1070.9	−882.7	1.108	1.590
− 65	69.70	55.00	1.801	31.50	−1067.9	−881.7	1.116	1.587
−60	77.37	62.67	1.634	31.26	−1065.0	−880.7	1.123	1.584
−55	85.60	71.90	1.485	31.02	−1061.9	−879.7	1.131	1.581
−50	94.13	79.43	1.352	30.78	−1058.8	−878.7	1.139	1.578
−45	103.50	88.80	1.234	30.53	−1055.6	−877.8	1.147	1.575
−40	113.4	98.7	1.127	30.28	−1052.4	−876.9	1.154	1.572
−35	123.2	108.5	1.030	30.02	−1049.0	−876.0	1.162	1.569
−30	135.5	120.8	0.9445	29.77	−1045.8	−875.2	1.169	1.566
−25	147.2	132.5	0.8665	29.51	−1042.5	−874.5	1.176	1.563
−20	160.6	145.9	.7960	29.22	−1039.2	−873.8	1.184	1.560
−15	172.8	158.1	.7302	28.92	−1035.8	−873.1	1.191	1.557
−10	189.0	174.3	.6722	28.62	−1032.3	−872.5	1.199	1.554
− 5	203.0	188.3	.6196	28.32	−1028.8	−871.9	1.207	1.552
0	220.8	206.1	0.5726	28.01	−1025.4	−871.5	1.215	1.549
5	238.4	223.7	.5286	27.69	−1021.8	−871.1	1.222	1.547
10	256.3	241.6	.4882	27.36	−1018.2	−870.8	1.230	1.544
15	275.6	260.9	.4512	27.01	−1014.5	−870.6	1.238	1.541
20	295.9	280.2	.4166	26.64	−1010.8	−870.5	1.246	1.538
25	317.1	302.4	0.3850	26.26	−1007.1	−870.4	1.254	1.535
30	339.7	325.0	.3554	25.87	−1002.6	−870.3	1.262	1.532
35	363.3	348.6	.3282	25.43	− 998.4	−870.5	1.269	1.528
40	388.1	373.4	.3025	25.03	− 994.0	−870.7	1.277	1.524
45	414.1	399.4	.2785	24.60	− 989.2	−871.0	1.286	1.520
50	441.3	426.6	0.2564	24.13	− 984.6	−871.6	1.294	1.516
55	469.8	455.1	.2353	23.57	− 979.4	−872.3	1.304	1.512
60	499.6	484.9	.2158	22.95	− 974.3	−873.2	1.313	1.507
65	530.8	516.1	.1973	22.30	− 968.7	−874.3	1.321	1.501
70	563.3	548.6	.1795	21.62	− 963.0	−876.0	1.331	1.495
75	597.2	582.5	0.1616	20.88	− 956.6	−878.2	1.342	1.488
80	632.7	618.0	.1441	19.75	− 949.6	−881.5	1.355	1.481
85	670.3	655.6	.1255	18.22	− 940.6	−886.0	1.372	1.472
90	709.7	695.0	0.0831	12.06	− 911.1	−910.9	1.421	1.423
90.01	709.8	695.1	.0830	12.05	− 911.0	−911.0	1.422	1.422

*Inches of mercury below one atmosphere

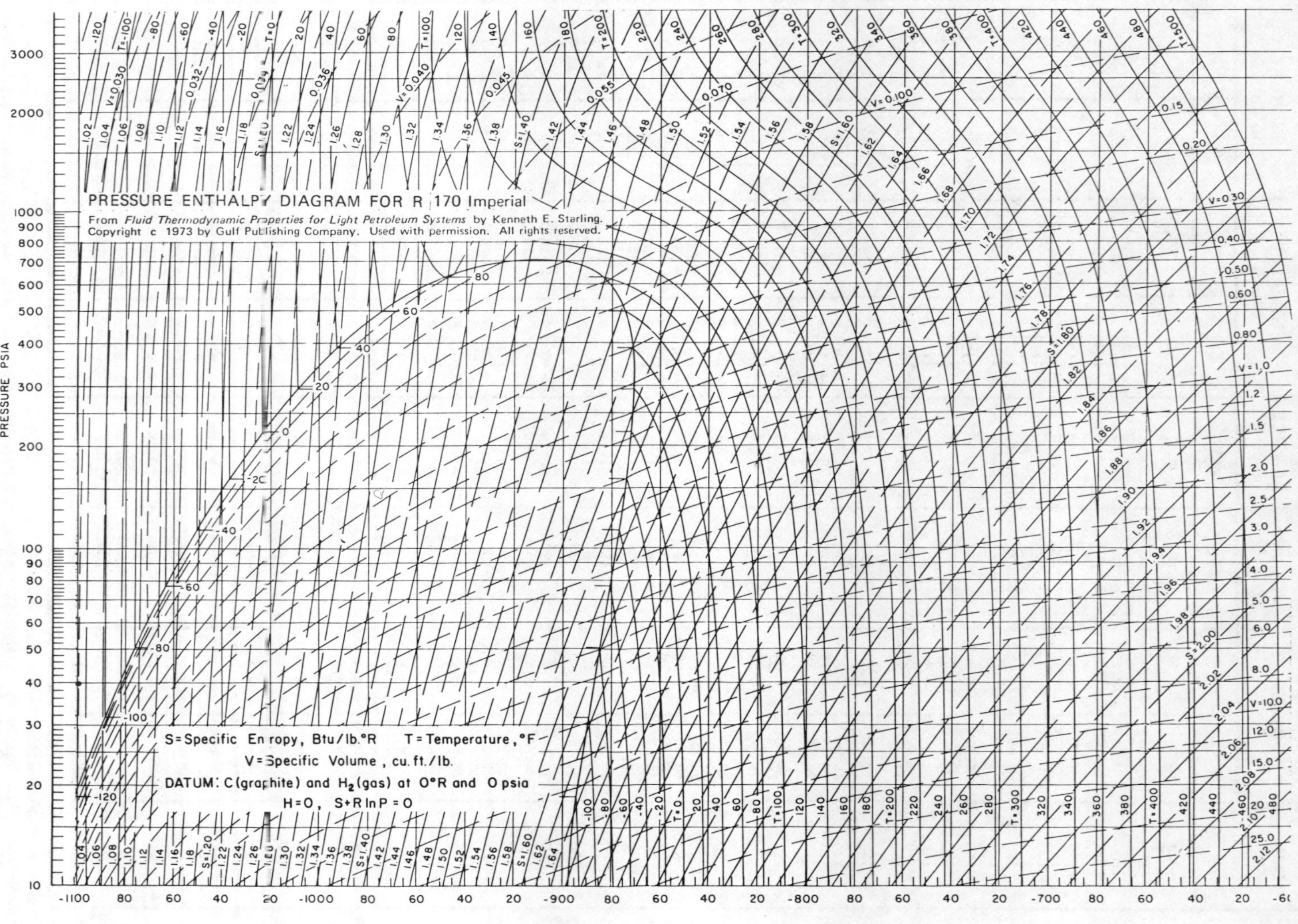
PRESSURE ENTHALPY DIAGRAM FOR R 170 Imperial
From Fluid Thermodynamic Properties for Light Petroleum Systems by Kenneth E. Starling.
Copyright c 1973 by Gulf Publishing Company. Used with permission. All rights reserved.
S=Specific Entropy, Btu/lb.°R
T=Temperature, °F
V=Specific Volume, cu. ft./lb.
DATUM: C(graphite) and H_2(gas) at 0°R and 0 psia
H=0, S+R lnP=0
PRESSURE PSIA

DATA SHEET FOR R 216

$C_2H_3ClF_2$	HEXAFLUORODICHLOROPROPANE	Molecular Wt. 221

Pressure		Temperature				Volume		Density	
28.1	kg/cm^2	18	°C	356	°F	1.75	l/kg	.572	kg/l.
400	psia	453	°K	816	°R	.028	ft^3/lb	35.7	lbs/ft^3
								7.6	Air=1

AT CRITICAL POINT

Discharge Pressure	
.85	kg/cm^2
12.1	psia

Inlet Pressure	
.112	kg/cm^2
1.6	psia

Discharge Temperature			
30	°C	86	°F
303	°K	546	°R

Normal Boiling Point			
35.5	°C	96	°F
308	°K	556	°R

Triple Point			
-92	°C	-134	°F
180	°K	326	°R

Refrigerating Effect			
25	kcal/m^3	2.83	Btu/ft^3

Latent Heat at NBP	28 kcal/kg	6213 /kg mol	50.6 Btu/lb	11183 /lb mol

Trouton's No. 20.11	Gas Constant 3.83 kg.m/kg/° K	6.99 ft.lbs/lb/° R

Specific Heat Liquid at 30° C 86° F	Gas C_p	C_p/C_v 1.05 at 0°C

Liquid at 30° C 86° F Density 1.54 kg/l.	96.5 lbs/ft^3 Viscosity cp

Reduced Form at NBP Pressure .037	1 / Temperature 1.47

Reduced value of 1/T	1.92	1.64	1.33
when Reduced Pressure is	a) .001	b) .01	c) .1

TEMP. °F	PRESSURE		VOLUME cu ft/lb	DENSITY lb/cu ft	ENTHALPY Btu/lb		ENTROPY Btu/(lb)(°R)	
	psia	psig	Vapor	Liquid	Liquid	Vapor	Liquid	Vapor
-40	0.339	29.231*	59.9567	107.93	-0.000	62.41	0.00000	0.14872
-35	.412	29.083*	49.9056	107.49	1.196	63.12	.00283	.14865
-30	.497	28.909*	41.7763	107.06	2.391	63.83	.00563	.14863
-25	.597	28.705*	35.1618	106.62	3.585	64.55	.00839	.14865
-20	.713	28.468*	29.7486	106.18	4.778	65.27	.01111	.14871
-15	0.848	28.195*	25.2943	105.74	5.969	66.00	0.01380	0.14881
-10	1.003	27.880*	21.6095	105.30	7.160	66.73	.01647	.14895
-5	1.180	27.519*	18.5455	104.86	8.351	67.46	.01910	.14912
0	1.382	27.106*	15.9857	104.42	9.540	68.20	.02170	.14932
5	1.612	26.638*	13.8366	103.97	10.730	68.95	.02427	.14956
10	1.873	26.108*	12.0243	103.53	11.91	69.69	0.02681	0.14983
15	2.167	25.510*	10.4895	103.09	13.10	70.44	.02933	.15012
20	2.497	24.838*	9.1842	102.64	14.29	71.19	.03182	.15044
25	2.866	24.085*	8.0695	102.19	15.48	71.95	.03428	.15079
30	3.279	23.246*	7.1141	101.74	16.67	72.71	.03672	.15116
35	3.738	22.311*	6.2920	101.29	17.86	73.47	0.03913	0.15155
40	4.247	21.275*	5.5822	100.84	19.05	74.23	.04152	.15196
45	4.810	20.129*	4.9672	100.38	20.24	75.00	.04389	.15240
50	5.430	18.865*	4.4325	99.93	21.43	75.77	.04623	.15285
55	6.113	17.474*	3.9662	99.47	22.62	76.54	.04856	.15332
60	6.862	15.949*	3.5582	99.00	23.82	77.31	0.05086	0.15381
65	7.682	14.280*	3.2002	98.54	25.01	78.09	.05314	.15431
70	8.577	12.458*	2.8852	98.07	26.20	78.87	.05540	.15483
75	9.552	10.473*	2.6072	97.60	27.40	79.64	.05764	.15536
80	10.612	8.316*	2.3613	97.13	28.59	80.42	.05986	.15590
85	11.761	5.976*	2.1431	96.65	29.79	81.21	0.06206	0.15646
90	13.004	3.444*	1.9490	96.17	30.99	81.99	.06425	.15703
95	14.348	0.709*	1.7760	95.69	32.19	82.77	.06641	.15761
100	15.797	1.101	1.6214	95.20	33.39	83.55	.06856	.15820
105	17.356	2.660	1.4830	94.71	34.59	84.34	.07069	.15880
110	19.031	4.335	1.3587	94.22	35.79	85.12	0.07281	0.15941
115	20.828	6.132	1.2469	93.72	36.99	85.91	.07490	.16002
120	22.753	8.057	1.1461	93.21	38.20	86.70	.07699	.16065
125	24.810	10.114	1.0551	92.70	39.41	87.48	.07905	.16128
130	27.008	12.312	0.9728	92.19	40.62	88.27	.08111	.16191
135	29.351	14.655	0.8981	91.67	41.83	89.05	.08314	0.16256
140	31.845	17.149	.8303	91.15	43.04	89.84	.08517	.16320
145	34.498	19.802	.7686	90.62	44.26	90.63	.08718	.16386
150	37.315	22.619	.7124	90.08	45.48	91.41	.08917	.16451
155	40.303	25.607	.6611	89.54	46.70	92.19	.09116	.16517
160	43.46	28.77	0.6142	88.99	47.93	92.98	0.09313	0.16583
165	46.81	32.12	.5712	88.43	49.15	93.76	.09509	.16650
170	50.36	35.66	.5317	87.87	50.38	94.54	.09704	.16717
175	54.10	39.40	.4955	87.30	51.62	95.32	.09898	.16784
180	58.04	43.35	.4622	86.72	52.86	96.09	.10091	.16851
185	62.20	47.50	0.4315	86.14	54.10	96.87	0.10283	0.16918
190	66.58	51.88	.4032	85.54	55.34	97.64	.10474	.16985
195	71.19	56.49	.3770	84.94	56.60	98.41	.10665	.17052
200	76.03	61.33	.3528	84.32	57.85	99.18	.10854	.17119
205	81.11	66.42	.3304	83.70	59.11	99.95	.11043	.17186
210	86.45	71.75	0.3096	83.06	60.38	100.71	0.11231	0.17253
215	92.05	77.35	.2903	82.42	61.65	101.47	.11419	.17320
220	97.91	83.21	.2724	81.76	62.93	102.22	.11606	.17386
225	104.05	89.35	.2557	81.09	64.22	102.97	.11792	.17452
230	110.4	95.78	.2402	80.40	65.51	103.72	.11978	.17517
235	117.1	102.49	0.2256	79.70	66.82	104.46	0.12164	0.17582
240	124.2	109.51	.2121	78.99	68.13	105.19	.12350	.17647
245	131.5	116.84	.1994	78.26	69.45	105.92	.12535	.17711
250	139.1	124.50	.1875	77.51	70.78	106.64	.12721	.17774
255	147.1	132.48	.1764	76.74	72.12	107.36	.12906	.17837
260	155.5	140.80	0.1650	75.95	73.47	108.06	0.13092	0.17898
265	164.1	149.47	.1561	75.14	74.83	108.76	.13278	.17959
270	173.2	158.51	.1469	74.30	76.21	109.45	.13464	.18019
275	182.6	167.91	.1382	73.44	77.60	110.12	.13651	.18077
280	192.4	177.70	.1300	72.54	79.01	110.78	.13838	.18134
285	202.5	187.89	0.1222	71.61	80.43	111.43	0.14027	0.18190
290	213.1	198.47	.1148	70.65	81.88	112.07	.14216	.18243
295	224.1	209.48	.1079	69.64	83.34	112.68	.14407	.18295
300	235.6	220.93	.1013	68.58	84.83	113.28	.14600	.18344
305	247.5	232.82	.0949	67.47	86.34	113.85	.14794	.18391
310	259.8	245.17	0.0889	66.29	87.89	114.39	0.14991	0.18434
315	272.7	258.00	.0831	65.03	89.47	114.90	.15190	.18473
320	286.0	271.33	.0776	63.68	91.08	115.37	.15394	.18508
325	299.8	285.18	.0722	62.22	92.75	115.79	.15602	.18537
330	314.2	299.57	.0669	60.60	94.48	116.14	.15816	.18558

*Inches of mercury below one atmosphere

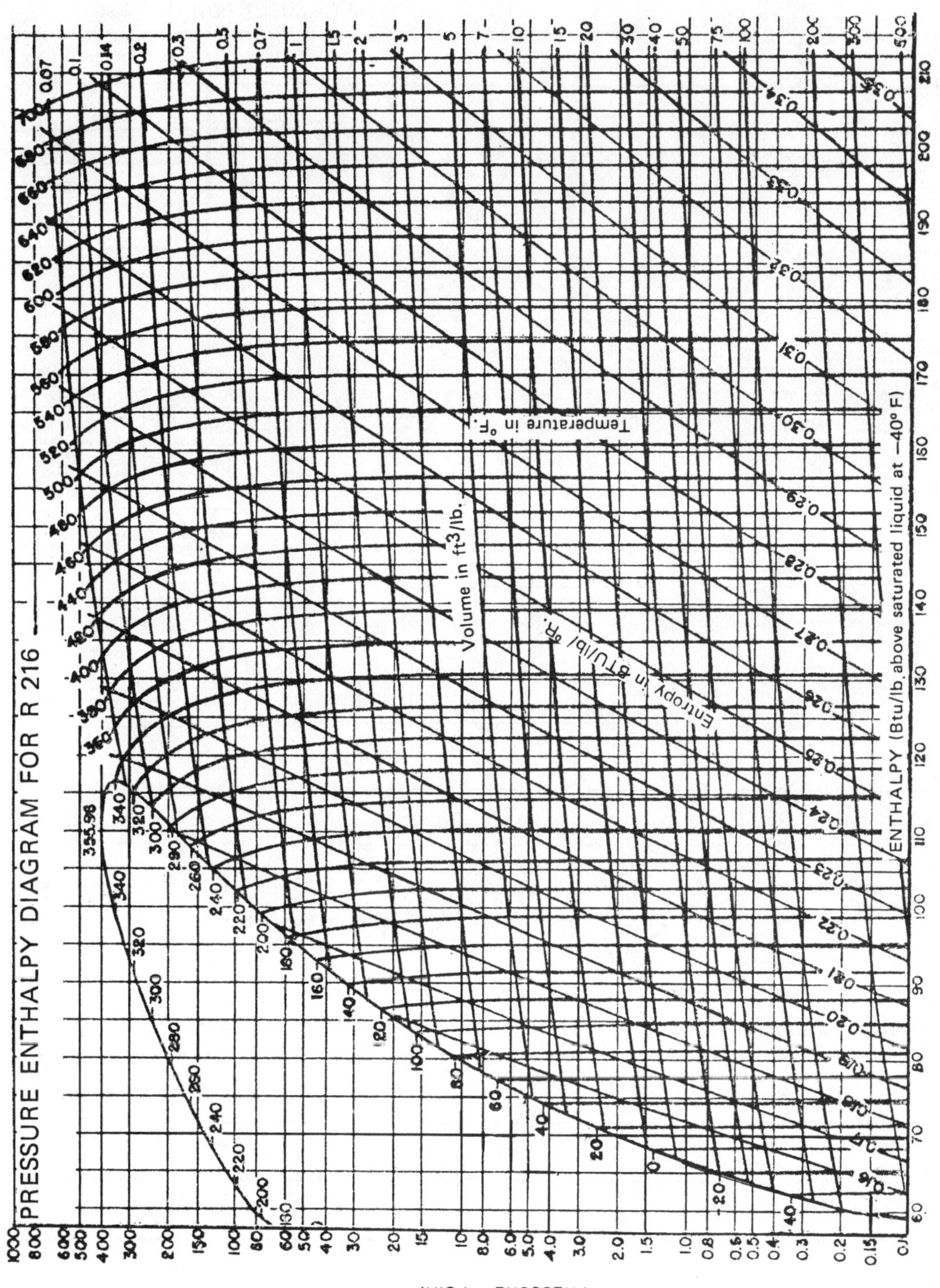
PRESSURE ENTHALPY DIAGRAM FOR R 216
Volume in ft3/lb.
Temperature in °F.
Entropy in BTU/lb/°R.
ENTHALPY (Btu/lb. above saturated liquid at −40° F)
PRESSURE PSIA.
355.98

DATA SHEET FOR R 290

C_3H_8	PROPANE	Molecular Wt. 44

Pressure		Temperature				Volume		Density	
43.4	kg/cm^2	94.4	°C	206	°F	4.58	l/kg	.218	kg/l.
617	psia	367	°K	666	°R	.0735	ft^3/lb	13.6	lbs/ft^3
								1.52	Air=1

AT CRITICAL POINT

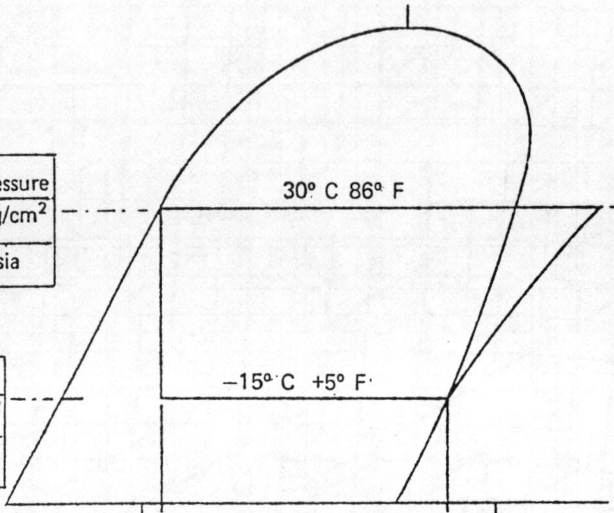

30° C 86° F

−15° C +5° F

Discharge Pressure	
10.9	kg/cm^2
155	psia

Inlet Pressure	
2.25	kg/cm^2
32.1	psia

Discharge Temperature			
36	°C	97	°F
309	°K	557	°R

Normal Boiling Point			
-42	°C	-44	°F
231	°K	416	°R

Triple Point			
-188	°C	-306	°F
85	°K	154	°R

Refrigerating Effect			
434	kcal/m^3	48.8	Btu/ft^3

Latent Heat at NBP	101 kcal/kg	4449 /kg mol	182 Btu/lb	8008 /lb mol

Trouton's No.	19.2	Gas Constant	19.3 kg.m/kg/° K	35.1 ft.lbs/lb/° R

Specific Heat Liquid at 30° C 86° F	Fig 17	Gas C_p	C_p C_v at

Liquid at 30° C 86° F Density	.476 kg/l.	30.4 lbs/ft^3	Viscosity cp

Reduced Form at NBP	Pressure .0238	1 / Temperature	1.6

Reduced value of 1/T		1.73	1.37
when Reduced Pressure is	a) .001	b) .01	c) .1

TEMP. °F	PRESSURE	VOLUME cu ft/lb	DENSITY lb/cu ft	ENTHALPY Btu/lb		ENTROPY Btu/(lb)(°R)	
	psia	Vapor	Liquid	Liquid	Vapor	Vapor	Vapor
-100	2.88	29.94	38.46	-32.04	164.1	-0.0821	0.4632
- 95	3.40	25.61	38.28	-29.43	165.5	- .0749	.4597
- 90	4.01	22.01	38.10	-26.81	166.9	- .0678	.4563
- 85	4.70	19.01	37.92	-24.17	168.3	- .0607	.4531
- 80	5.47	16.49	37.74	-21.53	169.7	- .0537	.4501
- 75	6.35	14.36	37.55	-18.88	171.2	-0.0468	0.4472
- 70	7.34	12.56	37.36	-16.22	172.6	- .0399	.4445
- 65	8.44	11.02	37.17	-13.54	174.0	- .0331	.4420
- 60	9.68	9.71	36.98	-10.86	175.4	- .0264	.4396
- 55	11.05	8.59	36.79	- 8.16	176.8	- .0197	.437
- 50	12.56	7.62	36.60	- 5.45	178.2	-0.0131	0.4352
- 45	14.24	6.78	36.40	- 2.73	179.6	- .0065	.4331
- 43.73	14.69	6.58	36.36	- 2.04	179.97	- .0050	.4327
- 40	16.09	6.05	36.21	0.00	181.0	.0000	.4312
- 35	18.11	5.41	36.01	2.74	182.3	.0064	.4294
- 30	20.33	4.86	35.80	5.50	183.7	0.0129	0.427
- 25	22.75	4.37	35.60	8.28	185.1	.0193	.426
- 20	25.39	3.95	35.39	11.07	186.5	.0256	.424
- 15	28.25	3.57	35.19	13.87	187.9	.0319	.423
- 10	31.36	3.23	34.98	16.69	189.2	.0382	.421
- 5	34.72	2.94	34.76	19.52	190.6	0.0444	0.4207
0	38.34	2.67	34.55	22.37	191.9	.0506	.4195
5	42.24	2.44	34.33	25.23	193.3	.0567	.4185
10	46.44	2.23	34.11	28.11	194.6	.0629	.4174
15	50.94	2.04	33.89	31.01	195.9	.0689	.4165
20	55.76	1.87	33.66	33.92	197.3	0.0750	0.4156
25	60.91	1.72	33.43	36.85	198.6	.0810	.4148
30	66.41	1.58	33.20	39.80	199.9	.0870	.4140
35	72.28	1.46	32.97	42.77	201.2	.0930	.4133
40	78.52	1.34	32.73	45.77	202.5	.0989	.4126
45	85.15	1.24	32.49	48.78	203.7	0.1048	0.4120
50	92.18	1.15	32.24	51.82	205.0	.1108	.4114
55	99.64	1.0674	32.00	54.88	206.2	0.1167	0.4108
60	107.53	0.9897	31.75	57.97	207.5	.1225	.4103
65	115.87	0.9186	31.49	61.09	208.7	.1284	.4098
70	124.68	0.8535	31.23	64.23	209.9	.1343	.4094
75	133.97	.7936	30.97	67.40	211.1	.1401	.4089
80	143.76	0.7386	30.70	70.60	212.2	0.1460	0.4085
85	154.07	.6879	30.42	73.84	213.4	.1518	.4081
90	164.91	.6411	30.15	77.11	214.5	.1577	.4077
95	176.29	.5979	29.86	80.41	215.6	.1635	.4073
100	188.25	.5578	29.57	83.76	216.6	.1694	.4069
105	200.79	0.5207	29.28	87.15	217.7	0.1753	0.4065
110	213.93	.4862	28.97	90.58	218.7	.1812	.4061
115	227.70	.4541	28.67	94.06	219.6	.1871	.4057
120	242.10	.4242	28.35	97.59	220.5	.1931	.4052
125	257.17	.3963	28.02	101.18	221.4	.1991	.4048
130	272.93	0.3702	27.68	104.83	222.2	0.2051	0.4042
135	289.38	.3458	27.34	108.54	222.9	.2112	.4037
140	306.57	.3228	26.98	112.33	223.6	.2173	.4030
145	324.50	.3012	26.61	116.19	224.2	.2236	.4023
150	343.21	.2808	26.22	120.14	224.7	.2299	.4015
155	302.72	0.2010	25.82	124.19	225.2	0.2362	0.4006
160	383.06	.2433	25.39	128.35	225.4	.2427	.3995
165	404.26	.2250	24.94	132.64	225.6	.2494	.3982
170	426.35	.2093	24.46	137.07	225.6	.2562	.3968
175	449.36	.1933	23.95	141.69	225.3	.2632	.3950
180	473.34	0.1778	23.38	146.52	224.8	0.2705	0.3929
185	498.34	.1626	22.75	151.63	223.9	.2781	.3902
190	524.39	.1475	22.03	157.13	222.4	.2863	.3869
195	551.58	.1319	21.14	163.23	220.2	.2953	.3823
200	579.97	.1149	19.94	170.44	216.3	.3059	.3755
205	609.68	.0911	17.58	181.49	207.1	.3221	.3608
206.26	617.40	.0735	13.73	197.56	197.5	.346	.3462

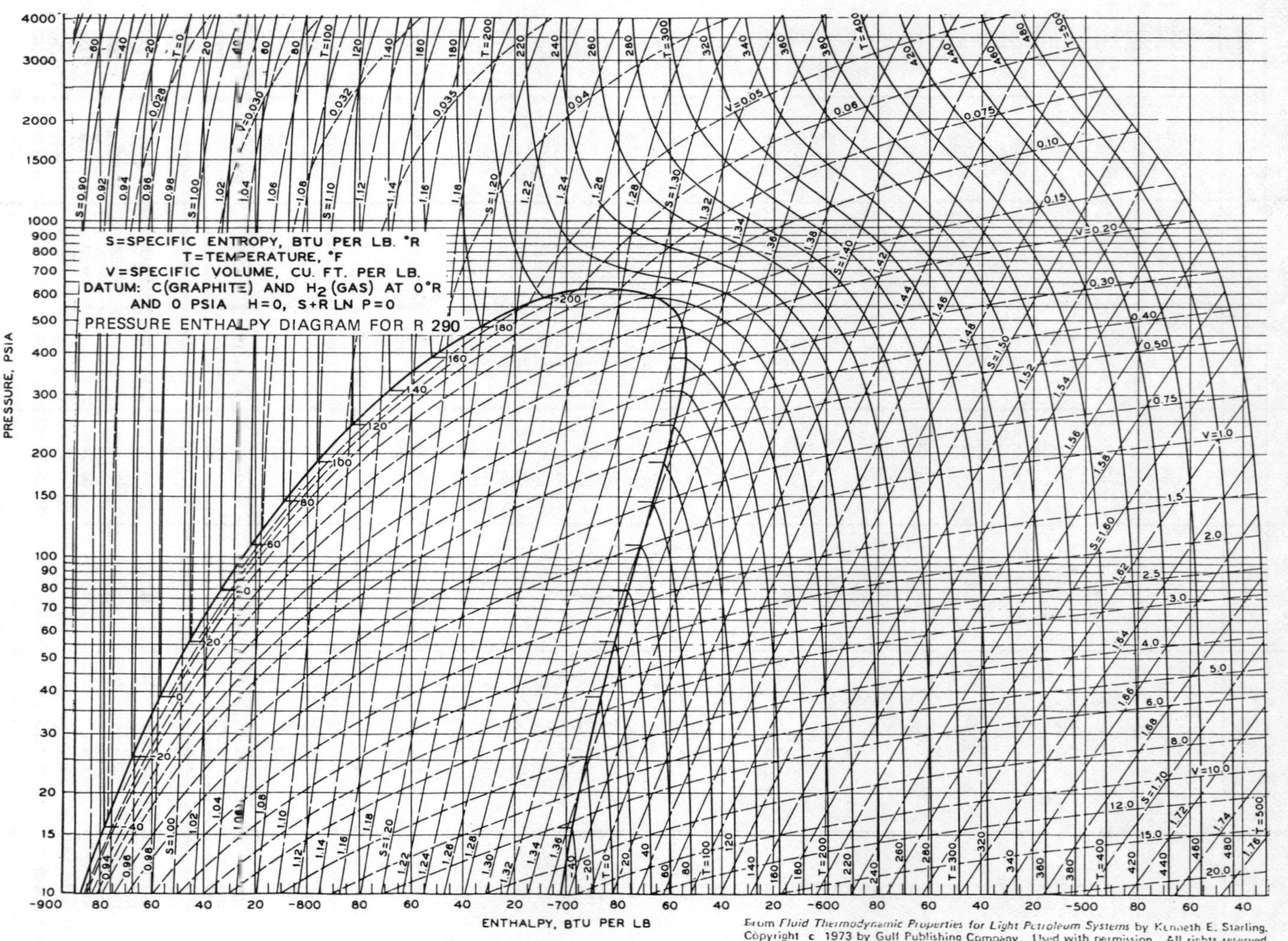

From *Fluid Thermodynamic Properties for Light Petroleum Systems* by Kenneth E. Starling.

DATA SHEET FOR R C318

C_4F_8	OCTAFLUOROCYCLOBUTANE	Molecular Wt. 200

Pressure		Temperature				Volume		Density	
28.33	kg/cm²	115	°C	240	°F	1.61	l/kg	.621	kg/l.
403.6	psia	388	°K	700	°R	.0258	ft³/lb	38.7	lbs/ft³
								6.9	Air=1

AT CRITICAL POINT

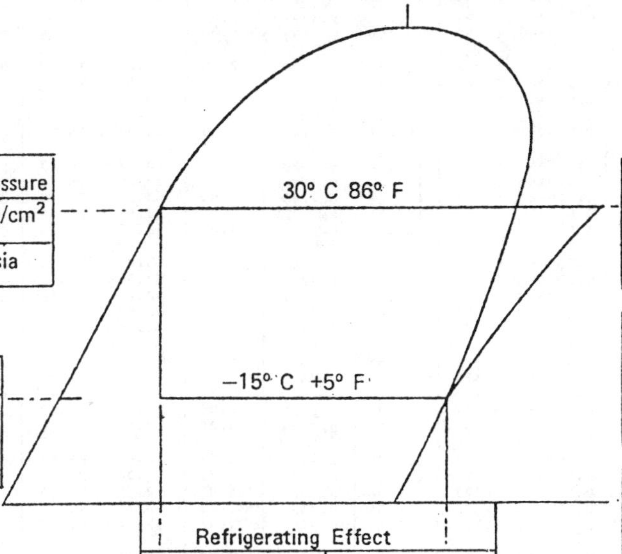

30° C 86° F

−15° C +5° F

Discharge Pressure	
3.76	kg/cm²
53.5	psia

Inlet Pressure	
.7	kg/cm²
9.9	psia

Discharge Temperature	
°C	°F
°K	°R

Normal Boiling Point			
-6	°C	21	°F
267	°K	481	°R

Triple Point			
-41	°C	-42	°F
232	°K	418	°R

Refrigerating Effect	
113 kcal/m³	12.7 Btu/ft³

Latent Heat at NBP	27.8 kcal/kg	5565 /kg mol	50.1 Btu/lb	10017 /lb mol

Trouton's No. 20.8	Gas Constant 4.24 kg.m/kg/°K	7.74 ft.lbs/lb/°R

Specific Heat Liquid at 30° C 86° F .266 Fig 15	Gas C_p .189	C_p/C_v 1.055 at 30°C

Liquid at 30° C 86° F Density 1.48 kg/l.	92.1 lbs/ft³ Viscosity .32 cp

Reduced Form at NBP Pressure .036	1 / Temperature 1.434

Reduced value of 1/T when Reduced Pressure is	a) .001	b) .01	c) .1
		1.62	1.32

TEMP.	PRESSURE		VOLUME cu ft/lb		DENSITY lb/cu ft		ENTHALPY Btu/lb			ENTROPY Btu/(lb)(°R)	
°F	PSIA	PSIG	LIQUID v_f	VAPOR v_g	LIQUID $1/v_f$	VAPOR $1/v_g$	LIQUID h_f	LATENT h_{fg}	VAPOR h_g	LIQUID s_f	VAPOR s_g
−40	2.79	24.23	0.00930	7.936	107.47	0.1260	0.	54.25	54.25	0.	0.1292
−39	2.88	24.05	0.00931	7.705	107.37	0.1297	0.21	54.19	54.40	0.0005	0.1293
−38	2.97	23.86	0.00932	7.483	107.27	0.1336	0.42	54.13	54.55	0.0010	0.1293
−37	3.06	23.67	0.00933	7.268	107.16	0.1375	0.63	54.08	54.71	0.0015	0.1294
−36	3.16	23.47	0.00934	7.060	107.06	0.1416	0.84	54.02	54.86	0.0020	0.1295
−35	3.26	23.27	0.00934	6.860	106.96	0.1457	1.05	53.96	55.02	0.0025	0.1295
−34	3.36	23.07	0.00935	6.666	106.85	0.1500	1.26	53.91	55.17	0.0030	0.1296
−33	3.46	22.85	0.00936	6.478	106.75	0.1543	1.48	53.85	55.33	0.0034	0.1297
−32	3.57	22.64	0.00937	6.297	106.64	0.1587	1.69	53.79	55.48	0.0039	0.1297
−31	3.68	22.41	0.00938	6.122	106.54	0.1633	1.90	53.73	55.64	0.0044	0.1298
−30	3.79	22.18	0.00939	5.952	106.43	0.1679	2.12	53.67	55.79	0.0049	0.1299
−29	3.91	21.95	0.00940	5.788	106.33	0.1727	2.33	53.61	55.95	0.0054	0.1299
−28	4.03	21.71	0.00941	5.630	106.22	0.1776	2.55	53.55	56.10	0.0059	0.1300
−27	4.15	21.46	0.00942	5.476	106.12	0.1825	2.76	53.49	56.26	0.0064	0.1301
−26	4.27	21.21	0.00943	5.328	106.01	0.1876	2.98	53.43	56.42	0.0069	0.1302
−25	4.40	20.96	0.00944	5.184	105.90	0.1928	3.20	53.37	56.57	0.0074	0.1302
−24	4.53	20.69	0.00945	5.045	105.80	0.1982	3.41	53.31	56.73	0.0079	0.1303
−23	4.66	20.42	0.00946	4.910	105.69	0.2036	3.63	53.25	56.88	0.0084	0.1304
−22	4.80	20.14	0.00947	4.779	105.58	0.2092	3.85	53.19	57.04	0.0089	0.1305
−21	4.93	19.86	0.00948	4.653	105.48	0.2149	4.07	53.12	57.19	0.0094	0.1305
−20	5.08	19.57	0.00949	4.530	105.37	0.2207	4.29	53.06	57.35	0.0099	0.1306
−19	5.22	19.27	0.00950	4.412	105.26	0.2266	4.51	53.00	57.51	0.0104	0.1307
−18	5.37	18.97	0.00950	4.297	105.15	0.2327	4.73	52.93	57.66	0.0109	0.1308
−17	5.53	18.66	0.00951	4.185	105.05	0.2389	4.95	52.87	57.82	0.0114	0.1309
−16	5.68	18.34	0.00952	4.077	104.94	0.2452	5.17	52.80	57.98	0.0119	0.1309
−15	5.84	18.01	0.00953	3.973	104.83	0.2516	5.39	52.74	58.13	0.0124	0.1310
−14	6.01	17.68	0.00954	3.871	104.72	0.2582	5.61	52.67	58.29	0.0129	0.1311
−13	6.17	17.34	0.00955	3.773	104.61	0.2650	5.83	52.61	58.45	0.0134	0.1312
−12	6.34	16.99	0.00956	3.677	104.50	0.2719	6.06	52.54	58.60	0.0139	0.1313
−11	6.52	16.64	0.00957	3.585	104.39	0.2789	6.28	52.47	58.76	0.0144	0.1314
−10	6.70	16.27	0.00958	3.495	104.28	0.2861	6.51	52.41	58.92	0.0149	0.1315
−9	6.88	15.90	0.00959	3.408	104.17	0.2934	6.73	52.34	59.07	0.0154	0.1316
−8	7.07	15.52	0.00960	3.323	104.06	0.3008	6.95	52.27	59.23	0.0159	0.1316
−7	7.26	15.13	0.00961	3.241	103.95	0.3085	7.18	52.20	59.39	0.0164	0.1317
−6	7.45	14.73	0.00962	3.161	103.84	0.3162	7.41	52.14	59.55	0.0169	0.1318
−5	7.65	14.33	0.00964	3.084	103.73	0.3242	7.63	52.07	59.70	0.0174	0.1319
−4	7.85	13.91	0.00965	3.009	103.62	0.3322	7.86	52.00	59.86	0.0179	0.1320
−3	8.06	13.49	0.00966	2.936	103.51	0.3405	8.09	51.93	60.02	0.0184	0.1321
−2	8.27	13.06	0.00967	2.865	103.40	0.3489	8.32	51.86	60.18	0.0189	0.1322
−1	8.49	12.62	0.00968	2.796	103.29	0.3575	8.54	51.79	60.33	0.0194	0.1323
0	8.71	12.17	0.00969	2.730	103.18	0.3662	8.77	51.71	60.49	0.0199	0.1324
1	8.94	11.71	0.00970	2.665	103.06	0.3752	9.00	51.64	60.65	0.0204	0.1325
2	9.17	11.24	0.00971	2.602	102.95	0.3843	9.23	51.57	60.81	0.0209	0.1326
3	9.40	10.76	0.00972	2.540	102.84	0.3935	9.46	51.50	60.96	0.0214	0.1327
4	9.64	10.27	0.00973	2.481	102.73	0.4030	9.69	51.42	61.12	0.0219	0.1328
5	9.89	9.77	0.00974	2.423	102.61	0.4126	9.93	51.35	61.28	0.0224	0.1329
6	10.14	9.26	0.00975	2.367	102.50	0.4224	10.16	51.28	61.44	0.0229	0.1330
7	10.39	8.75	0.00976	2.312	102.38	0.4324	10.39	51.20	61.60	0.0234	0.1331
8	10.65	8.22	0.00977	2.259	102.27	0.4426	10.62	51.13	61.75	0.0239	0.1332
9	10.92	7.68	0.00978	2.207	102.16	0.4530	10.86	51.05	61.91	0.0244	0.1333
10	11.19	7.12	0.00979	2.157	102.04	0.4635	11.09	50.97	62.07	0.0249	0.1334
11	11.47	6.56	0.00981	2.108	101.93	0.4743	11.33	50.90	62.23	0.0254	0.1335
12	11.75	5.99	0.00982	2.060	101.81	0.4853	11.56	50.82	62.39	0.0259	0.1336
13	12.03	5.41	0.00983	2.014	101.70	0.4965	11.80	50.74	62.55	0.0264	0.1337
14	12.33	4.81	0.00984	1.968	101.58	0.5078	12.03	50.67	62.70	0.0269	0.1338
15	12.62	4.21	0.00985	1.925	101.46	0.5194	12.27	50.59	62.86	0.0274	0.1339
16	12.93	3.59	0.00986	1.882	101.35	0.5312	12.51	50.51	63.02	0.0279	0.1341
17	13.24	2.96	0.00987	1.840	101.23	0.5432	12.74	50.43	63.18	0.0284	0.1342
18	13.55	2.32	0.00988	1.800	101.12	0.5554	12.98	50.35	63.34	0.0289	0.1343
19	13.87	1.66	0.00990	1.760	101.00	0.5679	13.22	50.27	63.49	0.0294	0.1344
20	14.20	1.00	0.00991	1.722	100.88	0.5805	13.46	50.19	63.65	0.0298	0.1345
21	14.53	0.32	0.00992	1.685	100.76	0.5934	13.70	50.11	63.81	0.0303	0.1346
22	14.87	0.17	0.00993	1.648	100.65	0.6065	13.94	50.03	63.97	0.0308	0.1347
23	15.22	0.52	0.00994	1.613	100.53	0.6199	14.18	49.94	64.13	0.0313	0.1348
24	15.57	0.87	0.00995	1.578	100.41	0.6334	14.42	49.86	64.29	0.0318	0.1349
25	15.93	1.23	0.00997	1.544	100.29	0.6473	14.66	49.78	64.45	0.0323	0.1351
26	16.29	1.59	0.00998	1.512	100.17	0.6613	14.90	49.70	64.60	0.0328	0.1352
27	16.67	1.97	0.00999	1.480	100.05	0.6756	15.15	49.61	64.76	0.0333	0.1353
28	17.04	2.34	0.01000	1.448	99.93	0.6901	15.39	49.53	64.92	0.0338	0.1354
29	17.43	2.73	0.01001	1.418	99.81	0.7049	15.63	49.44	65.08	0.0343	0.1355

TEMP. °F	PRESSURE		VOLUME cu ft/lb		DENSITY lb/cu ft		ENTHALPY Btu/lb			ENTROPY Btu/(lb)(°R)	
	PSIA	PSIG	LIQUID v_f	VAPOR v_g	LIQUID $1/v_f$	VAPOR $1/v_g$	LIQUID h_f	LATENT h_{fg}	VAPOR h_g	LIQUID s_f	VAPOR s_g
30	17.82	3.124	0.0100	1.3888	99.69	0.7200	15.88	49.36	65.24	0.034	0.1356
31	18.22	3.523	0.0100	1.3599	99.57	0.7352	16.12	49.27	65.40	0.035	0.1357
32	18.62	3.928	0.0100	1.3318	99.45	0.7508	16.37	49.18	65.55	0.035	0.1359
33	19.04	4.341	0.0100	1.3043	99.33	0.7666	16.61	49.10	65.71	0.036	0.1360
34	19.46	4.760	0.0100	1.2775	99.21	0.7827	16.86	49.01	65.87	0.036	0.1361
35	19.88	5.187	0.0100	1.2514	99.08	0.7990	17.10	48.92	66.03	0.037	0.1362
36	20.32	5.621	0.0101	1.2259	98.96	0.8156	17.35	48.83	66.19	0.037	0.1363
37	20.76	6.063	0.0101	1.2010	98.84	0.8325	17.60	48.74	66.35	0.038	0.1365
38	21.21	6.512	0.0101	1.1768	98.72	0.8497	17.85	48.65	66.51	0.038	0.1366
39	21.66	6.969	0.0101	1.1531	98.59	0.8671	18.10	48.56	66.66	0.039	0.1367
40	22.13	7.433	0.0101	1.1300	98.47	0.8849	18.34	48.47	66.82	0.039	0.1368
41	22.60	7.905	0.0101	1.1074	98.35	0.9029	18.59	48.38	66.98	0.040	0.1369
42	23.08	8.385	0.0101	1.0854	98.22	0.9212	18.84	48.29	67.14	0.040	0.1371
43	23.57	8.873	0.0101	1.0640	98.10	0.9398	19.09	48.20	67.30	0.041	0.1372
44	24.06	9.368	0.0102	1.0430	97.97	0.9587	19.35	48.11	67.46	0.041	0.1373
45	24.57	9.87	0.0102	1.0225	97.85	0.9779	19.60	48.01	67.61	0.042	0.1374
46	25.08	10.38	0.0102	1.0025	97.72	0.9974	19.85	47.92	67.77	0.042	0.1375
47	25.60	10.90	0.0102	0.9830	97.59	1.0172	20.10	47.82	67.93	0.043	0.1377
48	26.13	11.43	0.0102	0.9639	97.47	1.0374	20.35	47.73	68.09	0.043	0.1378
49	26.67	11.97	0.0102	0.9453	97.34	1.0578	20.61	47.63	68.25	0.044	0.1379
50	27.21	12.51	0.0102	0.9271	97.21	1.0786	20.86	47.54	68.40	0.044	0.1380
51	27.76	13.06	0.0103	0.9093	97.09	1.0997	21.12	47.44	68.56	0.045	0.1382
52	28.33	13.63	0.0103	0.8919	96.96	1.1211	21.37	47.34	68.72	0.045	0.1383
53	28.90	14.20	0.0103	0.8749	96.83	1.1428	21.63	47.25	68.88	0.046	0.1384
54	29.48	14.78	0.0103	0.8583	96.70	1.1649	21.88	47.15	69.04	0.046	0.1385
55	30.07	15.37	0.0103	0.8421	96.57	1.1873	22.14	47.05	69.19	0.047	0.1387
56	30.67	15.97	0.0103	0.8263	96.44	1.2101	22.40	46.95	69.35	0.047	0.1388
57	31.28	16.58	0.0103	0.8108	96.31	1.2332	22.65	46.85	69.51	0.048	0.1389
58	31.89	17.19	0.0103	0.7957	96.18	1.2567	22.91	46.75	69.67	0.048	0.1391
59	32.52	17.82	0.0104	0.7809	96.05	1.2805	23.17	46.65	69.83	0.049	0.1392
60	33.15	18.45	0.0104	0.7664	95.92	1.3047	23.43	46.55	69.98	0.049	0.1393
61	33.80	19.10	0.0104	0.7522	95.79	1.3293	23.69	46.45	70.14	0.050	0.1394
62	34.45	19.75	0.0104	0.7384	95.66	1.3542	23.95	46.35	70.30	0.050	0.1396
63	35.12	20.42	0.0104	0.7248	95.53	1.3795	24.21	46.24	70.46	0.051	0.1397
64	35.79	21.09	0.0104	0.7116	95.40	1.4051	24.47	46.14	70.61	0.051	0.1398
65	36.48	21.78	0.0104	0.6987	95.26	1.4312	24.73	46.04	70.77	0.052	0.1400
66	37.17	22.47	0.0105	0.6860	95.13	1.4576	24.99	45.93	70.93	0.052	0.1401
67	37.87	23.17	0.0105	0.6736	95.00	1.4844	25.25	45.83	71.09	0.053	0.1402
68	38.59	23.89	0.0105	0.6615	94.86	1.5116	25.52	45.72	71.24	0.053	0.1403
69	39.31	24.61	0.0105	0.6496	94.73	1.5392	25.78	45.61	71.40	0.054	0.1405
70	40.05	25.35	0.0105	0.6380	94.59	1.5673	26.04	45.51	71.56	0.054	0.1406
71	40.79	26.09	0.0105	0.6266	94.46	1.5957	26.31	45.40	71.71	0.055	0.1407
72	41.55	26.85	0.0106	0.6155	94.32	1.6245	26.57	45.29	71.87	0.055	0.1409
73	42.31	27.61	0.0106	0.6046	94.18	1.6538	26.84	45.18	72.03	0.056	0.1410
74	43.09	28.39	0.0106	0.5939	94.05	1.6835	27.10	45.07	72.18	0.056	0.1411
75	43.88	29.18	0.0106	0.5835	93.91	1.7136	27.37	44.96	72.34	0.057	0.1413
76	44.68	29.98	0.0106	0.5733	93.77	1.7441	27.64	44.85	72.50	0.057	0.1414
77	45.49	30.79	0.0106	0.5633	93.63	1.7751	27.90	44.74	72.65	0.058	0.1415
78	46.31	31.61	0.0106	0.5535	93.50	1.8065	28.17	44.63	72.81	0.058	0.1417
79	47.14	32.44	0.0107	0.5439	93.36	1.8384	28.44	44.52	72.96	0.059	0.1418
80	47.98	33.28	0.0107	0.5345	93.22	1.8707	28.71	44.41	73.12	0.059	0.1419
81	48.84	34.14	0.0107	0.5253	93.08	1.9035	28.98	44.29	73.28	0.060	0.1421
82	49.71	35.01	0.0107	0.5163	92.94	1.9367	29.25	44.18	73.43	0.060	0.1422
83	50.58	35.88	0.0107	0.5074	92.80	1.9704	29.52	44.07	73.59	0.061	0.1423
84	51.47	36.77	0.0107	0.4988	92.65	2.0046	29.79	43.95	73.74	0.061	0.1425
85	52.37	37.67	0.0108	0.4903	92.51	2.0393	30.06	43.84	73.90	0.062	0.1426
86	53.29	38.59	0.0108	0.4820	92.37	2.0744	30.33	43.72	74.05	0.062	0.1427
87	54.21	39.51	0.0108	0.4739	92.23	2.1101	30.60	43.60	74.21	0.063	0.1429
88	55.15	40.45	0.0108	0.4659	92.08	2.1462	30.87	43.49	74.37	0.063	0.1430
89	56.10	41.40	0.0108	0.4581	91.94	2.1828	31.15	43.37	74.52	0.064	0.1431
90	57.06	42.36	0.0108	0.4504	91.80	2.2200	31.42	43.25	74.68	0.064	0.1433
91	58.03	43.33	0.0109	0.4429	91.65	2.2576	31.69	43.13	74.83	0.065	0.1434
92	59.02	44.32	0.0109	0.4355	91.50	2.2958	31.97	43.01	74.98	0.065	0.1435
93	60.02	45.32	0.0109	0.4283	91.36	2.3345	32.24	42.89	75.14	0.066	0.1437
94	61.03	46.33	0.0109	0.4212	91.21	2.3738	32.52	42.77	75.29	0.066	0.1438
95	62.05	47.35	0.0109	0.4143	91.06	2.4136	32.79	42.65	75.45	0.067	0.1440
96	63.09	48.39	0.0109	0.4075	90.92	2.4539	33.07	42.53	75.60	0.067	0.1441
97	64.14	49.44	0.0110	0.4008	90.77	2.4948	33.35	42.40	75.76	0.068	0.1442
98	65.20	50.50	0.0110	0.3942	90.62	2.5362	33.62	42.28	75.91	0.068	0.1444
99	66.28	51.58	0.0110	0.3878	90.47	2.5782	33.90	42.16	76.06	0.069	0.1445

TEMP. °F	PRESSURE PSIA	PRESSURE PSIG	VOLUME cu ft/lb LIQUID v_f	VOLUME cu ft/lb VAPOR v_g	DENSITY lb/cu ft LIQUID $1/v_f$	DENSITY lb/cu ft VAPOR $1/v_g$	ENTHALPY Btu/lb LIQUID h_f	ENTHALPY Btu/lb LATENT h_{fg}	ENTHALPY Btu/lb VAPOR h_g	ENTROPY Btu/(lb)(°R) LIQUID s_f	ENTROPY Btu/(lb)(°R) VAPOR s_g
100	67.3	52.6	0.0110	0.3815	90.32	2.620	34.18	42.03	76.22	0.069	0.1446
101	68.4	53.7	0.0110	0.3753	90.17	2.663	34.46	41.91	76.37	0.070	0.1448
102	69.5	54.8	0.0111	0.3693	90.02	2.707	34.74	41.78	76.52	0.070	0.1449
103	70.7	56.0	0.0111	0.3633	89.87	2.752	35.02	41.65	76.68	0.071	0.1450
104	71.8	57.1	0.0111	0.3575	89.71	2.797	35.30	41.53	76.83	0.071	0.1452
105	73.0	58.3	0.0111	0.3517	89.56	2.842	35.58	41.40	76.98	0.072	0.1453
106	74.1	59.4	0.0111	0.3461	89.41	2.888	35.86	41.27	77.13	0.072	0.1454
107	75.3	60.6	0.0112	0.3406	89.25	2.935	36.14	41.14	77.29	0.073	0.1456
108	76.5	61.8	0.0112	0.3352	89.10	2.982	36.42	41.01	77.44	0.073	0.1457
109	77.7	63.0	0.0112	0.3299	88.94	3.031	36.70	40.88	77.59	0.074	0.1459
110	79.0	64.3	0.0112	0.3247	88.79	3.079	36.99	40.75	77.74	0.074	0.1460
111	80.2	65.5	0.0112	0.3195	88.63	3.129	37.27	40.62	77.89	0.074	0.1461
112	81.5	66.8	0.0113	0.3145	88.47	3.179	37.55	40.49	78.04	0.075	0.1463
113	82.7	68.0	0.0113	0.3096	88.31	3.229	37.84	40.35	78.20	0.075	0.1464
114	84.0	69.3	0.0113	0.3047	88.16	3.281	38.12	40.22	78.35	0.076	0.1465
115	85.3	70.6	0.0113	0.3000	88.00	3.333	38.41	40.08	78.50	0.076	0.1467
116	86.6	71.9	0.0113	0.2953	87.84	3.386	38.69	39.95	78.65	0.077	0.1468
117	87.9	73.2	0.0114	0.2907	87.67	3.439	38.98	39.81	78.80	0.077	0.1469
118	89.3	74.6	0.0114	0.2862	87.51	3.493	39.27	39.68	78.95	0.078	0.1471
119	90.6	75.9	0.0114	0.2817	87.35	3.548	39.55	39.54	79.10	0.078	0.1472
120	92.0	77.3	0.0114	0.2774	87.19	3.604	39.84	39.40	79.25	0.079	0.1474
121	93.4	78.7	0.0114	0.2731	87.02	3.660	40.13	39.26	79.40	0.079	0.1475
122	94.8	80.1	0.0115	0.2689	86.86	3.717	40.42	39.12	79.54	0.080	0.1476
123	96.2	81.5	0.0115	0.2648	86.69	3.775	40.71	38.98	79.69	0.080	0.1478
124	97.7	83.0	0.0115	0.2607	86.53	3.834	41.00	38.84	79.84	0.081	0.1479
125	99.1	84.4	0.0115	0.2568	86.36	3.893	41.29	38.70	79.99	0.081	0.1480
126	100.6	85.9	0.0116	0.2528	86.19	3.954	41.58	38.56	80.14	0.082	0.1482
127	102.1	87.4	0.0116	0.2490	86.02	4.015	41.87	38.41	80.29	0.082	0.1483
128	103.6	88.9	0.0116	0.2452	85.85	4.077	42.16	38.27	80.43	0.083	0.1484
129	105.1	90.4	0.0116	0.2415	85.68	4.140	42.45	38.13	80.58	0.083	0.1486
130	106.6	91.9	0.0116	0.2378	85.51	4.203	42.74	37.98	80.73	0.084	0.1487
131	108.2	93.5	0.0117	0.2342	85.34	4.268	43.03	37.83	80.87	0.084	0.1488
132	109.7	95.0	0.0117	0.2307	85.17	4.333	43.33	37.69	81.02	0.085	0.1490
133	111.3	96.6	0.0117	0.2272	84.99	4.399	43.62	37.54	81.17	0.085	0.1491
134	112.9	98.2	0.0117	0.2238	84.82	4.466	43.92	37.39	81.31	0.086	0.1492
135	114.5	99.8	0.0118	0.2205	84.64	4.534	44.21	37.24	81.46	0.086	0.1494
136	116.1	101.4	0.0118	0.2172	84.47	4.603	44.51	37.09	81.60	0.087	0.1495
137	117.8	103.1	0.0118	0.2139	84.29	4.673	44.80	36.94	81.75	0.087	0.1496
138	119.5	104.8	0.0118	0.2107	84.11	4.744	45.10	36.79	81.89	0.088	0.1498
139	121.1	106.4	0.0119	0.2076	83.93	4.816	45.39	36.64	82.03	0.088	0.1499
140	122.8	108.1	0.0119	0.2045	83.75	4.889	45.69	36.48	82.18	0.089	0.1500
141	124.5	109.8	0.0119	0.2014	83.57	4.962	45.99	36.33	82.32	0.089	0.1502
142	126.3	111.6	0.0119	0.1985	83.39	5.037	46.29	36.17	82.46	0.090	0.1503
143	128.0	113.3	0.0120	0.1955	83.20	5.113	46.58	36.02	82.61	0.090	0.1504
144	129.8	115.1	0.0120	0.1926	83.02	5.190	46.88	35.86	82.75	0.091	0.1505
145	131.6	116.9	0.0120	0.1898	82.83	5.268	47.18	35.70	82.89	0.091	0.1507
146	133.4	118.7	0.0120	0.1870	82.65	5.347	47.48	35.54	83.03	0.092	0.1508
147	135.2	120.5	0.0121	0.1842	82.46	5.427	47.78	35.38	83.17	0.092	0.1509
148	137.1	122.4	0.0121	0.1815	82.27	5.508	48.08	35.22	83.31	0.093	0.1511
149	138.9	124.2	0.0121	0.1788	82.08	5.590	48.39	35.06	83.45	0.093	0.1512
150	140.8	126.1	0.0122	0.1762	81.89	5.674	48.69	34.90	83.59	0.094	0.1513
151	142.7	128.0	0.0122	0.1736	81.69	5.759	48.99	34.74	83.73	0.094	0.1514
152	144.6	129.9	0.0122	0.1710	81.50	5.845	49.29	34.57	83.87	0.095	0.1516
153	146.5	131.8	0.0122	0.1685	81.31	5.932	49.60	34.41	84.01	0.095	0.1517
154	[illegible]	[illegible]	0.0123	0.1660	81.11	6.020	49.90	34.24	84.15	0.096	0.1518
155	150.5	135.8	0.0123	0.1636	80.91	6.110	50.21	34.07	84.28	0.096	0.1520
156	152.4	137.7	0.0123	0.1612	80.71	6.200	50.51	33.90	84.42	0.097	0.1521
157	154.5	139.8	0.0124	0.1589	80.51	6.293	50.82	33.73	84.56	0.097	0.1522
158	156.5	141.8	0.0124	0.1565	80.31	6.386	51.12	33.56	84.69	0.098	0.1523
159	158.5	143.8	0.0124	0.1542	80.11	6.481	51.43	33.39	84.83	0.098	0.1524
160	160.6	145.9	0.0125	0.1520	79.91	6.577	51.74	33.22	84.96	0.099	0.1526
161	162.7	148.0	0.0125	0.1498	79.70	6.675	52.05	33.04	85.09	0.099	0.1527
162	164.8	150.1	0.0125	0.1476	79.49	6.774	52.35	32.87	85.23	0.099	0.1528
163	166.9	152.2	0.0126	0.1454	79.28	6.875	52.66	32.69	85.36	0.100	0.1529
164	169.1	154.4	0.0126	0.1433	79.07	6.977	52.97	32.51	85.49	0.100	0.1531
165	171.2	156.5	0.0126	0.1412	78.86	7.081	53.28	32.34	85.62	0.101	0.1532
166	173.4	158.7	0.0127	0.1391	78.65	7.186	53.59	32.16	85.75	0.101	0.1533
167	175.6	160.9	0.0127	0.1371	78.43	7.293	53.91	31.97	85.88	0.102	0.1534
168	177.8	163.1	0.0127	0.1351	78.22	7.401	54.22	31.79	86.01	0.102	0.1535
169	180.1	165.4	0.0128	0.1331	78.00	7.511	54.53	31.61	86.14	0.103	0.1536

TEMP.	PRESSURE		VOLUME cu ft/lb		DENSITY lb/cu ft		ENTHALPY Btu/lb			ENTROPY Btu/(lb)(°R)	
°F	PSIA	PSIG	LIQUID v_f	VAPOR v_g	LIQUID $1/v_f$	VAPOR $1/v_g$	LIQUID h_f	LATENT h_{fg}	VAPOR h_g	LIQUID s_f	VAPOR s_g
170	182.4	167.7	0.0128	0.1311	77.78	7.623	54.84	31.42	86.27	0.1038	0.1538
171	184.7	170.0	0.0128	0.1292	77.56	7.736	55.16	31.24	86.40	0.1043	0.1539
172	187.0	172.3	0.0129	0.1273	77.33	7.852	55.47	31.05	86.52	0.1048	0.1540
173	189.3	174.6	0.0129	0.1254	77.11	7.969	55.79	30.86	86.65	0.1053	0.1541
174	191.6	176.9	0.0130	0.1236	76.88	8.088	56.10	30.67	86.78	0.1058	0.1542
175	194.0	179.3	0.0130	0.1218	76.65	8.209	56.42	30.47	86.90	0.1063	0.1543
176	196.4	181.7	0.0130	0.1200	76.42	8.332	56.74	30.28	87.02	0.1068	0.1544
177	198.8	184.1	0.0131	0.1182	76.19	8.457	57.06	30.09	87.14	0.1073	0.1545
178	201.3	186.6	0.0131	0.1164	75.95	8.584	57.37	29.89	87.27	0.1078	0.1546
179	203.7	189.0	0.0132	0.1147	75.71	8.713	57.69	29.69	87.39	0.1083	0.1548
180	206.2	191.5	0.0132	0.1130	75.47	8.844	58.01	29.49	87.51	0.1088	0.1549
181	208.7	194.0	0.0132	0.1113	75.23	8.978	58.33	29.29	87.62	0.1092	0.1550
182	211.3	196.6	0.0133	0.1097	74.99	9.114	58.66	29.08	87.74	0.1097	0.1551
183	213.8	199.1	0.0133	0.1080	74.74	9.252	58.98	28.88	87.86	0.1102	0.1552
184	216.4	201.7	0.0134	0.1064	74.49	9.392	59.30	28.67	87.97	0.1107	0.1553
185	219.0	204.3	0.0134	0.1048	74.24	9.535	59.63	28.46	88.09	0.1112	0.1554
186	221.6	206.9	0.0135	0.1032	73.99	9.681	59.95	28.25	88.20	0.1117	0.1555
187	224.3	209.6	0.0135	0.1017	73.73	9.829	60.28	28.03	88.31	0.1122	0.1556
188	226.9	212.2	0.0136	0.1001	73.47	9.980	60.60	27.82	88.42	0.1127	0.1556
189	229.6	214.9	0.0136	0.0986	73.20	10.134	60.93	27.60	88.53	0.1132	0.1557
190	232.3	217.6	0.0137	0.0971	72.94	10.291	61.26	27.38	88.64	0.1137	0.1558
191	235.1	220.4	0.0137	0.0956	72.67	10.451	61.59	27.15	88.75	0.1142	0.1559
192	237.8	223.1	0.0138	0.0942	72.40	10.613	61.92	26.93	88.85	0.1147	0.1560
193	240.6	225.9	0.0138	0.0927	72.12	10.779	62.25	26.70	88.96	0.1152	0.1561
194	243.4	228.7	0.0139	0.0913	71.84	10.949	62.58	26.47	89.06	0.1157	0.1562
195	246.3	231.6	0.0139	0.0899	71.56	11.121	62.92	26.24	89.16	0.1162	0.1563
196	249.2	234.5	0.0140	0.0885	71.28	11.298	63.25	26.00	89.26	0.1167	0.1563
197	252.0	237.3	0.0140	0.0871	70.99	11.478	63.59	25.76	89.36	0.1172	0.1564
198	255.0	240.3	0.0141	0.0857	70.69	11.662	63.93	25.52	89.45	0.1177	0.1565
199	257.9	243.2	0.0142	0.0843	70.39	11.849	64.27	25.28	89.55	0.1182	0.1566
200	260.9	246.2	0.0142	0.0830	70.09	12.041	64.61	25.03	89.64	0.1187	0.1566
201	263.9	249.2	0.0143	0.0817	69.79	12.238	64.95	24.77	89.73	0.1192	0.1567
202	266.9	252.2	0.0143	0.0803	69.48	12.438	65.29	24.52	89.81	0.1197	0.1568
203	269.9	255.2	0.0144	0.0790	69.16	12.644	65.63	24.26	89.90	0.1202	0.1568
204	273.0	258.3	0.0145	0.0777	68.84	12.854	65.98	24.00	89.98	0.1207	0.1569
205	276.1	261.4	0.0145	0.0765	68.51	13.070	66.33	23.73	90.06	0.1212	0.1569
206	279.2	264.5	0.0146	0.0752	68.18	13.291	66.68	23.46	90.14	0.1217	0.1570
207	282.4	267.7	0.0147	0.0739	67.84	13.517	67.03	23.18	90.22	0.1223	0.1570
208	285.6	270.9	0.0148	0.0727	67.50	13.750	67.38	22.90	90.29	0.1228	0.1571
209	288.8	274.1	0.0148	0.0714	67.15	13.989	67.74	22.62	90.36	0.1233	0.1571
210	292.0	277.3	0.0149	0.0702	66.80	14.234	68.09	22.32	90.42	0.1238	0.1571
211	295.3	280.6	0.0150	0.0690	66.43	14.486	68.45	22.03	90.49	0.1243	0.1572
212	298.6	283.9	0.0151	0.0678	66.06	14.746	68.82	21.72	90.55	0.1249	0.1572
213	302.0	287.3	0.0152	0.0666	65.68	15.014	69.18	21.42	90.60	0.1254	0.1572
214	305.3	290.6	0.0153	0.0654	65.30	15.289	69.55	21.10	90.65	0.1259	0.1572
215	308.7	294.0	0.0154	0.0642	64.90	15.574	69.92	20.78	90.70	0.1265	0.1573
216	312.1	297.4	0.0155	0.0630	64.49	15.868	70.29	20.45	90.74	0.1270	0.1573
217	315.6	300.9	0.0156	0.0618	64.08	16.173	70.67	20.11	90.78	0.1275	0.1573
218	319.1	304.4	0.0157	0.0606	63.65	16.488	71.05	19.76	90.81	0.1281	0.1572
219	322.6	307.9	0.0158	0.0594	63.21	16.815	71.43	19.40	90.84	0.1286	0.1572
220	326.2	311.5	0.0159	0.0582	62.76	17.155	71.82	19.03	90.86	0.1292	0.1572
221	329.8	315.1	0.0160	0.0571	62.30	17.509	72.22	18.65	90.87	0.1298	0.1572
222	333.4	318.7	0.0161	0.0559	61.82	17.878	72.62	18.26	90.88	0.1303	0.1571
223	337.0	322.3	0.0163	0.0547	61.32	18.263	73.02	17.85	90.88	0.1309	0.1571
224	340.7	326.0	0.0164	0.0535	60.80	18.667	73.43	17.43	90.87	0.1315	0.1570
225	344.5	329.8	0.0165	0.0523	60.27	19.091	73.85	17.00	90.85	0.1321	0.1569
226	348.2	333.5	0.0167	0.0511	59.71	19.538	74.27	16.54	90.82	0.1327	0.1568
227	352.0	337.3	0.0169	0.0499	59.13	20.011	74.71	16.06	90.77	0.1333	0.1567
228	355.9	341.2	0.0170	0.0487	58.52	20.512	75.15	15.56	90.72	0.1339	0.1566
229	359.8	345.1	0.0172	0.0475	57.87	21.048	75.61	15.03	90.64	0.1346	0.1564
230	363.7	349.0	0.0174	0.0462	57.19	21.622	76.08	14.47	90.55	0.1352	0.1562
231	367.7	353.0	0.0177	0.0449	56.47	22.243	76.56	13.88	90.44	0.1359	0.1560
232	371.7	357.0	0.0179	0.0436	55.69	22.920	77.06	13.23	90.30	0.1366	0.1558
233	375.7	361.0	0.0182	0.0422	54.84	23.666	77.59	12.53	90.13	0.1374	0.1555
234	379.8	365.1	0.0185	0.0408	53.91	24.501	78.15	11.77	89.92	0.1381	0.1551
235	384.0	369.3	0.0189	0.0392	52.87	25.452	78.74	10.90	89.65	0.1390	0.1547
236	388.1	373.4	0.0193	0.0376	51.68	26.568	79.39	9.92	89.31	0.1399	0.1542
237	392.4	377.7	0.0198	0.0357	50.26	27.937	80.12	8.73	88.86	0.1409	0.1535
238	396.7	382.0	0.0206	0.0335	48.43	29.764	80.99	7.21	88.20	0.1421	0.1525
239	401.0	386.3	0.0219	0.0304	45.57	32.797	82.20	4.82	87.03	0.1439	0.1508

DATA SHEET FOR R 500

A	AZEOTROPE R12 / R152a	Molecular Wt. 99.3

Pressure		Temperature				Volume		Density	
45.1	kg/cm^2	105	°C	222	°F	2.01	l/kg	.497	kg/l.
642	psia	378	°K	682	°R	.0322	ft^3/lb	30.9	lbs/ft^3
								3.4	Air=1

AT CRITICAL POINT

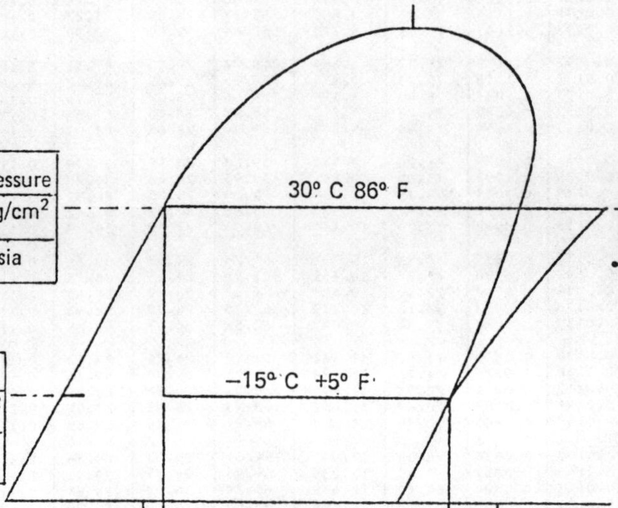

Discharge Pressure	
8.97	kg/cm^2
127.6	psia

Inlet Pressure	
2.18	kg/cm^2
31	psia

Refrigerating Effect	
359 $kcal/m^3$	40.4 Btu/ft^3

Discharge Temperature	
40.5 °C	105 °F
313 °K	565 °R

Normal Boiling Point	
-33 °C	-28 °F
240 °K	432 °R

Triple Point	
-159 °C	-254 °F
114 °K	206 °R

Latent Heat at NBP	48 kcal/kg	4754 /kg mol	86.4 Btu/lb	8558 /lb mol
Trouton's No. 19.8	Gas Constant 8.54 kg.m/kg/°K		15.55 ft.lbs/lb/°R	
Specific Heat Liquid at 30° C 86° F .29 Fig 16		Gas C_p .173	C_p/C_v 1.13 at 25°C	
Liquid at 30° C 86° F Density 1.14 kg/l.	71.1 lbs/ft^3	Viscosity .29 cp @15°C		
Reduced Form at NBP Pressure .023	1 / Temperature 1.58			

Reduced value of 1/T when Reduced Pressure is	a) .001	b) .01	c) .1
		1.696	1.38

A 73.8/26.2% by weight

TEMP. °F	PRESSURE lb per sq in.		VAPOR VOLUME cubic ft per lb	LIQUID DENSITY lb per cubic ft	ENTHALPY Btu per lb			ENTROPY Btu per (lb) (°R)	
t	Absolute psia	Gage psig	v_g	$\frac{1}{v_f}$	Liquid h_f	Latent h_{fg}	Vapor h_g	Liquid s_f	Vapor s_g
−100	1.735	*26.4	22.18	89.64	−13.34	93.11	79.78	−0.0341	0.2248
−90	2.469	*24.9	15.97	88.77	−11.27	92.39	81.12	−0.0285	0.2215
−80	3.433	*22.9	11.72	87.89	−9.06	91.52	82.46	−0.0226	0.2185
−70	4.713	*20.3	8.754	87.01	−6.82	90.61	83.79	−0.0168	0.2157
−60	6.340	*17.0	6.643	86.11	−4.61	89.73	85.12	−0.0112	0.2133
−50	8.395	*12.8	5.115	85.20	−2.30	88.74	86.44	−0.0055	0.2111
−48	8.864	*11.9	4.863	85.02	−1.87	88.57	86.70	−0.0045	0.2107
−46	9.353	*10.9	4.625	84.83	−1.39	88.35	86.96	−0.0033	0.2103
−44	9.865	* 9.84	4.401	84.65	−0.93	88.15	87.22	−0.0022	0.2099
−42	10.40	* 8.75	4.191	84.46	−0.46	87.94	87.48	−0.0011	0.2095
−40	10.95	* 7.62	3.992	84.28	0.00	87.74	87.74	0.0000	0.2091
−38	11.53	* 6.44	3.805	84.09	0.47	87.54	88.00	0.0011	0.2087
−36	12.14	* 5.21	3.628	83.91	0.96	87.31	88.26	0.0023	0.2083
−34	12.77	* 3.93	3.461	83.72	1.43	87.09	88.52	0.0034	0.2080
−32	13.42	* 2.60	3.303	83.53	1.89	86.88	88.78	0.0045	0.2076
−30	14.10	* 1.22	3.154	83.35	2.38	86.66	89.04	0.0056	0.2073
−28	14.81	0.110	3.013	83.16	2.85	86.44	89.29	0.0067	0.2069
−26	15.54	0.846	2.879	82.97	3.34	86.21	89.55	0.0078	0.2066
−24	16.31	1.61	2.753	82.78	3.81	86.00	89.80	0.0089	0.2062
−22	17.10	2.40	2.633	82.59	4.30	85.76	90.06	0.0100	0.2059
−20	17.92	3.23	2.520	82.40	4.79	85.52	90.31	0.0111	0.2056
−18	18.78	4.08	2.412	82.21	5.26	85.31	90.57	0.0122	0.2053
−16	19.66	4.96	2.310	82.02	5.75	85.07	90.82	0.0133	0.2050
−14	20.58	5.88	2.213	81.82	6.24	84.83	91.07	0.0144	0.2047
−12	21.53	6.83	2.121	81.63	6.74	84.58	91.32	0.0155	0.2044
−10	22.52	7.82	2.034	81.44	7.22	84.35	91.57	0.0165	0.2041
−8	23.54	8.84	1.951	81.24	7.72	84.10	91.82	0.0176	0.2038
−6	24.59	9.90	1.872	81.05	8.21	83.86	92.07	0.0187	0.2035
−4	25.68	11.0	1.797	80.85	8.71	83.61	92.32	0.0198	0.2033
−2	26.81	12.1	1.725	80.66	9.21	83.23	92.56	0.0209	0.2030
0	27.98	13.3	1.657	80.46	9.71	83.10	92.81	0.0220	0.2027
2	29.19	14.5	1.593	80.26	10.21	82.85	93.06	0.0230	0.2025
4	30.44	15.7	1.531	80.06	10.71	82.59	93.20	0.0241	0.2022
6	31.72	17.0	1.472	79.86	11.21	82.33	93.54	0.0252	0.2020
8	33.05	18.4	1.416	79.66	11.72	82.07	93.79	0.0263	0.2017
10	34.43	19.7	1.362	79.46	12.23	81.80	94.03	0.0274	0.2015
12	35.84	21.1	1.311	79.26	12.74	81.53	94.27	0.0284	0.2013
14	37.30	22.6	1.263	79.06	13.25	81.26	94.51	0.0295	0.2010
16	38.81	24.1	1.216	78.85	13.76	80.99	94.75	0.0306	0.2008
18	40.36	25.7	1.172	78.65	14.28	80.70	94.98	0.0316	0.2006
20	41.96	27.3	1.129	78.45	14.79	80.43	95.22	0.0327	0.2004
22	43.61	28.9	1.088	78.24	15.31	80.15	95.46	0.0338	0.2002
24	45.30	30.6	1.049	78.03	15.83	79.86	95.69	0.0348	0.2000
26	47.05	32.4	1.012	77.82	16.35	79.58	95.92	0.0359	0.1997
28	48.85	34.2	0.9763	77.62	16.88	79.28	96.16	0.0370	0.1995
30	50.70	36.0	0.9422	77.41	17.40	78.99	96.39	0.0380	0.1993
32	52.60	37.9	0.9095	77.20	17.94	78.68	96.62	0.0391	0.1991
34	54.55	39.9	0.8781	76.98	18.46	78.39	96.85	0.0402	0.1990
36	56.56	41.9	0.8481	76.77	18.99	78.08	97.08	0.0412	0.1988
38	58.63	43.9	0.8192	76.56	19.52	77.78	97.30	0.0423	0.1986
40	60.75	46.1	0.7916	76.34	20.05	77.47	97.53	0.0433	0.1984
42	62.94	48.2	0.7651	76.13	20.59	77.16	97.75	0.0444	0.1982
44	65.18	50.5	0.7392	75.91	21.13	76.85	97.98	0.0455	0.1980
46	67.47	52.8	0.7152	75.69	21.66	76.54	98.20	0.0465	0.1979
48	69.84	55.1	0.6917	75.48	22.20	76.22	98.42	0.0476	0.1977
50	72.26	57.6	0.6691	75.26	22.75	75.89	98.64	0.0486	0.1975
52	74.74	60.0	0.6475	75.03	23.29	75.56	98.86	0.0497	0.1974
54	77.29	62.6	0.6266	74.81	23.83	75.24	99.07	0.0507	0.1972
56	79.90	65.2	0.6066	74.59	24.38	74.90	99.29	0.0518	0.1970
58	82.58	67.9	0.5873	74.36	24.93	74.57	99.50	0.0528	0.1969

*Inches of Mercury Vacuum

TEMP. °F	PRESSURE lb per sq in.		VAPOR VOLUME cubic ft per lb	LIQUID DENSITY lb per cubic ft	ENTHALPY Btu per lb			ENTROPY Btu per (lb) (°R)	
t	Absolute psia	Gage psig	v_g	$\frac{1}{v_f}$	Liquid h_f	Latent h_{fg}	Vapor h_g	Liquid s_f	Vapor s_g
60	85.33	70.6	0.5687	74.14	25.48	74.23	99.71	0.0539	0.1967
62	88.14	73.4	0.5509	73.91	26.04	73.89	99.92	0.0549	0.1966
64	91.03	76.3	0.5337	73.68	26.59	73.54	100.13	0.0560	0.1964
66	93.98	79.3	0.5170	73.45	27.15	73.19	100.34	0.0570	0.1963
68	97.00	82.3	0.5011	73.22	27.72	72.83	100.55	0.0581	0.1961
70	100.1	85.4	0.4857	72.98	28.28	72.47	100.75	0.0591	0.1960
72	103.3	88.6	0.4709	72.75	28.84	72.12	100.95	0.0602	0.1958
74	106.5	91.8	0.4566	72.51	29.40	71.75	101.16	0.0612	0.1957
76	109.8	95.1	0.4428	72.28	29.97	71.38	101.36	0.0623	0.1955
78	113.2	98.5	0.4295	72.04	30.54	71.01	101.55	0.0633	0.1954
80	116.7	102.	0.4167	71.80	31.12	70.63	101.75	0.0644	0.1952
82	120.2	106.	0.4043	71.55	31.69	70.25	101.94	0.0654	0.1951
84	123.9	109.	0.3923	71.31	32.27	69.87	102.14	0.0664	0.1950
86	127.6	113.	0.3808	71.06	32.85	69.48	102.33	0.0675	0.1948
88	131.4	117.	0.3696	70.81	33.43	69.09	102.51	0.0685	0.1947
90	135.3	121.	0.3588	70.56	34.01	68.69	102.70	0.0696	0.1945
92	139.2	125.	0.3484	70.31	34.60	68.29	102.89	0.0706	0.1944
94	143.3	129.	0.3383	70.06	35.19	67.88	103.07	0.0717	0.1943
96	147.4	133.	0.3286	69.80	35.78	67.47	103.25	0.0727	0.1941
98	151.6	137.	0.3191	69.54	36.38	67.05	103.43	0.0738	0.1940
100	155.9	141.	0.3100	69.28	36.97	66.63	103.60	0.0748	0.1939
102	160.3	146.	0.3012	69.02	37.57	66.20	103.77	0.0759	0.1937
104	164.8	150.	0.2927	68.76	38.17	65.77	103.94	0.0769	0.1936
106	169.4	155.	0.2844	68.49	38.78	65.33	104.11	0.0780	0.1935
108	174.1	159.	0.2764	68.22	39.39	64.89	104.28	0.0790	0.1933
110	178.8	164.	0.2686	67.95	40.00	64.44	104.44	0.0801	0.1932
112	183.7	169.	0.2611	67.68	40.61	63.99	104.60	0.0811	0.1931
114	188.6	174.	0.2538	67.40	41.23	63.53	104.76	0.0822	0.1929
116	193.7	179.	0.2467	67.12	41.85	63.06	104.92	0.0832	0.1928
118	198.9	184.	0.2399	66.84	42.48	62.59	105.07	0.0843	0.1926
120	204.1	189.	0.2332	66.55	43.10	62.11	105.22	0.0853	0.1925
122	209.5	195.	0.2268	66.26	43.73	61.63	105.36	0.0864	0.1924
124	214.9	200.	0.2205	65.97	44.37	61.14	105.50	0.0875	0.1922
126	220.5	206.	0.2145	65.68	45.00	60.64	105.64	0.0885	0.1921
128	226.2	211.	0.2086	65.38	45.64	60.14	105.78	0.0896	0.1919
130	231.9	217.	0.2028	65.08	46.29	59.62	105.91	0.0907	0.1918
132	237.8	223.	0.1973	64.77	46.94	59.10	106.04	0.0917	0.1916
134	243.8	229.	0.1919	64.46	47.59	58.57	106.16	0.0928	0.1915
136	249.9	235.	0.1866	64.15	48.25	58.03	106.28	0.0939	0.1913
138	256.1	241.	0.1815	63.83	48.91	57.49	106.40	0.0950	0.1911
140	262.4	248.	0.1765	63.51	49.58	56.94	106.51	0.0960	0.1910
142	268.9	254.	0.1717	63.19	50.24	56.38	106.62	0.0971	0.1908
144	275.4	261.	0.1670	62.86	50.92	55.80	106.72	0.0982	0.1906
146	282.1	267.	0.1624	62.52	51.60	55.22	106.82	0.0993	0.1905
148	288.9	274.	0.1580	62.18	52.29	54.63	106.92	0.1004	0.1903
150	295.8	281.	0.1537	61.84	52.98	54.03	107.00	0.1015	0.1901
152	302.8	288.	0.1494	61.49	53.67	53.42	107.09	0.1026	0.1899
154	310.0	295.	0.1453	61.13	54.38	52.79	107.16	0.1037	0.1897
156	317.2	303.	0.1413	60.77	55.08	52.15	107.23	0.1048	0.1895
158	324.6	310.	0.1374	60.40	55.79	51.51	107.30	0.1059	0.1893
160	332.1	317.	0.1336	60.03	56.51	50.85	107.36	0.1071	0.1891
165	351.5	337.	0.1245	59.06	58.34	49.13	107.47	0.1099	0.1886
170	371.7	357.	0.1159	58.04	60.21	47.33	107.54	0.1128	0.1879
175	392.7	378.	0.1077	56.96	62.14	45.41	107.55	0.1157	0.1873
180	414.6	400.	0.1000	55.81	64.12	43.38	107.49	0.1187	0.1865
190	461.0	446.	0.0857	53.25	68.30	38.85	107.14	0.1250	0.1847
200	511.1	496.	0.0726	50.16	72.87	33.47	106.34	0.1317	0.1824
210	567.3	553.	0.0590	46.05	76.76	27.64	104.40	0.1372	0.1785
220	629.7	615.	0.0417	38.39	86.01	12.64	98.65	0.1505	0.1691
221.9	641.9	627.	0.0323	31.00	92.43	—	92.43	0.1598	0.1598

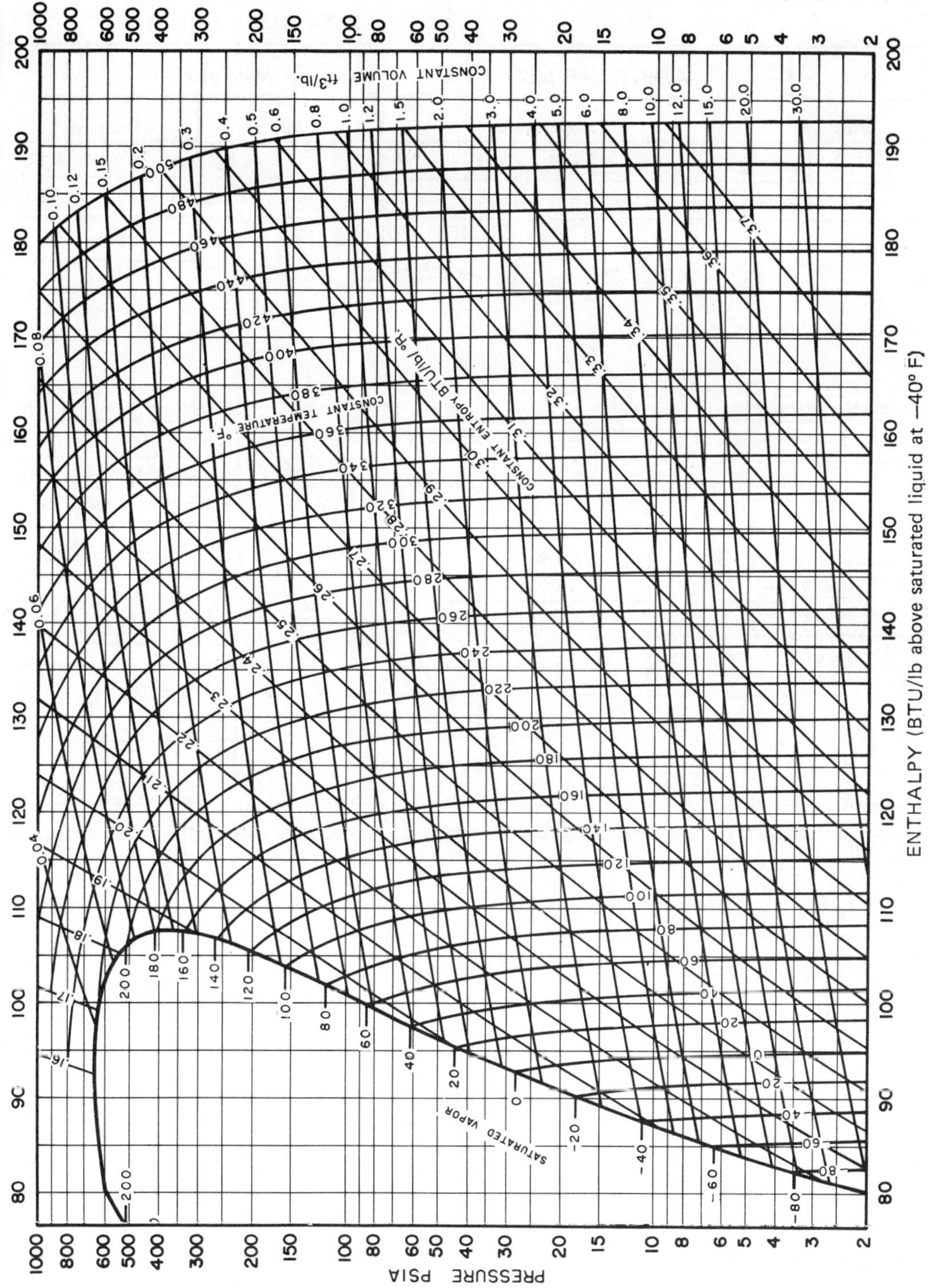
CONSTANT VOLUME ft3/lb.
CONSTANT TEMPERATURE °F
CONSTANT ENTROPY BTU/lb/°R
SATURATED VAPOR
PRESSURE PSIA
ENTHALPY (BTU/lb above saturated liquid at −40° F)

DATA SHEET FOR R 502

A	AZEOTROPE R22 / R115	Molecular Wt.	111.6

Pressure		Temperature				Volume		Density	
43.5	kg/cm^2	90	°C	194	°F	1.78	l/kg	.560	kg/l.
619	psia	363	°K	654	°R	.0286	ft^3/lb	34.9	lbs/ft^3
								3.83	Air=1

AT CRITICAL POINT

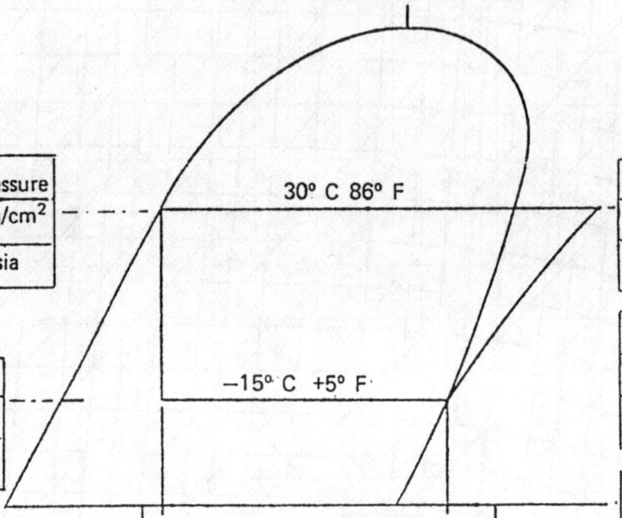

30° C 86° F

−15° C +5° F

Discharge Pressure	
12.3	kg/cm^2
175	psia

Inlet Pressure	
2.53	kg/cm^2
36	psia

Discharge Temperature			
37	°C	99	°F
310	°K	559	°R

Normal Boiling Point			
-46	°C	-50	°F
227	°K	410	°R

Triple Point			
	°C		°F
	°K		°R

Refrigerating Effect			
459	kcal/m^3	55.7	Btu/ft^3

Latent Heat at NBP	42.48 kcal/kg	4740 /kg mol	76.46 Btu/lb	8533 /lb mol

Trouton's No. 20.81	Gas Constant 7.6 kg.m/kg/° K	13.84 ft.lbs/lb/° R

Specific Heat Liquid at 30° C 86° F .260	Gas C_p .168	C_p/C_v 1.132 at 25°C

Liquid at 30° C 86° F Density 1.22 kg/l.	76.1 lbs/ft^3 Viscosity .15 cp

Reduced Form at NBP Pressure .0237	1 / Temperature 1.59

Reduced value of 1/T	2.056	1.72	1.377
when Reduced Pressure is	a) .001	b) .01	c) .1

A 48.8/51.2% by weight

TEMP. °F	PRESSURE		VOLUME cu ft/lb		DENSITY lb/cu ft		ENTHALPY Btu/lb			ENTROPY Btu/(lb)(°R)		TEMP. °F
	PSIA	PSIG	LIQUID	VAPOR	LIQUID	VAPOR	LIQUID	LATENT	VAPOR	LIQUID	VAPOR	
−150	0.42666	29.05250*	0.0097205	69.573	102.87	0.014373	−21.366	81.275	59.908	−0.05845	0.20401	−150
−149	0.44747	29.01013*	0.0097298	66.544	102.77	0.015027	−21.203	81.228	60.025	−0.05792	0.20354	−149
−148	0.46914	28.96601*	0.0097390	63.668	102.67	0.015706	−21.038	81.181	60.142	−0.05739	0.20307	−148
−147	0.49170	28.92008*	0.0097484	60.935	102.58	0.016410	−20.874	81.133	60.259	−0.05686	0.20262	−147
−146	0.51518	28.87227*	0.0097577	58.337	102.48	0.017141	−20.709	81.085	60.376	−0.05634	0.20216	−146
−145	0.53961	28.82253*	0.0097670	55.867	102.38	0.017899	−20.543	81.037	60.494	−0.05581	0.20172	−145
−144	0.56502	28.77080*	0.0097764	53.517	102.28	0.018685	−20.377	80.988	60.611	−0.05528	0.20127	−144
−143	0.59144	28.71700*	0.0097859	51.282	102.18	0.019499	−20.210	80.940	60.729	−0.05476	0.20084	−143
−142	0.61891	28.66108*	0.0097953	49.155	102.08	0.020343	−20.043	80.890	60.847	−0.05423	0.20040	−142
−141	0.64745	28.60297*	0.0098048	47.129	101.99	0.021218	−19.876	80.841	60.965	−0.05370	0.19997	−141
−140	0.67711	28.54259*	0.0098143	45.200	101.89	0.022123	−19.708	80.791	61.083	−0.05318	0.19955	−140
−139	0.70791	28.47988*	0.0098238	43.363	101.79	0.023060	−19.539	80.740	61.201	−0.05265	0.19913	−139
−138	0.73989	28.41477*	0.0098333	41.612	101.69	0.024031	−19.370	80.690	61.319	−0.05212	0.19872	−138
−137	0.77308	28.34718*	0.0098429	39.943	101.59	0.025035	−19.200	80.639	61.438	−0.05160	0.19831	−137
−136	0.80754	28.27703*	0.0098525	38.352	101.49	0.026074	−19.030	80.587	61.556	−0.05107	0.19790	−136
−135	0.84328	28.20426*	0.0098622	36.834	101.39	0.027148	−18.860	80.535	61.675	−0.05055	0.19750	−135
−134	0.88035	28.12878*	0.0098718	35.386	101.29	0.028259	−18.689	80.483	61.794	−0.05002	0.19711	−134
−133	0.91879	28.05052*	0.0098815	34.004	101.19	0.029407	−18.517	80.430	61.913	−0.04949	0.19671	−133
−132	0.95864	27.96938*	0.0098913	32.685	101.09	0.030594	−18.345	80.377	62.032	−0.04897	0.19633	−132
−131	0.99994	27.88530*	0.0099010	31.425	100.99	0.031821	−18.172	80.324	62.151	−0.04844	0.19594	−131
−130	1.0427	27.7981*	0.0099108	30.222	100.89	0.033088	−17.999	80.270	62.270	−0.04792	0.19556	−130
−129	1.0870	27.7079*	0.0099206	29.072	100.79	0.034396	−17.825	80.215	62.389	−0.04739	0.19519	−129
−128	1.1329	27.6144*	0.0099305	27.973	100.69	0.035747	−17.651	80.161	62.509	−0.04686	0.19482	−128
−127	1.1804	27.5177*	0.0099403	26.923	100.59	0.037142	−17.476	80.105	62.628	−0.04634	0.19445	−127
−126	1.2296	27.4175*	0.0099502	25.919	100.49	0.038581	−17.301	80.050	62.748	−0.04581	0.19409	−126
−125	1.2805	27.3139*	0.0099602	24.958	100.39	0.040066	−17.125	79.994	62.868	−0.04529	0.19373	−125
−124	1.3332	27.2067*	0.0099701	24.039	100.29	0.041598	−16.949	79.937	62.988	−0.04476	0.19338	−124
−123	1.3877	27.0958*	0.0099801	23.160	100.19	0.043177	−16.772	79.880	63.108	−0.04424	0.19303	−123
−122	1.4440	26.9811*	0.0099901	22.318	100.09	0.044806	−16.595	79.823	63.228	−0.04371	0.19268	−122
−121	1.5022	26.8626*	0.010000	21.511	99.997	0.046485	−16.417	79.765	63.348	−0.04318	0.19234	−121
−120	1.5624	26.7400*	0.010010	20.739	99.896	0.048216	−16.238	79.706	63.468	−0.04266	0.19200	−120
−119	1.6246	26.6134*	0.010020	20.000	99.795	0.049999	−16.059	79.648	63.588	−0.04213	0.19166	−119
−118	1.6888	26.4826*	0.010030	19.291	99.694	0.051836	−15.879	79.588	63.709	−0.04160	0.19133	−118
−117	1.7551	26.3476*	0.010040	18.612	99.593	0.053728	−15.699	79.528	63.829	−0.04108	0.19100	−117
−116	1.8237	26.2081*	0.010051	17.960	99.492	0.055677	−15.518	79.468	63.949	−0.04055	0.19068	−116
−115	1.8944	26.0641*	0.010061	17.336	99.391	0.057683	−15.337	79.407	64.070	−0.04003	0.19036	−115
−114	1.9674	25.9154*	0.010071	16.737	99.289	0.059747	−15.155	79.346	64.191	−0.03950	0.19004	−114
−113	2.0427	25.7621*	0.010081	16.162	99.188	0.061872	−14.972	79.284	64.311	−0.03897	0.18973	−113
−112	2.1204	25.6038*	0.010092	15.610	99.086	0.064058	−14.789	79.222	64.432	−0.03845	0.18942	−112
−111	2.2006	25.4406*	0.010102	15.081	98.984	0.066307	−14.606	79.159	64.553	−0.03792	0.18911	−111
−110	2.2833	25.2723*	0.010112	14.572	98.882	0.068621	−14.422	79.096	64.674	−0.03739	0.18881	−110
−109	2.3685	25.0988*	0.010123	14.084	98.780	0.070999	−14.237	79.032	64.795	−0.03686	0.18851	−109
−108	2.4563	24.9199*	0.010133	13.615	98.678	0.073444	−14.052	78.968	64.916	−0.03634	0.18821	−108
−107	2.5469	24.7356*	0.010144	13.165	98.576	0.075958	−13.866	78.903	65.037	−0.03581	0.18792	−107
−106	2.6402	24.5456*	0.010154	12.732	98.474	0.078541	−13.679	78.837	65.158	−0.03528	0.18763	−106
−105	2.7363	24.3499*	0.010165	12.315	98.371	0.081196	−13.492	78.771	65.279	−0.03475	0.18734	−105
−104	2.8353	24.1484*	0.010176	11.915	98.268	0.083922	−13.304	78.705	65.400	−0I03423	0.18706	−104
−103	2.9372	23.9409*	0.010186	11.530	98.166	0.086723	−13.116	78.638	65.521	−0.03370	0.18678	−103
−102	3.0421	23.7273*	0.010197	11.160	98.063	0.089599	−12.927	78.570	65.642	−0.03317	0.18650	−102
−101	3.1501	23.5074*	0.010208	10.804	97.960	0.092553	−12.738	78.502	65.764	−0.03264	0.18622	−101
−100	3.2613	23.2810*	0.010218	10.461	97.857	0.095584	−12.548	78.433	65.885	−0.03211	0.18595	−100
− 99	3.3757	23.0482*	0.010229	10.132	97.753	0.098696	−12.357	78.364	66.006	−0.03158	0.18568	− 99
− 98	3.4933	22.8086*	0.010240	9.8145	97.650	0.10188	−12.166	78.294	66.128	−0.03106	0.18542	− 98
− 97	3.6143	22.5623*	0.010251	9.5087	97.546	0.10516	−11.974	78.223	66.249	−0.03053	0.18516	− 97
− 96	3.7387	22.3089*	0.010262	9.2142	97.443	0.10852	−11.781	78.152	66.370	−0.03000	0.18490	− 96

*Inches of mercury below one atmosphere

TEMP. °F	PRESSURE		VOLUME cu ft/lb		DENSITY lb/cu ft		ENTHALPY Btu/lb			ENTROPY Btu/(lb)(°R)		TEMP. °F
	PSIA	PSIG	LIQUID	VAPOR	LIQUID	VAPOR	LIQUID	LATENT	VAPOR	LIQUID	VAPOR	
−95	3,8667	22.0484*	0.010273	8.9306	97.339	0.11197	−11.588	78.080	66.492	−0.02947	0.18464	−95
−94	3.9982	21.7807*	0.010284	8.6572	97.235	0.11550	−11.394	78.008	66.613	−0.02894	0.18439	−94
−93	4.1333	21.5055*	0.010295	8.3938	97.131	0.11913	−11.200	77.935	66.735	−0.02841	0.18414	−93
−92	4.2722	21.2228*	0.010306	8.1399	97.027	0.12285	−11.005	77.862	66.856	−0.02788	0.18389	−92
−91	4.4149	20.9323*	0.010317	7.8951	96.923	0.12665	−10.810	77.788	66.978	−0.02735	0.18365	−91
−90	4.5615	20.6339*	0.010328	7.6591	96.818	0.13056	−10.614	77.713	67.099	−0.02682	0.18340	−90
−89	4.7120	20.3274*	0.010339	7.4314	96.713	0.13456	−10.417	77.638	67.221	−0.02629	0.18316	−89
−88	4.8665	20.0128*	0.010350	7.2118	96.609	0.13866	−10.219	77.562	67.342	−0.02575	0.18293	−88
−87	5.0252	19.6897*	0.010362	6.9999	96.504	0.14285	−10.021	77.485	67.463	−0.02522	0.18269	−87
−86	5.1881	19.3581*	0.010373	6.7954	96.399	0.14715	−9.823	77.408	67.585	−0.02469	0.18246	−86
−85	5.3552	19.0178*	0.010384	6.5980	96.294	0.15156	−9.624	77.330	67.706	−0.02416	0.18223	−85
−84	5.5267	18.6686*	0.010396	6.4074	96.188	0.15606	−9.424	77.252	67.828	−0.02363	0.18201	−84
−83	5.7026	18.3104*	0.010407	6.2234	96.083	0.16068	−9.223	77.173	67.949	−0.02310	0.18178	−83
−82	5.8831	17.9429*	0.010419	6.0457	95.977	0.16540	−9.022	77.093	68.070	−0.02256	0.18156	−82
−81	6.0682	17.5661*	0.010430	5.8740	95.872	0.17024	−8.820	77.013	68.192	−0.02203	0.18134	−81
−80	6.2580	17.1797*	0.010442	5.7081	95.766	0.17518	−8.618	76.932	68.313	−0.02150	0.18113	−80
−79	6.4526	16.7835*	0.010453	5.5479	95.660	0.18024	−8.415	76.850	68.435	−0.02097	0.18091	−79
−78	6.6521	16.3774*	0.010465	5.3930	95.553	0.18542	−8.211	76.768	68.556	−0.02043	0.18070	−78
−77	6.8565	15.9611*	0.010476	5.2432	95.447	0.19071	−8.007	76.685	68.677	−0.01990	0.18049	−77
−76	7.0660	15.5346	0.010488	5.0985	95.341	0.19613	−7.802	76.601	68.798	−0.01937	0.18028	−76
−75	7.2806	15.0976*	0.010500	4.9585	95.234	0.20167	−7.597	76.517	68.919	−0.01883	0.18008	−75
−74	7.5005	14.6499*	0.010512	4.8231	95.127	0.20733	−7.391	76.432	69.041	−0.01830	0.17988	−74
−73	7.7257	14.1914*	0.010524	4.6921	95.020	0.21312	−7.184	76.346	69.162	−0.01776	0.17968	−73
−72	7.9563	13.7218*	0.010535	4.5654	94.913	0.21903	−6.976	76.260	69.283	−0.01723	0.17948	−72
−71	8.1925	13.2410*	0.010547	4.4428	94.806	0.22508	−6.768	76.172	69.404	−0.01669	0.17928	−71
−70	8.4343	12.7488*	0.010559	4.3241	94.698	0.23126	−6.560	76.085	69.525	−0.01616	0.17909	−70
−69	8.6818	12.2449*	0.010571	4.2092	94.591	0.23757	−6.350	75.996	69.645	−0.01562	0.17890	−69
−68	8.9350	11.7292*	0.010583	4.0980	94.483	0.24402	−6.140	75.907	69.766	−0.01509	0.17871	−68
−67	9.1942	11.2015*	0.010595	3.9903	94.375	0.25060	−5.930	75.817	69.887	−0.01455	0.17853	−67
−66	9.4594	10.6616*	0.010608	3.8859	94.267	0.25733	−5.718	75.727	70.008	−0.01402	0.17834	−66
−65	9.7307	10.1092*	0.010620	3.7849	94.159	0.26420	−5.506	75.635	70.128	−0.01348	0.17816	−65
−64	10.008	9.544*	0.010632	3.6870	94.050	0.27121	−5.294	75.543	70.249	−0.01294	0.17798	−64
−63	10.292	8.966*	0.010644	3.5922	93.942	0.27838	−5.081	75.451	70.369	−0.01241	0.17780	−63
−62	10.582	8.375*	0.100657	3.5002	93.833	0.28569	−4.867	75.357	70.490	−0.01187	0.17762	−62
−61	10.879	7.771*	0.010669	3.4111	93.724	0.29315	−4.652	75.263	70.610	−0.01133	0.17745	−61
−60	11.182	7.153*	0.010682	3.3248	93.615	0.30076	−4.437	75.168	70.730	−0.01079	0.17728	−60
−59	11.492	6.522*	0.010694	3.2410	93.505	0.30853	−4.221	75.073	70.851	−0.01026	0.17711	−59
−58	11.809	5.877*	0.010707	3.1598	93.396	0.31646	−4.005	74.976	70.971	−0.00972	0.17694	−58
−57	12.133	5.218*	0.010719	3.0811	93.286	0.32455	−3.788	74.879	71.091	−0.00918	0.17677	−57
−56	12.463	4.544*	0.010732	3.0047	93.176	0.33281	−3.570	74.781	71.211	−0.00864	0.17661	−56
−55	12.801	3.856*	0.010744	2.9305	93.066	0.34122	−3.352	74.683	71.330	−0.00810	0.17644	−55
−54	13.147	3.153*	0.010757	2.8586	92.956	0.34981	−3.133	74.584	71.450	−0.00757	0.17628	−54
−53	13.499	2.435*	0.010770	2.7888	92.846	0.35856	−2.913	74.484	71.570	−0.00703	0.17612	−53
−52	13.860	1.702*	0.010783	2.7211	92.735	0.36749	−2.693	74.383	71.689	−0.00649	0.17597	−52
−51	14.227	0.954*	0.010796	2.6553	92.624	0.37659	−2.472	74.281	71.809	−0.00595	0.17581	−51
−50	14.602	0.190*	0.010809	2.5915	92.513	0.38587	−2.251	74.179	71.928	−0.00541	0.17566	−50
−49	14.985	0.289	0.010822	2.5295	92.402	0.39532	−2.028	74.076	72.047	−0.00487	0.17551	−49
−48	15.376	0.680	0.010835	2.4693	92.291	0.40496	−1.805	73.972	72.166	−0.00433	0.17536	−48
−47	15.775	1.079	0.010848	2.4108	92.179	0.41478	−1.582	73.868	72.285	−0.00379	0.17521	−47
−46	16.182	1.486	0.010861	2.3540	92.068	0.42479	−1.358	73.762	72.404	−0.00325	0.17506	−46
−45	16.597	1.901	0.010874	2.2988	91.956	0.43499	−1.133	73.656	72.523	−0.00271	0.17492	−45
−44	17.021	2.325	0.010888	2.2452	91.844	0.44538	−0.908	73.549	72.641	−0.00217	0.17477	−44
−43	17.453	2.757	0.010901	2.1931	91.731	0.45597	−0.682	73.442	72.760	−0.00162	0.17463	−43
−42	17.894	3.198	0.010914	2.1424	91.619	0.46675	−0.455	73.333	72.878	−0.00108	0.17449	−42
−41	18.344	3.648	0.010928	2.0932	91.506	0.47773	−0.227	73.224	72.996	−0.00054	0.17435	−41

*Inches of mercury below one atmosphere

TEMP. °F	PRESSURE		VOLUME cu ft/lb		DENSITY lb/cu ft		ENTHALPY Btu/lb			ENTROPY Btu/(lb)(°R)		TEMP. °F
	PSIA	PSIG	LIQUID	VAPOR	LIQUID	VAPOR	LIQUID	LATENT	VAPOR	LIQUID	VAPOR	
−40	18.802	4.106	0.010941	2.0453	91.393	0.48891	0.000	73.114	73.114	0.00000	0.17421	−40
−39	19.269	4.573	0.010955	1.9987	91.280	0.50030	0.228	73.003	73.232	0.00054	0.17408	−39
−38	19.746	5.050	0.010968	1.9535	91.167	0.51189	0.457	72.892	73.350	0.00108	0.17395	−38
−37	20.231	5.535	0.010982	1.9094	91.053	0.52370	0.687	72.780	73.468	0.00163	0.17381	−37
−36	20.726	6.030	0.010996	1.8666	90.939	0.53571	0.918	72.667	73.585	0.00217	0.17368	−36
−35	21.230	6.534	0.011010	1.8249	90.825	0.54795	1.149	72.553	73.702	0.00271	0.17355	−35
−34	21.744	7.048	0.011023	1.7844	90.711	0.56040	1.381	72.438	73.819	0.00325	0.17342	−34
−33	22.267	7.571	0.011037	1.7449	90.597	0.57307	1.613	72.323	73.936	0.00380	0.17330	−33
−32	22.800	8.104	0.011051	1.7065	90.482	0.58597	1.846	72.206	74.053	0.00434	0.17317	−32
−31	23.343	8.647	0.011065	1.6691	90.367	0.59909	2.080	72.089	74.170	0.00488	0.17305	−31
−30	23.897	9.201	0.011080	1.6328	90.252	0.61244	2.315	71.971	74.286	0.00543	0.17293	−30
−29	24.460	9.764	0.011094	1.5973	90.136	0.62602	2.550	71.853	74.403	0.00597	0.17281	−29
−28	25.033	10.337	0.011108	1.5628	90.021	0.63984	2.785	71.733	74.519	0.00651	0.17269	−28
−27	25.617	10.921	0.011122	1.5292	89.905	0.65390	3.021	71.613	74.635	0.00706	0.17257	−27
−26	26.212	11.516	0.011137	1.4965	89.789	0.66819	3.258	71.492	74.751	0.00760	0.17245	−26
−25	26.817	12.121	0.011151	1.4646	89.673	0.68273	3.496	71.370	74.866	0.00815	0.17234	−25
−24	27.433	12.737	0.011166	1.4336	89.556	0.69752	3.734	71.247	74.982	0.00869	0.17222	−24
−23	28.059	13.363	0.011180	1.4033	89.439	0.71256	3.973	71.124	75.097	0.00923	0.17211	−23
−22	28.697	14.001	0.011195	1.3739	89.322	0.72785	4.212	71.000	75.212	0.00978	0.17200	−22
−21	29.346	14.650	0.011210	1.3451	89.205	0.74340	4.452	70.875	75.327	0.01032	0.17189	−21
−20	30.006	15.310	0.011224	1.3171	89.088	0.75920	4.693	70.749	75.442	0.01087	0.17178	−20
−19	30.678	15.982	0.011239	1.2898	88.970	0.77527	4.934	70.622	75.556	0.01141	0.17167	−19
−18	31.361	16.665	0.011254	1.2632	88.852	0.79160	5.176	70.494	75.671	0.01196	0.17156	−18
−17	32.056	17.360	0.011269	1.2373	88.734	0.80820	5.418	70.366	75.785	0.01250	0.17146	−17
−16	32.762	18.066	0.011284	1.2120	88.615	0.82507	5.661	70.237	75.899	0.01305	0.17135	−16
−15	33.480	18.784	0.011299	1.1873	88.496	0.84222	5.905	70.107	76.012	0.01359	0.17125	−15
−14	34.211	19.515	0.011315	1.1632	88.377	0.85964	6.149	69.976	76.126	0.01414	0.17115	−14
−13	34.954	20.258	0.011330	1.1397	88.258	0.87734	6.394	69.844	76.239	0.01469	0.17105	−13
−12	35.709	21.013	0.011345	1.1168	88.138	0.89533	6.640	69.712	76.352	0.01523	0.17095	−12
−11	36.476	21.780	0.011361	1.0945	88.018	0.91361	6.886	69.578	76.465	0.01578	0.17085	−11
−10	37.256	22.560	0.011376	1.0727	87.898	0.93218	7.133	69.444	76.577	0.01632	0.17075	−10
− 9	38.049	23.353	0.011392	1.0514	87.778	0.95104	7.380	69.309	76.690	0.01687	0.17066	− 9
− 8	38.854	24.158	0.011408	1.0307	87.657	0.97020	7.628	69.173	76.802	0.01741	0.17056	− 8
− 7	39.673	24.977	0.011423	1.0104	87.536	0.98966	7.877	69.037	76.914	0.01796	0.17047	− 7
− 6	40.504	25.808	0.011439	0.99065	87.415	1.0094	8.126	68.899	77.025	0.01851	0.17037	− 6
− 5	41.349	26.653	0.011455	0.97134	87.293	1.0295	8.376	68.761	77.137	0.01905	0.17028	− 5
− 4	42.207	27.511	0.011471	0.95247	87.172	1.0498	8.626	68.622	77.248	0.01960	0.17019	− 4
− 3	43.078	28.382	0.011487	0.93406	87.050	1.0705	8.877	68.482	77.359	0.02014	0.17010	− 3
− 2	43.964	29.268	0.011503	0.91607	86.927	1.0916	9.129	68.341	77.470	0.02069	0.17001	− 2
− 1	44.863	30.167	0.011520	0.89851	86.804	1.1129	9.381	68.199	77.580	0.02124	0.16992	− 1
0	45.775	31.079	0.011536	0.88135	86.681	1.1346	9.633	68.056	77.690	0.02178	0.16983	0
1	46.702	32.006	0.011552	0.86458	86.558	1.1566	9.887	67.913	77.800	0.02233	0.16975	1
2	47.643	32.947	0.011569	0.84821	86.434	1.1789	10.141	67.769	77.910	0.02287	0.16966	2
3	48.599	33.903	0.011586	0.83220	86.310	1.2016	10.395	67.624	78.019	0.02342	0.16958	3
4	49.568	34.872	0.011602	0.81657	86.186	1.2246	10.650	67.478	78.128	0.02397	0.16949	4
5	50.553	35.857	0.011619	0.80129	86.062	1.2479	10.906	67.331	78.237	0.02451	0.16941	5
6	51.552	36.856	0.011636	0.78635	85.937	1.2716	11.162	67.183	78.346	0.02506	0.16933	6
7	52.565	37.869	0.011653	0.77175	85.811	1.2957	11.419	67.035	78.454	0.02561	0.16925	7
8	53.594	38.898	0.011670	0.75748	85.686	1.3201	11.676	66.885	78.562	0.02615	0.16917	8
9	54.638	39.942	0.011687	0.74353	85.560	1.3449	11.934	66.735	78.670	0.02670	0.16909	9
10	55.697	41.001	0.011704	0.72988	85.434	1.3700	12.193	66.584	78.777	0.02724	0.16901	10
11	56.771	42.075	0.011722	0.71654	85.307	1.3955	12.452	66.432	78.885	0.02779	0.16893	11
12	57.861	43.165	0.011739	0.70349	85.180	1.4214	12.711	66.280	78.991	0.02834	0.16885	12
13	58.967	44.271	0.011757	0.69073	85.053	1.4477	12.972	66.126	79.098	0.02888	0.16878	13
14	60.088	45.392	0.011774	0.67825	84.925	1.4743	13.232	65.971	79.204	0.02943	0.16870	14

TEMP.	PRESSURE		VOLUME cu ft/lb		DENSITY lb/cu ft		ENTHALPY Btu/lb			ENTROPY Btu/(lb)(°R)		TEMP.
°F	PSIA	PSIG	LIQUID v_f	VAPOR v_g	LIQUID $1/v_f$	VAPOR $1/v_g$	LIQUID h_f	LATENT h_{fg}	VAPOR h_g	LIQUID s_f	VAPOR s_g	°F
15	61.225	46.529	0.011792	0.66604	84.797	1.5013	13.494	65.816	79.310	0.02997	0.16863	15
16	62.379	47.683	0.011810	0.65410	84.669	1.5288	13.756	65.660	79.416	0.03052	0.16855	16
17	63.548	48.852	0.011828	0.64241	84.540	1.5566	14.018	65.503	79.521	0.03107	0.16848	17
18	64.734	50.038	0.011846	0.63098	84.411	1.5848	14.281	65.345	79.626	0.03161	0.16841	18
19	65.936	51.240	0.011864	0.61979	84.282	1.6134	14.545	65.186	79.731	0.03216	0.16833	19
20	67.155	52.459	0.011883	0.60884	84.152	1.6424	14.809	65.026	79.836	0.03270	0.16826	20
21	68.391	53.695	0.011901	0.59812	84.022	1.6719	15.073	64.866	79.940	0.03325	0.16819	21
22	69.643	54.947	0.011920	0.58762	83.891	1.7017	15.339	64.704	80.043	0.03379	0.16812	22
23	70.913	56.217	0.011938	0.57735	83.760	1.7320	15.604	64.542	80.147	0.03434	0.16805	23
24	72.199	57.503	0.011957	0.56730	83.629	1.7627	15.871	64.379	80.250	0.03488	0.16799	24
25	73.503	58.807	0.011976	0.55745	83.497	1.7938	16.138	64.215	80.353	0.03543	0.16792	25
26	74.824	60.128	0.011995	0.54781	83.365	1.8254	16.405	64.050	80.455	0.03597	0.16785	26
27	76.163	61.467	0.012014	0.53837	83.232	1.8574	16.673	63.884	80.557	0.03652	0.16778	27
28	77.520	62.824	0.012033	0.52912	83.099	1.8898	16.941	63.717	80.659	0.03706	0.16772	28
29	78.894	64.198	0.012053	0.52007	82.966	1.9228	17.210	63.550	80.760	0.03761	0.16765	29
30	80.286	65.590	0.012072	0.51120	82.832	1.9561	17.480	63.381	80.861	0.03815	0.16759	30
31	81.697	67.001	0.012092	0.50251	82.698	1.9900	17.750	63.212	80.962	0.03870	0.16752	31
32	83.126	68.430	0.012111	0.49399	82.563	2.0242	18.020	63.042	81.062	0.03924	0.16746	32
33	84.573	69.877	0.012131	0.48565	82.428	2.0590	18.291	62.871	81.162	0.03979	0.16739	33
34	86.038	71.342	0.012151	0.47748	82.292	2.0943	18.563	62.698	81.262	0.04033	0.16733	34
35	87.522	72.826	0.012171	0.46947	82.156	2.1300	18.835	62.526	81.361	0.04087	0.16727	35
36	89.026	74.330	0.012192	0.46162	82.020	2.1662	19.107	62.352	81.460	0.04142	0.16721	36
37	90.548	75.852	0.012212	0.45393	81.883	2.2029	19.381	62.177	81.558	0.04196	0.16715	37
38	92.089	77.393	0.012233	0.44638	81.746	2.2401	19.654	62.001	81.656	0.04250	0.16708	38
39	93.649	78.953	0.012253	0.43899	81.608	2.2779	19.928	61.825	81.754	0.04305	0.16702	39
40	95.229	80.533	0.012274	0.43175	81.469	2.3161	20.203	61.647	81.851	0.04359	0.16696	40
41	96.828	82.132	0.012295	0.42464	81.330	2.3549	20.478	61.469	81.948	0.04413	0.16690	41
42	98.447	83.751	0.012316	0.41767	81.191	2.3941	20.754	61.290	82.044	0.04468	0.16684	42
43	100.08	85.38	0.012337	0.41084	81.051	2.4339	21.030	61.110	82.140	0.04522	0.16678	43
44	101.74	87.04	0.012359	0.40414	80.911	2.4743	21.307	60.928	82.235	0.04576	0.16673	44
45	103.42	88.72	0.012380	0.39757	80.770	2.5152	21.584	60.746	82.331	0.04630	0.16667	45
46	105.12	90.42	0.012402	0.39113	80.629	2.5566	21.861	60.563	82.425	0.04685	0.16661	46
47	106.84	92.14	0.012424	0.38481	80.487	2.5986	22.140	60.380	82.520	0.04739	0.16655	47
48	108.58	93.88	0.012446	0.37861	80.345	2.6412	22.418	60.195	82.613	0.04793	0.16649	48
49	110.34	95.64	0.012468	0.37252	80.202	2.6843	22.697	60.009	82.707	0.04847	0.16644	49
50	112.12	97.42	0.012490	0.36655	80.058	2.7280	22.977	59.822	82.800	0.04901	0.16638	50
51	113.92	99.22	0.012513	0.36070	79.914	2.7723	23.257	59.635	82.892	0.04955	0.16632	51
52	115.74	101.05	0.012536	0.35495	79.769	2.8172	23.538	59.446	82.984	0.05009	0.16627	52
53	117.59	102.89	0.012558	0.34931	79.624	2.8627	23.819	59.257	83.076	0.05063	0.16621	53
54	119.45	104.75	0.012581	0.34377	79.479	2.9088	24.100	59.066	83.167	0.05117	0.16616	54
55	121.34	106.64	0.012605	0.33834	79.332	2.9555	24.382	58.875	83.257	0.05171	0.16610	55
56	123.25	108.55	0.012628	0.33301	79.185	3.0028	24.665	58.682	83.347	0.05225	0.16605	56
57	125.18	110.48	0.012652	0.32777	79.038	3.0508	24.948	58.489	83.437	0.05279	0.16599	57
58	127.13	112.43	0.012675	0.32263	[illegible]	[illegible]	[illegible]	[illegible]	[illegible]	[illegible]	0.16594	58
59	129.10	114.41	0.012699	0.31759	78.741	3.1486	25.515	58.099	83.615	0.05387	0.16588	59
60	131.10	116.40	0.012723	0.31263	78.592	3.1985	25.799	57.903	83.703	0.05441	0.16583	60
61	133.12	118.42	0.012748	0.30777	78.441	3.2491	26.084	57.705	83.790	0.05495	0.16577	61
62	135.16	120.46	0.012772	0.30299	78.291	3.3004	26.370	57.507	83.877	0.05549	0.16572	62
63	137.22	122.53	0.012797	0.29830	78.140	3.3523	26.656	57.308	83.964	0.05602	0.16566	63
64	139.31	124.61	0.012822	0.29369	77.988	3.4049	26.942	57.107	84.050	0.05656	0.16561	64
65	141.42	126.72	0.012847	0.28916	77.835	3.4582	27.229	56.906	84.135	0.05710	0.16556	65
66	143.55	128.86	0.012872	0.28471	77.682	3.5122	27.516	56.704	84.220	0.05764	0.16550	66
67	145.71	131.01	0.012898	0.28034	77.528	3.5669	27.804	56.500	84.304	0.05817	0.16545	67
68	147.89	133.19	0.012924	0.27605	77.373	3.6224	28.092	56.296	84.388	0.05871	0.16539	68
69	150.09	135.39	0.012950	0.27183	77.217	3.6786	28.380	56.091	84.471	0.05925	0.16534	69

TEMP. °F	PRESSURE		VOLUME cu ft/lb		DENSITY lb/cu ft		ENTHALPY Btu/lb			ENTROPY Btu/(lb)(°R)		TEMP. °F
	PSIA	PSIG	LIQUID	VAPOR	LIQUID	VAPOR	LIQUID	LATENT	VAPOR	LIQUID	VAPOR	
70	152.32	137.62	0.012976	0.26769	77.061	3.7356	28.669	55.884	84.554	0.05978	0.16529	70
71	154.57	139.87	0.013003	0.26362	76.904	3.7933	28.959	55.676	84.636	0.06032	0.16523	71
72	156.84	142.15	0.013029	0.25961	76.747	3.8517	29.249	55.468	84.717	0.06085	0.16518	72
73	159.14	144.44	0.013056	0.25568	76.588	3.9110	29.539	55.258	84.798	0.06139	0.16512	73
74	161.46	146.77	0.013083	0.25181	76.429	3.9711	29.830	55.047	84.878	0.06193	0.16507	74
75	163.81	149.11	0.013111	0.24801	76.269	4.0319	30.122	54.835	84.958	0.06246	0.16502	75
76	166.18	151.49	0.013139	0.24428	76.108	4.0936	30.414	54.622	85.037	0.06299	0.16496	76
77	168.58	153.88	0.013167	0.24060	75.947	4.1561	30.706	54.408	85.115	0.06353	0.16491	77
78	171.00	156.30	0.013195	0.23699	75.784	4.2194	30.999	54.193	85.193	0.06406	0.16485	78
79	173.45	158.75	0.013223	0.23344	75.621	4.2836	31.292	53.977	85.270	0.06460	0.16480	79
80	175.92	161.22	0.013252	0.22995	75.457	4.3487	31.586	53.759	85.346	0.06513	0.16474	80
81	178.41	163.72	0.013281	0.22651	75.292	4.4146	31.880	53.541	85.421	0.06566	0.16469	81
82	180.94	166.24	0.013310	0.22313	75.126	4.4815	32.175	53.321	85.496	0.06620	0.16463	82
83	183.48	168.79	0.013340	0.21981	74.959	4.5492	32.470	53.100	85.570	0.06673	0.16458	83
84	186.06	171.36	0.013370	0.21654	74.791	4.6179	32.766	52.878	85.644	0.06726	0.16452	84
85	188.66	173.96	0.013400	0.21333	74.622	4.6874	33.062	52.654	85.717	0.06780	0.16446	85
86	191.28	176.59	0.013431	0.21017	74.453	4.7580	33.359	52.430	85.789	0.06833	0.16441	86
87	193.94	179.24	0.013462	0.20705	74.282	4.8295	33.656	52.204	85.860	0.06886	0.16435	87
88	196.62	181.92	0.013493	0.20399	74.111	4.9020	33.953	51.977	85.930	0.06939	0.16429	88
89	199.32	184.62	0.013524	0.20098	73.938	4.9755	34.251	51.748	86.000	0.06992	0.16423	89
90	202.05	187.36	0.013556	0.19801	73.764	5.0500	34.550	51.519	86.069	0.07045	0.16418	90
91	204.81	190.12	0.013588	0.19509	73.590	5.1255	34.849	51.288	86.137	0.07098	0.16412	91
92	207.60	192.90	0.013621	0.19222	73.414	5.2021	35.148	51.055	86.204	0.07151	0.16406	92
93	210.41	195.72	0.013654	0.18939	73.237	5.2798	35.448	50.822	86.271	0.07205	0.16400	93
94	213.25	198.56	0.013687	0.18661	73.059	5.3585	35.749	50.587	86.336	0.07258	0.16394	94
95	216.12	201.43	0.013721	0.18387	72.880	5.4384	36.050	50.350	86.401	0.07311	0.16388	95
96	219.02	204.32	0.013755	0.18117	72.700	5.5194	36.352	50.113	86.465	0.07364	0.16382	96
97	221.94	207.25	0.013789	0.17852	72.519	5.6015	36.654	49.874	86.528	0.07417	0.16375	97
98	224.90	210.20	0.013824	0.17590	72.336	5.6848	36.956	49.633	86.590	0.07470	0.16369	98
99	227.88	213.18	0.013859	0.17333	72.152	5.7693	37.259	49.391	86.651	0.07522	0.16363	99
100	230.89	216.19	0.013895	0.17079	71.967	5.8550	37.563	49.147	86.711	0.07575	0.16356	100
101	233.93	219.23	0.013931	0.16829	71.781	5.9419	37.867	48.902	86.770	0.07628	0.16350	101
102	237.00	222.30	0.013967	0.16583	71.593	6.0301	38.172	48.656	86.828	0.07681	0.16343	102
103	240.09	225.40	0.014004	0.16340	71.405	6.1196	38.477	48.407	86.885	0.07734	0.16337	103
104	243.22	228.52	0.014042	0.16101	71.214	6.2104	38.783	48.158	86.941	0.07787	0.16330	104
105	246.38	231.68	0.014079	0.15866	71.023	6.3026	39.090	47.906	86.997	0.07840	0.16323	105
106	249.56	234.87	0.014118	0.15634	70.829	6.3961	39.398	47.653	87.051	0.07893	0.16316	106
107	252.78	238.08	0.014157	0.15405	70.635	6.4910	39.705	47.398	87.103	0.07946	0.16309	107
108	256.02	241.33	0.014196	0.15180	70.439	6.5873	40.013	47.142	87.155	0.07998	0.16302	108
109	259.30	244.60	0.014236	0.14958	70.241	6.6851	40.322	46.883	87.206	0.08051	0.16295	109
110	262.61	247.91	0.014277	0.14739	70.042	6.7843	40.631	46.623	87.255	0.08104	0.16288	110
111	265.94	251.25	0.014318	0.14523	69.841	6.8851	40.942	46.361	87.303	0.08157	0.16281	111
112	269.31	254.62	0.014359	0.14311	69.639	6.9875	41.252	46.098	87.350	0.08210	0.16273	112
113	272.71	258.02	0.014401	0.14101	69.435	7.0914	41.564	45.832	87.396	0.08263	0.16265	113
114	276.15	261.45	0.014444	0.13894	69.229	7.1970	41.876	45.564	87.441	0.08316	0.16258	114
115	279.61	264.91	0.014488	0.13690	69.022	7.3042	42.189	45.294	87.484	0.08368	0.16250	115
116	283.10	268.41	0.014532	0.13489	68.812	7.4132	42.503	45.022	87.525	0.08421	0.16242	116
117	286.63	271.94	0.014576	0.13290	68.601	7.5239	42.817	44.748	87.566	0.08474	0.16234	117
118	290.19	275.50	0.014622	0.13095	68.388	7.6364	43.132	44.472	87.605	0.08527	0.16225	118
119	293.78	279.09	0.014668	0.12902	68.173	7.7507	43.448	44.194	87.642	0.08580	0.16217	119
120	297.41	282.71	0.014715	0.12711	67.956	7.8669	43.765	43.913	87.678	0.08633	0.16208	120
121	301.07	286.37	0.014762	0.12523	67.737	7.9850	44.082	43.630	87.713	0.08686	0.16199	121
122	304.76	290.06	0.014811	0.12337	67.515	8.1052	44.401	43.344	87.746	0.08739	0.16190	122
123	308.49	293.79	0.014860	0.12154	67.292	8.2274	44.720	43.056	87.777	0.08792	0.16181	123
124	312.25	297.55	0.014910	0.11973	67.066	8.3516	45.040	42.766	87.807	0.08845	0.16172	124

TEMP. °F	PRESSURE		VOLUME cu ft/lb		DENSITY lb/cu ft		ENTHALPY Btu/lb			ENTROPY Btu/(lb)(°R)		TEMP. °F
	PSIA	PSIG	LIQUID	VAPOR	LIQUID	VAPOR	LIQUID	LATENT	VAPOR	LIQUID	VAPOR	
125	316.04	301.35	0.014961	0.11795	66.838	8.4781	45.361	42.472	87.834	0.08899	0.16163	125
126	319.87	305.17	0.015013	0.11618	66.608	8.6067	45.684	42.176	87.860	0.08952	0.16153	126
127	323.73	309.04	0.015065	0.11444	66.375	8.7377	46.007	41.878	87.885	0.09005	0.16143	127
128	327.63	312.94	0.015119	0.11272	66.140	8.8710	46.331	41.576	87.907	0.09058	0.16133	128
129	331.57	316.87	0.015174	0.11102	65.901	9.0067	46.656	41.271	87.928	0.09112	0.16122	129
130	335.54	320.84	0.015229	0.10934	65.661	9.1449	46.983	40.963	87.946	0.09165	0.16112	130
131	339.55	324.85	0.015286	0.10769	65.417	9.2858	47.310	40.652	87.962	0.09219	0.16101	131
132	343.59	328.89	0.015344	0.10605	65.171	9.4293	47.639	40.337	87.977	0.09273	0.16090	132
133	347.67	332.97	0.015403	0.10443	64.921	9.5755	47.969	40.019	87.989	0.09326	0.16078	133
134	351.79	337.09	0.015463	0.10283	64.669	9.7246	48.300	39.697	87.998	0.09380	0.16067	134
135	355.94	341.24	0.015524	0.10124	64.413	9.8767	48.633	39.372	88.006	0.09434	0.16055	135
136	360.13	345.44	0.015587	0.099682	64.154	10.031	48.967	39.043	88.011	0.09488	0.16042	136
137	364.36	349.67	0.015651	0.098133	63.892	10.190	49.303	38.709	88.013	0.09543	0.16030	137
138	368.63	353.94	0.015716	0.096601	63.626	10.351	49.641	38.372	88.013	0.09597	0.16017	138
139	372.94	358.25	0.015783	0.095086	63.356	10.516	49.980	38.030	88.010	0.09652	0.16004	139
140	377.29	362.60	0.015852	0.093586	63.082	10.685	50.320	37.683	88.004	0.09706	0.15990	140
141	381.68	366.98	0.015922	0.092101	62.805	10.857	50.663	37.331	87.995	0.09761	0.15976	141
142	386.11	371.41	0.015993	0.090630	62.523	11.033	51.008	36.975	87.983	0.09817	0.15962	142
143	390.58	375.88	0.016067	0.089174	62.237	11.213	51.354	36.613	87.968	0.09877	0.15947	143
144	395.09	380.39	0.016142	0.087732	61.946	11.398	51.703	36.245	87.949	0.09928	0.15931	144
145	399.64	384.95	0.016220	0.086303	61.650	11.587	52.054	35.872	87.927	0.09983	0.15916	145
146	404.24	389.54	0.016299	0.084886	61.350	11.780	52.408	35.493	87.901	0.10039	0.15899	146
147	408.88	394.18	0.016381	0.083481	61.044	11.978	52.764	35.107	87.871	0.10096	0.15882	147
148	413.56	398.86	0.016465	0.082088	60.732	12.181	53.123	34.714	87.838	0.10152	0.15865	148
149	418.29	403.59	0.016551	0.080706	60.415	12.390	53.485	34.314	87.800	0.10209	0.15847	149
150	423.06	408.35	0.016641	0.079335	60.092	12.604	53.850	33.907	87.757	0.10267	0.15828	150
151	427.87	413.18	0.016732	0.077973	59.762	12.824	54.218	33.491	87.710	0.10325	0.15809	151
152	432.74	418.04	0.016827	0.076620	59.425	13.051	54.590	33.067	87.657	0.10383	0.15789	152
153	437.65	422.95	0.016925	0.075276	59.082	13.284	54.966	32.633	87.599	0.10442	0.15769	153
154	442.60	427.91	0.017026	0.073939	58.730	13.524	55.346	32.190	87.536	0.10501	0.15747	154
155	447.61	432.91	0.017131	0.072610	58.371	13.772	55.730	31.736	87.467	0.10561	0.15725	155
156	452.66	437.97	0.017240	0.071286	58.002	14.027	56.119	31.271	87.391	0.10622	0.15701	156
157	457.77	443.07	0.017353	0.069968	57.625	14.292	56.513	30.794	87.308	0.10683	0.15677	157
158	462.92	448.22	0.017470	0.068655	57.237	14.565	56.913	30.304	87.218	0.10746	0.15652	158
159	468.13	453.43	0.017593	0.067345	56.839	14.848	57.320	29.800	87.120	0.10809	0.15625	159
160	473.38	458.69	0.017721	0.066037	56.429	15.142	57.732	29.281	87.013	0.10872	0.15598	160
161	478.70	464.00	0.017854	0.064730	56.007	15.448	58.153	28.744	86.898	0.10937	0.15569	161
162	484.06	469.37	0.017994	0.063423	55.571	15.767	58.581	28.190	86.772	0.11004	0.15538	162
163	489.48	474.79	0.018141	0.062114	55.121	16.099	59.019	27.616	86.635	0.11071	0.15506	163
164	494.96	480.27	0.018296	0.060802	54.654	16.446	59.466	27.019	86.486	0.11140	0.15472	164
165	500.50	485.80	0.018460	0.059484	54.169	16.811	59.925	26.398	86.324	0.11210	0.15436	165
166	506.10	491.40	0.018634	0.058158	53.665	17.194	60.397	25.749	86.146	0.11283	0.15398	166
167	511.75	497.06	0.018818	0.056822	53.138	17.598	60.883	25.069	85.953	0.11357	0.15358	167
168	517.47	502.78	0.019016	0.055472	52.586	18.026	61.385	24.355	85.740	0.11434	0.15314	168
169	523.26	508.56	0.019228	0.054104	52.005	18.482	61.906	23.599	85.506	0.11514	0.15268	169
170	529.11	514.41	0.019458	0.052714	51.392	18.970	62.449	22.798	85.248	0.11597	0.15217	170
171	535.03	520.33	0.019708	0.051295	50.739	19.495	63.019	21.941	84.960	0.11684	0.15163	171
172	541.01	526.32	0.019983	0.049839	50.041	20.064	63.619	21.018	84.638	0.11775	0.15103	172
173	547.07	532.38	0.020288	0.048335	49.288	20.688	64.258	20.016	84.274	0.11873	0.15037	173
174	553.20	538.51	0.020633	0.046769	48.465	21.381	64.944	18.913	83.858	0.11978	0.14962	174
175	559.41	544.72	0.021029	0.045118	47.552	22.163	65.693	17.681	83.374	0.12092	0.14878	175
176	565.70	551.01	0.021496	0.043348	46.518	23.069	66.525	16.273	82.798	0.12219	0.14779	176
177	572.08	557.38	0.022071	0.041396	45.307	24.156	67.480	14.607	82.087	0.12365	0.14659	177
178	578.54	563.84	0.022829	0.039137	43.803	25.551	68.635	12.515	81.151	0.12542	0.14505	178
179	585.09	570.39	0.023993	0.036203	41.677	27.621	70.210	9.518	79.729	0.12784	0.14275	179
179.889	591.00	576.30	0.028571	0.028571	35.000	35.000	74.654	0.000	74.654	0.13476	0.13476	179.889

From Published data of E.I. du Pont de Nemours & Co. Inc. Used by Permission.

TEMP °C	PRESSURE bar	VOLUME m³/kg · 10³		DENSITY kg/m³		ENTHALPY kcal kg			ENTROPY kcal (kg) (°K)	
		LIQUID v_f	VAPOR v_g	LIQUID $1/v_f$	VAPOR $1/v_g$	LIQUID h_f	LATENT h_{fg}	VAPOR h_g	LIQUID s_f	VAPOR s_g
-100	0.032	0.607	3974	1.644	0.00025	109.1	188.8	297.9	0.595	1.685
-99	0.035	0.609	3673	1.641	0.00027	109.8	188.6	298.4	0.599	1.682
-98	0.038	0.610	3394	1.639	0.00029	110.5	188.4	298.9	0.603	1.679
-97	0.041	0.611	3147	1.636	0.00032	111.2	188.2	299.4	0.607	1.675
-96	0.045	0.612	2917	1.633	0.00034	111.9	188.0	299.9	0.611	1.672
-95	0.048	0.613	2707	1.630	0.00037	112.6	187.8	300.4	0.615	1.669
-94	0.052	0.614	2514	1.627	0.00040	113.3	187.5	300.9	0.619	1.666
-93	0.057	0.615	2336	1.624	0.00043	114.0	187.3	301.4	0.623	1.663
-92	0.061	0.616	2174	1.621	0.00046	114.7	187.1	301.9	0.627	1.660
-91	0.066	0.617	2024	1.619	0.00049	115.4	186.9	302.4	0.631	1.657
-90	0.071	0.618	1886	1.616	0.00053	116.2	186.7	302.9	0.635	1.654
-89	0.077	0.619	1759	1.613	0.00057	116.9	186.4	303.4	0.639	1.651
-88	0.083	0.620	1642	1.610	0.00061	117.6	186.2	303.9	0.643	1.648
-87	0.089	0.622	1534	1.607	0.00065	118.4	186.0	304.4	0.646	1.646
-86	0.096	0.623	1435	1.604	0.00070	119.1	185.7	304.9	0.650	1.643
-85	0.103	0.624	1342	1.601	0.00074	119.8	185.5	305.4	0.654	1.640
-84	0.111	0.625	1257	1.598	0.00080	120.6	185.2	305.9	0.658	1.638
-83	0.119	0.626	1178	1.595	0.00085	121.4	185.0	306.4	0.662	1.635
-82	0.127	0.627	1105	1.593	0.00090	122.1	184.7	306.9	0.666	1.633
-81	0.136	0.628	1037	1.590	0.00096	122.9	184.5	307.4	0.670	1.631
-80	0.146	0.630	974.5	1.587	0.00103	123.6	184.2	307.9	0.674	1.628
-79	0.156	0.631	915.9	1.584	0.00109	124.4	184.0	308.4	0.678	1.626
-78	0.166	0.632	861.5	1.581	0.00116	125.2	183.7	308.9	0.682	1.624
-77	0.178	0.633	810.9	1.578	0.00123	126.0	183.4	309.4	0.686	1.622
-76	0.190	0.634	763.7	1.575	0.00131	126.7	183.1	309.9	0.690	1.619
-75	0.202	0.635	719.8	1.572	0.00139	127.5	182.9	310.4	0.694	1.617
-74	0.215	0.637	678.8	1.569	0.00147	128.3	182.6	310.9	0.698	1.615
-73	0.229	0.638	640.6	1.566	0.00156	129.1	182.3	311.5	0.702	1.613
-72	0.244	0.639	604.9	1.563	0.00165	129.9	182.0	312.0	0.706	1.611
-71	0.259	0.640	571.6	1.560	0.00175	130.7	181.7	312.5	0.710	1.609
-70	0.275	0.642	540.4	1.557	0.00185	131.5	181.4	313.0	0.714	1.607
-69	0.292	0.643	511.2	1.554	0.00196	132.3	181.1	313.5	0.718	1.605
-68	0.310	0.644	483.9	1.551	0.00207	133.2	180.8	314.0	0.722	1.603
-67	0.329	0.645	458.3	1.548	0.00218	134.0	180.5	314.5	0.726	1.602
-66	0.348	0.647	434.4	1.545	0.00230	134.8	180.1	315.0	0.730	1.600
-65	0.369	0.648	411.9	1.542	0.00243	135.6	179.8	315.5	0.734	1.598
-64	0.390	0.649	390.7	1.539	0.00256	136.5	179.5	316.0	0.738	1.596
-63	0.413	0.650	370.9	1.536	0.00270	137.3	179.2	316.5	0.742	1.595
-62	0.436	0.652	352.3	1.533	0.00284	138.2	178.8	317.0	0.746	1.593
-61	0.461	0.653	334.7	1.530	0.00299	139.0	178.5	317.6	0.750	1.592
-60	0.487	0.654	318.2	1.527	0.00314	139.9	178.1	318.1	0.754	1.590
-59	0.514	0.656	302.7	1.524	0.00330	140.7	177.8	318.6	0.758	1.588
-58	0.542	0.657	288.1	1.521	0.00347	141.6	177.4	319.1	0.762	1.587
-57	0.571	0.658	274.3	1.517	0.00364	142.5	177.0	319.6	0.766	1.586
-56	0.602	0.660	261.3	1.514	0.00383	143.4	176.7	320.1	0.770	1.584
-55	0.633	0.661	249.1	1.511	0.00401	144.2	176.3	320.6	0.774	1.583
-54	0.667	0.662	237.5	1.508	0.00421	145.1	175.9	321.1	0.778	1.581
-53	0.701	0.664	226.6	1.505	0.00441	146.0	175.5	321.6	0.782	1.580
-52	0.737	0.665	216.2	1.502	0.00462	146.9	175.1	322.1	0.786	1.579
-51	0.775	0.667	206.5	1.499	0.00484	147.8	174.7	322.6	0.790	1.577
-50	0.814	0.668	197.2	1.496	0.00507	148.7	174.3	323.1	0.795	1.576
-49	0.854	0.669	188.5	1.492	0.00530	149.6	173.9	323.6	0.799	1.575
-48	0.896	0.671	180.2	1.489	0.00555	150.5	173.5	324.1	0.803	1.574
-47	0.940	0.672	172.3	1.486	0.00580	151.5	173.1	324.6	0.807	1.572
-46	0.986	0.674	164.9	1.483	0.00606	152.4	172.7	325.1	0.811	1.571
-45	1.033	0.675	157.9	1.480	0.00633	153.3	172.3	325.6	0.815	1.570
-44	1.082	0.677	151.2	1.476	0.00661	154.3	171.8	326.1	0.819	1.569
-43	1.132	0.678	144.8	1.473	0.00690	155.2	171.4	326.6	0.823	1.568
-42	1.185	0.680	138.8	1.470	0.00720	156.1	170.9	327.1	0.827	1.567
-41	1.239	0.681	133.1	1.467	0.00751	157.1	170.5	327.6	0.831	1.566
-40	1.296	0.683	127.6	1.463	0.00783	158.0	170.0	328.1	0.835	1.565
-39	1.354	0.684	122.5	1.460	0.00816	159.0	169.5	328.6	0.839	1.564
-38	1.415	0.686	117.5	1.457	0.00850	160.0	169.1	329.1	0.843	1.563
-37	1.477	0.687	112.9	1.454	0.00886	160.9	168.6	329.6	0.847	1.562
-36	1.542	0.689	108.4	1.450	0.00922	161.9	168.1	330.1	0.852	1.561
-35	1.609	0.690	104.2	1.447	0.00960	162.9	167.6	330.6	0.856	1.560
-34	1.678	0.692	100.1	1.444	0.00998	163.9	167.1	331.0	0.860	1.559
-33	1.750	0.694	96.30	1.440	0.01038	164.8	166.6	331.5	0.864	1.558
-32	1.823	0.695	92.62	1.437	0.01080	165.8	166.1	332.0	0.868	1.557
-31	1.900	0.697	89.11	1.434	0.01122	166.8	165.6	332.5	0.872	1.556

TEMP. °C	PRESSURE bar	VOLUME $m^3/kg \cdot 10^3$		DENSITY kg/m^3		ENTHALPY kcal kg			ENTROPY kcal (kg) (°K)	
		LIQUID v_f	VAPOR v_g	LIQUID $1/v_f$	VAPOR $1/v_g$	LIQUID h_f	LATENT h_{fg}	VAPOR h_g	LIQUID s_f	VAPOR s_g
-30	1.978	0.698	85.76	1.430	0.0116	167.8	165.1	333.0	0.876	1.555
-29	2.059	0.700	82.57	1.427	0.0121	168.8	164.6	333.5	0.880	1.555
-28	2.143	0.702	79.52	1.424	0.0125	169.8	164.0	333.9	0.884	1.554
-27	2.229	0.703	76.60	1.420	0.0130	170.9	163.5	334.4	0.888	1.553
-26	2.318	0.705	73.81	1.417	0.0135	171.9	163.0	334.9	0.893	1.552
-25	2.410	0.707	71.15	1.413	0.0140	172.9	162.4	335.4	0.897	1.551
-24	2.504	0.709	68.60	1.410	0.0145	173.9	161.9	335.8	0.901	1.551
-23	2.601	0.710	66.16	1.406	0.0151	175.0	161.3	336.3	0.905	1.550
-22	2.701	0.712	63.83	1.403	0.0156	176.0	160.7	336.8	0.909	1.549
-21	2.804	0.714	61.60	1.399	0.0162	177.1	160.1	337.2	0.913	1.548
-20	2.910	0.716	59.46	1.395	0.0168	178.1	159.6	337.7	0.917	1.548
-19	3.018	0.717	57.41	1.392	0.0174	179.2	159.0	338.2	0.921	1.547
-18	3.130	0.719	55.44	1.389	0.0180	180.2	158.4	338.6	0.925	1.546
-17	3.245	0.721	53.56	1.385	0.0186	181.3	157.8	339.1	0.930	1.546
-16	3.364	0.723	51.75	1.382	0.0193	182.3	157.2	339.6	0.934	1.545
-15	3.485	0.725	50.02	1.378	0.0199	183.4	156.6	340.0	0.938	1.545
-14	3.610	0.727	48.35	1.374	0.0206	184.5	155.9	340.5	0.942	1.544
-13	3.738	0.729	46.76	1.371	0.0213	185.6	155.3	340.9	0.946	1.543
-12	3.869	0.731	45.23	1.367	0.0221	186.6	154.7	341.4	0.950	1.543
-11	4.004	0.733	43.75	1.364	0.0228	187.7	154.0	341.8	0.954	1.542
-10	4.143	0.735	42.34	1.360	0.0236	188.8	153.4	342.3	0.958	1.542
-9	4.284	0.737	40.99	1.356	0.0244	189.9	152.7	342.7	0.963	1.541
-8	4.430	0.739	39.67	1.352	0.0252	191.0	152.1	343.1	0.967	1.540
-7	4.579	0.741	38.41	1.349	0.0260	192.1	151.4	343.6	0.971	1.540
-6	4.732	0.743	37.20	1.345	0.0268	193.2	150.8	344.0	0.975	1.539
-5	4.889	0.745	36.04	1.341	0.0277	194.3	150.1	344.5	0.979	1.539
-4	5.049	0.747	34.92	1.337	0.0286	195.4	149.4	344.9	0.983	1.538
-3	5.214	0.749	33.84	1.334	0.0295	196.6	148.7	345.3	0.987	1.538
-2	5.382	0.751	32.80	1.330	0.0304	197.7	148.0	345.7	0.991	1.537
-1	5.554	0.753	31.80	1.326	0.0314	198.8	147.3	346.2	0.995	1.537
0	5.731	0.756	30.83	1.322	0.0324	200.0	146.6	346.6	0****	1.536
1	5.911	0.758	29.90	1.318	0.0334	201.1	145.9	347.0	1.004	1.536
2	6.096	0.760	29.01	1.314	0.0344	202.2	145.1	347.4	1.008	1.535
3	6.285	0.762	28.14	1.310	0.0355	203.4	144.4	347.8	1.012	1.535
4	6.478	0.765	27.31	1.306	0.0366	204.5	143.7	348.2	1.016	1.534
5	6.676	0.767	26.50	1.302	0.0377	205.7	142.9	348.6	1.020	1.534
6	6.878	0.769	25.73	1.298	0.0388	206.8	142.2	349.0	1.024	1.534
7	7.084	0.772	24.98	1.294	0.0400	208.0	141.4	349.4	1.028	1.533
8	7.295	0.774	24.25	1.290	0.0412	209.1	140.6	349.8	1.032	1.533
9	7.510	0.777	23.55	1.286	0.0424	210.3	139.9	350.2	1.036	1.532
10	7.730	0.779	22.88	1.282	0.0437	211.5	139.1	350.6	1.040	1.532
11	7.955	0.782	22.23	1.278	0.0449	212.7	138.3	351.0	1.044	1.531
12	8.184	0.784	21.59	1.274	0.0463	213.8	137.5	351.4	1.049	1.531
13	8.418	0.787	20.98	1.269	0.0476	215.0	136.7	351.8	1.053	1.531
14	8.657	0.790	20.39	1.265	0.0490	216.2	135.9	352.1	1.057	1.530
15	8.90	0.792	19.82	1.261	0.0504	217.4	135.1	352.5	1.061	1.530
16	9.15	0.795	19.27	1.257	0.0518	218.6	134.3	352.9	1.065	1.529
17	9.40	0.798	18.73	1.252	0.0532	219.8	133.4	353.3	1.069	1.529
18	9.66	0.801	18.22	1.248	0.0548	221.0	132.6	353.6	1.073	1.529
19	9.92	0.803	17.71	1.243	0.0564	222.2	131.7	354.0	1.077	1.528
20	10.1	0.806	17.23	1.239	0.0580	223.4	130.9	354.3	1.081	1.528
21	10.4	0.809	16.76	1.234	0.0596	224.6	130.0	354.7	1.085	1.527
22	10.7	0.812	16.30	1.230	0.0613	225.8	129.2	355.0	1.089	1.527
23	11.0	0.815	[illegible]	1.225	0.0630	227.0	128.3	355.3	1.093	1.526
24	11.3	0.818	15.43	1.221	0.0647	228.2	127.4	355.7	1.097	1.526
25	11.6	0.822	15.02	1.216	0.0665	229.5	126.5	356.0	1.101	1.526
26	11.9	0.825	14.61	1.211	0.0684	230.7	125.6	356.3	1.105	1.525
27	12.2	0.828	14.22	1.207	0.0702	231.9	124.7	356.7	1.109	1.525
28	12.5	0.831	13.84	1.202	0.0722	233.1	123.8	357.0	1.113	1.524
29	12.8	0.835	13.47	1.197	0.0741	234.4	122.8	357.3	1.117	1.524
30	13.1	0.838	13.12	1.192	0.0762	235.6	121.9	357.6	1.121	1.524
31	13.5	0.841	12.77	1.187	0.0782	236.9	121.0	357.9	1.125	1.523
32	13.8	0.845	12.43	1.182	0.0804	238.1	120.0	358.2	1.129	1.523
33	14.1	0.849	12.10	1.177	0.0825	239.4	119.0	358.5	1.133	1.522
34	14.5	0.852	11.78	1.172	0.0848	240.6	118.1	358.7	1.137	1.522
35	14.9	0.856	11.47	1.167	0.0871	241.9	117.1	359.0	1.141	1.521
36	15.2	0.860	11.17	1.162	0.0894	243.2	116.1	359.3	1.145	1.521
37	15.6	0.864	10.88	1.156	0.0918	244.4	115.1	359.5	1.149	1.520
38	16.0	0.868	10.59	1.151	0.0943	245.7	114.0	359.8	1.153	1.520
39	16.3	0.872	10.32	1.146	0.0968	247.0	113.0	360.0	1.157	1.519

TEMP °C	PRESSURE bar	VOLUME $m^3/kg \cdot 10^3$ LIQUID v_f	VOLUME VAPOR v_g	DENSITY kg/m^3 LIQUID $1/v_f$	DENSITY VAPOR $1/v_g$	ENTHALPY kcal kg LIQUID h_f	ENTHALPY LATENT h_{fg}	ENTHALPY VAPOR h_g	ENTROPY kcal (kg) (°K) LIQUID s_f	ENTROPY VAPOR s_g
40	16.7	0.876	10.05	1.140	0.0994	248.2	112.0	360.3	1.161	1.519
41	17.1	0.880	9.789	1.135	0.1021	249.5	110.9	360.5	1.165	1.518
42	17.5	0.885	9.532	1.129	0.1049	250.8	109.8	360.7	1.169	1.518
43	17.9	0.889	9.283	1.123	0.1077	252.1	108.8	360.9	1.173	1.517
44	18.3	0.894	9.040	1.118	0.1106	253.4	107.7	361.1	1.177	1.517
45	18.8	0.899	8.803	1.112	0.1135	254.7	106.6	361.3	1.181	1.516
46	19.2	0.903	8.572	1.106	0.1166	256.0	105.4	361.5	1.185	1.516
47	19.6	0.908	8.346	1.100	0.1198	257.3	104.3	361.7	1.189	1.515
48	20.1	0.913	8.126	1.094	0.1230	258.7	103.1	361.8	1.193	1.514
49	20.5	0.919	7.911	1.087	0.1263	260.0	102.0	362.0	1.197	1.514
50	21.0	0.924	7.702	1.081	0.1298	261.3	100.8	362.1	1.201	1.513
51	21.4	0.930	7.497	1.075	0.1333	262.6	99.60	362.3	1.205	1.512
52	21.9	0.935	7.297	1.068	0.1370	264.0	98.37	362.4	1.209	1.512
53	22.4	0.941	7.101	1.061	0.1408	265.3	97.12	362.5	1.213	1.511
54	22.9	0.947	6.910	1.054	0.1447	266.7	95.85	362.6	1.217	1.510
55	23.4	0.954	6.722	1.047	0.1487	268.1	94.55	362.6	1.221	1.509
56	23.9	0.960	6.539	1.040	0.1529	269.5	93.23	362.7	1.225	1.508
57	24.4	0.967	6.360	1.033	0.1572	270.8	91.88	362.7	1.229	1.508
58	24.9	0.974	6.184	1.025	0.1617	272.2	90.50	362.7	1.233	1.507
59	25.4	0.982	6.011	1.018	0.1663	273.7	89.09	362.8	1.237	1.506
60	26.0	0.989	5.842	1.010	0.1711	275.1	87.65	362.7	1.242	1.505
61	26.5	0.997	5.676	1.002	0.1761	276.5	86.17	362.7	1.246	1.504
62	27.1	1.005	5.512	0.994	0.1813	278.0	84.65	362.6	1.250	1.502
63	27.6	1.014	5.352	0.985	0.1868	279.4	83.08	362.5	1.254	1.501
64	28.2	1.023	5.194	0.976	0.1925	280.9	81.47	362.4	1.258	1.500
65	28.8	1.033	5.038	0.967	0.1984	282.4	79.81	362.3	1.263	1.499
66	29.4	1.043	4.884	0.958	0.2047	284.0	78.09	362.1	1.267	1.497
67	30.0	1.054	4.732	0.948	0.2112	285.5	76.31	361.8	1.271	1.496
68	30.6	1.065	4.582	0.938	0.2182	287.1	74.45	361.6	1.276	1.494
69	31.2	1.077	4.433	0.927	0.2255	288.7	72.51	361.3	1.280	1.492
70	31.9	1.090	4.286	0.916	0.233	290.4	70.48	360.9	1.285	1.491
71	32.5	1.104	4.138	0.905	0.241	292.1	68.35	360.5	1.290	1.489
72	33.2	1.119	3.992	0.892	0.250	293.9	66.09	360.0	1.295	1.486
73	33.9	1.136	3.844	0.879	0.260	295.7	63.68	359.4	1.300	1.484
74	34.5	1.154	3.697	0.866	0.270	297.6	61.10	358.7	1.305	1.481
75	35.2	1.174	3.547	0.851	0.281	299.6	58.31	358.0	1.311	1.478
76	35.9	1.197	3.394	0.834	0.294	301.8	55.25	357.0	1.317	1.475
77	36.7	1.223	3.237	0.817	0.308	304.1	51.84	355.9	1.323	1.471
78	37.4	1.254	3.074	0.796	0.325	306.6	47.98	354.6	1.330	1.466
79	38.2	1.292	2.899	0.773	0.344	309.4	43.44	352.9	1.338	1.461
80	39.0	1.342	2.706	0.745	0.369	312.8	37.85	350.6	1.347	1.454
81	39.7	1.414	2.473	0.706	0.404	317.1	30.19	347.3	1.359	1.444
82	40.6	1.582	2.084	0.631	0.479	325.0	14.72	339.7	1.381	1.422
82.2	40.7	1.783	1.783	0.560	0.560	331.8	0.00	331.8	1.400	1.400

The reference point for enthalpy is 200 kJ/kg for the liquid at 0°C and for the entropy 1 kJ/(kg) (K) for the liquid at 0°C.

From Published data of E.I. du Pont de Nemours & Co. Inc. Used by permission.

PRESSURE ENTHALPY DIAGRAM FOR R 502 Imperial

TEMPERATURE IN °F VOLUME IN CU FT PER LB ENTROPY IN BTU PER LB PER °F

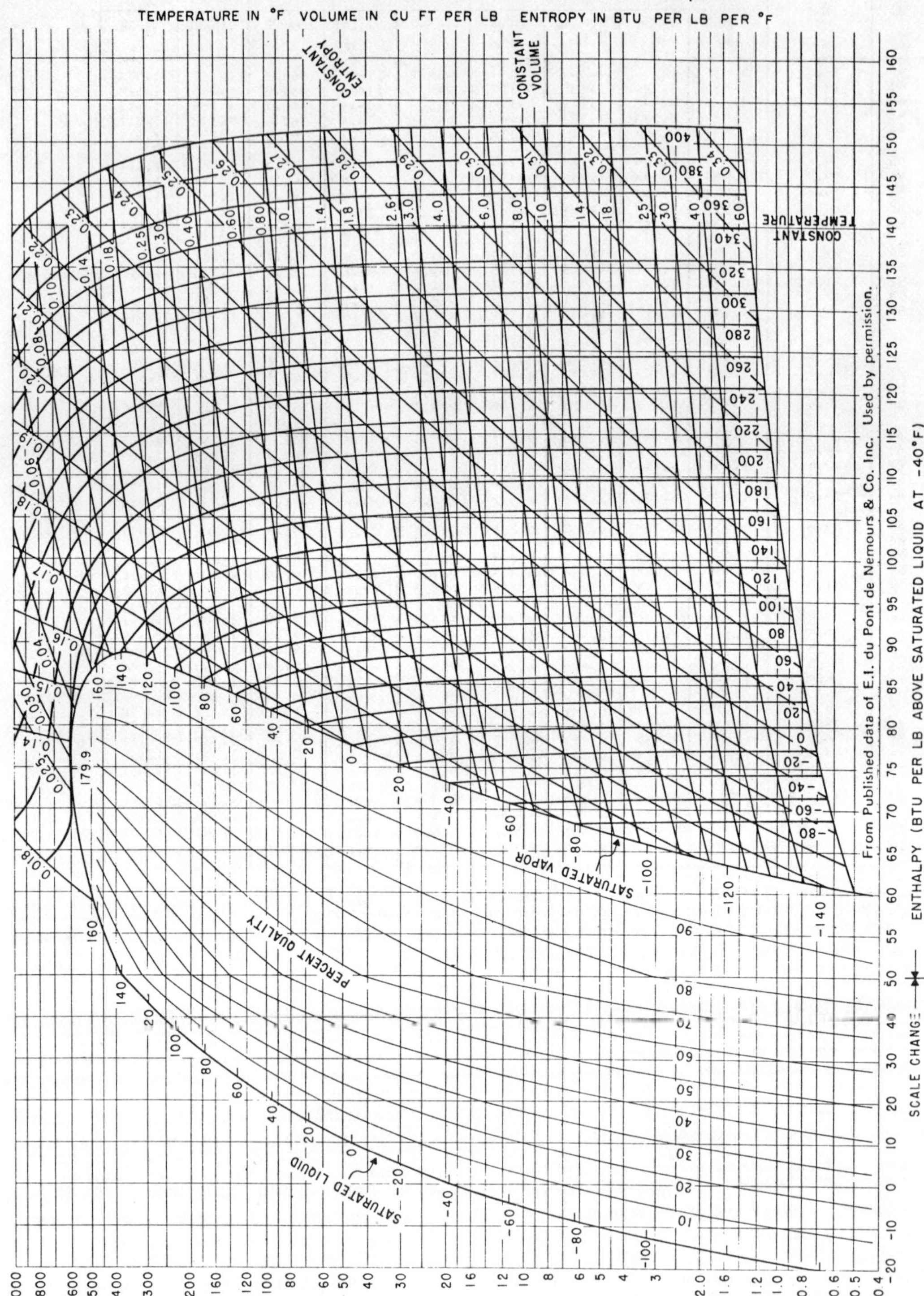

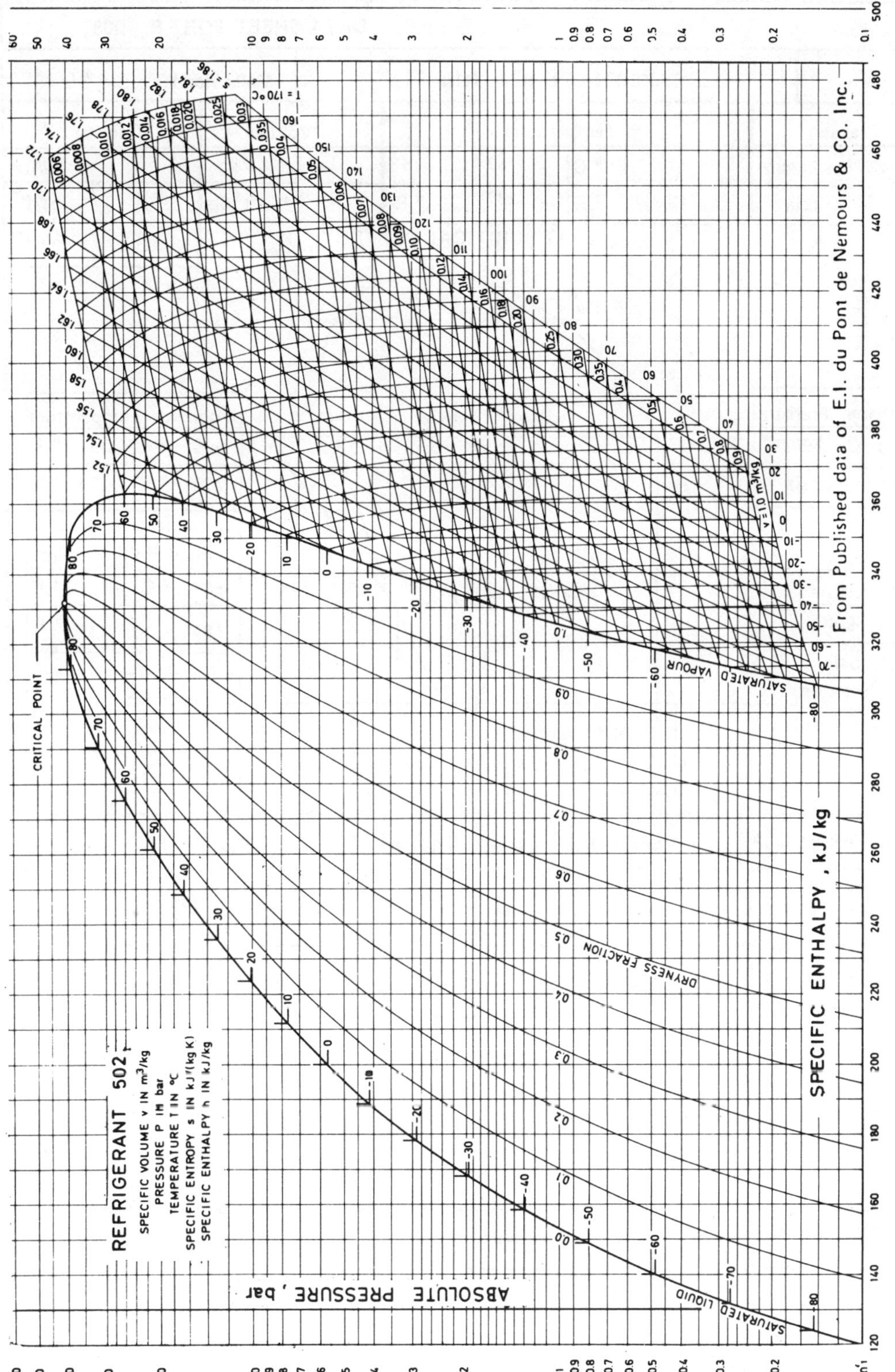
REFRIGERANT 502
SPECIFIC VOLUME v IN m3/kg
PRESSURE P IN bar
TEMPERATURE T IN °C
SPECIFIC ENTROPY s IN kJ/(kg K)
SPECIFIC ENTHALPY h IN kJ/kg
CRITICAL POINT
SATURATED LIQUID
SATURATED VAPOUR
DRYNESS FRACTION
ABSOLUTE PRESSURE, bar
SPECIFIC ENTHALPY, kJ/kg
From Published data of E.I. du Pont de Nemours & Co. Inc.
T = 170°C
v = 1.0 m3/kg

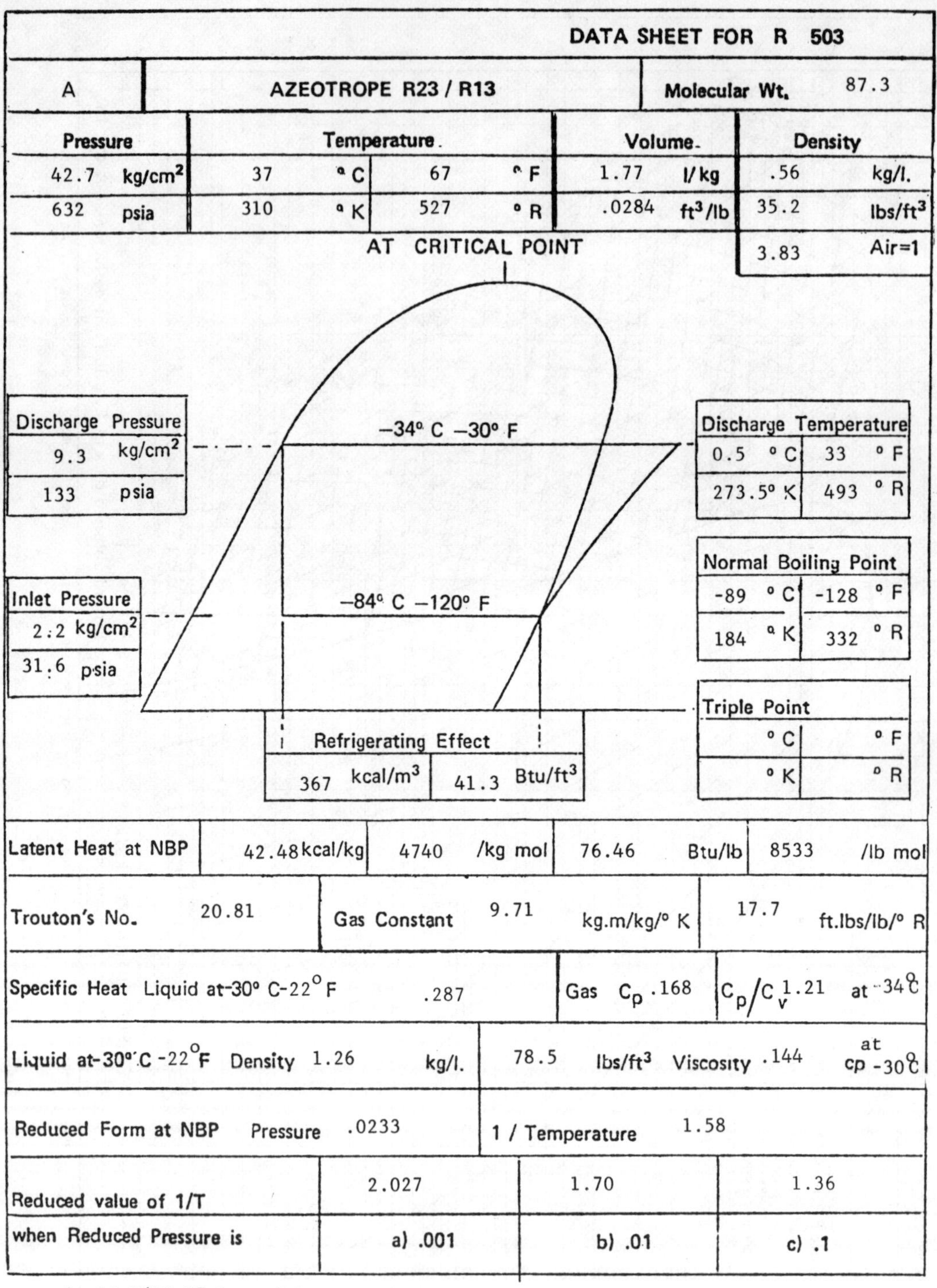

DATA SHEET FOR R 503

A	AZEOTROPE R23 / R13	Molecular Wt.	87.3

Pressure		Temperature				Volume		Density	
42.7	kg/cm²	37	°C	67	°F	1.77	l/kg	.56	kg/l.
632	psia	310	°K	527	°R	.0284	ft³/lb	35.2	lbs/ft³
								3.83	Air=1

AT CRITICAL POINT

−34° C −30° F

−84° C −120° F

Discharge Pressure	
9.3	kg/cm²
133	psia

Inlet Pressure	
2.2	kg/cm²
31.6	psia

Discharge Temperature			
0.5	°C	33	°F
273.5°	K	493	°R

Normal Boiling Point			
-89	°C	-128	°F
184	°K	332	°R

Triple Point			
	°C		°F
	°K		°R

Refrigerating Effect			
367	kcal/m³	41.3	Btu/ft³

Latent Heat at NBP	42.48 kcal/kg	4740 /kg mol	76.46 Btu/lb	8533 /lb mol

Trouton's No.	20.81	Gas Constant	9.71 kg.m/kg/°K	17.7 ft.lbs/lb/°R

Specific Heat Liquid at -30° C -22° F	.287	Gas C_p .168	C_p/C_v 1.21 at -34°C

Liquid at -30° C -22° F Density	1.26 kg/l.	78.5 lbs/ft³	Viscosity .144 cp at -30°C

Reduced Form at NBP	Pressure .0233	1 / Temperature 1.58

Reduced value of 1/T	2.027	1.70	1.36
when Reduced Pressure is	a) .001	b) .01	c) .1

A 40.1/59.9% by weight

TEMP. °F	PRESSURE		VOLUME cu ft/lb	DENSITY lb/cu ft	ENTHALPY Btu/lb			ENTROPY Btu/(lb)(°R)	
	psia	psig	Vapor	Liquid	Liquid	Latent	Vapor	Liquid	Vapor
—200	0.618	*28.7	51.44	98.52	—40.84	87.56	46.72	—0.1209	0.2163
—195	0.818	*28.3	39.58	98.17	—39.70	86.94	47.25	—0.1166	0.2119
—190	1.070	*27.7	30.80	97.80	—38.54	86.32	47.77	—0.1122	0.2078
—185	1.383	*27.1	24.23	97.43	—37.38	85.68	48.30	—0.1080	0.2039
—180	1.770	*26.3	19.25	97.04	—36.20	85.02	48.82	—0.1037	0.2003
—175	2.241	*25.4	15.44	96.64	—35.02	84.36	49.34	—0.0995	0.1968
—170	2.812	*24.2	12.50	96.23	—33.82	83.54	49.96	—0.0945	0.1929
—165	3.497	*22.8	10.20	95.81	—32.62	82.99	50.37	—0.0912	0.1904
—160	4.311	*21.1	8.390	95.37	—31.40	82.29	50.89	—0.0872	0.1874
—155	5.273	*19.2	6.954	94.92	—30.18	81.57	51.39	—0.0831	0.1846
—150	6.401	*16.9	5.804	94.46	—28.94	80.84	51.89	—0.0791	0.1819
—149	6.648	*16.4	5.602	94.36	—28.70	80.69	51.99	—0.0783	0.1814
—148	6.903	*15.9	5.409	94.27	—28.45	80.54	52.09	—0.0775	0.1809
—147	7.165	*15.3	5.224	94.17	—28.20	80.39	52.19	—0.0767	0.1804
—146	7.435	*14.8	5.047	94.08	—27.95	80.24	52.29	—0.0759	0.1799
—145	7.714	*14.2	4.876	93.98	—27.70	80.09	52.39	—0.0751	0.1794
—144	8.001	*13.6	4.713	93.88	—27.45	79.94	52.49	—0.0743	0.1789
—143	8.296	*13.0	4.556	93.79	—27.20	79.79	52.58	—0.0736	0.1784
—142	8.600	*12.4	4.406	93.69	—26.95	79.63	52.68	—0.0728	0.1779
—141	8.912	*11.8	4.261	83.59	—26.70	79.48	52.78	—0.0720	0.1774
—140	9.234	*11.1	4.123	93.49	—26.45	79.33	52.88	—0.0712	0.1769
—139	9.564	*10.4	3.989	93.39	—26.20	79.17	52.97	—0.0704	0.1765
—138	9.904	* 9.76	3.861	93.29	—25.95	79.02	53.07	—0.0696	0.1760
—137	10.25	* 9.04	3.738	93.19	—25.70	78.86	53.17	—0.0689	0.1755
—136	10.61	* 8.31	3.620	93.09	—25.44	78.71	53.26	—0.0681	0.1751
—135	10.98	* 7.56	3.506	92.99	—25.19	78.55	53.36	—0.0673	0.1746
—134	11.36	* 6.79	3.397	92.89	—24.94	78.40	53.46	—0.0665	0.1742
—133	11.75	* 6.00	3.291	92.78	—24.69	78.24	53.55	—0.0658	0.1737
—132	12.15	* 5.18	3.190	92.68	—24.43	78.08	53.65	—0.0650	0.1733
—131	12.56	* 4.35	3.092	92.58	—24.18	77.92	53.74	—0.0642	0.1729
—130	12.98	* 3.49	2.998	92.47	—23.93	77.76	53.84	—0.0634	0.1724
—129	13.41	* 2.61	2.908	92.37	—23.67	77.60	53.93	—0.0627	0.1720
—128	13.86	* 1.71	2.821	92.26	—23.42	77.44	54.03	—0.0619	0.1716
—127	14.31	* 0.78	2.737	92.15	—23.16	77.28	54.12	—0.0611	0.1711
—126	14.78	0.08	2.656	92.05	—22.91	77.12	54.21	—0.0604	0.1707
—125	15.26	0.56	2.577	91.94	—22.64	76.95	54.31	—0.0596	0.1703
—124	15.75	1.05	2.502	91.83	—22.39	76.79	54.40	—0.0588	0.1699
—123	16.25	1.55	2.430	91.72	—22.13	76.62	54.49	—0.0581	0.1695
—122	16.76	2.07	2.360	91.61	—21.87	76.46	54.58	—0.0573	0.1691
—121	17.29	2.59	2.292	91.50	—21.62	76.29	54.67	—0.0566	0.1687
—120	17.83	3.13	2.227	91.39	—21.36	76.13	54.77	—0.0558	0.1683
—119	18.38	3.69	2.164	91.28	—21.10	75.96	54.86	—0.0551	0.1679
—118	18.95	4.25	2.103	91.17	—20.85	75.79	54.95	—0.0543	0.1675
—117	19.53	4.83	2.044	91.06	—20.59	75.63	55.04	—0.0536	0.1671
—116	20.12	5.42	1.988	90.94	—20.33	75.46	55.13	—0.0528	0.1668
—115	20.73	6.03	1.933	90.83	—20.07	75.29	55.22	—0.0521	0.1664
—114	21.35	6.65	1.880	90.72	—19.81	75.12	55.31	—0.0513	0.1660
—113	21.98	7.29	1.829	90.60	—19.55	74.95	55.40	—0.0506	0.1656
—112	22.63	7.94	1.779	90.48	19.29	74.78	55.49	0.0498	0.1653
—111	23.30	8.60	1.731	90.37	—19.03	74.61	55.58	—0.0491	0.1649
—110	23.98	9.28	1.685	90.25	—18.77	74.44	55.66	—0.0483	0.1645
—109	24.67	9.98	1.640	90.13	—18.51	74.26	55.75	—0.0476	0.1642
—108	25.38	10.7	1.597	90.02	—18.25	74.09	55.84	—0.0469	0.1638
—107	26.11	11.4	1.555	89.90	—17.99	73.92	55.93	—0.0461	0.1635
—106	26.85	12.2	1.514	89.78	—17.73	73.43	56.01	—0.0454	0.1631
—105	27.61	12.9	1.475	89.66	—17.47	73.57	56.10	—0.0447	0.1628
—104	28.38	13.7	1.437	89.54	—17.21	73.39	56.18	—0.0439	0.1624
—103	29.17	14.5	1.400	89.42	—16.95	73.22	56.27	—0.0432	0.1621
—102	29.98	15.3	1.365	89.29	—16.69	73.04	56.36	—0.0425	0.1617
—101	30.80	16.1	1.330	89.17	—16.42	72.86	56.44	—0.0417	0.1614

*Inches of mercury vacuum.

TEMP. °F	PRESSURE		VOLUME cu ft/lb	DENSITY lb/cu ft	ENTHALPY Btu/lb			ENTROPY Btu/(lb)(°R)	
	psia	psig	Vapor	Liquid	Liquid	Latent	Vapor	Liquid	Vapor
—100	31.64	16.9	1.296	89.05	—16.16	72.69	56.52	—0.0410	0.1611
—99	32.50	17.8	1.264	88.92	—15.90	72.51	56.61	—0.0403	0.1607
—98	33.38	18.7	1.232	88.80	—15.64	72.33	56.69	—0.0396	0.1604
—97	34.27	19.6	1.202	88.67	—15.37	72.15	56.78	—0.0389	0.1601
—96	35.18	20.5	1.172	88.55	—15.11	71.97	56.86	—0.0381	0.1598
—95	36.11	21.4	1.144	88.42	—14.85	71.79	56.94	—0.0374	0.1594
—94	37.06	22.4	1.116	88.29	—14.59	71.61	57.02	—0.0367	0.1591
—93	38.03	23.3	1.089	88.17	—14.32	71.43	57.10	—0.0360	0.1588
—92	39.01	24.3	1.062	88.04	—14.06	71.24	57.19	—0.0353	0.1585
—91	40.02	25.3	1.037	87.91	—13.79	71.06	57.27	—0.0346	0.1582
—90	41.05	26.3	1.012	87.78	—13.53	70.88	57.35	—0.0339	0.1579
—88	43.15	28.5	0.9650	87.52	—13.00	70.51	57.56	—0.0324	0.1573
—86	45.34	30.6	0.9204	87.25	—12.47	70.14	57.66	—0.0310	0.1567
—84	47.62	32.9	0.8782	86.98	—11.93	69.75	57.82	—0.0296	0.1561
—82	49.97	35.3	0.8384	86.71	—11.40	69.37	57.97	—0.0282	0.1555
—80	52.42	37.7	0.8008	86.44	—10.87	68.99	58.13	—0.0268	0.1549
—78	54.95	40.3	0.7653	86.16	—10.33	68.61	58.28	—0.0254	0.1543
—76	57.57	42.9	0.7316	85.88	—9.80	68.22	58.43	—0.0241	0.1538
—74	60.29	45.6	0.6998	85.60	—9.26	67.84	58.57	—0.0227	0.1532
—72	63.10	48.4	0.6696	85.31	—8.73	67.44	58.72	—0.0213	0.1527
—70	66.00	51.3	0.6409	85.02	—8.19	67.05	58.86	—0.0199	0.1521
—68	69.01	54.3	0.6138	84.73	—7.65	66.65	59.00	—0.0186	0.1516
—66	72.11	57.4	0.5880	84.43	—7.11	66.25	59.14	—0.0172	0.1511
—64	75.32	60.6	0.5636	84.13	—6.57	65.85	59.27	—0.0159	0.1505
—62	78.64	63.9	0.5403	83.83	—6.03	65.44	59.41	—0.0145	0.1500
—60	82.05	67.4	0.5182	83.52	—5.49	65.03	59.54	—0.0132	0.1495
—58	85.58	70.9	0.4972	83.21	—4.95	64.62	59.67	—0.0118	0.1490
—56	89.22	74.5	0.4772	82.90	—4.40	64.19	59.80	—0.0105	0.1485
—54	92.97	78.3	0.4581	82.58	—3.85	63.77	59.92	—0.0092	0.1480
—52	96.84	82.1	0.4400	82.30	—3.31	63.35	60.04	—0.0079	0.1475
—50	100.8	86.1	0.4227	81.93	—2.76	62.92	60.16	—0.0065	0.1471
—48	104.9	90.2	0.4062	81.60	—2.21	62.49	60.28	—0.0052	0.1466
—46	109.1	94.5	0.3904	81.27	—1.66	62.05	60.39	—0.0039	0.1461
—44	113.5	98.8	0.3754	80.94	—1.11	61.61	60.51	—0.0026	0.1456
—42	118.0	103	0.3611	80.59	—0.55	61.17	60.62	—0.0013	0.1452
—40	122.6	108	0.3474	80.25	0.00	60.72	60.72	0.0000	0.1447
—38	127.3	113	0.3343	79.90	0.56	60.27	60.82	0.0013	0.1442
—36	132.2	117	0.3217	79.55	1.12	59.81	60.92	0.0026	0.1438
—34	137.2	122	0.3097	79.19	1.68	59.34	61.02	0.0039	0.1433
—32	142.3	128	0.2982	78.83	2.24	58.87	61.11	0.0052	0.1428
—30	147.6	133	0.2872	78.46	2.81	58.40	61.20	0.0065	0.1424
—28	153.0	138	0.2767	78.09	3.37	57.92	61.29	0.0078	0.1419
—26	158.6	144	0.2666	77.72	3.94	57.43	61.37	0.0090	0.1415
—24	164.3	150	0.2569	77.33	4.51	56.94	61.45	0.0103	0.1410
—22	170.2	155	0.2476	76.95	5.08	56.45	61.53	0.0116	0.1406
—20	176.2	161	0.2387	76.56	5.66	55.94	61.60	0.0129	0.1401
—15	191.9	177	0.2179	75.56	7.11	54.64	61.76	0.0161	0.1390
—10	208.6	194	0.1991	74.52	8.59	53.30	61.89	0.0193	0.1378
—5	226.4	212	0.1820	73.44	10.08	51.90	61.98	0.0225	0.1366
0	245.3	231	0.1664	72.33	11.60	50.44	62.05	0.0257	0.1354
5	265.3	251	0.1521	71.16	13.16	48.91	62.07	0.0290	0.1342
10	286.4	272	0.1391	69.95	14.74	47.30	62.04	0.0322	0.1329
15	308.9	294	0.1271	68.68	16.38	45.59	61.96	0.0356	0.1316
20	332.6	318	0.1160	67.35	18.05	43.77	61.82	0.0390	0.1302
25	357.6	343	0.1058	65.94	19.79	41.82	61.61	0.0424	0.1287
30	384.1	369	0.0962	64.45	21.60	39.71	61.31	0.0460	0.1271
40	440.6	426	0.0793	61.12	26.03	34.42	60.45	0.0546	0.1235
50	503.3	489	0.0640	57.09	29.69	29.25	58.95	0.0615	0.1189
60	574.8	560	0.0485	51.40	34.32	21.44	55.77	0.0700	0.1112
67.1	632.2	618	0.0284	35.21	45.36	0.00	45.36	0.0906	0.0906

From Published data of E. I. du Pont de Nemours & Co., Inc. Used by permission.

TEMPERATURE IN °F. PRESSURE IN LBS/SQ.IN.
VOLUME IN FT3/LB. ENTROPY IN BTU/LB/°R

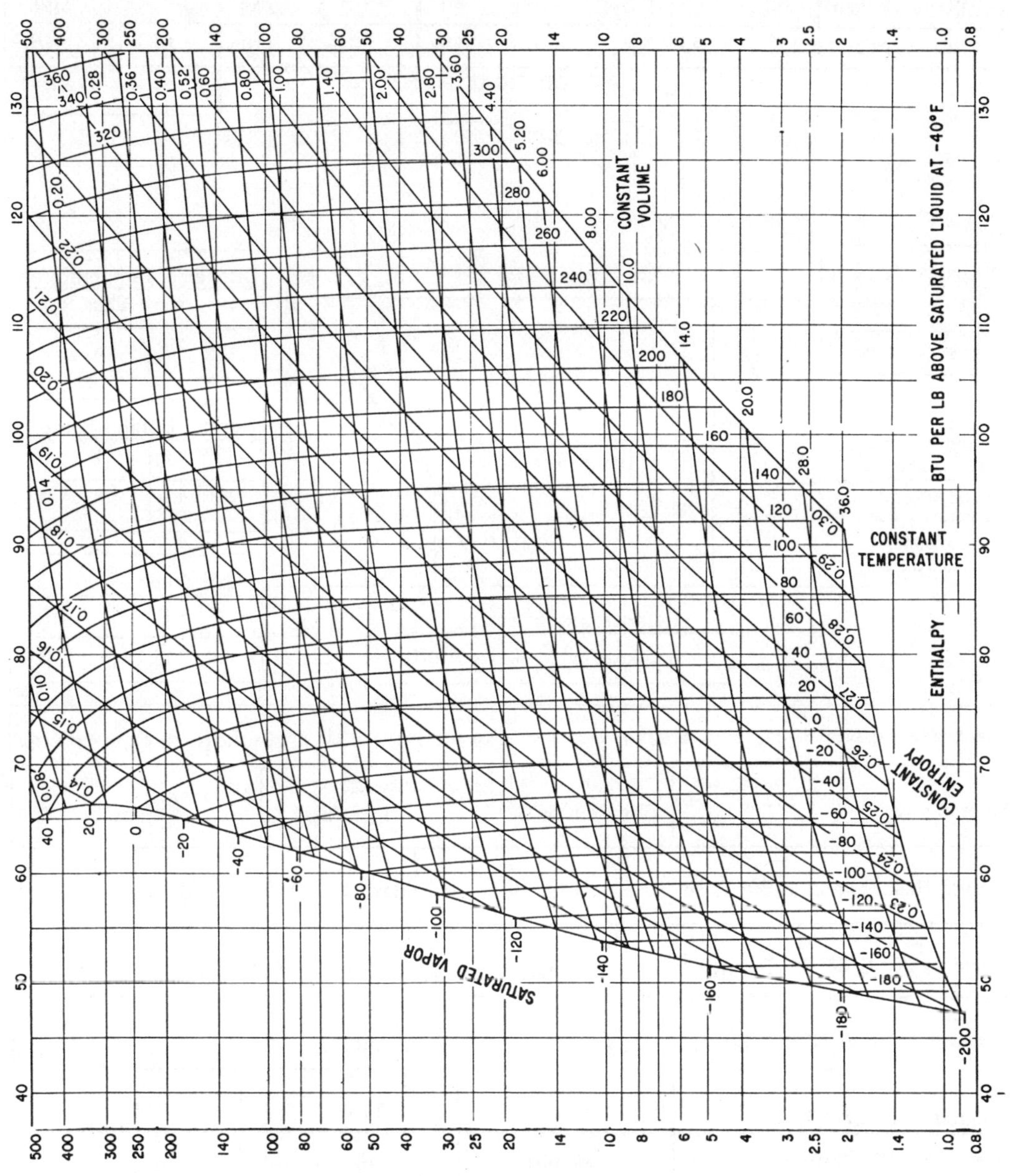

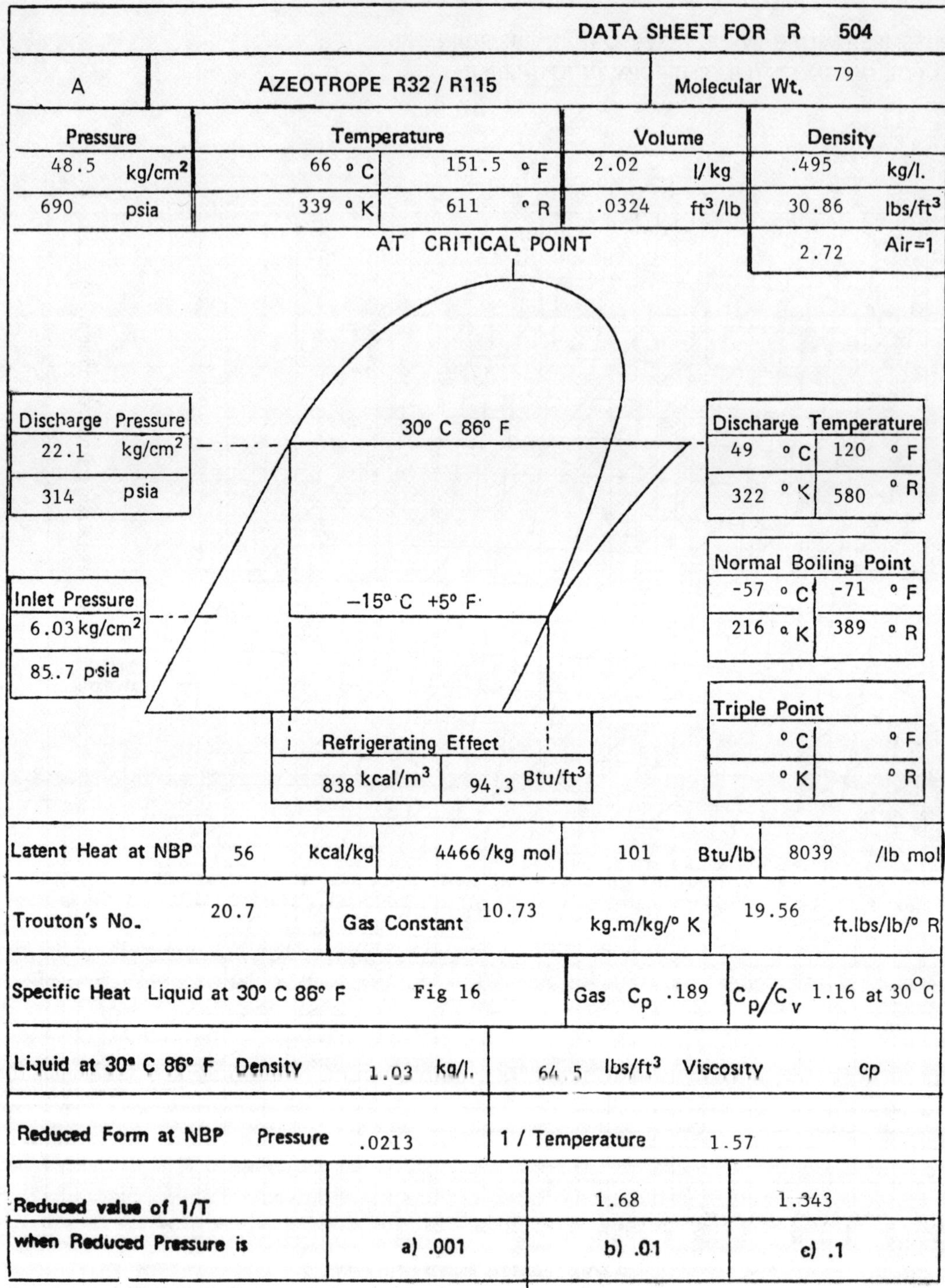

DATA SHEET FOR R 504

A	AZEOTROPE R32 / R115	Molecular Wt. 79

Pressure		Temperature				Volume		Density	
48.5	kg/cm²	66	°C	151.5	°F	2.02	l/kg	.495	kg/l.
690	psia	339	°K	611	°R	.0324	ft³/lb	30.86	lbs/ft³
								2.72	Air=1

AT CRITICAL POINT

Discharge Pressure	
22.1	kg/cm²
314	psia

Discharge Temperature			
49	°C	120	°F
322	°K	580	°R

Inlet Pressure	
6.03	kg/cm²
85.7	psia

Normal Boiling Point			
-57	°C	-71	°F
216	°K	389	°R

Refrigerating Effect	
838 kcal/m³	94.3 Btu/ft³

Triple Point			
	°C		°F
	°K		°R

Latent Heat at NBP	56 kcal/kg	4466 /kg mol	101 Btu/lb	8039 /lb mol

Trouton's No. 20.7	Gas Constant 10.73 kg.m/kg/° K	19.56 ft.lbs/lb/° R

Specific Heat Liquid at 30° C 86° F	Fig 16	Gas C_p .189	C_p/C_v 1.16 at 30°C

Liquid at 30° C 86° F Density	1.03 kg/l.	64.5 lbs/ft³	Viscosity cp

Reduced Form at NBP Pressure	.0213	1 / Temperature	1.57

Reduced value of 1/T when Reduced Pressure is		1.68	1.343
	a) .001	b) .01	c) .1

A 48.2/51.8% by weight

TEMP. °F	PRESSURE lb per sq in.		VAPOR VOLUME cubic ft per lb	LIQUID DENSITY lb per cubic ft	ENTHALPY Btu per lb			ENTROPY Btu per (lb) (°R)	
t	Absolute psia	Gage psig	v_g	$\frac{1}{v_f}$	Liquid h_f	Latent h_{fg}	Vapor h_g	Liquid s_f	Vapor s_g
—120	2.964	* 23.9	15.31	91.33	—21.48	108.17	86.69	—0.0565	0.2619
—110	4.280	* 21.2	10.86	90.34	—18.96	106.97	88.01	—0.0492	0.2567
—100	6.042	* 17.6	7.874	89.33	—16.39	105.71	89.31	—0.0420	0.2519
—90	8.356	* 12.9	5.817	88.31	—13.78	104.37	90.59	—0.0348	0.2475
—80	11.338	* 6.84	4.372	87.27	—11.12	102.96	91.84	—0.0277	0.2435
—78	12.026	* 5.44	4.138	87.06	—10.59	102.67	92.09	—0.0263	0.2427
—76	12.747	* 3.97	3.918	86.85	—10.05	102.38	92.33	—0.0249	0.2419
—74	13.503	* 2.43	3.712	86.63	—9.50	102.08	92.58	—0.0235	0.2412
—72	14.293	* 0.82	3.519	86.42	—8.96	101.78	92.82	—0.0221	0.2404
—70	15.121	0.42	3.338	86.21	—8.42	101.48	93.06	—0.0207	0.2397
—68	15.99	1.29	3.168	85.99	—7.86	101.16	93.30	—0.0193	0.2390
—66	16.89	2.19	3.009	85.77	—7.31	100.85	93.54	—0.0179	0.2383
—64	17.83	3.14	2.859	85.56	—6.76	100.54	93.78	—0.0165	0.2376
—62	18.82	4.12	2.718	85.34	—6.21	100.22	94.01	—0.0151	0.2369
—60	19.85	5.15	2.585	85.12	—5.65	99.90	94.25	—0.0137	0.2362
—58	20.92	6.22	2.460	84.90	—5.10	99.57	94.48	—0.0123	0.2355
—56	22.03	7.34	2.342	84.68	—4.54	99.25	94.71	—0.0110	0.2349
—54	23.20	8.50	2.231	84.46	—3.98	98.91	94.94	—0.0096	0.2342
—52	24.41	9.71	2.126	84.24	—3.41	98.58	95.17	—0.0082	0.2336
—50	25.67	11.0	2.028	84.01	—2.85	98.24	95.39	—0.0068	0.2330
—48	26.98	12.3	1.934	83.8	—2.28	97.90	95.62	—0.0055	0.2323
—46	28.34	13.6	1.846	83.6	—1.72	97.55	95.84	—0.0041	0.2317
—44	29.76	15.1	1.763	83.3	—1.15	97.21	96.06	—0.0027	0.2311
—42	31.23	16.5	1.684	83.1	—0.57	96.85	96.28	—0.0014	0.2305
—40	32.76	18.1	1.609	82.9	0.00	96.50	96.50	0.0000	0.2299
—38	34.34	19.6	1.539	82.6	0.58	96.14	96.71	0.0014	0.2293
—36	35.99	21.3	1.472	82.4	1.15	95.77	96.93	0.0027	0.2288
—34	37.69	23.0	1.408	82.2	1.73	95.41	97.14	0.0041	0.2282
—32	39.46	24.8	1.348	81.9	2.31	95.04	97.35	0.0054	0.2276
—30	41.29	26.6	1.291	81.7	2.90	94.66	97.56	0.0068	0.2271
—28	43.18	28.5	1.237	81.5	3.48	94.28	97.77	0.0081	0.2265
—26	45.14	30.4	1.185	81.2	4.07	93.90	97.97	0.0095	0.2260
—24	47.17	32.5	1.136	81.0	4.66	93.52	98.18	0.0108	0.2255
—22	49.27	34.6	1.090	80.8	5.26	93.12	98.38	0.0122	0.2249
—20	51.44	36.7	1.045	80.5	5.85	92.72	98.58	0.0135	0.2244
—18	53.68	39.0	1.003	80.3	6.45	92.32	98.77	0.0149	0.2239
—16	56.00	41.3	0.9631	80.0	7.05	91.92	98.97	0.0162	0.2234
—14	58.39	43.7	0.9249	79.8	7.65	91.51	99.16	0.0175	0.2229
—12	60.86	46.2	0.8885	79.5	8.25	91.10	99.35	0.0189	0.2224
—10	63.41	48.7	0.8539	79.3	8.86	90.69	99.54	0.0202	0.2219
—8	66.04	51.3	0.8208	79.0	9.46	90.27	99.73	0.0215	0.2214
—6	68.76	54.1	0.7893	78.8	10.07	89.84	99.91	0.0229	0.2209
—4	71.56	56.9	0.7591	78.5	10.68	89.41	100.10	0.0242	0.2204
—2	74.44	59.7	0.7304	78.3	11.30	88.98	100.28	0.0255	0.2199
0	77.41	62.7	0.7029	78.0	11.91	88.54	100.45	0.0269	0.2195
2	80.47	65.8	0.6767	77.8	12.53	88.10	100.63	0.0282	0.2190
4	83.62	68.9	0.6516	77.5	13.15	87.65	100.80	0.0295	0.2185
6	86.87	72.2	0.6276	77.2	13.77	87.20	100.97	0.0308	0.2181
8	90.21	75.5	0.6047	77.0	14.40	86.74	101.14	0.0322	0.2176
10	93.64	78.9	0.5827	76.7	15.03	86.28	101.30	0.0335	0.2172
12	97.18	82.5	0.5616	76.4	15.67	85.80	101.47	0.0348	0.2167
14	100.8	86.1	0.5415	76.2	16.30	85.33	101.63	0.0361	0.2163
16	104.5	89.8	0.5222	75.9	16.94	84.85	101.78	0.0375	0.2158
18	108.4	93.7	0.5037	75.6	17.57	84.36	101.94	0.0388	0.2154
20	112.3	97.6	0.4859	75.3	18.22	83.87	102.09	0.0401	0.2150
22	116.4	102	0.4689	75.1	18.86	83.38	102.24	0.0414	0.2145
24	120.5	106	0.4526	74.8	19.51	82.88	102.39	0.0427	0.2141
26	124.8	110	0.4369	74.5	20.16	82.37	102.53	0.0441	0.2137
28	129.2	114	0.4219	74.2	20.81	81.86	102.67	0.0454	0.2132
30	133.7	119	0.4074	73.9	21.47	81.34	102.81	0.0467	0.2128

*Inches of mercury vacuum

TEMP. °F	PRESSURE lb per sq in.		VAPOR VOLUME cubic ft per lb	LIQUID DENSITY lb per cubic ft	ENTHALPY Btu per lb			ENTROPY Btu per (lb) (°R)	
t	Absolute psia	Gage psig	v_g	$\frac{1}{v_f}$	Liquid h_f	Latent h_{fg}	Vapor h_g	Liquid s_f	Vapor s_g
32	138.3	124	0.3936	73.6	22.13	80.81	102.94	0.0480	0.2124
34	143.0	128	0.3802	73.4	22.79	80.28	103.07	0.0494	0.2120
36	147.9	133	0.3673	73.1	23.47	79.73	103.19	0.0507	0.2115
38	152.9	138	0.3550	72.8	24.14	79.18	103.32	0.0520	0.2111
40	158.0	143	0.3431	72.5	24.81	78.63	103.44	0.0533	0.2107
42	163.2	148	0.3317	72.2	25.49	78.06	103.55	0.0547	0.2103
44	168.5	154	0.3207	71.8	26.17	77.49	103.66	0.0560	0.2099
46	174.0	159	0.3101	71.5	26.86	76.92	103.77	0.0573	0.2094
48	179.6	165	0.2999	71.2	27.54	76.33	103.88	0.0587	0.2090
50	185.4	171	0.2901	70.9	28.24	75.74	103.98	0.0600	0.2086
52	191.3	177	0.2806	70.6	28.93	75.14	104.07	0.0613	0.2082
54	197.3	183	0.2714	70.3	29.64	74.52	104.16	0.0627	0.2077
56	203.5	189	0.2626	69.9	30.35	73.90	104.25	0.0640	0.2073
58	209.8	195	0.2541	69.6	31.06	73.27	104.33	0.0654	0.2069
60	216.2	202	0.2458	69.3	31.78	72.63	104.41	0.0667	0.2065
62	222.8	208	0.2379	68.9	32.50	71.98	104.48	0.0681	0.2060
64	229.6	215	0.2302	68.6	33.22	71.32	104.54	0.0694	0.2056
66	236.5	222	0.2228	68.2	33.95	70.65	104.60	0.0708	0.2052
68	243.5	229	0.2157	67.9	34.70	69.96	104.66	0.0721	0.2047
70	250.7	236	0.2087	67.5	35.44	69.27	104.70	0.0735	0.2043
72	258.1	243	0.2020	67.2	36.19	68.56	104.75	0.0749	0.2038
74	265.6	251	0.1956	66.8	36.94	67.84	104.78	0.0763	0.2034
76	273.3	259	0.1893	66.4	37.70	67.11	104.81	0.0776	0.2029
78	281.2	266	0.1832	66.0	38.47	66.37	104.83	0.0790	0.2025
80	289.2	274	0.1773	65.7	39.25	65.60	104.85	0.0804	0.2020
82	297.4	283	0.1717	65.3	40.03	64.83	104.85	0.0818	0.2015
84	305.7	291	0.1661	64.9	40.82	64.04	104.85	0.0832	0.2010
86	314.3	300	0.1608	64.5	41.61	63.23	104.84	0.0847	0.2005
88	323.0	308	0.1556	64.1	42.42	62.40	104.82	0.0861	0.2000
90	331.9	317	0.1505	63.6	43.23	61.56	104.80	0.0875	0.1995
92	340.9	326	0.1457	63.2	44.05	60.71	104.76	0.0890	0.1990
94	350.2	335	0.1409	62.8	44.88	59.83	104.71	0.0904	0.1985
96	359.6	345	0.1363	62.3	45.73	58.92	104.65	0.0919	0.1979
98	369.2	355	0.1318	61.9	46.57	58.00	104.58	0.0934	0.1974
100	379.1	364	0.1274	61.4	47.43	57.06	104.49	0.0948	0.1968
102	389.1	374	0.1232	61.0	48.31	56.09	104.39	0.0964	0.1962
104	399.3	385	0.1190	60.5	49.19	55.09	104.28	0.0979	0.1956
106	409.7	395	0.1150	60.0	50.09	54.06	104.15	0.0994	0.1950
108	420.3	406	0.1111	59.5	51.00	53.01	104.01	0.1009	0.1943
110	431.1	416	0.1072	59.0	51.92	51.93	103.85	0.1025	0.1937
112	442.1	427	0.1035	58.4	52.86	50.81	103.67	0.1041	0.1930
114	453.3	439	0.0998	57.9	53.81	49.65	103.47	0.1057	0.1923
116	464.8	450	0.0963	57.3	54.79	48.46	103.24	0.1073	0.1915
118	476.4	462	0.0928	56.7	55.77	47.22	102.99	0.1090	0.1907
120	488.3	474	0.0893	56.1	56.78	45.94	102.72	0.1107	0.1899
122	500.4	486	0.0859	55.5	57.81	44.60	102.41	0.1124	0.1890
124	512.7	498	0.0826	54.8	58.87	43.20	102.07	0.1141	0.1881
126	525.2	511	0.0793	54.1	59.95	41.74	101.69	0.1159	0.1871
128	538.0	523	0.0761	53.4	61.06	40.21	101.27	0.1177	0.1861
130	551.0	536	0.0729	52.7	62.20	38.60	100.80	0.1195	0.1850
132	564.2	550	0.0697	51.9	63.38	36.89	100.27	0.1215	0.1838
134	577.7	563	0.0666	51.0	64.60	35.07	99.67	0.1234	0.1825
136	591.2	577	0.0635	50.1	66.77	32.25	99.03	0.1270	0.1812
138	604.6	590	0.0606	49.1	68.59	29.79	98.38	0.1300	0.1798
140	618.1	603	0.0578	48.0	69.97	27.73	97.70	0.1322	0.1784
142	631.7	617	0.0550	46.8	71.04	25.90	96.94	0.1339	0.1769
144	645.8	631	0.0520	45.5	71.97	24.05	96.02	0.1353	0.1752
146	660.4	646	0.0488	43.8	72.95	21.86	94.81	0.1369	0.1729
148	675.9	661	0.0450	41.8	74.31	18.73	93.04	0.1390	0.1698
150	692.2	678	0.0394	38.5	76.96	12.81	89.76	0.1432	0.1642

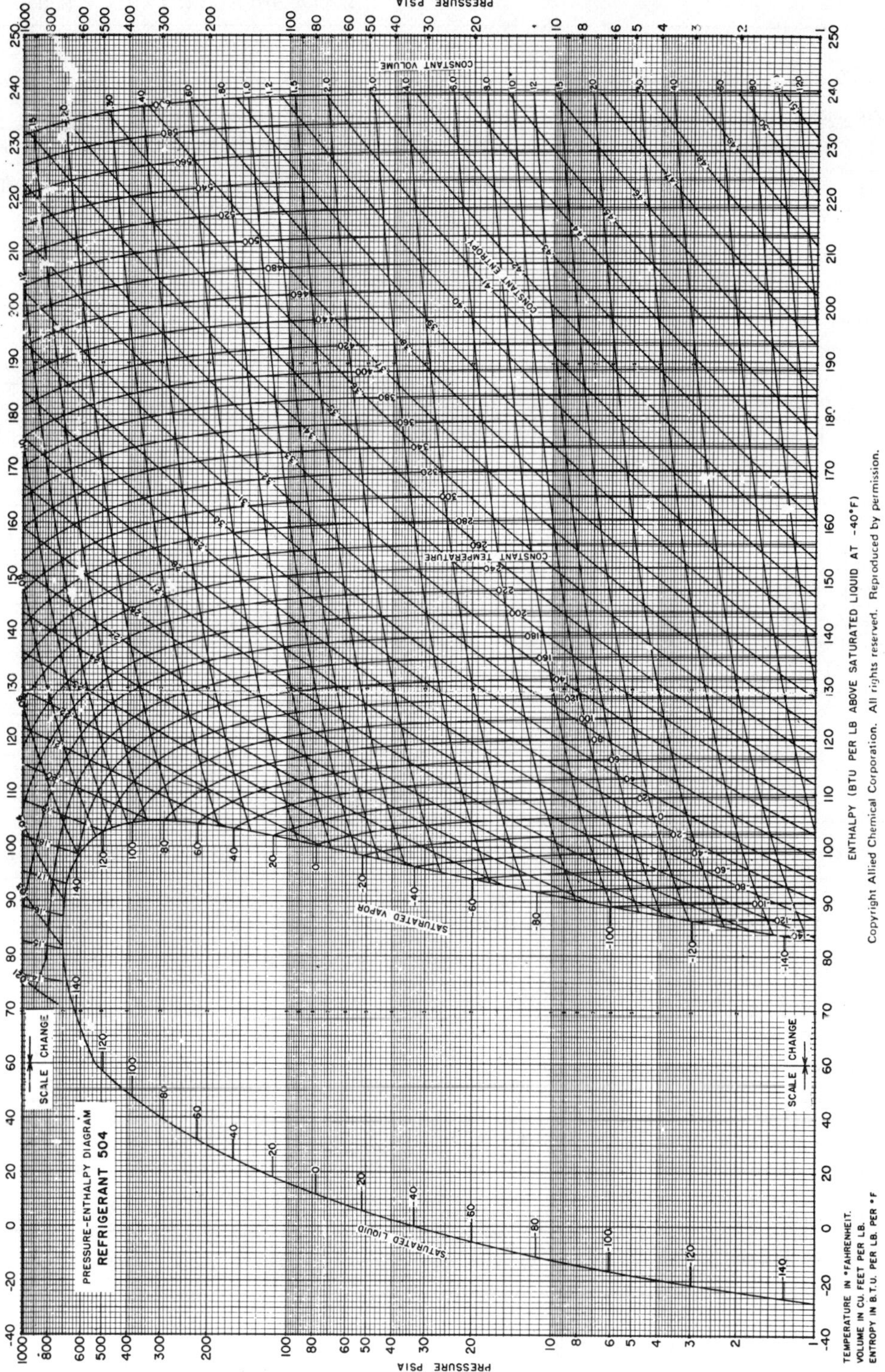
PRESSURE-ENTHALPY DIAGRAM
REFRIGERANT 504
PRESSURE PSIA
ENTHALPY (BTU PER LB ABOVE SATURATED LIQUID AT -40°F)
CONSTANT VOLUME
CONSTANT ENTROPY
CONSTANT TEMPERATURE
SATURATED VAPOR
SATURATED LIQUID
SCALE CHANGE
TEMPERATURE IN °FAHRENHEIT.
VOLUME IN CU. FEET PER LB.
ENTROPY IN B.T.U. PER LB. PER °F
Copyright Allied Chemical Corporation. All rights reserved. Reproduced by permission.

DATA SHEET FOR R 600

C_4H_{10}	BUTANE	Molecular Wt. 58

Pressure		Temperature				Volume		Density	
38.7	kg/cm²	152	°C	306	°F	4.38	l/kg	.228	kg/l.
551	psia	425	°K	766	°R	.0702	ft³/lb	14.2	lbs/ft³
								2	Air=1

AT CRITICAL POINT

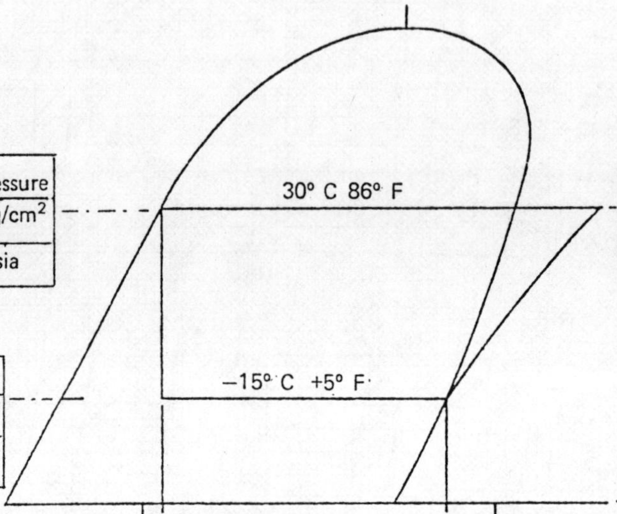

30° C 86° F

–15° C +5° F

Discharge Pressure	
2.9	kg/cm²
41.2	psia

Inlet Pressure	
.577	kg/cm²
8.2	psia

Discharge Temperature			
31	°C	88	°F
304	°K	548	°R

Normal Boiling Point			
-0.5	°C	31	°F
272.5	°K	491	°R

Triple Point			
-138	°C	-217	°F
135	°K	243	°R

Refrigerating Effect			
114	kcal/m³	12.8	Btu/ft³

Latent Heat at NBP	92.2 kcal/kg	5346 /kg mol	166 Btu/lb	9628 /lb mol

Trouton's No. 19.[illegible]	Gas Constant 14.62 kg.m/kg/°K	26.6 ft.lbs/lb/°R

Specific Heat Liquid at 30° C 86° F .58 Fig 16	Gas C_p	C_p C_v at

Liquid at 30° C 86° F Density .567 kg/l.	35.4 lbs/ft³ Viscosity cp

Reduced Form at NBP Pressure .027	1 / Temperature 1.56

Reduced value of 1/T when Reduced Pressure is	a) .001	b) .01	c) .1
		1.67	1.36

TEMP. °F	PRESSURE		VOLUME cu ft/lb	DENSITY lb/cu ft	ENTHALPY Btu/lb		ENTROPY Btu/(lb)(°R)	
	psia	psig	Vapor	Liquid	Liquid	Vapor	Liquid	Vapor
0	7.30	15.09*	11.100	38.59	−791	−619	0.8640	1.239
5	8.20	13.26*	10.030	38.42	−788	−618	.8697	1.238
10	9.21	11.21*	9.050	38.26	−786	−616	.8753	1.238
15	10.33	8.91*	8.150	38.09	−783	−614	.8810	1.238
20	11.56	6.42*	7.350	37.92	−780	−613	.8866	1.237
25	12.90	2.65*	6.640	37.76	−778	−611	0.8923	1.237
30	14.35	.82*	6.018	37.59	−775	−610	.8980	1.237
31.1	14.696	0	5.917	37.57	−775	−609	.8990	1.2369
35	15.95	1.25	5.488	37.41	−773	−608	0.9033	1.2369
40	17.62	2.92	4.998	37.22	−770	−606	.9085	1.2369
45	19.52	4.82	4.546	37.02	−768	−605	.9139	1.2369
50	21.55	6.85	4.141	36.83	−765	−603	.9193	1.2369
55	23.70	9.00	3.787	36.64	−762	−601	.9244	1.2369
60	26.02	11.32	3.474	36.44	−760	−600	0.9295	1.2369
65	28.52	13.82	3.186	36.25	−757	−598	.9346	1.2370
70	31.20	16.50	2.928	36.05	−754	−597	.9397	1.2370
75	34.00	19.20	2.695	35.86	−751	−595	.9446	1.2371
80	37.14	22.44	2.481	35.65	−749	−594	.9495	1.2371
85	40.51	25.81	2.284	35.45	−746	−592	0.9545	1.2372
90	43.91	29.21	2.109	35.25	−743	−590	.9594	1.2374
95	47.37	32.67	1.954	35.04	−740	−589	.9645	1.2376
100	51.37	36.67	1.808	34.83	−737	−587	.9696	1.2378
105	55.72	41.02	1.675	34.61	−734	−585	.9745	1.2381
110	60.27	46.57	1.553	34.40	−731	−584	0.9795	1.2384
115	64.98	50.28	1.442	34.19	−728	−582	0.9847	1.2388
120	69.98	55.28	1.341	33.97	−725	−581	0.9899	1.2392
125	75.23	60.53	1.248	33.75	−722	−579	0.9953	1.2398
130	80.83	66.13	1.163	33.52	−719	−578	1.0008	1.2404
135	86.83	72.13	1.085	33.29	−716	−576	1.0062	1.2410
140	92.87	78.17	1.012	33.06	−713	−575	1.0116	1.2417
145	99.37	84.67	0.945	32.83	−710	−574	1.0170	1.2423
150	106.2	91.5	0.8831	32.61	−707	−572	1.0224	1.2430
155	113.4	98.7	0.8264	32.38	−703	−571	1.0278	1.2437
160	120.9	106.2	0.7734	32.16	−700	−569	1.0332	1.2444
165	128.7	114.0	.7244	31.94	−697	−568	1.0386	1.2451
170	137.0	122.3	.6790	31.71	−694	−566	1.0441	1.2459
175	145.7	131.0	.6366	31.44	−690	−565	1.0495	1.2466
180	154.7	140.0	.5976	31.17	−687	−564	1.0547	1.2473
185	164.0	149.3	0.5617	30.87	−684	−562	1.0599	1.2480
190	173.8	159.1	.5285	30.58	−680	−561	1.0650	1.2488
195	184.0	169.3	.4969	30.26	−677	−560	1.0702	1.2495
200	194.7	180.0	.4672	29.94	−674	−558	1.0755	1.2503
205	205.9	191.2	.4395	29.63	−670	−557	1.0803	1.2510
210	217.4	202.7	0.4135	29.33	−666	−555	1.0851	1.2518
215	229.2	214.5	.3892	29.02	−663	−554	1.0908	1.2525
220	241.6	226.9	.3665	28.72	−659	−553	1.0965	1.2532
225	254.0	239.9	.3450	28.38	−655	−551	1.1018	1.2539
230	268.2	253.5	.3247	28.05	−652	−550	1.1071	1.2547
235	282.4	267.7	0.3054	27.69	−647	−548	1.1123	1.2553
240	297.2	282.5	.2868	27.32	−644	−547	1.1176	1.2559
245	312.5	297.8	.2692	26.87	−639	−545	1.1234	1.2565
250	328.3	313.6	.2525	26.42	−635	−544	1.1292	1.2572
255	344.8	330.1	.2364	25.91	−630	−543	1.1354	1.2580
260	361.9	347.2	0.2210	25.43	−626	−541	1.1417	1.2588
265	379.7	365.0	.2063	24.97	−621	−540	1.1488	1.2596
270	398.2	383.5	.1920	24.52	−615	−539	1.1559	1.2605
275	417.5	402.8	.1781	23.89	−610	−538	1.1630	1.2608
280	437.3	422.6	.1651	23.29	−605	−537	1.1701	1.2611
285	457.8	443.1	0.1521	22.61	−598	−537	1.1785	1.2609
290	479.3	464.6	.1384	21.82	−592	−537	1.1872	1.2607
295	501.6	486.9	.1224	20.83	−584	−538	1.1970	1.2590
300	524.9	510.2	.1005	18.50	−575	−540	1.2093	1.2554
305.62	550.7	536.0	.0702	14.245	−554	−554	1.2372	1.2372

* Inches of mercury below one standard atmosphere.

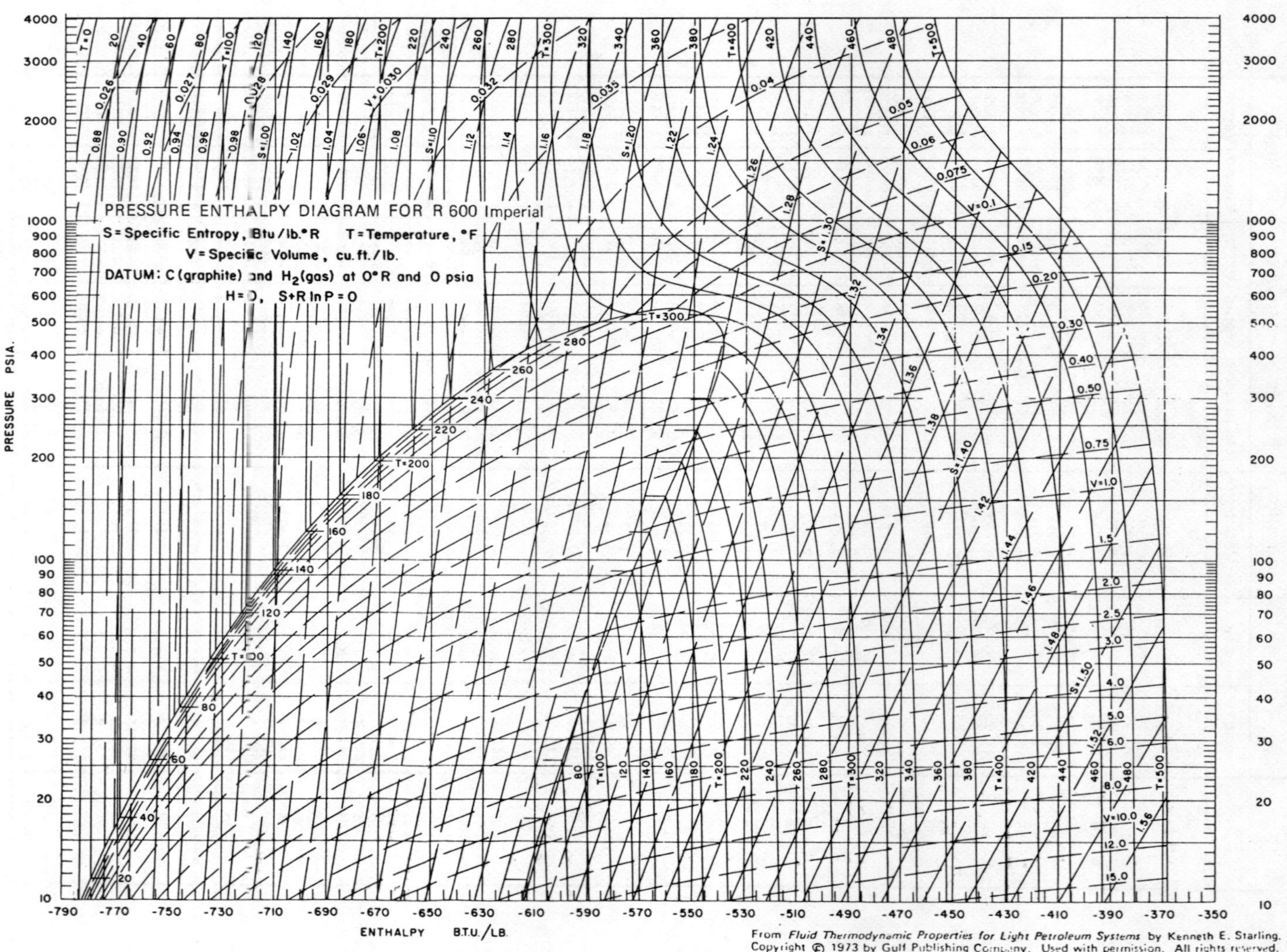

From *Fluid Thermodynamic Properties for Light Petroleum Systems* by Kenneth E. Starling.

DATA SHEET FOR R 601

C_4H_{10}	ISOBUTANE	Molecular Wt.	58

Pressure		Temperature				Volume		Density	
37.2	kg/cm²	135	°C	275	°F	4.52	l/kg	.221	kg/l.
529	psia	408	°K	735	°R	.0725	ft³/lb	13.8	lbs/ft³
								2	Air=1

AT CRITICAL POINT

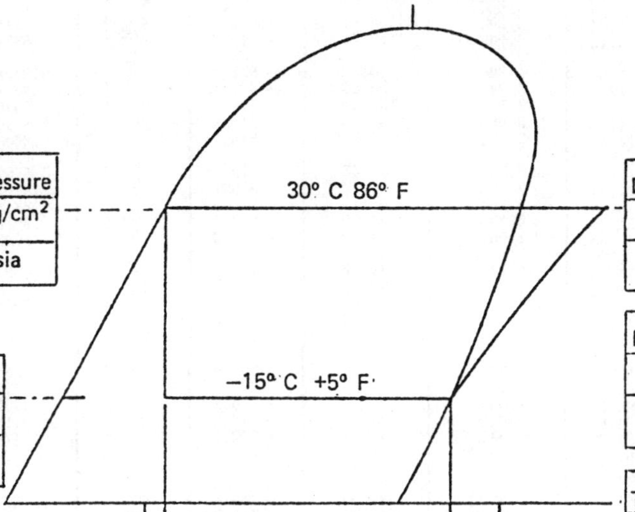

30° C 86° F

−15° C +5° F

Discharge Pressure	
4.2	kg/cm²
59.5	psia

Inlet Pressure	
.9	kg/cm²
12.8	psia

Discharge Temperature			
27	°C	80	°F
300	°K	540	°R

Normal Boiling Point			
-11	°C	10.9	°F
262	°K	471	°R

Triple Point			
-159	°C	-255	°F
114	°K	205	°R

Refrigerating Effect			
155	kcal/m³	17.4	Btu/ft³

Latent Heat at NBP	87 kcal/kg	5043 /kg mol	156.5 Btu/lb	9077 /lb mol

Trouton's No. 20.2	Gas Constant 14.62 kg.m/kg/° K	26.6 ft.lbs/lb/° R

Specific Heat Liquid at 30° C 86° F	Gas C_p	C_p C_v at

Liquid at 30° C 86° F Density .545 kg/l.	33.9 lbs/ft³ Viscosity cp

Reduced Form at NBP Pressure .0278	1 / Temperature 1.56

Reduced value of 1/T		1.72	1.36
when Reduced Pressure is	a) .001	b) .01	c) .1

TEMP. °F	PRESSURE		VOLUME cu ft/lb	DENSITY lb/cu ft	ENTHALPY Btu/lb		ENTROPY Btu/(lb)(°R)	
	psia	psig	Vapor	Liquid	Vapor	Liquid	Vapor	Liquid
−20	7.50	14.67*	10.70	38.36	−851.2	−688.4	0.795	1.166
−15	8.30	13.04*	9.66	38.11	−848.6	−686.7	.800	1.164
−10	9.28	11.04*	8.72	38.01	−846.0	−685.1	.804	1.162
−5	10.36	8.84*	7.88	37.68	−843.3	−683.4	.809	1.161
0	11.53	6.45*	7.116	37.48	−840.7	−681.8	.8146	1.1602
5	12.91	3.64*	6.416	37.29	−838.0	−680.2	0.8203	1.1599
10	14.41	0.58*	5.781	37.11	−835.3	−678.5	.8260	1.1596
10.89	14.69	0	5.685	37.05	−834.7	−678.2	.8272	1.1595
15	16.05	1.35	5.231	36.91	−832.5	−676.8	.8316	1.1593
20	17.82	3.12	4.740	36.72	−829.8	−675.2	.8372	1.1590
25	19.75	5.05	4.320	36.52	−827.0	−673.6	0.8428	1.1590
30	21.83	7.13	3.918	36.32	−824.1	−671.9	.8484	1.1590
35	24.07	9.37	3.575	36.13	−821.2	−670.3	.8540	1.1590
40	26.48	11.78	3.265	35.93	−818.4	−668.7	.8596	1.1590
45	29.07	14.37	2.985	35.73	−815.5	−667.1	.8652	1.1590
50	31.86	17.16	2.740	35.52	−812.7	−665.5	0.8707	1.1590
55	34.84	20.14	2.520	35.32	−809.7	−663.8	.8763	1.1592
60	38.04	23.34	2.314	35.11	−806.8	−662.2	.8818	1.1595
65	41.44	26.74	2.130	34.90	−803.9	−660.7	.8873	1.1598
70	45.06	30.36	1.967	34.70	−801.0	−659.1	.8928	1.1602
75	48.90	34.20	1.819	34.48	−798.0	−657.5	0.8983	1.1606
80	52.98	38.28	1.683	34.27	−795.1	−656.0	.9038	1.1611
85	57.33	42.63	1.559	34.05	−792.1	−654.5	.9093	1.1616
90	61.95	47.25	1.446	33.83	−789.0	−652.9	.9147	1.1621
95	66.86	52.16	1.343	33.60	−786.0	−651.4	.9202	1.1626
100	72.04	57.34	1.248	33.38	−782.9	−649.8	0.9256	1.1632
105	77.47	62.77	1.162	33.13	−779.9	−648.3	.9311	1.1639
110	83.22	68.52	1.083	32.91	−776.8	−646.7	.9365	1.1646
115	89.30	74.60	1.010	32.66	−773.8	−645.2	.9418	1.1653
120	95.73	81.03	0.9423	32.43	−770.7	−643.7	.9471	1.1660
125	102.55	87.85	0.8802	32.19	−767.5	−642.2	.9524	1.1667
130	109.72	95.02	0.8222	31.94	−764.3	−640.6	0.9577	1.1675
135	117.26	102.56	.7684	31.69	−761.0	−639.0	.9632	1.1682
140	125.18	110.48	.7189	31.43	−757.7	−637.5	.9687	1.1690
145	133.39	118.69	.6738	31.16	−754.4	−636.1	.9742	1.1697
150	141.98	127.28	.6315	30.89	−751.0	−634.6	.9797	1.1705
155	151.04	136.34	0.5917	30.62	−747.5	−633.1	0.9852	1.1712
160	160.55	145.85	.5545	30.35	−744.1	−631.8	.9908	1.1720
165	170.49	155.79	.5199	30.07	−740.7	−630.4	.9963	1.1727
170	180.89	166.19	.4876	29.78	−737.2	−629.0	1.0018	1.1735
175	191.73	177.03	.4577	29.47	−733.7	−627.7	1.0073	1.1742
180	203.04	188.34	0.4295	29.17	−730.1	−626.3	1.0128	1.1749
185	214.88	200.18	.4029	28.84	−726.5	−625.1	1.0183	1.1756
190	227.20	212.50	.3780	28.51	−722.8	−623.8	1.0239	1.1762
195	240.02	225.32	.3547	28.17	−719.1	−622.6	1.0294	1.1768
200	253.41	238.71	.3328	27.83	−715.3	−621.4	1.0349	1.1773
205	267.35	252.65	0.3121	27.46	−711.4	−620.2	1.0404	1.1777
210	281.85	267.15	.2925	27.09	−707.4	−619.0	1.0460	1.1780
215	296.94	282.24	.2742	26.67	−703.4	−618.0	1.0516	1.1781
220	312.60	297.90	.2567	26.25	−699.4	−617.0	1.0572	1.1783
225	328.88	314.18	.2400	25.77	−695.3	−616.3	1.0630	1.1783
230	345.82	331.12	0.2240	25.32	−691.2	−615.6	1.0688	1.1784
235	363.37	348.67	.2088	24.81	−687.0	−615.1	1.0747	1.1782
240	381.57	366.87	.1943	24.27	−682.8	−614.7	1.0806	1.1779
245	400.49	385.79	.1803	23.58	−678.2	−614.5	1.0867	1.1773
250	420.11	405.41	.1666	22.94	−673.5	−614.2	1.0929	1.1765
255	440.46	425.76	0.1533	22.08	−668.4	−614.3	1.0994	1.1752
260	461.54	444.84	.1403	21.28	−663.2	−614.5	1.1059	1.1735
265	483.35	468.65	.1272	19.80	−656.9	−615.5	1.1137	1.1711
270	505.85	481.15	.1142	18.52	−650.6	−616.7	1.1217	1.1681
274.96	529.06	514.36	.0724	13.81	−631.2	−631.2	1.1475	1.1475

* Inches of mercury below one standard atmosphere.

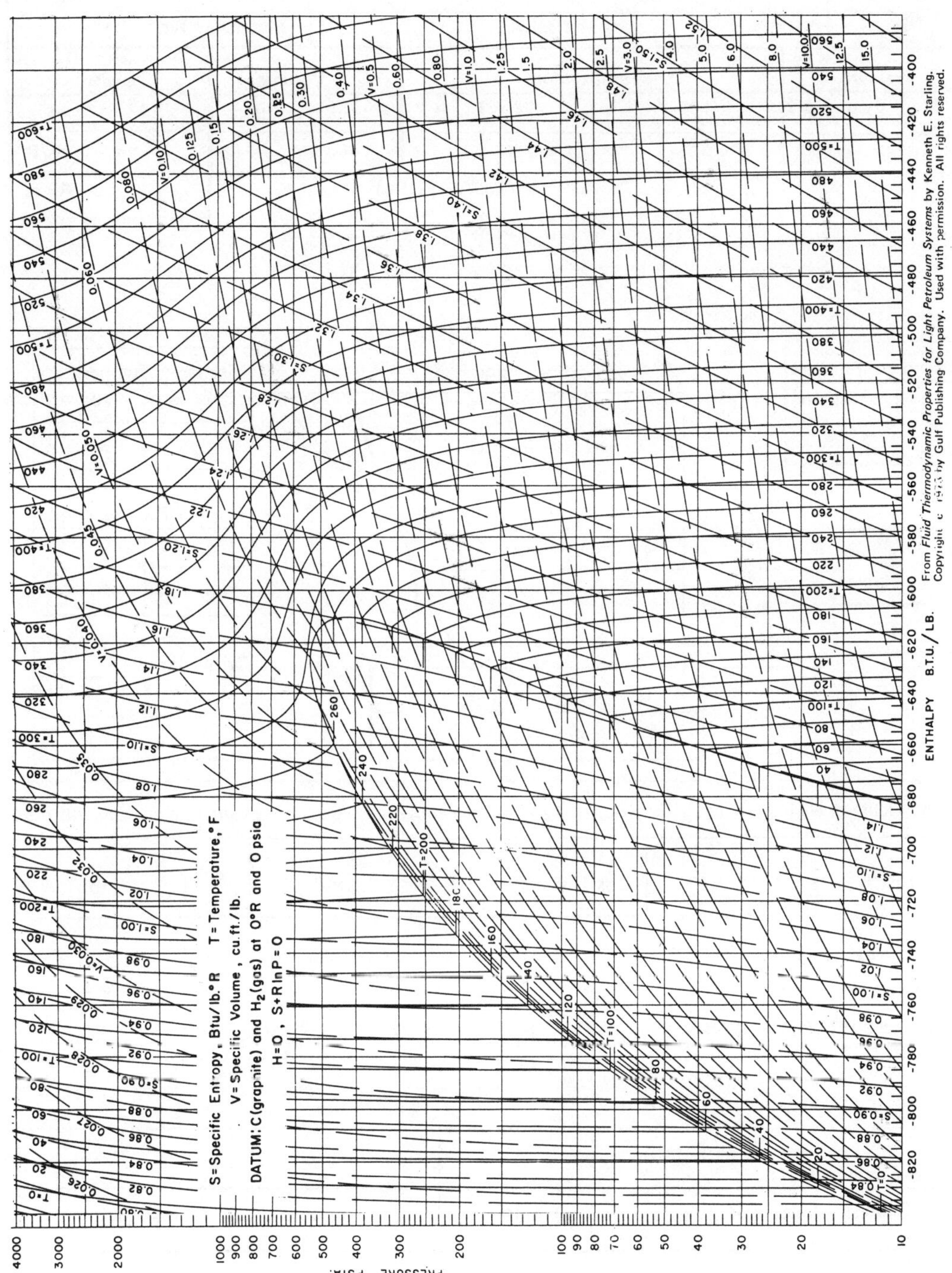

From *Fluid Thermodynamic Properties for Light Petroleum Systems* by Kenneth E. Starling.

DATA SHEET FOR R 717

NH_3	AMMONIA	Molecular Wt. 17

Pressure		Temperature				Volume		Density	
115.2	kg/cm²	132.4	°C	270.3	°F	4.26	l/kg	.235	kg/l.
1638	psia	405	°K	730	°R	.0683	ft³/lb	14.64	lbs/ft³
								.586	Air=1

AT CRITICAL POINT

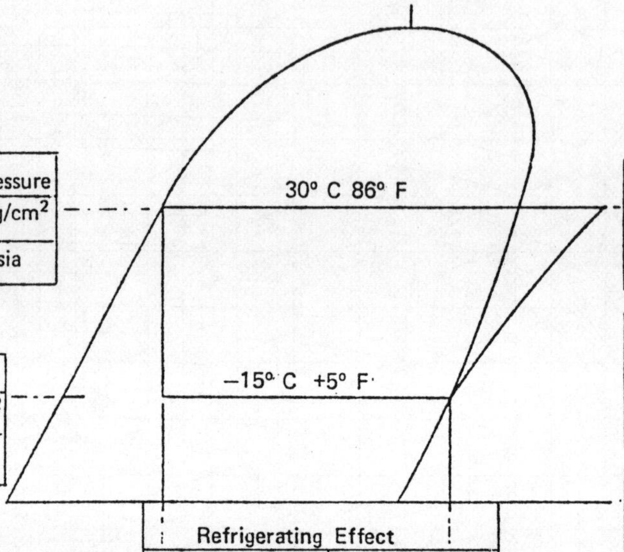

30° C 86° F

−15° C +5° F

Discharge Pressure	
11.9	kg/cm²
169	psia

Inlet Pressure	
2.4	kg/cm²
34.3	psia

Discharge Temperature			
99	°C	210	°F
373	°K	670	°R

Normal Boiling Point			
-33.3	°C	-28	°F
240	°K	432	°R

Triple Point			
-77.7	°C	-108	°F
240	°K	352	°R

Refrigerating Effect	
517 kcal/m³	58.2 Btu/ft³

Latent Heat at NBP	327 kcal/kg	5575 /kg mol	589 Btu/lb	10013 /lb mol

Trouton's No. 23.2	Gas Constant 49.8 kg.m/kg/° K	90.7 ft.lbs/lb/° R

Specific Heat Liquid at 30° C 86° F 1.15	Gas C_p	C_p/C_v 1.312 at 0°C

Liquid at 30° C 86° F Density 595 kg/l.	37.16 lbs/ft³ Viscosity cp

Reduced Form at NBP Pressure .0089	1 / Temperature 1.69

Reduced value of 1/T	1.99	1.67	1.34
when Reduced Pressure is	a) .001	b) .01	c) .1

TEMP. °F	PRESSURE		VOLUME cu ft/lb		DENSITY lb/cu ft		ENTHALPY Btu/lb			ENTROPY Btu/(lb) (°R)	
	PSIA	PSIG	LIQUID v_f	VAPOR v_g	LIQUID $1/v_f$	VAPOR $1/v_g$	LIQUID h_f	LATENT h_{fg}	VAPOR h_g	LIQUID s_f	VAPOR s_g
†−107·86	0·878	*28·1	0·02182	251·3	45·83	0·00398	−71·6	640·6	569·0	−0·1859	1·6354
−107	0·912	*28·1	0·02184	242·4	45·80	0·00413	−70·6	640·0	569·4	−0·1833	1·6320
−106	0·954	*28·0	0·02186	232·5	45·76	0·00430	−69·6	639·4	569·8	−0·1803	1·6281
−105	0·996	*27·9	0·02187	223·2	45·72	0·00448	−68·5	638·8	570·3	−0·1774	1·6243
−104	1·041	*27·8	0·02189	214·2	45·68	0·00467	−67·5	638·2	570·7	−0·1744	1·6205
−103	1·087	*27·7	0·02191	205·7	45·64	0·00486	−66·4	637·6	571·2	−0·1714	1·6167
−102	1·135	*27·6	0·02193	197·6	45·60	0·00506	−65·4	637·0	571·6	−0·1685	1·6129
−101	1·184	*27·5	0·02195	189·8	45·56	0·00527	−64·3	636·4	572·1	−0·1655	1·6092
−100	1·24	*27·4	0·02197	182·4	45·52	0·00548	−63·3	635·8	572·5	−0·1626	1·6055
− 99	1·29	*27·3	0·02199	175·3	45·48	0·00571	−62·2	635·2	572·9	−0·1597	1·6018
− 98	1·34	*27·2	0·02201	168·5	45·44	0·00593	−61·2	634·6	573·4	−0·1568	1·5982
− 97	1·40	*27·1	0·02203	162·1	45·40	0·00617	−60·1	633·9	573·8	−0·1539	1·5945
− 96	1·46	*26·9	0·02205	155·9	45·36	0·00641	−59·1	633·3	574·3	−0·1510	1·5910
− 95	1·52	*26·8	0·02207	150·0	45·32	0·00667	−58·0	632·7	574·7	−0·1481	1·5874
− 94	1·59	*26·7	0·02209	144·3	45·28	0·00693	−57·0	632·1	575·1	−0·1452	1·5838
− 93	1·65	*26·6	0·02211	138·9	45·24	0·00720	−55·9	631·5	575·6	−0·1423	1·5803
− 92	1·72	*26·4	0·02212	133·8	45·20	0·00747	−54·9	630·9	576·0	−0·1395	1·5768
− 91	1·79	*26·3	0·02214	128·9	45·16	0·00776	−53·8	630·3	576·5	−0·1366	1·5734
− 90	1·86	*26·1	0·02216	124·1	45·12	0·00806	−52·8	629·7	576·9	−0·1338	1·5699
− 89	1·94	*26·0	0·02218	119·6	45·08	0·00836	−51·7	629·0	577·3	−0·1309	1·5665
− 88	2·02	*25·8	0·02220	115·3	45·04	0·00868	−50·7	628·4	577·8	−0·1281	1·5631
− 87	2·10	*25·6	0·02222	111·1	45·00	0·00900	−49·6	627·8	578·2	−0·1253	1·5597
− 86	2·18	*25·5	0·02224	107·1	44·96	0·00933	−48·6	627·2	578·6	−0·1225	1·5564
− 85	2·27	*25·3	0·02226	103·33	44·92	0·00968	−47·5	626·6	579·1	−0·1197	1·5531
− 84	2·35	*25·1	0·02228	99·68	44·88	0·01003	−46·5	625·9	579·5	−0·1169	1·5498
− 83	2·45	*24·9	0·02230	96·17	44·84	0·01040	−45·4	625·3	579·9	−0·1141	1·5465
− 82	2·54	*24·7	0·02232	92·81	44·80	0·01077	−44·4	624·7	580·4	−0·1113	1·5432
− 81	2·64	*24·5	0·02234	89·59	44·76	0·01116	−43·3	624·1	580·8	−0·1085	1·5400
−80	2·74	*24·3	0·02236	86·50	44·72	0·01156	−42·2	623·5	581·2	−0·1057	1·5368
−79	2·84	*24·1	0·02238	83·54	44·68	0·01197	−41·2	622·8	581·6	−0·1030	1·5336
−78	2·95	*23·9	0·02240	80·69	44·64	0·01239	−40·1	622·2	582·1	−0·1002	1·5304
−77	3·06	*23·7	0·02242	77·96	44·60	0·01283	−39·1	621·6	582·5	−0·0975	1·5273
−76	3·18	*23·5	0·02244	75·33	44·56	0·01327	−38·0	621·0	582·9	−0·0947	1·5242
−75	3·29	*23·2	0·02246	72·81	44·52	0·01373	−37·0	620·3	583·3	−0·0920	1·5211
−74	3·42	*23·0	0·02248	70·39	44·48	0·01421	−35·9	619·7	583·8	−0·0892	1·5180
−73	3·54	*22·7	0·02250	68·06	44·44	0·01469	−34·9	619·1	584·2	−0·0865	1·5149
−72	3·67	*22·4	0·02252	65·82	44·40	0·01519	−33·8	618·4	584·6	−0·0838	1·5119
−71	3·80	*22·2	0·02254	63·67	44·36	0·01571	−32·8	617·8	585·0	−0·0811	1·5089
−70	3·94	*21·9	0·02256	61·60	44·32	0·01623	−31·7	617·2	585·5	−0·0784	1·5059
−69	4·08	*21·6	0·02258	59·61	44·28	0·01678	−30·7	616·6	585·9	−0·0757	1·5029
−68	4·23	*21·3	0·02260	57·69	44·24	0·01733	−29·6	615·9	586·3	−0·0730	1·4999
−67	4·38	*21·0	0·02262	55·85	44·20	0·01791	−28·6	615·3	586·7	−0·0703	1·4970
−66	4·53	*20·7	0·02265	54·08	44·16	0·01840	−27·5	614·6	587·1	−0·0676	1·4940
−65	4·69	*20·4	0·02267	52·37	44·11	0·01910	−26·5	614·0	587·5	−0·0650	1·4911
−64	4·85	*20·0	0·02269	50·73	44·07	0·01971	−25·4	613·4	588·0	−0·0623	1·4883
−63	5·02	*19·7	0·02271	49·14	44·03	0·02035	−24·4	612·7	588·4	−0·0596	1·4854
−62	5·19	*19·4	0·02274	47·62	43·99	0·02100	−23·3	612·1	588·8	−0·0570	1·4826
−61	5·37	*19·0	0·02276	46·15	43·95	0·02167	−22·2	611·4	589·2	−0·0543	1·4797
−60	5·55	*18·6	0·02278	44·73	43·91	0·02235	−21·2	610·8	589·6	−0·0517	1·4769
−59	5·74	*18·2	0·02280	43·37	43·86	0·02306	−20·1	610·1	590·0	−0·0490	1·4741
−58	5·93	*17·8	0·02282	42·05	43·82	0·02378	−19·1	609·5	590·4	−0·0464	1·4713
−57	6·13	*17·4	0·02284	40·79	43·78	0·02452	−18·0	608·8	590·8	−0·0438	1·4686
−56	6·33	*17·0	0·02286	39·56	43·74	0·02528	−17·0	608·2	591·2	−0·0412	1·4658
−55	6·54	*16·6	0·02288	38·38	43·70	0·02605	−15·9	607·5	591·6	−0·0386	1·4631
−54	6·75	*16·2	0·02290	37·24	43·66	0·02685	−14·8	606·9	592·1	−0·0360	1·4604
−53	6·97	*15·7	0·02292	36·15	43·62	0·02766	−13·8	606·2	592·4	−0·0334	1·4577
−52	7·20	*15·3	0·02295	35·09	43·57	0·02850	−12·7	605·6	592·9	−0·0307	1·4551
−51	7·43	*14·8	0·02297	34·06	43·53	0·02936	−11·7	604·9	593·2	−0·0281	1·4524

*Inches of mercury below one atmosphere

TEMP.	PRESSURE		VOLUME cu ft/lb		DENSITY lb/cu ft		ENTHALPY Btu/lb			ENTROPY Btu/(lb)(°R)	
°F	PSIA	PSIG	LIQUID v_f	VAPOR v_g	LIQUID $1/v_f$	VAPOR $1/v_g$	LIQUID h_f	LATENT h_{fg}	VAPOR h_g	LIQUID s_f	VAPOR s_g
−50	7·67	*14·3	0·02299	33·08	43·49	0·0302	−10·6	604·3	593·7	−0·0256	1·4497
−49	7·91	*13·8	0·02301	32·12	43·45	0·0311	− 9·6	603·6	594·0	−0·0230	1·4471
−48	8·16	*13·3	0·02303	31·20	43·41	0·0320	− 8·5	602·9	594·4	−0·0204	1·4445
−47	8·42	*12·8	0·02306	30·31	43·37	0·0329	− 7·4	602·3	594·9	−0·0179	1·4419
−46	8·68	*12·2	0·02308	29·45	43·33	0·0339	− 6·4	601·6	595·2	−0·0153	1·4393
−45	8·95	*11·7	0·02310	28·62	43·28	0·0349	−5·3	600·9	595·6	−0·0127	1·4368
−44	9·23	*11·1	0·02312	27·82	43·24	0·0359	−4·3	600·3	596·0	−0·0102	1·4342
−43	9·51	*10·6	0·02315	27·04	43·20	0·0369	−3·2	599·6	596·4	−0·0076	1·4317
−42	9·81	*10·0	0·02317	26·29	43·16	0·0380	−2·1	598·9	596·8	−0·0051	1·4292
−41	10·10	*9·3	0·02320	25·56	43·12	0·0391	−1·1	598·3	597·2	−0·0025	1·4267
−40	10·41	*8·7	0·02322	24·86	43·08	0·0402	0·0	597·6	597·6	0·0000	1·4242
−39	10·72	*8·1	0·02324	24·18	43·03	0·0413	1·1	596·9	598·0	0·0025	1·4217
−38	11·04	*7·4	0·02326	23·53	42·99	0·0425	2·1	596·2	598·3	0·0051	1·4193
−37	11·37	*6·8	0·02329	22·89	42·94	0·0436	3·2	595·5	598·7	0·0076	1·4169
−36	11·71	*6·1	0·02331	22·27	42·90	0·0448	4·3	594·8	599·1	0·0101	1·4144
−35	12·05	*5·4	0·02333	21·68	42·86	0·0461	5·3	594·2	599·5	0·0126	1·4120
−34	12·41	*4·7	0·02335	21·10	42·82	0·0473	6·4	593·5	599·9	0·0151	1·4096
−33	12·77	*3·9	0·02338	20·54	42·78	0·0486	7·4	592·8	600·2	0·0176	1·4072
−32	13·14	*3·2	0·02340	20·00	42·74	0·0499	8·5	592·1	600·6	0·0201	1·4048
−31	13·52	*2·4	0·02343	19·48	42·69	0·0513	9·6	591·4	601·0	0·0226	1·4025
−30	13·90	*1·6	0·02345	18·97	42·65	0·0527	10·7	590·7	601·4	0·0250	1·4001
−29	14·30	*0·8	0·02347	18·48	42·61	0·0441	11·7	590·0	601·7	0·0275	1·3978
−28	14·71	0·0	0·02350	18·00	42·57	0·0555	12·8	589·3	602·1	0·0300	1·3955
−27	15·12	0·4	0·02352	17·54	42·53	0·0570	13·9	588·6	602·5	0·0325	1·3932
−26	15·55	0·8	0·02355	17·09	42·48	0·0585	14·9	587·9	602·8	0·0350	1·3909
−25	15·98	1·3	0·02357	16·66	42·44	0·0600	16·0	587·2	603·2	0·0374	1·3886
−24	16·42	1·7	0·02359	16·24	42·39	0·0615	17·1	586·5	603·6	0·0399	1·3863
−23	16·88	2·2	0·02362	15·83	42·35	0·0631	18·1	585·8	603·9	0·0423	1·3840
−22	17·34	2·6	0·02364	15·43	42·31	0·0647	19·2	585·1	604·3	0·0448	1·3818
−21	17·81	3·1	0·02367	15·05	42·26	0·0664	20·3	584·3	604·6	0·0472	1·3796
−20	18·30	3·6	0·02369	14·68	42·22	0·0681	21·4	583·6	605·0	0·0497	1·3774
−19	18·79	4·1	0·02371	14·32	42·18	0·0698	22·4	582·9	605·3	0·0521	1·3752
−18	19·30	4·6	0·02374	13·97	42·13	0·0716	23·5	582·2	605·7	0·0545	1·3729
−17	19·81	5·1	0·02376	13·62	42·09	0·0734	24·6	581·5	606·1	0·0570	1·3708
−16	20·34	5·6	0·02379	13·29	42·04	0·0752	25·6	580·8	606·4	0·0594	1·3686
−15	20·88	6·2	0·02381	12·97	42·00	0·0770	26·7	580·0	606·7	0·0618	1·3664
−14	21·43	6·7	0·02383	12·66	41·96	0·0789	27·8	579·3	607·1	0·0642	1·3643
−13	21·99	7·3	0·02386	12·36	41·91	0·0809	28·9	578·6	607·5	0·0666	1·3621
−12	22·56	7·9	0·02388	12·06	41·87	0·0828	30·0	577·8	607·8	0·0690	1·3600
−11	23·15	8·5	0·02391	11·78	41·82	0·0849	31·0	577·1	608·1	0·0714	1·3579
−10	23·74	9·0	0·02393	11·50	41·78	0·0869	32·1	576·4	608·5	0·0738	1·3558
− 9	24·35	9·7	0·02396	11·23	41·74	0·0890	33·2	575·6	608·8	0·0762	1·3537
− 8	24·97	10·3	0·02398	10·97	41·70	0·0911	34·3	574·9	609·2	0·0786	1·3516
− 7	25·61	10·9	0·02401	10·71	41·65	0·0933	35·4	574·1	609·5	0·0809	1·3495
− 6	26·26	11·6	0·02403	10·47	41·61	0·0955	36·4	573·4	609·8	0·0833	1·3474
− 5	26·92	12·2	0·02406	10·23	41·56	0·0978	37·5	572·6	610·1	0·0857	1·3454
− 4	27·59	12·9	0·02409	9·991	41·52	0·1001	38·6	571·9	610·5	0·0880	1·3433
− 3	28·28	13·6	0·02411	9·763	41·48	0·1024	39·7	571·1	610·8	0·0904	1·3413
− 2	28·98	14·3	0·02414	9·541	41·43	0·1048	40·7	570·4	611·1	0·0928	1·3393
− 1	29·69	15·0	0·02416	9·326	41·39	0·1072	41·8	569·6	611·4	0·0951	1·3372
0	30·42	15·7	0·02419	9·116	41·34	0·1097	42·9	568·9	611·8	0·0975	1·3352
1	31·16	16·5	0·02422	8·912	41·29	0·1122	44·0	568·1	612·1	0·0998	1·3332
2	31·92	17·2	0·02424	8·714	41·25	0·1148	45·1	567·3	612·4	0·1022	1·3312
3	32·69	18·0	0·02427	8·521	41·21	0·1174	46·2	566·5	612·7	0·1045	1·3292
4	33·47	18·8	0·02429	8·333	41·16	0·1200	47·2	565·8	613·0	0·1069	1·3273
5	34·27	19·6	0·02432	8·150	41·11	0·1227	48·3	565·0	613·3	0·1092	1·3253
6	35·09	20·4	0·02435	7·971	41·07	0·1254	49·4	564·2	613·6	0·1115	1·3234
7	35·92	21·2	0·02438	7·798	41·03	0·1282	50·5	563·4	613·9	0·1138	1·3214
8	36·77	22·1	0·02440	7·629	40·99	0·1311	51·6	562·7	614·3	0·1162	1·3195
9	37·63	22·9	0·02443	7·464	40·94	0·1340	52·7	561·9	614·6	0·1185	1·3176

*Inches of mercury below one atmosphere

TEMP. °F	PRESSURE		VOLUME cu ft/lb		DENSITY lb/cu ft		ENTHALPY Btu/lb			ENTROPY Btu/(lb) (°R)	
	PSIA	PSIG	LIQUID v_f	VAPOR v_g	LIQUID $1/v_f$	VAPOR $1/v_g$	LIQUID h_f	LATENT h_{fg}	VAPOR h_g	LIQUID s_f	VAPOR s_g
10	38·51	23·8	0·02446	7·304	40·89	0·1369	53·8	561·1	614·9	0·1208	1·3157
11	39·40	24·7	0·02449	7·148	40·84	0·1399	54·9	560·3	615·2	0·1231	1·3137
12	40·31	25·6	0·02452	6·996	40·79	0·1429	56·0	559·5	615·5	0·1254	1·3118
13	41·24	26·5	0·02454	6·847	40·75	0·1460	57·1	558·7	615·8	0·1277	1·3099
14	42·18	27·5	0·02457	6·703	40·70	0·1492	58·2	557·9	616·1	0·1300	1·3081
15	43·14	28·4	0·02460	6·562	40·66	0·1524	59·2	557·1	616·3	0·1323	1·3062
16	44·12	29·4	0·02463	6·425	40·61	0·1556	60·3	556·3	616·6	0·1346	1·3043
17	45·12	30·4	0·02466	6·291	40·56	0·1590	61·4	555·5	616·9	0·1369	1·3025
18	46·13	31·4	0·02468	6·161	40·52	0·1623	62·5	554·7	617·2	0·1392	1·3006
19	47·16	32·5	0·02471	6·034	40·47	0·1657	63·6	553·9	617·5	0·1415	1·2988
20	48·21	33·5	0·02474	5·910	40·43	0·1692	64·7	553·1	617·8	0·1437	1·2969
21	49·28	34·6	0·02477	5·789	40·38	0·1728	65·8	552·2	618·0	0·1460	1·2951
22	50·36	35·7	0·02480	5·671	40·33	0·1763	66·9	551·4	618·3	0·1483	1·2933
23	51·47	36·8	0·02482	5·556	40·28	0·1800	68·0	550·6	618·6	0·1505	1·2915
24	52·59	37·9	0·02485	5·443	40·24	0·1837	69·1	549·8	618·9	0·1528	1·2897
25	53·73	39·0	0·02488	5·334	40·20	0·1875	70·2	548·9	619·1	0·1551	1·2879
26	54·90	40·2	0·02491	5·227	40·15	0·1913	71·3	548·1	619·4	0·1573	1·2861
27	56·08	41·4	0·02494	5·123	40·10	0·1952	72·4	547·3	619·7	0·1596	1·2843
28	57·28	42·6	0·02497	5·021	40·05	0·1992	73·5	546·4	619·9	0·1618	1·2825
29	58·50	43·8	0·02500	4·922	40·00	0·2032	74·6	545·6	620·2	0·1641	1·2808
30	59·74	45·0	0·02503	4·825	39·96	0·2073	75·7	544·8	620·5	0·1663	1·2790
31	61·00	46·3	0·02506	4·730	39·91	0·2114	76·8	543·9	620·7	0·1686	1·2773
32	62·29	47·6	0·02509	4·637	39·86	0·2156	77·9	543·1	621·0	0·1708	1·2755
33	63·59	48·9	0·02512	4·547	39·81	0·2199	79·0	542·2	621·2	0·1730	1·2738
34	64·91	50·2	0·02515	4·459	39·76	0·2243	80·1	541·4	621·5	0·1753	1·2721
35	66·26	51·6	0·02518	4·373	39·72	0·2287	81·2	540·5	621·7	0·1775	1·2704
36	67·63	52·9	0·02521	4·289	39·67	0·2332	82·3	539·7	622·0	0·1797	1·2686
37	69·02	54·3	0·02524	4·207	39·62	0·2377	83·4	538·8	622·2	0·1819	1·2669
38	70·43	55·7	0·02527	4·126	39·58	0·2423	84·6	537·9	622·5	0·1841	1·2652
39	71·87	57·2	0·02530	4·048	39·54	0·2470	85·7	537·0	622·7	0·1863	1·2635
40	73·32	58·6	0·02533	3·971	39·49	0·2518	86·8	536·2	623·0	0·1885	1·2618
41	74·80	60·1	0·02536	3·897	39·44	0·2566	87·9	535·3	623·2	0·1908	1·2602
42	76·31	61·6	0·02539	3·823	39·40	0·2616	89·0	534·4	623·4	0·1930	1·2585
43	77·83	63·1	0·02542	3·752	39·34	0·2665	90·1	533·6	623·7	0·1952	1·2568
44	79·38	64·7	0·02545	3·682	39·29	0·2716	91·2	532·7	623·9	0·1974	1·2552
45	80·96	66·3	0·02548	3·614	39·24	0·2767	92·3	531·8	624·1	0·1996	1·2535
46	82·55	67·9	0·02551	3·547	39·20	0·2819	93·5	530·9	624·4	0·2018	1·2519
47	84·18	69·5	0·02554	3·481	39·15	0·2872	94·6	530·0	624·6	0·2040	1·2502
48	85·82	71·1	0·02558	3·418	39·10	0·2926	95·7	529·1	624·8	0·2062	1·2486
49	87·49	72·8	0·02561	3·355	39·05	0·2981	96·8	528·2	625·0	0·2083	1·2469
50	89·19	74·5	0·02564	3·294	39·00	0·3036	97·9	527·3	625·2	0·2105	1·2453
51	90·91	76·2	0·02567	3·234	38·95	0·3092	99·1	526·4	625·5	0·2127	1·2437
52	92·66	78·0	0·02571	3·176	38·90	0·3149	100·2	525·5	625·7	0·2149	1·2421
53	94·43	79·7	0·02574	3·119	38·85	0·3207	101·3	524·6	625·9	0·2171	1·2405
54	96·23	81·5	0·02578	3·063	38·80	0·3265	102·4	523·7	626·1	0·2192	1·2389
55	98·06	83·4	0·02581	3·008	38·75	0·3325	103·5	522·8	626·3	0·2214	1·2373
56	99·91	85·2	0·02584	2·954	38·70	0·3385	104·7	521·8	626·5	0·2236	1·2357
57	101·8	87·1	0·02587	2·902	38·65	0·3446	105·8	520·9	626·7	0·2257	1·2341
58	103·7	89·0	0·02591	2·851	38·60	0·3508	106·9	520·0	626·9	0·2279	1·2325
59	105·6	90·9	0·02594	2·800	38·55	0·3571	108·1	519·0	627·1	0·2301	1·2310
60	107·6	92·9	0·02597	2·751	38·50	0·3635	109·2	518·1	627·3	0·2322	1·2294
61	109·6	94·9	0·02600	2·703	38·45	0·3700	110·3	517·2	627·5	0·2344	1·2278
62	111·6	96·9	0·02604	2·656	38·40	0·3765	111·5	516·2	627·7	0·2365	1·2262
63	113·6	98·9	0·02607	2·610	38·35	0·3832	112·6	515·3	627·9	0·2387	1·2247
64	115·7	101·0	0·02611	2·565	38·30	0·3899	113·7	514·3	628·0	0·2408	1·2231
65	117·8	103·1	0·02614	2·520	38·25	0·3968	114·8	513·4	628·2	0·2430	1·2216
66	120·0	105·3	0·02618	2·477	38·20	0·4037	116·0	512·4	628·4	0·2451	1·2201
67	122·1	107·4	0·02621	2·435	38·15	0·4108	117·1	511·5	628·6	0·2473	1·2186
68	124·3	109·6	0·02625	2·393	38·10	0·4179	118·3	510·5	628·8	0·2494	1·2170
69	126·5	111·8	0·02628	2·352	38·05	0·4251	119·4	509·5	628·9	0·2515	1·2155

TEMP. °F	PRESSURE		VOLUME cu ft/lb		DENSITY lb/cu ft		ENTHALPY Btu/lb			ENTROPY Btu/(lb) (°R)	
	PSIA	PSIG	LIQUID v_f	VAPOR v_g	LIQUID $1/v_f$	VAPOR $1/v_g$	LIQUID h_f	LATENT h_{fg}	VAPOR h_g	LIQUID s_f	VAPOR s_g
70	128·8	114·1	0·02632	2·312	38·00	0·4325	120·5	508·6	629·1	0·2537	1·2140
71	131·1	116·4	0·02636	2·273	37·94	0·4399	121·7	507·6	629·3	0·2558	1·2125
72	133·4	118·7	0·02639	2·235	37·89	0·4474	122·8	506·6	629·4	0·2579	1·2110
73	135·7	121·0	0·02643	2·197	37·84	0·4551	124·0	505·6	629·6	0·2601	1·2095
74	138·1	123·4	0·02646	2·161	37·79	0·4628	125·1	504·7	629·8	0·2622	1·2080
75	140·5	125·8	0·02650	2·125	37·74	0·4707	126·2	503·7	629·9	0·2643	1·2065
76	143·0	128·3	0·02654	2·089	37·69	0·4786	127·4	502·7	630·1	0·2664	1·2050
77	145·4	130·7	0·02657	2·055	37·64	0·4867	128·5	501·7	630·2	0·2685	1·2035
78	147·9	133·2	0·02661	2·021	37·58	0·4949	129·7	500·7	630·4	0·2706	1·2020
79	150·5	135·8	0·02664	1·988	37·53	0·5031	130·8	499·7	630·5	0·2728	1·2006
80	153·0	138·3	0·02668	1·955	37·48	0·5115	132·0	498·7	630·7	0·2749	1·1991
81	155·6	140·9	0·02672	1·923	37·43	0·5200	133·1	497·7	630·8	0·2769	1·1976
82	158·3	143·6	0·02676	1·892	37·38	0·5287	134·3	496·7	631·0	0·2791	1·1962
83	161·0	146·3	0·02679	1·861	37·33	0·5374	135·4	495·7	631·1	0·2812	1·1947
84	163·7	149·0	0·02683	1·831	37·27	0·5462	136·6	494·7	631·3	0·2833	1·1933
85	166·4	151·7	0·02687	1·801	37·21	0·5552	137·8	493·6	631·4	0·2854	1·1918
86	169·2	154·5	0·02691	1·772	37·15	0·5643	138·9	492·6	631·5	0·2875	1·1904
87	172·0	157·3	0·02695	1·744	37·10	0·5735	140·1	491·6	631·7	0·2895	1·1889
88	174·8	160·1	0·02699	1·716	37·05	0·5828	141·2	490·6	631·8	0·2917	1·1875
89	177·7	163·0	0·02703	1·688	37·00	0·5923	142·4	489·5	631·9	0·2937	1·1860
90	180·6	165·9	0·02707	1·661	36·95	0·6019	143·5	488·5	632·0	0·2958	1·1846
91	183·6	168·9	0·02711	1·635	36·89	0·6116	144·7	487·4	632·1	0·2979	1·1832
92	186·6	171·9	0·02715	1·609	36·83	0·6214	145·8	486·4	632·2	0·3000	1·1818
93	189·6	174·9	0·02719	1·584	36·78	0·6314	147·0	485·3	632·3	0·3021	1·1804
94	192·7	178·0	0·02723	1·559	36·72	0·6415	148·2	484·3	632·5	0·3041	1·1789
95	195·8	181·1	0·02727	1·534	36·67	0·6517	149·4	483·2	632·6	0·3062	1·1775
96	198·9	184·2	0·02731	1·510	36·61	0·6620	150·5	482·1	632·6	0·3083	1·1761
97	202·1	187·4	0·02735	1·487	36·56	0·6725	151·7	481·1	632·8	0·3104	1·1747
98	205·3	190·6	0·02739	1·464	36·51	0·6832	152·9	480·0	632·9	0·3125	1·1733
99	208·6	193·9	0·02743	1·441	36·45	0·6939	154·0	478·9	632·9	0·3145	1·1719
100	211·9	197·2	0·02747	1·419	36·40	0·7048	155·2	477·8	633·0	0·3166	1·1705
101	215·2	200·5	0·02751	1·397	36·35	0·7159	156·4	476·7	633·1	0·3187	1·1691
102	218·6	203·9	0·02756	1·375	36·29	0·7270	157·6	475·6	633·2	0·3207	1·1677
103	222·0	207·3	0·02760	1·354	36·23	0·7384	158·7	474·6	633·3	0·3228	1·1663
104	225·4	210·7	0·02765	1·334	36·17	0·7498	159·9	473·5	633·4	0·3248	1·1649
105	228·9	214·2	0·02769	1·313	36·12	0·7615	161·1	472·3	633·4	0·3269	1·1635
106	232·5	217·8	0·02773	1·293	36·06	0·7732	162·3	471·2	633·5	0·3289	1·1621
107	236·0	221·3	0·02777	1·274	36·00	0·7852	163·5	470·1	633·6	0·3310	1·1607
108	239·7	225·0	0·02782	1·254	35·95	0·7972	164·6	469·0	633·6	0·3330	1·1593
109	243·3	228·6	0·02786	1·235	35·90	0·8095	165·8	467·9	633·7	0·3351	1·1580
110	247·0	232·3	0·02790	1·217	35·84	0·8219	167·0	466·7	633·7	0·3372	1·1566
111	250·8	236·1	0·02795	1·198	35·78	0·8344	168·2	465·6	633·8	0·3392	1·1552
112	254·5	239·8	0·02799	1·180	35·73	0·8471	169·4	464·4	633·8	0·3413	1·1538
113	258·4	243·7	0·02804	1·163	35·67	0·8600	170·6	463·3	633·9	0·3433	1·1524
114	262·2	247·5	0·02808	1·145	35·61	0·8730	171·8	462·1	633·9	0·3453	1·1510
115	266·2	251·5	0·02813	1·128	35·55	0·8862	173·0	460·9	633·9	0·3474	1·1497
116	270·1	255·4	0·02818	1·112	35·49	0·8996	174·2	459·8	634·0	0·3495	1·1483
117	274·1	259·4	0·02822	1·095	35·44	0·9132	175·4	458·6	634·0	0·3515	1·1469
118	278·2	263·5	0·02827	1·079	35·38	0·9269	176·6	457·4	634·0	0·3535	1·1455
119	282·3	267·6	0·02831	1·063	35·32	0·9408	177·8	456·2	634·0	0·3556	1·1441
120	286·4	271·7	0·02836	1·047	35·26	0·9549	179·0	455·0	634·0	0·3576	1·1427
121	290·6	275·9	0·02841	1·032	35·20	0·9692	180·2	453·8	634·0	0·3597	1·1414
122	294·8	280·1	0·02846	1·017	35·14	0·9837	181·4	452·6	634·0	0·3618	1·1400
123	299·1	284·4	0·02850	1·002	35·08	0·9983	182·6	451·4	634·0	0·3638	1·1386
124	303·4	288·7	0·02855	0·987	35·02	1·0132	183·9	450·1	634·0	0·3659	1·1372
125	307·8	293·1	0·02860	0·973	34·96	1·0280	185·1	448·9	634·0	0·3679	1·1358

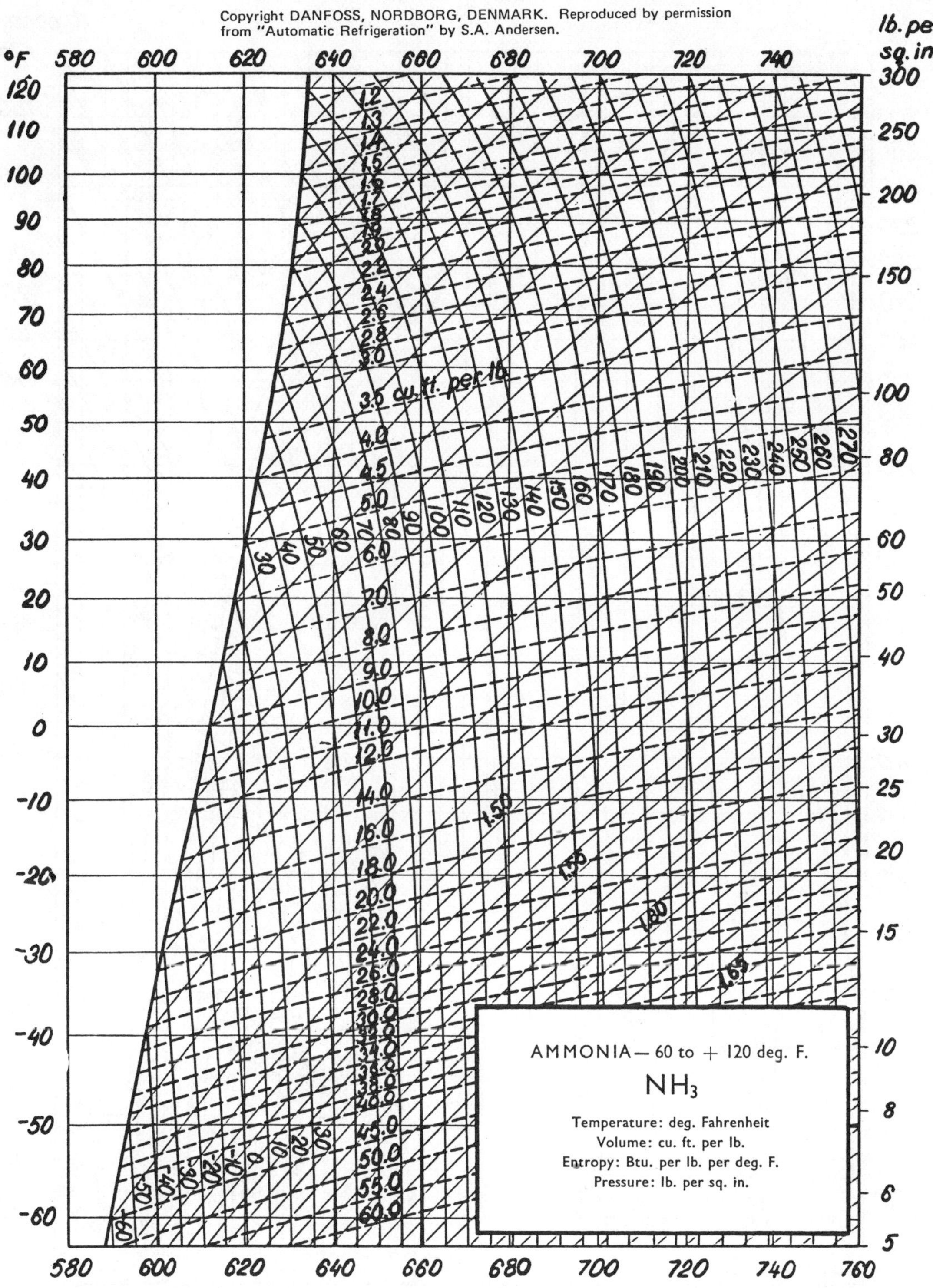
°F
580
600
620
640
660
680
700
720
740
760
lb. per sq. in.
300
250
200
150
100
80
60
50
40
30
25
20
15
10
8
6
5
3.5 cu. ft. per lb.
AMMONIA — 60 to + 120 deg. F.
NH3
Temperature: deg. Fahrenheit
Volume: cu. ft. per lb.
Entropy: Btu. per lb. per deg. F.
Pressure: lb. per sq. in.

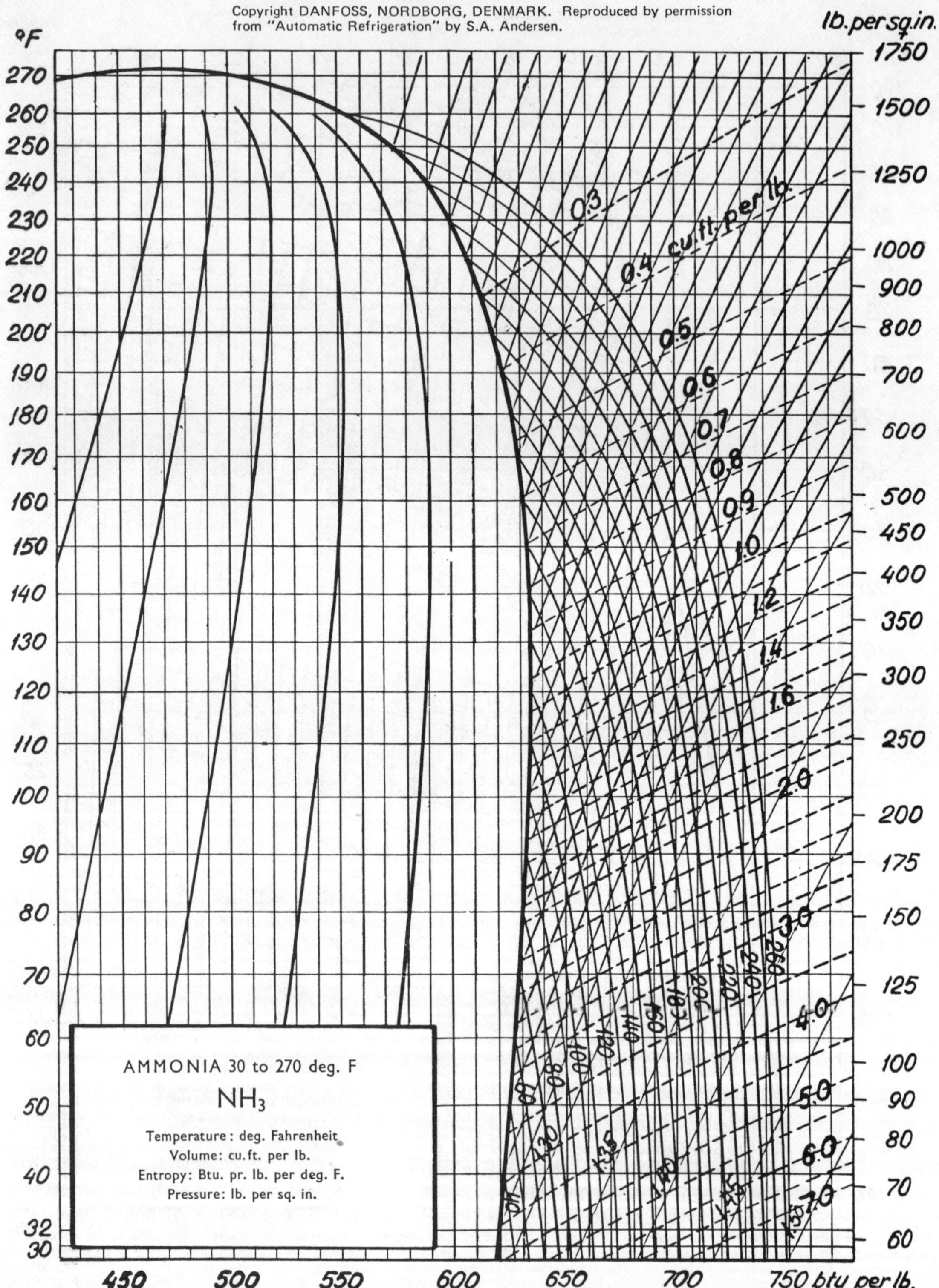
°F
lb. per sq. in.
AMMONIA 30 to 270 deg. F
NH_3
Temperature: deg. Fahrenheit
Volume: cu.ft. per lb.
Entropy: Btu. pr. lb. per deg. F.
Pressure: lb. per sq. in.
0.3
0.4 cu.ft. per lb.
btu per lb.

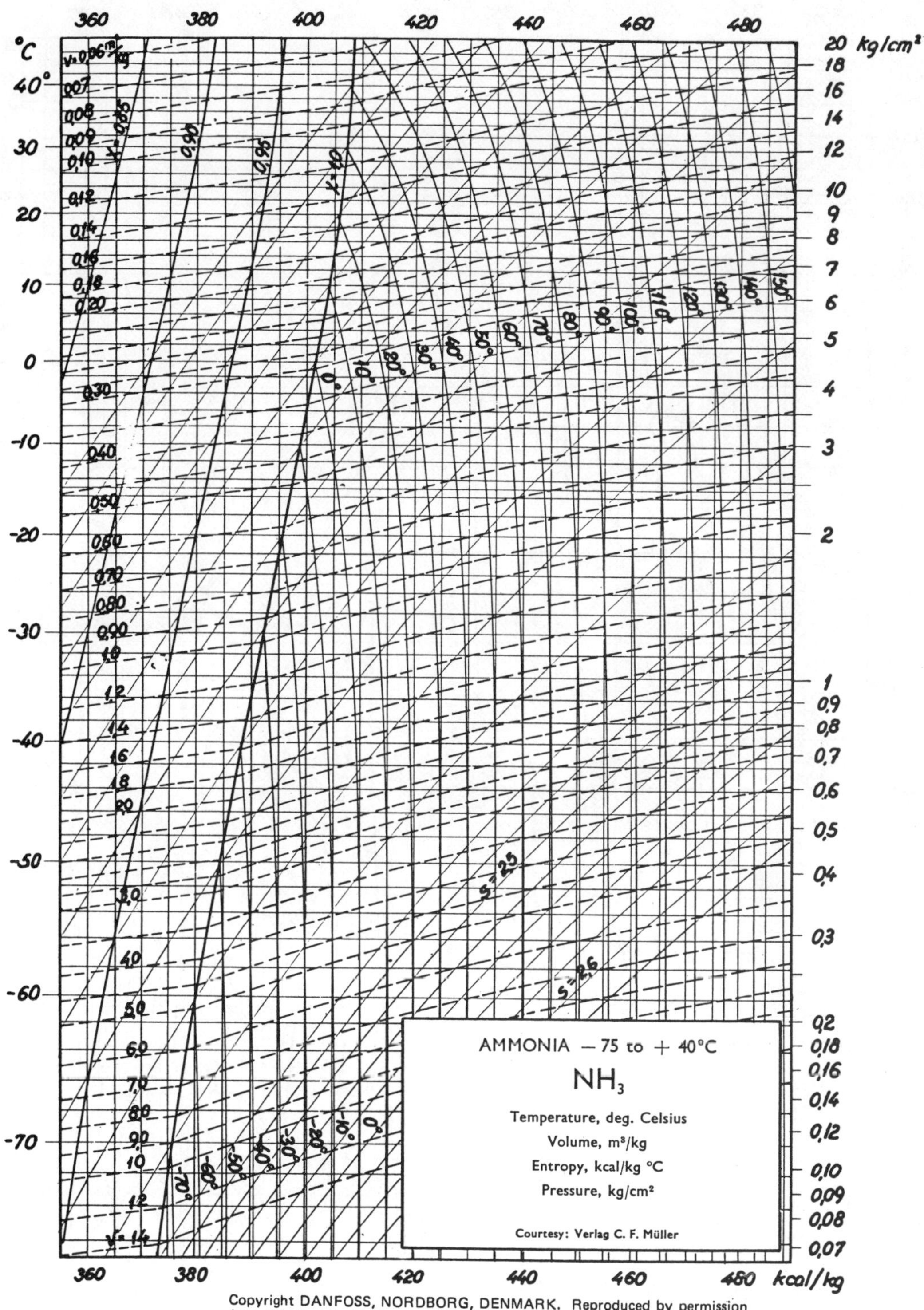
AMMONIA −75 to + 40°C
NH_3
Temperature, deg. Celsius
Volume, m³/kg
Entropy, kcal/kg °C
Pressure, kg/cm²
Courtesy: Verlag C. F. Müller
°C
kg/cm²
kcal/kg

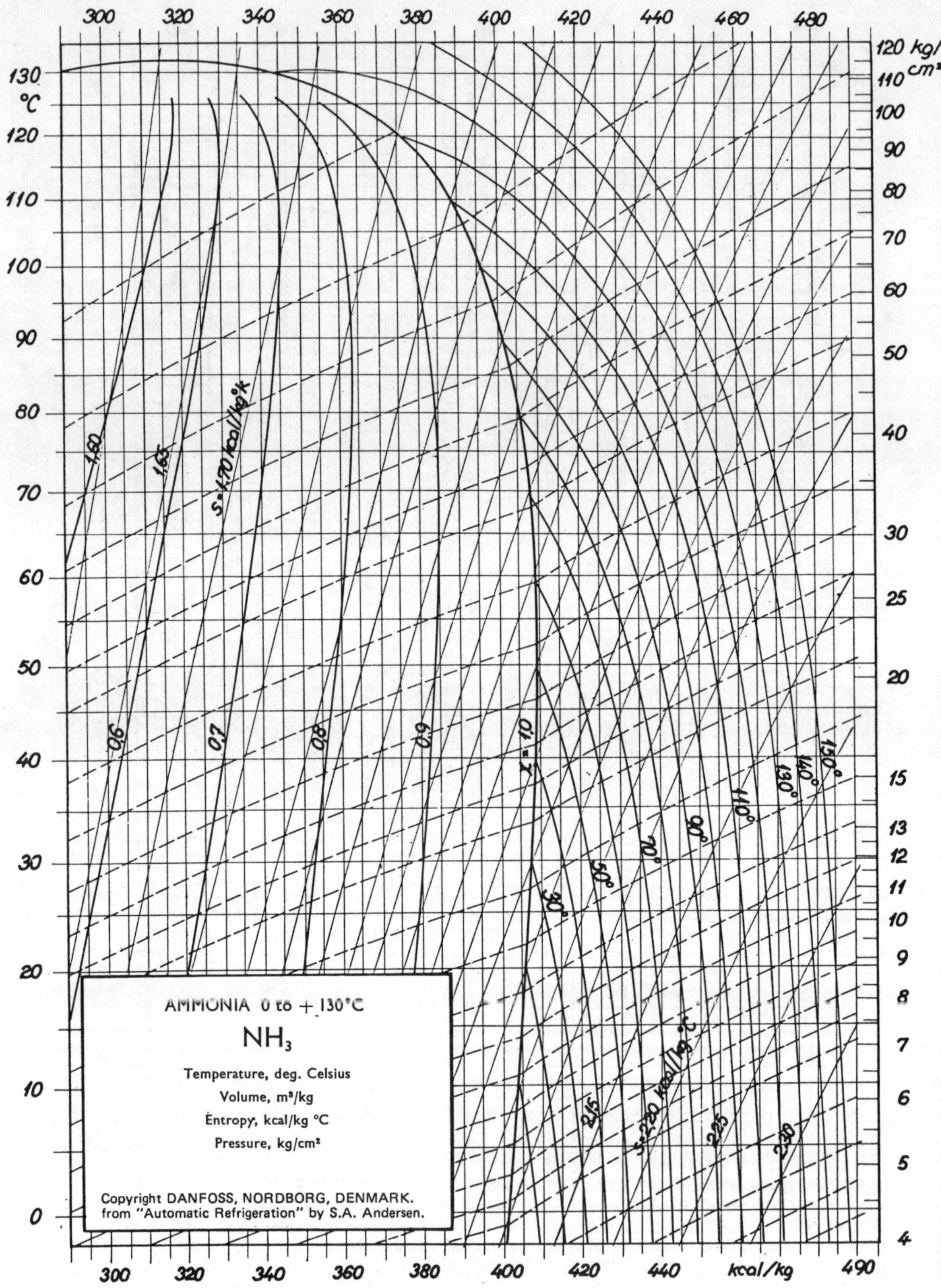
AMMONIA 0 to + 130°C
NH_3
Temperature, deg. Celsius
Volume, m³/kg
Entropy, kcal/kg °C
Pressure, kg/cm²
Copyright DANFOSS, NORDBORG, DENMARK.
from "Automatic Refrigeration" by S.A. Andersen.
300 320 340 360 380 400 420 440 460 480
130 °C 120 110 100 90 80 70 60 50 40 30 20 10 0
120 kg/cm² 110 100 90 80 70 60 50 40 30 25 20 15 13 12 11 10 9 8 7 6 5 4
1,60 1,65 s = 1,70 kcal/kg°K
0,6 0,7 0,8 0,9 x = 1,0
30° 50° 70° 90° 110° 130° 140° 150°
2,15 s = 2,20 kcal/kg °C 2,25 2,30
300 320 340 360 380 400 420 440 kcal/kg 490

DATA SHEET FOR R 744

CO_2	CARBON DIOXIDE	Molecular Wt. 44

Pressure		Temperature				Volume		Density	
75.2	kg/cm^2	31	°C	87.8	°F	2.15	l/kg	.465	kg/l.
1069	psia	304	°K	548	°R	.0345	ft^3/lb	28.98	lbs/ft^3
								1.52	Air=1

AT CRITICAL POINT

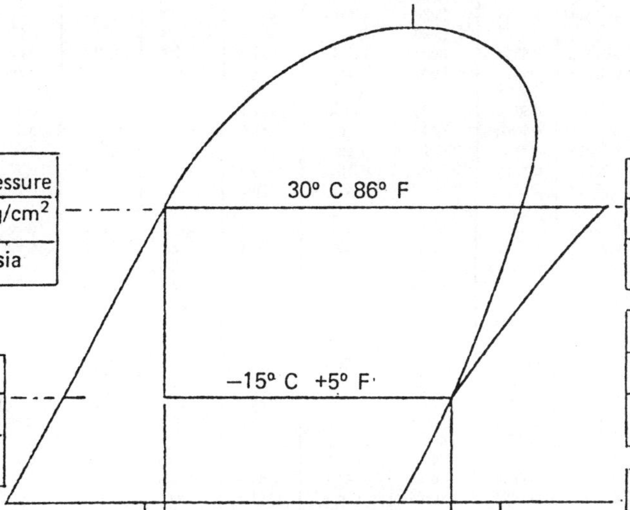

30° C 86° F

−15° C +5° F

Discharge Pressure	
73.3	kg/cm^2
1043	psia

Inlet Pressure	
23.2	kg/cm^2
332	psia

Discharge Temperature			
66	°C	150	°F
339	°K	610	°R

Normal Boiling Point			
-78	°C	-109	°F
195	°K	351	°R

Triple Point			
-56	°C	-70	°F
217	°K	390	°R

Refrigerating Effect	
1827 kcal/m^3	205.5 Btu/ft^3

Latent Heat at NBP	90 kcal/kg	3960 /kg mol	162.1 Btu/lb	7132 /lb mol

Trouton's No. 20.3	Gas Constant 19.3 kg.m/kg/° K	35.1 ft.lbs/lb/° R

Specific Heat Liquid at 30° C 86° F	Gas C_p .184	C_p/C_v 1.3 at -40°C

Liquid at 30° C 86° F Density .059 kg/l.	3.7 lbs/ft^3 Viscosity cp

Reduced Form at NBP Pressure .0137	1 / Temperature 1.56

Reduced value of 1/T			1.53
when Reduced Pressure is	a) .001	b) .01	c) .1

TEMP.	PRESSURE		VOLUME cu ft/lb		DENSITY lb/cu ft		ENTHALPY Btu/lb			ENTROPY Btu/(lb)(°R)	
°F	PSIA	PSIG	LIQUID v_f	VAPOR v_g	LIQUID $1/v_f$	VAPOR $1/v_g$	LIQUID h_f	LATENT h_{fg}	VAPOR h_g	LIQUID s_f	VAPOR s_g
†−69·9	75·1	60·4	0·01360	1·157	73·53	0·8640	129·6	149·6	279·2	0·8885	1·2724
−69	76·7	62·0	0·01362	1·133	73·42	0·8826	130·0	149·3	279·3	0·8895	1·2716
−68	78·6	63·9	0·01365	1·108	73·26	0·9025	130·4	148·9	279·3	0·8906	1·2708
−67	80·5	65·8	0·01367	1·083	73·16	0·9233	130·9	148·5	279·4	0·8917	1·2700
−66	82·4	67·7	0·01369	1·0589	73·05	0·9443	131·4	148·1	279·5	0·8929	1·2692
−65	84·4	69·7	0·01372	1·0354	72·88	0·9658	131·8	147·8	279·6	0·8941	1·2685
−64	86·4	71·7	0·01374	1·0125	72·78	0·9876	132·3	147·3	279·6	0·8952	1·2677
−63	88·4	73·7	0·01377	0·9903	72·62	1·010	132·7	147·0	279·7	0·8963	1·2670
−62	90·5	75·8	0·01379	0·9686	72·52	1·032	133·2	146·6	279·8	0·8975	1·2662
−61	92·6	77·9	0·01381	0·9475	72·41	1·055	133·6	146·2	279·8	0·8986	1·2654
−60	94·7	80·0	0·01384	0·9270	72·26	1·079	134·1	145·8	279·9	0·8997	1·2647
−59	96·9	82·2	0·01386	0·9070	72·15	1·102	134·5	145·4	279·9	0·9009	1·2639
−58	99·1	84·4	0·01389	0·8875	72·00	1·127	135·0	145·0	280·0	0·9020	1·2631
−57	101·4	86·7	0·01391	0·8686	71·89	1·151	135·5	144·6	280·1	0·9031	1·2623
−56	103·7	89·0	0·01393	0·8502	71·79	1·176	135·9	144·2	280·1	0·9042	1·2616
−55	106·0	91·3	0·01396	0·8323	71·63	1·201	136·4	143·8	280·2	0·9053	1·2609
−54	108·4	93·7	0·01398	0·8149	71·53	1·227	136·8	143·4	280·2	0·9065	1·2601
−53	110·8	96·1	0·01401	0·7978	71·38	1·253	137·3	143·0	280·3	0·9076	1·2594
−52	113·2	98·5	0·01404	0·7812	71·23	1·280	137·7	142·7	280·4	0·9087	1·2587
−51	115·7	101·0	0·01406	0·7650	71·12	1·307	138·2	142·2	280·4	0·9099	1·2579
−50	118·2	103·5	0·01409	0·7492	70·98	1·335	138·6	141·9	280·5	0·9110	1·2572
−49	120·8	106·1	0·01411	0·7338	70·87	1·363	139·1	141·5	280·6	0·9120	1·2565
−48	123·4	108·7	0·01414	0·7188	70·72	1·391	139·5	141·1	280·6	0·9131	1·2558
−47	126·0	111·3	0·01417	0·7042	70·57	1·420	140·0	140·7	280·7	0·9142	1·2551
−46	128·7	114·0	0·01420	0·6899	70·42	1·449	140·5	140·3	280·8	0·9153	1·2544
−45	131·4	116·7	0·01423	0·6760	70·28	1·479	141·0	139·9	280·9	0·9165	1·2537
−44	134·2	119·5	0·01425	0·6624	70·18	1·510	141·4	139·5	280·9	0·9175	1·2530
−43	137·0	122·3	0·01428	0·6491	70·03	1·540	141·9	139·1	281·0	0·9186	1·2524
−42	139·9	125·2	0·01431	0·6362	69·88	1·572	142·3	138·7	281·0	0·9196	1·2517
−41	142·8	128·1	0·01434	0·6236	69·74	1·604	142·8	138·2	281·0	0·9207	1·2509
−40	145·8	131·1	0·01437	0·6113	69·60	1·635	143·3	137·8	281·1	0·9218	1·2503
−39	148·8	134·1	0·01440	0·5993	69·45	1·669	143·7	137·4	281·1	0·9229	1·2496
−38	151·8	137·1	0·01442	0·5876	69·35	1·702	144·2	137·0	281·2	0·9239	1·2490
−37	154·9	140·2	0·01445	0·5761	69·20	1·736	144·6	136·6	281·2	0·9250	1·2483
−36	158·0	143·3	0·01448	0·5649	69·06	1·770	145·1	136·2	281·3	0·9261	1·2476
−35	161·2	146·5	0·01451	0·5540	68·92	1·805	145·5	135·8	281·3	0·9272	1·2470
−34	164·4	149·7	0·01454	0·5433	68·78	1·840	146·0	135·4	281·4	0·9282	1·2463
−33	167·7	153·0	0·01457	0·5329	68·64	1·876	146·5	134·9	281·4	0·9293	1·2456
−32	171·0	156·3	0·01460	0·5227	68·49	1·913	146·9	134·5	281·4	0·9303	1·2449
−31	174·4	159·7	0·01463	0·5127	68·35	1·950	147·4	134·1	281·5	0·9314	1·2443
−30	177·8	163·1	0·01466	0·5029	68·22	1·988	147·8	133·7	281·5	0·9325	1·2436
−29	181·3	166·6	0·01469	0·4934	68·08	2·027	148·3	133·3	281·6	0·9335	1·2430
−28	184·8	170·1	0·01472	0·4841	67·93	2·066	148·7	132·9	281·6	0·9345	1·2424
−27	188·4	173·7	0·01476	0·4750	67·75	2·105	149·2	132·5	281·7	0·9356	1·2417
−26	192·0	177·3	0·01479	0·4661	67·61	2·145	149·7	132·0	281·7	0·9366	1·2410
−25	195·7	181·0	0·01482	0·4574	67·48	2·186	150·1	131·6	281·7	0·9377	1·2404
−24	199·4	184·7	0·01485	0·4489	67·34	2·228	150·6	131·2	281·8	0·9387	1·2398
−23	203·2	188·5	0·01488	0·4406	67·20	2·270	151·1	130·7	281·8	0·9397	1·2391
−22	207·0	192·3	0·01491	0·4325	67·07	2·312	151·5	130·3	281·8	0·9408	1·2385
−21	210·9	196·2	0·01494	0·4246	66·94	2·355	152·0	129·8	281·8	0·9418	1·2378
−20	214·9	200·2	0·01498	0·4168	66·76	2·399	152·4	129·4	281·8	0·9430	1·2372
−19	218·9	204·2	0·01501	0·4092	66·62	2·444	152·9	129·0	281·9	0·9439	1·2366
−18	223·0	208·3	0·01504	0·4017	66·49	2·489	153·4	128·5	281·9	0·9449	1·2359
−17	227·1	212·4	0·01508	0·3944	66·31	2·535	153·8	128·1	281·9	0·9460	1·2353
−16	231·2	216·5	0·01511	0·3872	66·18	2·583	154·3	127·6	281·9	0·9470	1·2347

Base : Enthalpy of liquid 180 BThU/lb at 32°F.
Entropy of liquid 1.0 BThU/lb/°R at 32°F.

† Triple Point

TEMP.	PRESSURE		VOLUME cu ft/lb		DENSITY lb/cu ft		ENTHALPY Btu/lb			ENTROPY Btu/(lb)(°R)	
°F	PSIA	PSIG	LIQUID v_f	VAPOR v_g	LIQUID $1/v_f$	VAPOR $1/v_g$	LIQUID h_f	LATENT h_{fg}	VAPOR h_g	LIQUID s_f	VAPOR s_g
−15	235·4	220·7	0·01515	0·3802	66·00	2·630	154·8	127·2	282·0	0·9480	1·2340
−14	239·6	224·9	0·01518	0·3733	65·88	2·679	155·3	126·7	282·0	0·9491	1·2334
−13	243·9	229·2	0·01522	0·3666	65·70	2·728	155·8	126·2	282·0	0·9501	1·2328
−12	248·3	233·6	0·01525	0·3600	65·58	2·778	156·2	125·8	282·0	0·9511	1·2321
−11	252·8	238·1	0·01529	0·3535	65·40	2·829	156·7	125·3	282·0	0·9522	1·2315
−10	257·3	242·6	0·01532	0·3472	65·28	2·880	157·2	124·8	282·0	0·9532	1·2309
− 9	261·9	247·2	0·01536	0·3410	65·10	2·932	157·7	124·3	282·0	0·9542	1·2303
− 8	266·5	251·8	0·01540	0·3349	64·94	2·986	158·1	123·9	282·0	0·9552	1·2297
− 7	271·2	256·5	0·01544	0·3289	64·77	3·040	158·6	123·4	282·0	0·9563	1·2291
− 6	275·9	261·2	0·01547	0·3231	64·64	3·095	159·1	122·9	282·0	0·9573	1·2284
− 5	280·6	265·9	0·01552	0·3174	64·43	3·150	159·6	122·5	282·1	0·9584	1·2278
− 4	285·4	270·7	0·01555	0·3118	64·31	3·207	160·1	122·0	282·1	0·9594	1·2272
− 3	290·3	275·6	0·01559	0·3063	64·14	3·264	160·6	121·5	282·1	0·9604	1·2265
− 2	295·3	280·6	0·01563	0·3009	63·98	3·323	161·1	121·0	282·1	0·9614	1·2259
− 1	300·4	285·7	0·01566	0·2956	63·86	3·383	161·5	120·6	282·1	0·9625	1·2253
0	305·5	290·8	0·01570	0·2904	63·69	3·443	162·1	120·1	282·1	0·9636	1·2247
1	310·7	296·0	0·01575	0·2853	63·49	3·505	162·5	119·6	282·1	0·9646	1·2241
2	315·9	301·2	0·01579	0·2803	63·33	3·567	163·1	119·0	282·1	0·9657	1·2235
3	321·2	306·5	0·01584	0·2755	63·13	3·629	163·6	118·5	282·1	0·9668	1·2229
4	326·5	311·8	0·01588	0·2707	62·97	3·694	164·1	118·0	282·1	0·9679	1·2224
5	331·9	317·2	0·01592	0·2660	62·81	3·759	164·6	117·5	282·1	0·9690	1·2218
6	337·4	322·7	0·01596	0·2614	62·66	3·825	165·1	116·9	282·0	0·9701	1·2212
7	343·0	328·3	0·01600	0·2568	62·50	3·894	165·7	116·3	282·0	0·9711	1·2205
8	348·7	334·0	0·01605	0·2524	62·31	3·962	166·2	115·8	282·0	0·9722	1·2199
9	354·4	339·7	0·01610	0·2480	62·11	4·032	166·7	115·3	282·0	0·9733	1·2194
10	360·2	345·5	0·01614	0·2437	61·96	4·103	167·3	114·7	282·0	0·9744	1·2188
11	366·0	351·3	0·01618	0·2395	61·80	4·175	167·8	114·2	282·0	0·9754	1·2182
12	371·9	357·2	0·01623	0·2354	61·62	4·248	168·3	113·6	281·9	0·9765	1·2176
13	377·9	363·2	0·01628	0·2314	61·43	4·321	168·9	113·0	281·9	0·9776	1·2170
14	383·9	369·2	0·01632	0·2274	61·27	4·397	169·4	112·5	281·9	0·9787	1·2163
15	390·0	375·3	0·01637	0·2235	61·09	4·474	169·9	111·9	281·8	0·9798	1·2157
16	396·2	381·5	0·01642	0·2197	60·90	4·551	170·5	111·3	281·8	0·9810	1·2151
17	402·5	387·8	0·01647	0·2159	60·72	4·632	171·1	110·7	281·8	0·9821	1·2145
18	408·9	394·2	0·01652	0·2121	60·53	4·715	171·6	110·1	281·7	0·9833	1·2139
19	415·3	400·6	0·01658	0·2085	60·31	4·796	172·2	109·5	281·7	0·9844	1·2133
20	421·8	407·1	0·01663	0·2049	60·13	4·880	172·7	108·9	281·6	0·9856	1·2127
21	428·3	413·6	0·01668	0·2014	59·95	4·965	173·3	108·3	281·6	0·9867	1·2121
22	434·9	420·2	0·01673	0·1979	59·78	5·053	173·8	107·7	281·5	0·9879	1·2115
23	441·6	426·9	0·01679	0·1945	59·56	5·141	174·4	107·1	281·5	0·9890	1·2109
24	448·4	433·7	0·01684	0·1912	59·38	5·230	175·0	106·4	281·4	0·9902	1·2103
25	455·3	440·6	0·01689	0·1879	59·21	5·322	175·6	105·8	281·4	0·9915	1·2097
26	462·2	447·5	0·01695	0·1846	59·00	5·417	176·2	105·1	281·3	0·9927	1·2091
27	469·2	454·5	0·01701	0·1814	58·79	5·513	176·8	104·5	281·3	0·9939	1·2085
28	476·3	461·6	0·01707	0·1783	58·58	5·608	177·4	103·8	281·2	0·9951	1·2079
29	483·5	468·8	0·01713	0·1752	58·38	5·708	178·1	103·1	281·2	0·9963	1·2073
30	490·8	476·1	0·01719	0·1722	58·17	5·807	178·7	102·4	281·1	0·9976	1·2067
31	498·1	483·4	0·01725	0·1692	57·97	5·910	179·4	101·7	281·1	0·9988	1·2061
32	505·5	490·8	0·01731	0·1663	57·77	6·013	180·0	101·0	281·0	1·0000	1·2055
33	513·0	498·3	0·01738	0·1634	57·54	6·120	180·6	100·3	280·9	1·0011	1·2047
34	520·6	505·9	0·01744	0·1605	57·34	6·231	181·2	99·5	280·7	1·0023	1·2039
35	528·3	513·6	0·01751	0·1557	57·11	6·341	181·8	98·8	280·6	1·0034	1·2033
36	536·0	521·3	0·01759	0·1550	56·85	6·452	182·4	98·1	280·5	1·0046	1·2024
37	543·8	529·1	0·01766	0·1523	56·62	6·566	183·0	97·4	280·4	1·0058	1·2017
38	551·7	537·0	0·01773	0·1496	56·40	6·685	183·7	96·5	280·2	1·0069	1·2008
39	559·7	545·0	0·01780	0·1470	56·18	6·802	184·3	95·8	280·1	1·0081	1·2002
40	567·8	553·1	0·01787	0·1444	55·96	6·925	185·0	95·0	280·0	1·0092	1·1994
41	576·0	561·3	0·01794	0·1418	55·75	7·054	185·6	94·2	279·8	1·0103	1·1985
42	584·3	569·6	0·01801	0·1393	55·52	7·178	186·2	93·4	279·6	1·0115	1·1978
43	592·7	578·0	0·01809	0·1368	55·28	7·310	186·9	92·6	279·5	1·0128	1·1970
44	601·1	586·4	0·01817	0·1344	55·02	7·440	187·6	91·8	279·4	1·0140	1·1962

TEMP	PRESSURE		VOLUME cu ft/lb		DENSITY lb/cu ft		ENTHALPY Btu/lb			ENTROPY Btu/(lb)(°R)	
°F	PSIA	PSIG	LIQUID v_f	VAPOR v_g	LIQUID $1/v_f$	VAPOR $1/v_g$	LIQUID h_f	LATENT h_{fg}	VAPOR h_g	LIQUID s_f	VAPOR s_g
45	609·6	594·9	0·01825	0·1320	54·79	7·576	188·2	91·0	279·2	1·0153	1·1955
46	618·2	603·5	0·01834	0·1297	54·52	7·710	188·9	90·1	279·0	1·0166	1·1947
47	626·9	612·2	0·01842	0·1273	54·29	7·856	189·6	89·2	278·8	1·0179	1·1939
48	635·7	621·0	0·01851	0·1250	54·02	8·000	190·3	88·4	278·7	1·0192	1·1932
49	644·6	629·9	0·01859	0·1227	53·79	8·150	191·0	87·5	278·5	1·0204	1·1924
50	653·6	638·9	0·01868	0·1205	53·56	8·298	191·7	86·6	278·3	1·0218	1·1917
51	662·7	648·0	0·01878	0·1183	53·25	8·453	192·4	85·6	278·0	1·0231	1·1908
52	671·9	657·2	0·01887	0·1161	52·99	8·614	193·1	84·7	277·8	1·0244	1·1900
53	681·2	666·5	0·01896	0·1139	52·74	8·780	193·8	83·7	277·5	1·0258	1·1891
54	690·6	675·9	0·01906	0·1117	52·46	8·953	194·5	82·7	277·2	1·0272	1·1882
55	700·0	685·3	0·01916	0·1096	52·19	9·124	195·2	81·8	277·0	1·0283	1·1874
56	709·5	694·8	0·01927	0·1075	51·90	9·302	195·9	80·8	276·7	1·0299	1·1865
57	719·1	704·4	0·01938	0·1055	51·60	9·479	196·6	79·8	276·4	1·0312	1·1855
58	728·8	714·1	0·01948	0·1034	51·34	9·671	197·3	78·7	276·0	1·0326	1·1846
59	738·6	723·9	0·01959	0·1014	51·06	9·863	198·0	77·7	275·7	1·0340	1·1835
60	748·6	733·9	0·01970	0·0994	50·76	10·06	198·8	76·6	275·4	1·0353	1·1826
61	758·7	744·0	0·01982	0·0975	50·46	10·26	199·5	75·5	275·0	1·0368	1·1816
62	768·9	754·2	0·01994	0·0956	50·15	10·46	200·3	74·4	274·7	1·0382	1·1805
63	779·1	764·4	0·02006	0·0937	49·85	10·67	201·1	73·2	274·3	1·0396	1·1794
64	789·4	774·7	0·02020	0·0918	49·50	10·89	201·9	72·0	273·9	1·0410	1·1783
65	799·8	785·1	0·02033	0·0899	49·19	11·12	202·7	70·7	273·4	1·0422	1·1773
66	810·3	795·6	0·02048	0·0880	48·83	11·36	203·5	69·5	273·0	1·0438	1·1760
67	820·9	806·2	0·02064	0·0861	48·45	11·61	204·4	68·1	272·5	1·0453	1·1747
68	831·6	816·9	0·02079	0·0842	48·11	11·87	205·2	66·8	272·0	1·0468	1·1734
69	842·5	827·8	0·02094	0·0823	47·75	12·15	206·1	65·4	271·5	1·0482	1·1719
70	853·4	838·7	0·02112	0·0804	47·35	12·44	207·0	63·8	270·8	1·0500	1·1705
71	864·4	849·7	0·02131	0·0784	46·93	12·74	207·9	62·3	270·2	1·0515	1·1689
72	875·5	860·8	0·02150	0·0765	46·51	13·06	208·8	60·7	269·5	1·0532	1·1674
73	886·8	872·1	0·02172	0·0746	46·04	13·40	209·7	59·0	268·7	1·0549	1·1657
74	898·2	883·5	0·02192	0·0726	45·62	13·76	210·6	57·2	267·8	1·0568	1·1640
75	909·7	895·0	0·02216	0·0707	45·12	14·14	211·6	55·4	267·0	1·0589	1·1622
76	921·3	906·6	0·02242	0·0687	44·60	14·55	212·7	53·4	266·1	1·0607	1·1604
77	933·0	918·3	0·02270	0·0667	44·06	14·98	213·8	51·4	265·2	1·0628	1·1585
78	944·8	930·1	0·02300	0·0647	43·48	15·45	214·9	49·3	264·2	1·0649	1·1565
79	956·7	942·0	0·02333	0·0626	42·86	15·95	216·0	47·2	263·2	1·0671	1·1544
80	968·7	954·0	0·02370	0·0606	42·19	16·49	217·2	44·8	262·0	1·0694	1·1522
81	980·8	966·1	0·02413	0·0585	41·44	17·07	218·4	42·6	261·0	1·0716	1·1503
82	993·0	978·3	0·02456	0·0564	40·72	17·70	219·7	40·2	259·9	1·0740	1·1479
83	1005·3	990·6	0·02502	0·0543	39·97	18·39	221·1	37·5	258·6	1·0764	1·1451
84	1017·7	1003·0	0·02553	0·0522	39·17	19·15	222·7	34·5	257·2	1·0790	1·1423
85	1030·3	1015·6	0·02613	0·0500	38·27	19·97	224·4	31·2	255·6	1·0820	1·1390
86	1043·0	1028·3	0·02686	0·0478	37·23	20·88	226·6	27·1	253·7	1·0854	1·1351
87	1055·6	1040·9	0·02806	0·0441	35·64	22·66	230·0	20·2	250·2	1·0907	1·1285
87·8	1066·2	1051·5	0·03454	0·0345	28·96	28·96	240·3	0	240·3	1·1098	1·1098

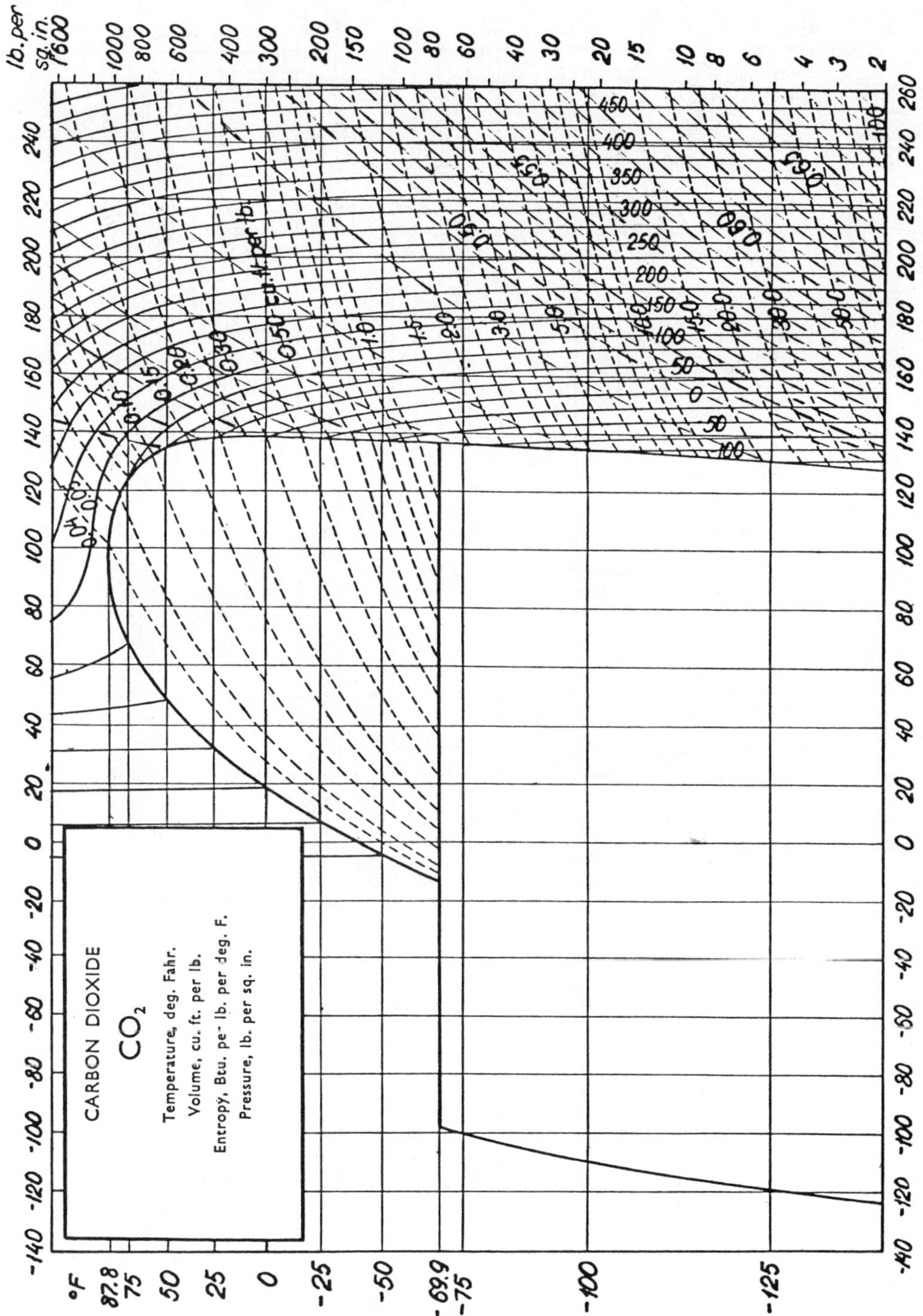
CARBON DIOXIDE
CO_2
Temperature, deg. Fahr.
Volume, cu. ft. per lb.
Entropy, Btu. pe- lb. per deg. F.
Pressure, lb. per sq. in.

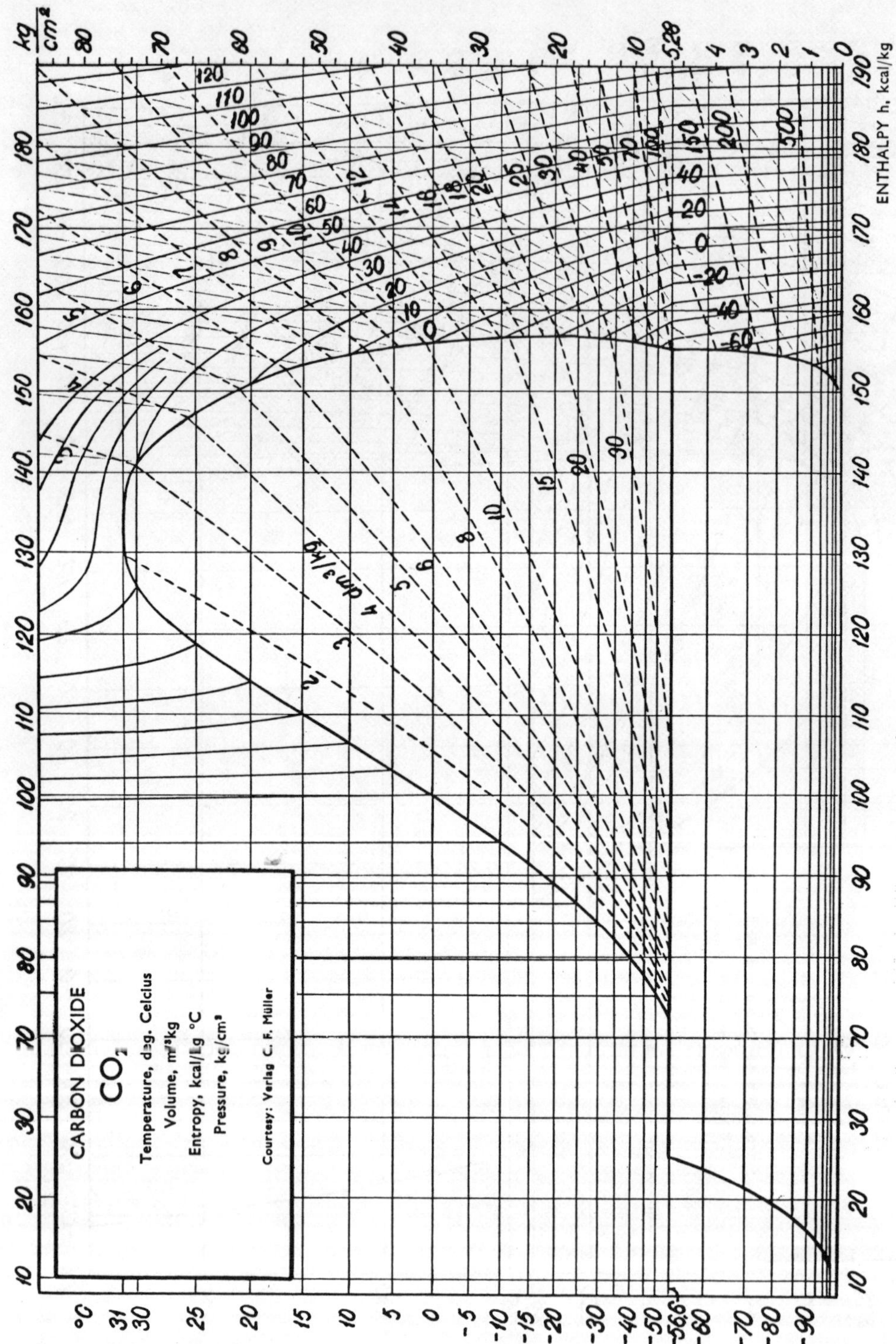
CARBON DIOXIDE
CO2
Temperature, deg. Celcius
Volume, m³/kg
Entropy, kcal/kg °C
Pressure, kg/cm²
Courtesy: Verlag C. F. Müller
ENTHALPY h, kcal/kg

DATA SHEET FOR R 764

SO_2	SULPHUR DIOXIDE	Molecular Wt. 64

Pressure	Temperature		Volume	Density
80.3 kg/cm^2	157 °C	315 °F	1.91 l/kg	.524 kg/l.
1142 psia	430 °K	775 °R	.0306 ft^3/lb	32.7 lbs/ft^3
				2.21 Air=1

AT CRITICAL POINT

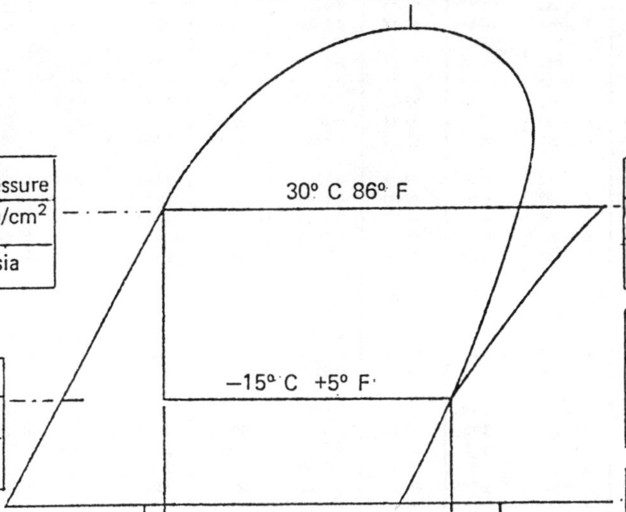

Discharge Pressure	
4.71	kg/cm^2
67	psia

30° C 86° F

Discharge Temperature	
93.7° C	200 °F
367 °K	660 °R

Normal Boiling Point	
-10 °C	14 °F
263 °K	474 °R

Inlet Pressure
.823 kg/cm^2
11.7 psia

−15° C +5° F

Triple Point	
-75.5° C	-104 °F
197 °K	356 °R

Refrigerating Effect	
195 kcal/m^3	22.0 Btu/ft^3

Latent Heat at NBP	93 kcal/kg	5965 /kg mol	167.5 Btu/lb	10737 /lb mol

Trouton's No. 22.65	Gas Constant 13.2 kg.m/kg/° K	24.1 ft.lbs/lb/° R

Specific Heat Liquid at 30° C 86° F	Gas C_p	C_p/C_v 1.27 at 0°C

Liquid at 30° C 86° F Density 1.35 kg/l.	84.5 lbs/ft^3 Viscosity cp

Reduced Form at NBP Pressure .0129	1 / Temperature 1.63

Reduced value of 1/T	1.99	1.67	1.34
when Reduced Pressure is	a) .001	b) .01	c) .1

TEMP.	PRESSURE		VOLUME cu ft/lb		DENSITY lb/cu ft		ENTHALPY Btu/lb			ENTROPY Btu/(lb)(°R)	
°F	PSIA	PSIG	LIQUID v_f	VAPOR v_g	LIQUID $1/v_f$	VAPOR $1/v_g$	LIQUID h_f	LATENT h_{fg}	VAPOR h_g	LIQUID s_f	VAPOR s_g
−100	0·29	*29·33	0·00985	204·7	101·46	0·00488	0	190·1	190·1	0	0·5285
− 90	0·46	*28·98	0·00995	132·9	100·46	0·00752	3·3	188·1	191·4	0·0090	0·5179
− 80	0·71	*28·48	0·01005	89·3	99·46	0·01120	6·5	186·2	192·7	0·0177	0·5081
− 70	1·05	*27·78	0·01015	61·0	98·52	0·01639	9·8	184·2	194·0	0·0262	0·4990
− 60	1·55	*26·77	0·01025	42·6	97·56	0·02347	13·0	182·3	195·3	0·0344	0·4905
− 50	2·22	*25·40	0·01034	30·4	96·71	0·03289	16·3	180·4	196·7	0·0424	0·4827
− 40	3·12	*23·58	0·01044	22·2	95·79	0·04504	19·6	178·4	198·0	0·0502	0·4753
− 30	4·33	*21·11	0·01053	16·4	94·97	0·06098	22·8	176·4	199·2	0·0579	0·4684
− 25	5·06	*19·63	0·01057	14·20	94·61	0·07042	24·4	175·4	199·8	0·0617	0·4651
− 20	5·88	*17·96	0·01062	12·33	94·16	0·08110	26·0	174·4	200·4	0·0654	0·4620
− 15	6·80	*16·08	0·01067	10·80	93·72	0·09260	27·7	173·4	201·1	0·0690	0·4590
− 10	7·83	*13·99	0·01072	9·48	93·28	0·1055	29·3	172·4	201·7	0·0726	0·4560
− 5	8·97	*11·67	0·01077	8·33	92·85	0·1201	30·9	171·4	202·3	0·0762	0·4531
0	10·26	* 9·04	0·01082	7·35	92·42	0·1361	32·5	170·3	202·8	0·0797	0·4503
5	11·71	* 6·09	0·01087	6·50	92·00	0·1538	34·1	169·3	203·4	0·0832	0·4476
10	13·3	* 2·85	0·01092	5·77	91·58	0·1733	35·7	168·3	204·0	0·0867	0·4450
15	15·0	*0·3	0·01097	5·14	91·16	0·1945	37·4	167·3	204·7	0·0901	0·4426
20	16·9	2·2	0·01103	4·59	90·66	0·2179	39·0	166·3	205·3	0·0935	0·4402
25	19·0	4·3	0·01108	4·11	90·25	0·2433	40·6	165·2	205·8	0·0969	0·4379
30	21·3	6·6	0·01114	3·70	89·77	0·2703	42·2	164·2	206·4	0·1002	0·4356
35	23·9	9·2	0·01120	3·34	89·29	0·2944	43·8	163·2	207·0	0·1035	0·4334
40	26·6	11·9	0·01125	3·02	88·89	0·3311	45·5	162·2	207·7	0·1067	0·4313
45	29·6	14·9	0·01131	2·74	88·42	0·3650	47·1	161·1	208·2	0·1100	0·4292
50	32·9	18·2	0·01137	2·48	87·95	0·4032	48·7	160·0	208·7	0·1132	0·4271
55	36·5	21·8	0·01143	2·25	87·49	0·4444	50·3	158·9	209·2	0·1164	0·4251
60	40·3	25·6	0·01149	2·05	87·03	0·4878	52·0	157·8	209·8	0·1195	0·4232
65	44·5	29·8	0·01156	1·87	86·50	0·5348	53·6	156·7	210·3	0·1227	0·4213
70	49·1	34·4	0·01162	1·70	86·06	0·5883	55·3	155·5	210·8	0·1258	0·4194
75	54·0	39·3	0·01168	1·55	85·62	0·6452	56·9	154·3	211·2	0·1289	0·4176
80	59·3	44·6	0·01175	1·42	85·11	0·7042	58·6	153·1	211·7	0·1320	0·4158
85	65·0	50·3	0·01182	1·30	84·60	0·7693	60·3	151·9	212·2	0·1351	0·4140
90	71·0	56·3	0·01189	1·20	84·10	0·8333	61·9	150·7	212·6	0·1382	0·4123
95	77·3	62·6	0·01196	1·11	83·61	0·9010	63·6	149·4	213·0	0·1413	0·4106
100	84·1	69·4	0·01204	1·02	83·06	0·9804	65·3	148·2	213·5	0·1443	0·4090
105	91·4	76·7	0·01212	0·946	82·51	1·057	67·0	147·0	214·0	0·1473	0·4075
110	99·1	84·4	0·01219	0·868	82·04	1·152	68·8	145·7	214·5	0·1503	0·4060
115	107·5	92·8	0·01227	0·805	81·50	1·243	70·5	144·4	214·9	0·1534	0·4045
120	116·3	101·6	0·01235	0·746	80·97	1·341	72·2	143·0	215·2	0·1564	0·4030
125	125·6	110·9	0·01243	0·690	80·45	1·450	74·0	141·5	215·5	0·1594	0·4015
130	135·8	121·1	0·01251	0·640	79·94	1·563	75·8	140·0	215·8	0·1624	0·4001
135	146·3	131·6	0·01260	0·596	79·37	1·678	77·5	138·6	216·1	0·1654	0·3986
140	157·7	143·0	0·01269	0·554	78·80	1·805	79·3	137·1	216·4	0·1684	0·3972
145	169·6	154·9	0·01278	0·516	78·25	1·938	81·1	135·7	216·8	0·1714	0·3958
150	182	167·3	0·01288	0·481	77·64	2·079	83·0	134·2	217·2	0·1744	0·3944
155	195	180·3	0·01299	0·448	76·98	2·232	84·8	132·6	217·4	0·1775	0·3930
160	[illegible]	[illegible]	0·01309	0·418	[illegible]	2·392	86·7	130·9	217·6	0·1805	0·3917
165	224	209·3	0·01319	0·390	75·82	2·564	88·6	129·3	217·9	0·1835	0·3903
170	238	223·3	0·01330	0·364	75·19	2·748	90·5	127·6	218·1	0·1865	0·3892

*Inches of mercury below one atmosphere

Index